Lecture Notes in Computer Science　　10761

Commenced Publication in 1973
Founding and Former Series Editors:
Gerhard Goos, Juris Hartmanis, and Jan van Leeuwen

Editorial Board

More information about this series at http://www.springer.com/series/7407

Alexander Gelbukh (Ed.)

Computational Linguistics and Intelligent Text Processing

18th International Conference, CICLing 2017
Budapest, Hungary, April 17–23, 2017
Revised Selected Papers, Part I

 Springer

Editor
Alexander Gelbukh
CIC, Instituto Politécnico Nacional
Mexico City, Mexico

ISSN 0302-9743   ISSN 1611-3349   (electronic)
Lecture Notes in Computer Science
ISBN 978-3-319-77112-0   ISBN 978-3-319-77113-7   (eBook)
https://doi.org/10.1007/978-3-319-77113-7

Library of Congress Control Number: 2018934347

LNCS Sublibrary: SL1 – Theoretical Computer Science and General Issues

This Springer imprint is published by the registered company Springer Nature Switzerland AG
The registered company address is: Gewerbestrasse 11, 6330 Cham, Switzerland

Preface

CICLing 2017 was the 18th International Conference on Computational Linguistics and Intelligent Text Processing. The CICLing conferences provide a wide-scope forum for discussion of the art and craft of natural language processing research, as well as the best practices in its applications.

This set of two books contains four invited papers and a selection of regular papers accepted for presentation at the conference. Since 2001, the proceedings of the CICLing conferences have been published in Springer's *Lecture Notes in Computer Science* series as volumes 2004, 2276, 2588, 2945, 3406, 3878, 4394, 4919, 5449, 6008, 6608, 6609, 7181, 7182, 7816, 7817, 8403, 8404, 9041, 9042, 9623, and 9624.

The set has been structured into 18 sections representative of the current trends in research and applications of natural language processing:

General
Morphology and Text Segmentation
Syntax and Parsing
Word Sense Disambiguation
Reference and Coreference Resolution
Named Entity Recognition
Semantics and Text Similarity
Information Extraction
Speech Recognition
Applications to Linguistics and the Humanities
Sentiment Analysis
Opinion Mining
Author Profiling and Authorship Attribution
Social Network Analysis
Machine Translation
Text Summarization
Information Retrieval and Text Classification
Practical Applications

This year our invited speakers were Marco Baroni (Facebook Artificial Intellgence Research), Iryna Gurevych (Ubiquitous Knowledge Processing Lab, TU Darmstadt), Björn W. Schuller (University of Passau, Imperial College London, Harbin Institute of Technology, University of Geneva, Joanneum Research, and EERING GmbH), and Hinrich Schuetze (Center for Information and Language Processing, University of Munich). They delivered excellent extended lectures and organized lively discussions. Full contributions of these invited talks are included in this book set.

After careful reviewing, the Program Committee selected 86 papers for presentation, out of 356 submissions from 60 countries.

To encourage providing algorithms and data along with the published papers, we selected three winners of our Verifiability, Reproducibility, and Working Description Award. The main factors in choosing the awarded submission were technical correctness and completeness, readability of the code and documentation, simplicity of installation and use, and exact correspondence to the claims of the paper. Unnecessary sophistication of the user interface was discouraged; novelty and usefulness of the results were not evaluated, instead, they were evaluated for the paper itself and not for the data.

The following papers received the Best Paper Awards, the Best Student Paper Award, as well as the Verifiability, Reproducibility, and Working Description Awards, respectively:

Best Verifiability Award, First Place:
"Label-Dependencies Aware Recurrent Neural Networks"
by Yoann Dupont, Marco Dinarelle, and Isabelle Tellier

Best Paper Award, Second Place, and Best Presentation Award:
"Idioms: Humans or Machines, It's All About Context"
by Manali Pradhan, Jing Peng, Anna Feldman, and Bianca Wright

Best Student Paper Award:
"Dialogue Act Taxonomy Interoperability Using a Meta-Model"
by Soufian Salim, Nicolas Hernandez, and Emmanuel Morin

Best Paper Award, First Place:
"Gold Standard Online Debates Summaries and First Experiments Towards Automatic Summarization of Online Debate Data"
by Nattapong Sanchan, Ahmet Aker, and Kalina Bontcheva

Best Paper Award, Third Place:
"Efficient Semantic Search over Structured Web Data: A GPU Approach" by Ha-Hguyen Tran, Erik Cambria, and Hoang Giang Do.

A conference is the result of the work of many people. First of all I would like to thank the members of the Program Committee for the time and effort they devoted to the reviewing of the submitted articles and to the selection process. Obviously I thank the authors for their patience in the preparation of the papers, not to mention the very development of their scientific results that form this book. I also express my most cordial thanks to the members of the local Organizing Committee for their considerable contribution to making this conference become a reality.

January 2018   Alexander Gelbukh

Organization

CICLing 2017 was hosted by the Pázmány Péter Catholic University, Faculty of Information Technology and Bionics, Budapest, Hungary, and organized by the CICLing 2017 Organizing Committee in conjunction with the Pázmány Péter Catholic University, Faculty of Information Technology and Bionics, the Natural Language and Text Processing Laboratory of the CIC, IPN, and the Mexican Society of Artificial Intelligence (SMIA).

Organizing Committee

Attila Novák (Chair)   MTA-PPKE Language Technology Research Group,
                              Pázmány Péter Catholic University
Gábor Prószéky   MTA-PPKE Language Technology Research Group,
                              Pázmány Péter Catholic University
Borbála Siklósi   MTA-PPKE Language Technology Research Group,
                              Pázmány Péter Catholic University

Program Committee

Bayan Abushawar
Galia Angelova
Alexandra Balahur
Sivaji Bandyopadhyay
Leslie Barrett
Roberto Basili
Pushpak Bhattacharyya
Christian Boitet
Nicoletta Calzolari
Nick Campbell
Michael Carl
Violetta Cavalli-Sforza
Niladri Chatterjee
Dan Cristea
Walter Daelemans
Mike Dillinger
Samhaa El-Beltagy
Michael Elhadad
Anna Feldman
Robert Gaizauskas
Alexander Gelbukh
Dafydd Gibbon

Gregory Grefenstette
Tunga Gungor
Eva Hajicova
Yasunari Harada
Karin Harbusch
Koiti Hasida
Ales Horak
Veronique Hoste
Diana Inkpen
Hitoshi Isahara
Aminul Islam
Guillaume Jacquet
Milos Jakubicek
Sylvain Kahane
Alma Kharrat
Philipp Koehn
Valia Kordoni
Mathieu Lafourcade
Elena Lloret
Bente Maegaard
Cerstin Mahlow
Suresh Manandhar

Inderjeet Mani
Alexander Mehler
Farid Meziane
Rada Mihalcea
Evangelos Milios
Ruslan Mitkov
Dunja Mladenic
Marie-Francine Moens
Hermann Moisl
Masaki Murata
Preslav Nakov
Costanza Navarretta
Joakim Nivre
Kjetil Norvag
Attila Novák
Nir Ofek
Kemal Oflazer
Constantin Orasan
Ivandre Paraboni
Saint-Dizier Patrick
Maria Teresa Pazienza
Ted Pedersen

Software Reviewing Committee

Best Paper Award Committee

Contents – Part I

Syntax and Parsing

Word Sense Disambiguation

Reference and Coreference Resolution

Invited Paper:

Named Entity Recognition

Semantics and Text Similarity

Best Paper Award, Second Place, and Best Presentation Award:

Best Student Paper Award:

Information Extraction

Speech Recognition

Applications to Linguistics and the Humanities

Invited Papers:

Contents – Part II

Opinion Mining

Author Profiling and Authorship Attribution

Invited Paper:

Social Network Analysis

Machine Translation

Text Summarization

Best Paper Award, First Place:

Information Retrieval and Text Classification

Best Paper Award, Third Place:

Practical Applications

General

Overview of Character-Based Models
for Natural Language Processing

Heike Adel[1], Ehsaneddin Asgari[1,2], and Hinrich Schütze[1(✉)]

[1] Center for Information and Language Processing, LMU Munich, Munich, Germany
heike@cis.lmu.de, inquiries@cislmu.org
[2] Applied Science and Technology, University of California, Berkeley, CA, USA
asgari@berkeley.edu

Abstract. Character-based models become more and more popular for different natural language processing task, especially due to the success of neural networks. They provide the possibility of directly model text sequences without the need of tokenization and, therefore, enhance the traditional preprocessing pipeline. This paper provides an overview of character-based models for a variety of natural language processing tasks. We group existing work in three categories: tokenization-based approaches, bag-of-n-gram models and end-to-end models. For each category, we present prominent examples of studies with a particular focus on recent character-based deep learning work.

Keywords: Natural language processing · Neural networks
Document representation · Feature selection
Natural language generation · Language models
Structured prediction · Supervised learning by classification

1 Introduction

Traditionally, natural language processing (NLP) relies on a preprocessing pipeline, such as the one described in [33] and depicted in Fig. 1. First, the document is tokenized. This step needs language-specific tokenization tools. The token sequence is then segmented into sentences. Afterwards, syntactic and semantic analysis is performed (usually sentence-wise). Syntactic analysis outputs part-of-speech tags, syntactic dependencies, etc. Semantic analysis extracts named entity tags, semantic roles, etc. The actual natural language processing/understanding (NLP/NLU) task, e.g., question answering or information extraction, uses features from those preprocessing steps.

Since every preprocessing step can have deficiencies, the whole pipeline of modules is prone to subsequent errors. Usually, it is hard, inefficient or even impossible to recover from those errors, especially when they occur during tokenization, i.e., in the first step of the pipeline. Although tokenization is easy for many cases in English,[1] it can be very hard for other languages, e.g., for Chinese

---

[1] There are also difficult cases in English, such as "Yahoo!" or "San Francisco-Los Angeles flights".

© Springer Nature Switzerland AG 2018
A. Gelbukh (Ed.): CICLing 2017, LNCS 10761, pp. 3–16, 2018.
https://doi.org/10.1007/978-3-319-77113-7_1

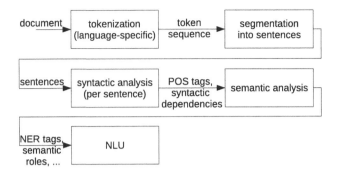

Fig. 1. Traditional NLP preprocessing pipeline

because tokens are not separated by spaces, for German because of compounds and for agglutinative languages like Turkish. Therefore, character-based models have the potential of being more robust for natural language processing. Futhermore, they support end-to-end approaches for text that do not require manual definitions of features, similar to pixel-based models in vision or acoustic signal-based approaches in speech recognition.

In the following, we will present an overview of work on character-based models for a variety of tasks from different NLP areas.[2]

2   Character-Level Models for NLP

The history of character-based research in NLP is long and spans a broad array of tasks. Here we make an attempt to categorize the literature of character-level work into three classes based on the way they incorporate character-level information into their computational models. The three classes we identified are: **tokenization-based models, bag-of-n-gram models** and **end-to-end models** [80]. However, there are also mixtures possible, such as tokenization-based bag-of-n-gram models or bag-of-n-gram models trained end-to-end.

On top of the categorization based on the underlying representation model, we sub-categorize the work within each group into six abstract types of NLP tasks (if possible) to be able to compare them more directly. These task types are the following:

1. **Representation learning for character sequences:** Work in this category attempts to learn a generic representation for sequences of characters in an unsupervised fashion on large corpora. Learning such representations has been shown to be useful for solving downstream NLP tasks [23,52,65].

---

[2] In our view, morpheme-based models are not true instances of character-level models as linguistically motivated morphological segmentation is an equivalent step to tokenization, but on a different level. We therefore do not cover most work on morphological segmentation in this paper.

2. **Sequence-to-sequence generation:** This category includes a variety of NLP tasks mapping variable-length input sequences to variable-length output sequences. Tasks in this category include those that are naturally suited for character-based modeling, such as grapheme-to-phoneme conversion [9,46, 81], transliteration [37,51,75], spelling normalization for historical text [71], or diacritics restauration [64]. Machine translation and question answering are other major examples of this category.

3. **Sequence labeling:** NLP tasks that assign a categorical label to a part of a sequence (a character, a sequence of characters or a token) are included within this group. Part-of-speech tagging, named entity recognition, morphological segmentation and word alignment are exemplary instances of sequence labeling.

4. **Language modeling:** The other type of tasks for that character-based modeling has been important for a very long time is language modeling. In 1951, Shannon [85] proposed a guessing game asking "How well can the next letter of a text be predicted when the preceding N letters are known?" This is basically the task of character-based language modeling.

5. **Information retrieval:** The information retrieval task is to retrieve the most relevant character sequence to a given character sequence (the query) from a set of existing character sequences.

6. **Sequence classification:** In this type of NLP tasks, a categorical label will be assigned to a character sequence (e.g., a document). Instances of this type are language identification, sentiment classification, authorship attribution, topic classification and word sense disambiguation.

2.1   Tokenization-Based Approaches

We group character-level models that are based on tokenization as a necessary preprocessing step in the category of tokenization-based approaches. Those can be either models with tokenized text as input or models that operate only on individual tokens (such as studies on morphological inflection of words).

In the following paragraphs, we cover a subset of tokenization-based models that are used for representation learning, sequence-to-sequence generation, sequence labeling, language modeling, and sequence classification tasks.

Representation Learning for Character Sequences. Creating word representations based on characters has attracted much attention recently. Such representations can model rare words, complex words, out-of-vocabulary words and noisy texts. In comparison to traditional word representation models that learn separate vectors for word types, character-level models are more compact as they only need vector representations for characters as well as a compositional model.

Various neural network architectures have been proposed for learning token representations based on characters. Examples of such architectures are depicted in Fig. 2 (from left to right): averaging character embeddings, (bidirectional) recurrent neural networks (RNNs) (with or without gates) over character embeddings and convolutional neural networks (CNNs) over character embeddings.

Studies on the general task of learning word representations from characters include [17,59,61,94]. These character-based word representations are often combined with word embeddings and integrated into a hierarchical system, such as hierarchical RNNs (see Fig. 3) or CNNs (see Fig. 4) or combinations of both (see Fig. 5) to solve other task types. We will provide more concrete examples in the following paragraphs.

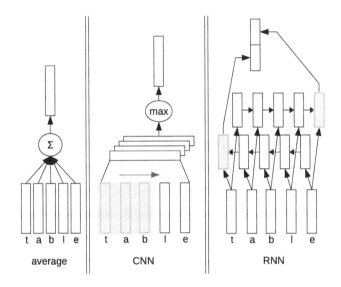

Fig. 2. Models to calculate word embeddings based on characters

Sequence-to-Sequence Generation (Machine Translation). Character-based machine translation is no new topic. Using character-based methods has been a natural way to overcome challenges like rare words or out-of-vocabulary words in machine translation. Traditional machine translation models based on characters or character n-grams have been investigated by [57,90,93]. Neural machine translation with character-level and subword units has become popular recently [24,60,82,94]. In such neural models, using a joint attention/translation model makes joint learning of alignment and translation possible [59].

Both hierarchical RNNs [59,60] (similar to Fig. 3) and combinations of CNNs and RNNs networks have been proposed for neural machine translation [24,94] (similar to Fig. 5).

Sequence Labeling. Examples of early efforts on sequence labeling using tokenization-based models include: bilingual character-level alignment extraction [21]; unsupervised multilingual part-of-speech induction based on characters [22]; part-of-speech tagging with subword/character-level information [2,38,74]; morphological segmentation and tagging [25,68]; and identification of language inclusion with character-based features [1].

Recently, various hierarchical character-level neural networks have been applied to a variety of sequence labeling tasks.

– Recurrent neural networks are used for part-of-speech tagging (depicted in Fig. 3) [58,72,101], named entity recognition [55,101], chunking [101] and morphological segmentation/inflection generation [13,31,43–45,73,95,102]. Such hierarchical RNNs are also used for dependency parsing [7]. This work has shown that morphologically rich languages benefit from character-level models in dependency parsing.
– Convolutional neural networks are used for part-of-speech tagging (shown in Fig. 4) [78] and named entity recognition [77].
– The combination of RNNs and CNNs is used, for instance, for named entity recognition (shown in Fig. 5) [18,91].

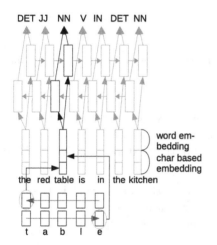

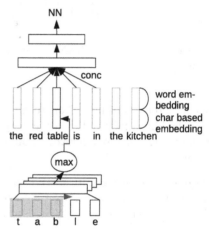

Fig. 3. Hierarchical RNN for part-of-speech tagging with character embeddings as input

Fig. 4. Hierarchical CNN + MLP for part-of-speech tagging with character embeddings as input

Language Modeling. Earlier work on sub-word language modeling has used morpheme-level features for language models [8,40,49,84,92]. In addition, hybrid word/n-gram language models for out-of-vocabulary words have been applied to speech recognition [39,53,69,83]. Furthermore, characters and character n-grams have been used as input to restricted boltzmann machine-based language models for machine translation [86].

More recently, character-level neural language modeling has been proposed by a large body of work [11,12,48,58,66,84,86]. Although most of this work is using RNNs, there exist architectures that combine CNNs and RNNs [48]. While most of these studies combine the output of the character model with word embeddings, the authors of [48] report that this does not help them for

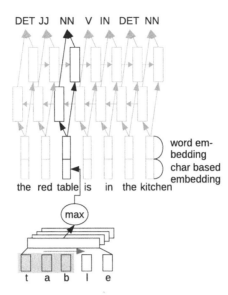

Fig. 5. Hierarchical CNN + RNN for part-of-speech tagging with character embeddings as input

their character-aware neural language model. They use convolution over character embeddings followed by a highway network [87] and feed its output into a long short-term memory network that predicts the next word using a softmax function.

Sequence Classification. Examples of tokenization-based models that perform sequence classification are CNNs used for sentiment classification [76] and combinations of RNNs and CNNs used for language identification [41].

2.2 Bag-of-n-gram Models

Character n-grams have a long history as features for specific NLP applications, such as information retrieval. However, there is also work on representing words or larger input units, such as phrases, with character n-gram embeddings. Those embeddings can be within-token or cross-token, i.e., there is no tokenization necessary.

Although such models learn/use character n-gram embeddings from tokenized text or short text segments, to represent a piece of text, the occurring character n-grams are usually summed without the need for tokenization. For example, the phrase "Berlin is located in Germany" is represented with character 4-grams as follows: "Berl erli rlin lin_ in_i n_is _is_ is_l s_lo _loc loca ocat cate ated ted_ ed_i d_in _in_ in_G n_Ge _Ger Germ erma rman many any." Note that the input has not been tokenized and there are n-grams spanning token boundaries. We also include non-embedding approaches using bag-of-n-grams within this group as they go beyond word and token representations.

In the following, we explore a subset of bag-of-ngram models that are used for representation learning, information retrieval, and sequence classification tasks.

Representation Learning for Character Sequences. An early study in this category of character-based models is [79]. Its goal is to create corpus-based fixed-length distributed semantic representations for text. To train k-gram embeddings, the top character k-grams are extracted from a corpus along with their cooccurrence counts. Then, singular value decomposition (SVD) is used to create low dimensional k-gram embeddings given their cooccurrence matrix. To apply them to a piece of text, the k-grams of the text are extracted and their corresponding embeddings are summed. The study evaluates the k-gram embeddings in the context of word sense disambiguation.

A more recent study [96] trains character n-gram embeddings in an end-to-end fashion with a neural network. They are evaluated on word similarity, sentence similarity and part-of-speech tagging.

Training character n-gram embeddings has also been proposed for biological sequences [3,4] for a variety of bioinformatics tasks.

Information Retrieval. As mentioned before, character n-gram features are widely used in the area of information retrieval [14, 16, 26, 27, 47, 63].

Sequence Classification. Bag-of-n-gram models are used for language identification [6,28], topic labeling [54], authorship attribution [70], word/text similarity [10,30,96] and word sense disambiguation [79].

2.3   End-to-end Models

Similar to bag-of-n-gram models, end-to-end models are tokenization-free. Their input is a sequence of characters or bytes and they are directly optimized on a (task-specific) objective. Thus, they learn their own, task-specific representation of the input sequences. Recently, character-based end-to-end models have gained a lot of popularity due to the success of neural networks.

We explore the subset of these models that are used for sequence generation, sequence labeling, language modeling and sequence classification tasks.

Sequence-to-Sequence Generation. In 2011, the authors of [88] already proposed an end-to-end model for generating text. They train RNNs with multiplicative connections on the task of character-level language modeling. Afterwards, they use the model to generate text and find that the model captures linguistic structure and a large vocabulary. It produces only a few uncapitalized non-words and is able to balance parantheses and quotes even over long distances (e.g., 30 characters). A similar study by [35] uses a long short-term memory network to create character sequences.

Recently, character-based neural network sequence-to-sequence models have been applied to instances of generation tasks like machine translation [20, 42, 56, 97, 100] (which was previously proposed on the token-level [89]), question answering [34] and speech recognition [5, 15, 29, 36].

Sequence Labeling. Character and character n-gram-based features were already proposed in 2003 for named entity recognition in an end-to-end manner using a hidden markov model [50]. More recently, the authors of [62] have proposed an end-to-end neural network based model for named entity recognition and part-of-speech tagging. An end-to-end model is also suggested for unsupervised, language-independent identification of phrases or words [32].

A prominent recent example of neural end-to-end sequence labeling is the paper by [33] about multilingual language processing from bytes. A window is slid over the input sequence, which is represented by its byte string. Thus, the segments in the window can begin and end mid-word or even mid-character. The authors apply the same model for different languages and evaluate it on part-of-speech tagging and named entity recognition.

Language Modeling. The authors of [19] propose a hierarchical multiscale recurrent neural network for language modeling. The model uses different timescales to encode temporal dependencies and is able to discover hierarchical structures in a character sequence without explicit tokenization. Other studies on end-to-end language models include [42,67].

Sequence Classification. Another recent end-to-end model uses character-level inputs for document classification [98,103,104]. To capture long-term dependencies of the input, the authors combine convolutional layers with recurrent layers. The model is evaluated on sentiment analysis, ontology classification, question type classification and news categorization.

End-to-end models are also used for entity typing based on the character sequence of the entity's name [99].

3   Conclusion

Characters or character n-grams have a long history as features for specific applications in natural language processing. Nowadays, character-based models become increasingly popular. This development is promoted especially by the success and popularity of neural networks. In this paper, we grouped studies on character-based models for NLP into three categories: Tokenization-based models, bag-of-n-gram models and end-to-end models. For each category, we provided examples for a variety of NLP tasks. While tokenization-based models still require tokenization of the input sequence into words, bag-of-n-gram models and end-to-end models are tokenization-free. Thus, they overcome the challenge of dealing with tokenization errors and provide the possibility of enhancing or even completely replacing the traditional NLP preprocessing pipeline.

References

1. Alex, B.: An unsupervised system for identifying english inclusions in german text. In: Annual Meeting of the Association for Computational Linguistics (2005)
2. Andor, D., et al.: Globally normalized transition-based neural networks. In: Annual Meeting of the Association for Computational Linguistics (2016)
3. Asgari, E., Mofrad, M.R.K.: Continuous distributed representation of biological sequences for deep proteomics and genomics. PLoS ONE **10**(11), 1–15 (2015)
4. Asgari, E., Mofrad, M.R.K.: Comparing fifty natural languages and twelve genetic languages using word embedding language divergence (WELD) as a quantitative measure of language distance. In: Workshop on Multilingual and Cross-lingual Methods in NLP, pp. 65–74 (2016)
5. Bahdanau, D., Chorowski, J., Serdyuk, D., Brakel, P., Bengio, Y.: End-to-end attention-based large vocabulary speech recognition. In: IEEE International Conference on Acoustics, Speech and Signal Processing, pp. 4945–4949 (2016)
6. Baldwin, T., Lui, M.: Language identification: the long and the short of the matter. In: Conference of the North American Chapter of the Association for Computational Linguistics/Human Language Technologies, pp. 229–237 (2010)
7. Ballesteros, M., Dyer, C., Smith, N.A.: Improved transition-based parsing by modeling characters instead of words with LSTMS. In: Conference on Empirical Methods in Natural Language Processing (2015)
8. Bilmes, J., Kirchhoff, K.: Factored language models and generalized parallel backoff. In: Conference of the North American Chapter of the Association for Computational Linguistics/Human Language Technologies (2003)
9. Bisani, M., Ney, H.: Joint-sequence models for grapheme-to-phoneme conversion. Speech Commun. **50**(5), 434–451 (2008)
10. Bojanowski, P., Grave, E., Joulin, A., Mikolov, T.: Enriching word vectors with subword information. Transactions of the Association for Computational Linguistics (2017)
11. Bojanowski, P., Joulin, A., Mikolov, T.: Alternative structures for character-level RNNS. In: Workshop at International Conference on Learning Representations (2016)
12. Botha, J.A., Blunsom, P.: Compositional morphology for word representations and language modelling. In: International Conference on Machine Learning (2014)
13. Cao, K., Rei, M.: A joint model for word embedding and word morphology. In: Annual Meeting of the Association for Computational Linguistics, pp. 18–26 (2016)
14. Cavnar, W.: Using an n-gram-based document representation with a vector processing retrieval model. NIST SPECIAL PUBLICATION SP, pp. 269–269 (1995)
15. Chan, W., Jaitly, N., Le, Q.V., Vinyals, O.: Listen, attend and spell: A neural network for large vocabulary conversational speech recognition. In: IEEE International Conference on Acoustics, Speech and Signal Processing, pp. 4960–4964 (2016)
16. Chen, A., He, J., Xu, L., Gey, F.C., Meggs, J.: Chinese text retrieval without using a dictionary. ACM SIGIR Forum **31**(SI), 42–49 (1997)
17. Chen, X., Xu, L., Liu, Z., Sun, M., Luan, H.: Joint learning of character and word embeddings. In: International Joint Conference on Artificial Intelligence, pp. 1236–1242 (2015)
18. Chiu, J.P.C., Nichols, E.: Named entity recognition with bidirectional LSTM-CNNS. Trans. Assoc. Comput. Linguist. **4**, 357–370 (2016)

19. Chung, J., Ahn, S., Bengio, Y.: Hierarchical multiscale recurrent neural networks. In: Proceedings of International Conference on Learning Representations (2017)

20. Chung, J., Cho, K., Bengio, Y.: A character-level decoder without explicit segmentation for neural machine translation. In: Annual Meeting of the Association for Computational Linguistics (2016)

21. Church, K.W.: Char_align: a program for aligning parallel texts at the character level. In: Annual Meeting of the Association for Computational Linguistics, pp. 1–8 (1993)

22. Clark, A.: Combining distributional and morphological information for part of speech induction. In: Conference of the European Chapter of the Association for Computational Linguistics, pp. 59–66 (2003)

23. Collobert, R., Weston, J., Bottou, L., Karlen, M., Kavukcuoglu, K., Kuksa, P.: Natural language processing (almost) from scratch. J. Mach. Learn. Res. **12**, 2493–2537 (2011)

24. Costa-Jussà, M.R., Fonollosa, J.A.R.: Character-based neural machine translation. In: Annual Meeting of the Association for Computational Linguistics (2016)

25. Cotterell, R., Vieira, T., Schütze, H.: A joint model of orthography and morphological segmentation. In: Conference of the North American Chapter of the Association for Computational Linguistics/Human Language Technologies (2016)

26. Damashek, M.: Gauging similarity with n-grams: language-independent categorization of text. Science **267**, 843–848 (1995)

27. De Heer, T.: Experiments with syntactic traces in information retrieval. Inf. Storage Retr. **10**(3–4), 133–144 (1974)

28. Dunning, T.: Statistical identification of language. Technical Report MCCS 940–273, Computing Research Laboratory, New Mexico State (1994)

29. Eyben, F., Wöllmer, M., Schuller, B.W., Graves, A.: From speech to letters - using a novel neural network architecture for grapheme based ASR. In: IEEE Workshop on Automatic Speech Recognition & Understanding (ASRU), pp. 376–380 (2009)

30. Eyecioglu, A., Keller, B.: ASOBEK at SemEval-2016 task 1: sentence representation with character n-gram embeddings for semantic textual similarity. In: SemEval-2016: The 10th International Workshop on Semantic Evaluation, pp. 1320–1324 (2016)

31. Faruqui, M., Tsvetkov, Y., Neubig, G., Dyer, C.: Morphological inflection generation using character sequence to sequence learning. In: Conference of the North American Chapter of the Association for Computational Linguistics/Human Language Technologies (2016)

32. Gerdjikov, S., Schulz, K.U.: Corpus analysis without prior linguistic knowledge-unsupervised mining of phrases and subphrase structure. CoRR abs/1602.05772 (2016)

33. Gillick, D., Brunk, C., Vinyals, O., Subramanya, A.: Multilingual language processing from bytes. In: North American Chapter of the Association for Computational Linguistics, pp. 1296–1306, June 2016

34. Golub, D., He, X.: Character-level question answering with attention. In: Conference on Empirical Methods in Natural Language Processing (2016)

35. Graves, A.: Generating sequences with recurrent neural networks. CoRR abs/1308.0850 (2013)

36. Graves, A., Jaitly, N.: Towards end-to-end speech recognition with recurrent neural networks. In: International Conference on Machine Learning, pp. 1764–1772 (2014)

37. Haizhou, L., Min, Z., Jian, S.: A joint source-channel model for machine transliteration. In: Annual Meeting of the Association for Computational Linguistics, p. 159 (2004)
38. Hardmeier, C.: A neural model for part-of-speech tagging in historical texts. In: International Conference on Computational Linguistics, pp. 922–931 (2016)
39. Hirsimäki, T., Creutz, M., Siivola, V., Kurimo, M., Virpioja, S., Pylkkönen, J.: Unlimited vocabulary speech recognition with morph language models applied to finnish. Comput. Speech Lang. **20**(4), 515–541 (2006)
40. Ircing, P., et al.: On large vocabulary continuous speech recognition of highly inflectional language-czech. In: Proceedings of the 7th European Conference on Speech Communication and Technology, vol. 1, pp. 487–490. ISCA: International Speech Communication Association (2001)
41. Jaech, A., Mulcaire, G., Hathi, S., Ostendorf, M., Smith, N.A.: Hierarchical character-word models for language identification. In: Proceedings of The Fourth International Workshop on Natural Language Processing for Social Media, pp. 84–93 (2016)
42. Kalchbrenner, N., Espeholt, L., Simonyan, K., van den Oord, A., Graves, A., Kavukcuoglu, K.: Neural machine translation in linear time. CoRR abs/1610.10099 (2016)
43. Kann, K., Cotterell, R., Schütze, H.: Neural morphological analysis: encoding-decoding canonical segments. In: Conference on Empirical Methods in Natural Language Processing (2016)
44. Kann, K., Schütze, H.: MED: The LMU system for the SIGMORPHON 2016 shared task on morphological reinflection. In: SIGMORPHON Workshop (2016)
45. Kann, K., Schütze, H.: Single-model encoder-decoder with explicit morphological representation for reinflection. In: Annual Meeting of the Association for Computational Linguistics (2016)
46. Kaplan, R.M., Kay, M.: Regular models of phonological rule systems. Comput. Linguist. **20**(3), 331–378 (1994)
47. Kettunen, K., McNamee, P., Baskaya, F.: Using syllables as indexing terms in full-text information retrieval. In: Human Language Technologies - The Baltic Perspective - Proceedings of the Fourth International Conference Baltic HLT 2010, Riga, Latvia, October 7–8, 2010, pp. 225–232 (2010)
48. Kim, Y., Jernite, Y., Sontag, D., Rush, A.M.: Character-aware neural language models. In: AAAI Conference on Artificial Intelligence, pp. 2741–2749 (2016)
49. Kirchhoff, K., Vergyri, D., Bilmes, J., Duh, K., Stolcke, A.: Morphology-based language modeling for conversational arabic speech recognition. Comput. Speech Lang. **20**(4), 589–608 (2006)
50. Klein, D., Smarr, J., Nguyen, H., Manning, C.D.: Named entity recognition with character-level models. In: Computational Natural Language Learning, pp. 180–183 (2003)
51. Knight, K., Graehl, J.: Machine transliteration. Comput. Linguist. **24**(4), 599–612 (1998)
52. Kocmi, T., Bojar, O.: SubGram: extending skip-gram word representation with substrings. In: Sojka, P., Horák, A., Kopeček, I., Pala, K. (eds.) TSD 2016. LNCS (LNAI), vol. 9924, pp. 182–189. Springer, Cham (2016). https://doi.org/10.1007/978-3-319-45510-5_21
53. Kombrink, S., Hannemann, M., Burget, L., Heřmanský, H.: Recovery of rare words in lecture speech. In: Sojka, P., Horák, A., Kopeček, I., Pala, K. (eds.) TSD 2010. LNCS (LNAI), vol. 6231, pp. 330–337. Springer, Heidelberg (2010). https://doi.org/10.1007/978-3-642-15760-8_42

54. Kou, W., Li, F., Baldwin, T.: Automatic labelling of topic models using word vectors and letter trigram vectors. In: Asia Information Retrieval Societies Conference (AIRS), pp. 253–264 (2015)
55. Lample, G., Ballesteros, M., Subramanian, S., Kawakami, K., Dyer, C.: Neural architectures for named entity recognition. In: Conference of the North American Chapter of the Association for Computational Linguistics/Human Language Technologies (2016)
56. Lee, J., Cho, K., Hofmann, T.: Fully character-level neural machine translation without explicit segmentation. CoRR abs/1610.03017 (2016)
57. Lepage, Y., Denoual, E.: Purest ever example-based machine translation: detailed presentation and assessment. Mach. Transl. **19**(3–4), 251–282 (2005)
58. Ling, W., et al.: Finding function in form: compositional character models for open vocabulary word representation. In: Conference on Empirical Methods in Natural Language Processing, pp. 1520–1530 (2015)
59. Ling, W., Trancoso, I., Dyer, C., Black, A.W.: Character-based neural machine translation. CoRR abs/1511.04586 (2015)
60. Luong, M., Manning, C.D.: Achieving open vocabulary neural machine translation with hybrid word-character models. In: Annual Meeting of the Association for Computational Linguistics (2016)
61. Luong, M.T., Socher, R., Manning, C.D.: Better word representations with recursive neural networks for morphology. In: Computational Natural Language Learning (2013)
62. Ma, X., Hovy, E.H.: End-to-end sequence labeling via bi-directional LSTM-CNNS-CRF. In: Annual Meeting of the Association for Computational Linguistics (2016)
63. McNamee, P., Mayfield, J.: Character n-gram tokenization for european language text retrieval. Inf. Retr. **7**(1–2), 73–97 (2004)
64. Mihalcea, R., Nastase, V.: Letter level learning for language independent diacritics restoration. In: Computational Natural Language Learning (2002)
65. Mikolov, T., Sutskever, I., Chen, K., Corrado, G.S., Dean, J.: Distributed representations of words and phrases and their compositionality. In: Advances in Neural Information Processing Systems, pp. 3111–3119 (2013)
66. Mikolov, T., Sutskever, I., Deoras, A., Le, H.S., Kombrink, S., Cernocky, J.: Subword language modeling with neural networks (2012)
67. Miyamoto, Y., Cho, K.: Gated word-character recurrent language model. In: Conference on Empirical Methods in Natural Language Processing, pp. 1992–1997 (2016)
68. Müller, T., Schmid, H., Schütze, H.: Efficient higher-order CRFs for morphological tagging. In: Conference on Empirical Methods in Natural Language Processing, pp. 322–332 (2013)
69. Parada, C., Dredze, M., Sethy, A., Rastrow, A.: Learning sub-word units for open vocabulary speech recognition. In: Proceedings of the 49th Annual Meeting of the Association for Computational Linguistics: Human Language Technologies, vol. 1, pp. 712–721 (2011)
70. Peng, F., Schuurmans, D., Wang, S., Keselj, V.: Language independent authorship attribution using character level language models. In: Conference of the European Chapter of the Association for Computational Linguistics, pp. 267–274 (2003)
71. Pettersson, E., Megyesi, B., Nivre, J.: A multilingual evaluation of three spelling normalisation methods for historical text. In: Proceedings of the 8th Workshop on Language Technology for Cultural Heritage, Social Sciences, and Humanities, pp. 32–41 (2014)

72. Plank, B., Søgaard, A., Goldberg, Y.: Multilingual part-of-speech tagging with bidirectional long short-term memory models and auxiliary loss. In: Annual Meeting of the Association for Computational Linguistics (2016)
73. Rastogi, P., Cotterell, R., Eisner, J.: Weighting finite-state transductions with neural context. In: Conference of the North American Chapter of the Association for Computational Linguistics/Human Language Technologies, pp. 623–633 (2016)
74. Ratnaparkhi, A., et al.: A maximum entropy model for part-of-speech tagging. In: Conference on Empirical Methods in Natural Language Processing, vol. 1, pp. 133–142. Philadelphia, USA (1996)
75. Sajjad, H.: Statistical models for unsupervised, semi-supervised and supervised transliteration mining. In: Computational Linguistics (2012)
76. dos Santos, C.N., Gatti, M.: Deep convolutional neural networks for sentiment analysis of short texts. In: International Conference on Computational Linguistics. pp. 69–78 (2014)
77. dos Santos, C.N., Guimarães, V.: Boosting named entity recognition with neural character embeddings. In: Fifth Named Entity Workshop, pp. 25–33 (2015)
78. dos Santos, C.N., Zadrozny, B.: Learning character-level representations for part-of-speech tagging. In: International Conference on Machine Learning, pp. 1818–1826 (2014)
79. Schütze, H.: Word space. In: Advances in Neural Information Processing Systems, pp. 895–902 (1992)
80. Schütze, H.: Nonsymbolic text representation. CoRR abs/1610.00479 (2016)
81. Sejnowski, T.J., Rosenberg, C.R.: Parallel networks that learn to pronounce english text. Complex Syst. 1(1), 145–168 (1987)
82. Sennrich, R., Haddow, B., Birch, A.: Neural machine translation of rare words with subword units. In: Annual Meeting of the Association for Computational Linguistics (2016)
83. Shaik, M.A.B., Mousa, A.E.D., Schlüter, R., Ney, H.: Hybrid language models using mixed types of sub-lexical units for open vocabulary german lvcsr. In: Annual Conference of the International Speech Communication Association, pp. 1441–1444 (2011)
84. Shaik, M.A.B., Mousa, A.E., Schlüter, R., Ney, H.: Feature-rich sub-lexical language models using a maximum entropy approach for german LVCSR. In: Annual Conference of the International Speech Communication Association, pp. 3404–3408 (2013)
85. Shannon, C.E.: Prediction and entropy of printed english. Bell Labs Tech. J. 30(1), 50–64 (1951)
86. Sperr, H., Niehues, J., Waibel, A.: Letter n-gram-based input encoding for continuous space language models. In: Workshop on Continuous Vector Space Models and their Compositionality, pp. 30–39 (2013)
87. Srivastava, R.K., Greff, K., Schmidhuber, J.: Highway networks. In: ICML 2015 Deep Learing Workshop (2015)
88. Sutskever, I., Martens, J., Hinton, G.E.: Generating text with recurrent neural networks. In: International Conference on Machine Learning, pp. 1017–1024 (2011)
89. Sutskever, I., Vinyals, O., Le, Q.V.: Sequence to sequence learning with neural networks. In: Advances in Neural Information Processing Systems, pp. 3104–3112 (2014)

90. Tiedemann, J., Nakov, P.: Analyzing the use of character-level translation with sparse and noisy datasets. In: Recent Advances in Natural Language Processing, RANLP 2013, 9–11 September, 2013, Hissar, Bulgaria, pp. 676–684 (2013)
91. Murthy, V., Khapra, M.M., Bhattacharyya, P.: Sharing network parameters for crosslingual named entity recognition. CoRR abs/1607.00198 (2016)
92. Vergyri, D., Kirchhoff, K., Duh, K., Stolcke, A.: Morphology-based language modeling for arabic speech recognition. In: Annual Conference of the International Speech Communication Association, **4**, 2245–2248 (2004)
93. Vilar, D., Peter, J.T., Ney, H.: Can we translate letters? In: Workshop on Statistical Machine Translation (2007)
94. Vylomova, E., Cohn, T., He, X., Haffari, G.: Word representation models for morphologically rich languages in neural machine translation. CoRR abs/1606.04217 (2016)
95. Wang, L., Cao, Z., Xia, Y., de Melo, G.: Morphological segmentation with window LSTM neural networks. In: AAAI Conference on Artificial Intelligence (2016)
96. Wieting, J., Bansal, M., Gimpel, K., Livescu, K.: Charagram: embedding words and sentences via character n-grams. In: Conference on Empirical Methods in Natural Language Processing (2016)
97. Wu, Y., et al.: Google's neural machine translation system: Bridging the gap between human and machine translation. CoRR abs/1609.08144 (2016)
98. Xiao, Y., Cho, K.: Efficient character-level document classification by combining convolution and recurrent layers. CoRR abs/1602.00367 (2016)
99. Yaghoobzadeh, Y., Schütze, H.: Multi-level representations for fine-grained typing of knowledge base entities. In: Conference of the European Chapter of the Association for Computational Linguistics (2017)
100. Yang, Z., Chen, W., Wang, F., Xu, B.: A character-aware encoder for neural machine translation. In: International Conference on Computational Linguistics, pp. 3063–3070 (2016)
101. Yang, Z., Salakhutdinov, R., Cohen, W.W.: Multi-task cross-lingual sequence tagging from scratch. CoRR abs/1603.06270 (2016)
102. Yu, L., Buys, J., Blunsom, P.: Online segment to segment neural transduction. In: Conference on Empirical Methods in Natural Language Processing, pp. 1307–1316 (2016)
103. Zhang, X., LeCun, Y.: Text understanding from scratch. CoRR abs/1502.01710 (2015)
104. Zhang, X., Zhao, J., LeCun, Y.: Character-level convolutional networks for text classification. In: Advances in Neural Information Processing Systems, pp. 649–657 (2015)

Pooling Word Vector Representations Across Models

Rajendra Banjade[1]($\boxtimes$), Nabin Maharjan[1], Dipesh Gautam[1], Frank Adrasik[2],
Arthur C. Graesser[2], and Vasile Rus[1]

[1] Department of Computer Science/Institute for Intelligent Systems,
The University of Memphis, Memphis, TN 38152, USA
{rbanjade,nmharjan,dgautam,vrus}@memphis.edu
[2] Department of Psychology/Institute for Intelligent Systems,
The University of Memphis, Memphis, TN 38152, USA

Abstract. Vector based word representation models are typically developed from very large corpora with the hope that the representations are reliable and have wide coverage, i.e. they cover, ideally, all words. However, we often encounter words in real world applications that are not available in a single vector-based model. In this paper, we present a novel Neural Network (NN) based approach for obtaining representations for words that are missing in a *target* model from another model, called the *source* model, where representations for these words are available, effectively pooling together their vocabularies and the corresponding representations. Our experiments with three different types of pre-trained models (*Word2vec*, *GloVe*, and *LSA*) show that the representations obtained using our transformation approach can substantially and effectively extend the word coverage of existing models. The increase in the number of unique words covered by a model varies from few to several times depending on which model vocabulary is taken as reference. The transformed word representations are also well correlated (average correlation up to 0.801 for words in Simlex-999 dataset) with the native target model representations indicating that the transformed vectors can effectively be used as substitutes of native word representations. Furthermore, an extrinsic evaluation based on a word-to-word similarity task using the Simlex-999 dataset leads to results close to those obtained using native target model representations.

Keywords: Semantics · Word representations
Handling missing words

1 Introduction

Different approaches have been proposed over the years to represent the meaning of words, phrases, sentences, or even larger texts in continuous vector forms (also called embeddings; [4,7,10,11,15,21,26,27]). These vector based representations have been used in many Natural Language Processing (NLP) applications [7,12,

© Springer Nature Switzerland AG 2018
A. Gelbukh (Ed.): CICLing 2017, LNCS 10761, pp. 17–29, 2018.
https://doi.org/10.1007/978-3-319-77113-7_2

14, 22, 24]. Preferably, and which is often the case, the representations are derived in an unsupervised way from extremely large collections of texts following the distributional semantics principle according to which the meaning of words is derived from its usage, i.e. by computing co-occurrence statistics from large collections of texts. For instance, the pre-trained Word2vec [15][1] and GloVe [21][2] word vector representations were developed from texts containing billions of tokens covering millions of unique words: the pre-trained Word2vec model covers 3 million unique words, the GloVe model has a coverage of 1.9 million words, and a Latent Semantic Analysis (LSA) model developed ourselves from the whole set of Wikipedia articles (LSA$_{wiki}$; [25])[3] contains representations for 1.1 million words[4].

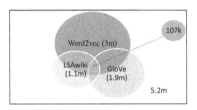

Fig. 1. Vocabulary size of three different pre-trained models (k - thousand, m - million).

While these are impressive numbers compared to manually created resources such as WordNet [17], it is important to note that the aforementioned word representation models share a limited number of words, as illustrated in Fig. 1. The GloVe and Word2vec have about 154,000 words in common. Only about 107,000 words are common to all three models, which equates to only 3 to 10% of the words depending which model's vocabulary size is used as a reference. This clearly indicates that a significant chunk of words in each of these models are unique to the respective models and that they are missing from the other models. For example, the word *"Totalizator"* is present in Word2vec model but not in other two models. Therefore, systems using the LSA$_{wiki}$ or GloVe model will have difficulty processing the word*"Totalizator"* because of the missing word representation whereas systems using Word2vec model will not encounter this situation, and so on. Even though these numbers and the overlap in vocabulary among models can vary depending on the source of data used to build the models and the nature of preprocessing steps performed (e.g., lemmatization keeps only the base or dictionary form of the words), we will not have any single model that covers all the words one may encounter in an application. On the other hand, by design, many (if not all) existing NLP algorithms do not work with multiple types of representations side by side. Using multiple heterogeneous representations, a

---

[1] https://code.google.com/p/word2vec/.

[2] http://nlp.stanford.edu/data/glove.42B.300d.zip.

[3] `Wiki_NVAR_f7` at http://semanticsimilarity.org/.

[4] We have used 'token' and 'word' interchangeably.

potential solution to the problem of missing words in a target word meaning representation, can greatly increase the complexity of such algorithms.

Yet another, better approach to the problem of missing word representations in vector based models, which we propose and explore in this paper, is to automatically map word vector representations from one model (where they are present; the *source* model) to another (where they are missing; the *target* model). We rely on a novel Neural Network (NN) based approach to obtain vector-based representations for missing words in a *target* model from the *source* model, where representations for these words are available. That is, we make use of existing word representation models in combination with the NN-based mapping approach to extend the coverage (i.e., expand the vocabulary) of a given target model. The benefit of our approach is that we extend the coverage of a target model without the need to collect any extra texts and re-train the model, which could be non-trivial, as already mentioned, because such representation models are generally developed by different groups or organizations using non comparable set of corpora and obtaining all these corpora is not always possible due to various reasons including copyright and privacy issues. Using our approach we can expand, for instance, the Word2vec, GloVe, and LSA models coverage to about 5.2 million unique words while showing that the transformed representations are well correlated (average correlation up to 0.801 for words in Simlex-999 dataset) with the native target model representations indicating that the transformed vectors can effectively be used as substitutes for native word representations of the target model. Also, the process can be automated if pre-trained word representation models are available.

We evaluate our approach *intrinsically* and *extrinsically* (see Sect. 4) on all possible *source → target* model permutations of three different pre-trained word vector models: word2vec, GloVe, and LSA$_{wiki}$. The results show that obtaining word representations for one model from another without much loss of representation power relative to the native target vectors is possible and indicate that the transformed vectors can be used to augment the target models.

2   Related Work

The issue of handling unknown and missing words has been previously explored to some extent. [1,6] proposed deriving continuous word representations for unknown or missing words in Neural Language Models (NLMs) based on the words in context. However, (full) context of a word is not always available. [16] demonstrated that Word2vec vectors capture enough syntactic and semantic linguistic regularities to derive vector representations of missing words based on simple vector operations. For example, the following expression illustrates a singular/plural relation: $v('cats') = v('dogs') - v('dog') + v('cat')$. However, such nice linguistic regularities might not hold for complex and rare words and their vector representations might not be properly estimated [13]. Furthermore, it's hard to automatically find out such relations, such as in the case of proper nouns. Also, it will not work if word representations that are needed on the right hand side of an expression like the one above are not available.

Recursive Neural Networks (RNNs) have also been used to construct missing word representations from the vectors of words' morphemes [13]. This approach works only if the missing word can be broken into morphemes, which in the case of some words such as proper nouns this is not possible, and representations for morphemes are available. In our previous work [3] we used the representation of a word's synonyms obtained from WordNet as a substitute representation for the target word. However, this only works if representations for the word's synonyms exist, which is not always the case.

Though, to some extent, these techniques can handle the issue of missing word representations, the processes are not very straightforward to automate. Also, the increase in coverage will be limited as these methods have difficulty handling named entities which constitutes a large chunk of the vocabulary derived from very large corpora. In our approach, we directly transform word representations from one model to another model effectively pooling together their vocabularies (i.e., expanding each model's vocabulary). Additionally, our approach has potential to be equally applicable to phrase level or sentence level representations which have much more acute missing representation issues as they are even sparser - dealing with missing representations for phrase level model is beyond the scope of this paper.

3   Our Approach

As discussed previously and illustrated in Fig. 1, words missing from a model may be present in another model. Therefore, by learning a word vector mapping model (or function) that can map one vector representation onto another, representations for missing words in the target model can be obtained from source model where the words are present. The schematic diagrams in Fig. 2 illustrate this approach.

$$TrV = T_{source \rightarrow target}(SrcV) \qquad (1)$$

$$Voc_{target}^{e} = Voc_{target} \cup (\cup_{i=1}^{S} Voc_{source(i)}) \qquad (2)$$

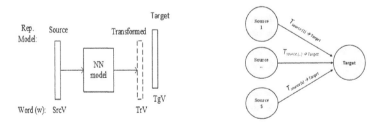

Fig. 2. Schematic diagram of (a) A transformation model, and (b) Multiple source-to-target transformations (*NN*- Neural Network, T - Transformation function/model, $SrcV$ - Source model vector, TrV - Transformed vector, TgV - Target model vector).

In Eq. (1), $T_{source \rightarrow target}$ is a transformation model (function) corresponding to the *source* → *target* transformation. The transformation model consists of a feed-forward Neural Network. The input to the model is in the form of source model vectors ($SrcV$) and the output of the transformation model (transformed vectors; TrV) is similar to target model vectors (TgV). That is, ideally, the TrV should be the mirror image of TgV. The source vectors and target vectors can be of different types. For instance, the $SrcV$ can be LSA vectors while the TgV can be Word2vec vectors and vice versa. Also, the dimensionality of the source and target vectors may be different. Using S different source models (as depicted in Fig. 2b), the effective size of the target model (Voc_{target}^e) will be increased greatly, particularly when there is less overlap among model vocabularies. For example, if one model is developed using a corpus containing academic text-books, such as Touchstone Applied Science Associates (TASA) corpus [11] and another model is built from Wikipedia articles, then many words in Wikipedia will be missing in the TASA-based model (the vocabulary of Wikipedia-based model is much larger than that of the TASA-based model). The proposed trans-formation model can map vectors derived using Wikipedia onto the TASA-based model, thus greatly increasing the coverage of the latter. Similarly, the TASA-based model's word coverage can be further increased by adding other source models.

Specifically, we developed Neural Network models to map between any two of the following vector-based word representation models: LSA, word2Vec, and GloVe. There are six different transformations such as LSA-to-GloVe or word2vec-to-GloVe. It is important to note that these models are quite different in their underlying principles to derive word representations and that they are all unsupervised. LSA is an algebraic method. Word2vec is a feed-forward neural network based language model and its bag of word model utilizes the context of four words (two before and two after). GloVe model is based on algebraic as well as probabilistic theories.

4   Evaluation Methods

We evaluated our transformation approach *intrinsically* and *extrinsically* using a simulation based approach in both cases, i.e. we simulated a set of missing words from an existing target model by removing them from it and then try-ing to project them back, comparing the transformed vectors with the origi-nal ones. This enables us to accurately assess the transformed vectors (TrVs) with respect to the native vectors in target model (TgVs). When comparing the obtained transformed representations with the native representations from the target model (the vectors that were purposely removed) we check if they are alike (intrinsic evaluation) and whether they perform similarly when used in NLP applications such as word-to-word similarity computations (extrinsic evaluation).

Intrinsic evaluation. We chose as our simulated missing words a set of N words that are present in both the source and target models so that the TrV could be directly compared to the TgV, i.e. the representations of the underlying target model itself. Then, we calculated an average correlation (r) score ($AvgCorr$) between the two vectors as shown in Eq. (3).

$$AvgCorr = \frac{1}{N} \sum_{i=1}^{N} r(TrV_i, TgV_i) \tag{3}$$

Extrinsic Evaluation. For an extrinsic evaluation of the transformed vectors, we used a word-to-word similarity task which is one of the approaches used to measure the quality of word representations. If word representations are good, then similar words will lead to high similarity scores whereas dissimilar words will lead to low similarity scores. Using a benchmark dataset containing pairs of words together with human-expert judgments of similarity, described in more detail in Sect. 5, we computed similarity scores between vector-based representations ($Sim(V_{w1}, V_{w2})$) of words using the standard cosine similarity measure (normalized dot-product) applied to the transformed vectors TrVs of those words. Then, an overall correlation (r) between the similarity scores and human judgments' scores were computed as shown below.

$$TrSim = r(\{(Sim(TrV_{i1}, TrV_{i2}), H_i)\}) \tag{4}$$

$$TgSim = r(\{(Sim(TgV_{i1}, TgV_{i2}), H_i)\}) \tag{5}$$

where $1 \leq i \leq K$, K: size (# word pairs) of word similarity evaluation dataset, H_i: human rated similarity score for i^{th} word pair in the word-to-word similarity dataset.

We repeated the process using TgVs. When using the transformed vectors we obtained an overall correlation similarity score denoted as $TrSim$ and when using the native vectors the overall correlation score across all word pairs in our benchmark dataset is denoted as $TgSim$. A comparable $TrSim$ score to the $TgSim$ score would indicate that the transformed vectors can act as a substitute for word representations of the target model.

Baselines. We also used two baseline approaches to obtain transformed representations. A baseline approach used randomly chosen word vectors from the source model to transform onto the target model (we denote this transformation as $RandV@Src$). A second baseline approach used randomly chosen word vector representations from the target model itself without using any transformation model ($RandV@Trg$). The $RandV@Src$ and $RandV@Trg$ vectors were then compared with the actual, native word vector representations. These baselines help detect whether the system is actually learning something or it simply does a random mapping.

5 Data

Selected Word Representation Models. We performed experiments with three different word representation models: (a) LSA model built using whole

Wikipedia articles, (b) Word2vec model, and (c) GloVe model. These models were developed independently by different groups and were downloaded "as-is" without any intervention on our part as our purpose was to take advantage of existing models without altering them in any way.

- **Word2vec**: This model is a pre-trained vector model based on the Google News dataset (about 100 billion words) and was developed by [16] at Google. The distributed word vectors were computed using feed forward neural network based on a skip-gram model.
- **GloVe**: The GloVe (Global Vector), developed at Stanford University, is an unsupervised learning model for representing words [21]. The model was trained on non-zero elements in a global word co-occurrence matrix. We used the pre-trained model GloVe-42B which was trained on 42 billion words of Common Crawl corpus and it contains about 1.9 million unique tokens.
- **LSA$_{wiki}$**: We used the LSA model generated from the whole set of English Wikipedia articles (an early-2013 snapshot) by [25]. The model was generated considering the lemmas of the content words that appeared at least 7 times in the corpus. This model contains 1.1 million unique entries.

All these models have 300-dimensional vectors, which, in the context of our research, is a pure coincidence as the dimensionality of various source and target models can be different.

Simlex-999. Simlex-999 (*Simlex*; [9]) is a recently released dataset for word-to-word similarity evaluation. In this dataset, the related but semantically less equivalent word pairs are rated with low similarity scores by human judges. For instance, *lemon* and *tea* are related but not similar, and therefore, they are rated with low similarity score. The dataset consists of 999 word pairs. But some of the words in Simlex-999 dataset were not available in LSA$_{wiki}$ and for consistency of our evaluation, we used only 955 word pairs that are available in all three word representation models. We used this dataset for both extrinsic and intrinsic evaluations.

Training, Validation, and Test Datasets. From the pre-trained Word2vec, GloVe, and LSA models, we extracted 107,813 vectors corresponding to the common words in all three models (only 107,813 words were common in all three models). For each pair of models, we set-aside 1,017 *Simlex* word vectors for intrinsic evaluation and the remaining ones were randomly assigned for training (95,000 pairs of vectors), validation (5,000 pairs), and intrinsic evaluation (5,000 pairs or *5k-test*). *Simlex* words were used for both intrinsic as well as extrinsic evaluation. That is, we used two datasets (*Simlex*, and *5k-test*) for intrinsic evaluation.

The difference between *Simlex* and *5k-test* is that the words in *Simlex* are curated words and contains only common and meaningful words whereas the words in *5k-test* are randomly selected from the vocabulary containing millions of words. All the words in *5k-test* are not necessarily meaningful ones (due to typos and other reasons) but this test set is bigger and practically more general. The remaining 1,796 vectors from the common vocabulary along with other

Table 1. Summary of training, validation, and test datasets. Pair of vectors correspond to the words common to both *source* and *target* model. The information in this table applies to each transformation model.

Data	Count (pairs)	Remarks
Training vectors	95,000	Used to build transformation model
Validation vectors	5,000	Used for validating transformation model
Simlex word vectors	1,017	Intrinsic evaluation (set 1)
5k-test vectors	5,000	Intrinsic evaluation (set 2)
Simlex words	955	Extrinsic (word-to-word similarity) evaluation
Baseline 1 (*Rand@Src*)	6,017	Source model vectors were randomly selected
Baseline 2 (*Rand@Trg*)	6,017	Randomly selected target model vectors used as TrVs

(excluding vectors in training, validation, and test: *Simlex* and *5k-test*) randomly selected vectors (6,017 in total) from the corresponding source/target model were used for evaluating the baseline transformations, *RandV@Src* and *RandV@Trg*. The vectors were normalized by their L^2-norms to be in the same scale. These datasets are summarized in Table 1.

6   Experiments and Results

We built NN models with a number of input units and output units equal to the size of the vectors in corresponding source and target models, respectively. They all were 300-dimension vectors (which was a pure coincidence and not a constraint of our mapping model). Therefore, the number of input units and output units were 300.

We added only one hidden layer keeping the model relatively simple and considering potential sparseness during training and performed experiments with varying number of hidden units. We developed those models using the neural network toolbox in Matlab (R2015a). The NN learning algorithm was set to the Scaled Conjugate Gradient [19] with logistic activation function and the number of iterations set to 1,000.

Each *source* → *target* transformation model was trained using the training dataset of 95k pairs of vectors. We did experiments with different number of hidden units from 100 to 800 incrementing by 100. The *AvgCorr* (see Eq. (3)) on the validation set was used to calibrate the number of hidden units in the NN models. The results were improving with the increasing number of hidden units up to 600. However, the differences among the results with 400–600 hidden units were very small in all models[5]. Therefore, we chose to use 600 hidden unit models for all pairs of *source* → *target* models. We then evaluated the learned models on the test data (*Simlex* and *5k-test*). The results are summarized in Table 2. The preliminary results using the unnormalized vectors only from GloVe and Word2vec models can be found in [2].

Table 2. Results of vector transformation models (Results for the *RandV@Src* and *RandV@Trg* are presented as a range, ↓ - same as next rows, Std - Standard deviation).

Source → Target	Word similarity		AvgCorr (TrV, TgV) (Std)	
	TgSim	TrSim	Simlex	5k-test
RandV@Src (baseline 1)	↓	∼0.0	0.0–0.251	0.0–0.187
RandV@Trg (baseline 2)		∼0.0	0.005–0.136	0.012–0.100
Word2vec→GloVe	0.427	0.446	0.748 (0.085)	0.488 (0.180)
LSA$_{wiki}$→GloVe		0.284	0.677 (0.092)	0.380 (0.160)
Word2vec→LSA$_{wiki}$	0.276	0.301	0.791 (0.104)	0.553 (0.214)
GloVe→LSA$_{wiki}$		0.292	**0.801** (0.103)	0.541 (0.217)
LSA$_{wiki}$→Word2vec	0.469	0.262	0.538 (0.089)	0.515 (0.147)
GloVe→Word2vec		0.369	0.676 (0.073)	**0.610** (0.116)

The *TgSim* column presents the correlations (r) between the word similarities computed using target vectors and the human annotated similarity scores (see Eq. (4)), for the word pairs in the *Simlex* dataset. The *TrSim* column shows the same correlations for word similarities but this time using transformed vectors (see Eq. (5)). The *TrSim* scores when compared with *TgSim* scores indicate how well the transformed vectors can act as a substitute for word representations in the target model. It is customary to interpret the word similarity results in *TrSim* with respect to *TgSim* as the goal here is to have the transformed vectors that perform as good as the native target model vectors. In fact, *Simlex* is considered as a difficult dataset when compared with other popular word similarity evaluation datasets such as WordSim-353 [8] because the related but not similar word pairs (e.g., *bread* and *butter*) in it are also assigned low similarity scores. And the correlation between similarity scores obtained using state-of-the-art word representation models and human judgment scores was found to be less than 0.5 for *Simlex* [9].

We can see in Table 2 that the word-to-word similarity results using the transformed vectors (*TrSim*) are comparable with or better in some cases than the results obtained using the native target model vectors (*TgSim*). For instance, the correlation between the similarity scores obtained using the native GloVe vectors and human judgments is 0.427 while the correlation (with human judgments) of similarity scores obtained using vectors transformed to GloVe from the

---

[5] In order to reduce the complexity of the model (or risk of overfitting), the number of hidden units could be set to 500 or 400 with small reduction in performance.

Word2vec model is better at 0.446 (see Word2vec→GloVe in Table 2). In some cases, particularly the results obtained using transformations from LSA$_{wiki}$ are relatively lower than those obtained by using the native target model vectors directly. For instance, $TgSim$ for GloVe is 0.427 but the results using the vectors transformed from LSA$_{wiki}$ to Glove is 0.284 (in LSA$_{wiki}$→Glove). Still, a correlation of 0.284 can be considered as good given the difficulty of the $Simlex$ dataset. Additionally, for each transformation we calculated a correlation score between word-to-word similarity scores calculated for the $Simlex$ dataset using the target model vectors and the similarity scores calculated using the transformed vectors. This correlation score was up to 0.842 (for the Word2vec→GloVe transformation), indicating that the transformed vectors behave similar to the target model vectors in calculating word-to-word similarity.

Moreover, the average correlation score of TrVs with corresponding TgVs (in $AvgCorr$ column) for $Simlex$ words was up to 0.801 (in Glove→LSA$_{wiki}$) and up to 0.610 (in Glove→Word2vec) for the $5k\text{-}test$ dataset. These correlation scores indicate that the transformed vectors closely resemble the target model vectors. On average, the $AvgCorr$ scores across all transformations were 0.705 and 0.514 for $Simlex$ words and words in $5k\text{-}test$, respectively. In some cases, particularly the LSA$_{wiki}$ model, the transformation process yielded relatively lower scores. We conclude that the Word2vec model and the GloVe model were more effective than LSA$_{wiki}$ for the task of word-to-word similarity. Nevertheless, it is important to note that the scientific community is still striving towards finding a common ground on what measures to use to evaluate the word representation models and which word representation models are more powerful than others [5] by organizing events, such as RepEval (Representation Evaluation) workshop [20].

Table 3. Examples of words in $5k\text{-}test$ set for which the correlation between Word2vec model representations and the representations obtained from GloVe by using our transformation model (GloVe→Word2vec) were high (on left), and low (on right).

Word	Correlation (TrV, TgV)	Word	Correlation (TrV, TgV)
whimsical	0.8428	poins	0.2225
fashionable	0.8421	witrh	0.2184
thinkers	0.8413	killingly	0.2159
kayaking	0.8411	poppermost	0.2120
likable	0.8403	pacifically	0.2046
estrategia	0.8400	witih	0.1945
nicer	0.8396	tasman	0.1887
whiny	0.8391	biolabs	0.1830

Correlations are relatively stronger and less spread (i.e, Std values are low) for $Simlex$ than $5k\text{-}test$. It seems that some of the words in $5k\text{-}test$ are not quite common (or meaningful) as compared to the *Simlex* words and the transformation of such words' representations was not very effective. Table 3 shows examples of words in *5k-test* for which the correlation between Word2vec model representations (TgVs) and the representations obtained from GloVe (TrVs) by using our transformation model GloVe→Word2vec are high (on left) and low (on right). We can see that the words on the right are rare words or misspelled words. Similarly, the correlation between TgV and TrV for misspelled word *"whihc"* found in *5k-test* is less than 0.35 in all *source → target* transformations.

Results for the *RandV@Src* and *RandV@Trg* baselines are presented as a range because the results were similar for all six different transformations. The highest average correlation ($AvgCorr$) was 0.251 for the GloVe to LSA$_{wiki}$ transformation of *Simlex* words. In all other cases, the correlations were below 0.2. The word similarity results ($TrSim$) were around zero. These mean that providing random vectors from the source model as input or using randomly selected words for missing words in the target model has no significant outcome. Additionally, we checked the direct correlation between native source and target vectors in each case but it was approximately zero when tested on *5k-test*. These indicate that learning a mapping function is needed and effective when properly done, as is the case in our work.

7   Conclusion

We have presented a novel approach of expanding the vocabularies of word representation models by mapping vectors from one model (*source*) to another (*target*). Our results with three different pre-trained models indicate that the Neural Network based vector mapping approach is effective as the resulting transformed vectors have an average correlation with the target model's vectors to 0.801 for the words in Simlex-999 dataset. The extrinsic evaluation using word-to-word similarity task with Simlex dataset shows the results obtained using the transformed vectors are comparable with that of using the target model's representations. The results indicate that the transformed vectors mostly behave similar to the target model vectors and, therefore, the transformed vectors can be used with confidence to augment the target model.

Such type of mappings that vastly increases the coverage of a target model can be very useful in many NLP applications which most likely need to handle missing words or phrases. Our experiments with pre-trained models (Word2vec, GloVe, and LSA$_{wiki}$) showed an increase in word coverage that varies from few to several times depending on which model vocabulary is taken as reference.

Nevertheless, finding out whether certain type of source or target model makes transformations more or less effective is a topic of future investigation. Additionally, the proposed solution can be used to obtain phrase representations which are even sparser than words. It is another interesting topic for future research.

The pooled representations will be an integral part of our career counseling system meant to guide users on career paths by holding conversational counseling sessions with them. One type of dialogue we have in mind is called motivational interviewing (MI; [18]), which could be useful when individuals are not happy with the pace of reaching their career goals. MI arose as a means of addressing a number of problems individuals face when, on the one hand they are aware that a change in behavior is needed, yet, at the same time are lacking in motivation and information and experiencing denial and ambivalence about changing. In these types of situations, approaches that are too directive, confrontational, or appeal to expert advice fall upon deaf ears. MI is best characterized as non-adversarial, non-confrontational, and non-judgmental. Our MI career counseling dialogues will use generic question asking, reflective listening statements, tactful advice giving and feedback sharing, and key summarizing statements ([23]). Users' response to the counseling system's prompts will be evaluated using the pooled representations presented in this paper.

Acknowledgments. The research was supported by the Office of Naval Research (N00014-00-1-0600, N00014-15-P-1184; N00014-12-C-0643; N00014-16-C-3027) and the National Science Foundation Data Infrastructure Building Blocks program (ACI-1443068). Any opinions, findings, and conclusions expressed are solely the authors'.

References

1. Alexandrescu, A., Kirchhoff, K.: Factored neural language models. In: Proceedings of the Human Language Technology Conference of the NAACL, Companion Volume: Short Papers, pp. 1–4. Association for Computational Linguistics (2006)
2. Banjade, R., Maharjan, N., Gautam, D., Rus, V.: Handling missing words by mapping across word vector representations. In: FLAIRS Conference, pp. 250–253 (2016)
3. Banjade, R., et al.: Nerosim: A system for measuring and interpreting semantic textual similarity. In: Proceedings of the 9th International Workshop on SemEval (Co-located with NAACL), pp. 164–171 (2015)
4. Baroni, M., Zamparelli, R.: Nouns are vectors, adjectives are matrices: representing adjective-noun constructions in semantic space. In: Proceedings of the 2010 Conference on Empirical Methods in Natural Language Processing, pp. 1183–1193. Association for Computational Linguistics (2010)
5. Batchkarov, M., Kober, T., Reffin, J., Weeds, J., Weir, D.: A critique of word similarity as a method for evaluating distributional semantic models (2016)
6. Bengio, Y., Ducharme, R., Vincent, P., Janvin, C.: A neural probabilistic language model. J. Mach. Learn. Res. **3**, 1137–1155 (2003)
7. Collobert, R., Weston, J., Bottou, L., Karlen, M., Kavukcuoglu, K., Kuksa, P.: Natural language processing (almost) from scratch. J. Mach. Learn. Res. **12**, 2493–2537 (2011)
8. Finkelstein, L., et al.: Placing search in context: the concept revisited. In: Proceedings of the 10th International Conference on World Wide Web, pp. 406–414. ACM (2001)
9. Hill, F., Reichart, R., Korhonen, A.: Simlex-999: evaluating semantic models with (genuine) similarity estimation. arXiv preprint arXiv:1408.3456 (2014)
10. Iacobacci, I., Pilehvar, M.T., Navigli, R.: Sensembed: learning sense embeddings for word and relational similarity. In: Proceedings of ACL, pp. 95–105 (2015)

11. Landauer, T.K., Foltz, P.W., Laham, D.: An introduction to latent semantic analysis. Discourse Process. **25**(2–3), 259–284 (1998)
12. Lei, T., Xin, Y., Zhang, Y., Barzilay, R., Jaakkola, T.: Low-rank tensors for scoring dependency structures. In: Proceedings of the 52nd Annual Meeting of the Association for Computational Linguistics, vol. 1, pp. 1381–1391 (2014)
13. Luong, M.T., Socher, R., Manning, C.D.: Better word representations with recursive neural networks for morphology. In: CoNLL-2013, p. 104 (2013)
14. Manning, C.D., Raghavan, P., Schütze, H.: Introduction to Information Retrieval, vol. 1. Cambridge University Press Cambridge, New York (2008)
15. Mikolov, T., Chen, K., Corrado, G., Dean, J.: Efficient estimation of word representations in vector space. arXiv preprint arXiv:1301.3781 (2013)
16. Mikolov, T., Sutskever, I., Chen, K., Corrado, G.S., Dean, J.: Distributed representations of words and phrases and their compositionality. In: Advances in Neural Information Processing Systems, pp. 3111–3119 (2013)
17. Miller, G.A.: Wordnet: a lexical database for english. Commun. ACM **38**(11), 39–41 (1995)
18. Miller, W.R.: Motivational interviewing with problem drinkers. Behav. Psychother. **11**, 147–172 (1983)
19. Møller, M.F.: A scaled conjugate gradient algorithm for fast supervised learning. Neural Netw. **6**(4), 525–533 (1993)
20. Nayak, N., Angeli, G., Manning, C.D.: Evaluating word embeddings using a representative suite of practical tasks. ACL **2016**, 19 (2016)
21. Pennington, J., Socher, R., Manning, C.D.: Glove: Global vectors for word representation. In: Proceedings of the Empiricial Methods in Natural Language Processing (EMNLP 2014) vol. 12, pp. 1532–1543 (2014)
22. Rus, V., Lintean, M.C., Banjade, R., Niraula, N.B., Stefanescu, D.: Semilar: the semantic similarity toolkit. In: ACL (Conference System Demonstrations), pp. 163–168. Association for Computational Linguistics (2013)
23. Sobell, L.C., Sobell, M.B.: Motivational interviewing strategies and techniques: rationales and examples (2008). http://www.nova.edu/gsc/forms/mi_rationale_techniques.pdf
24. Socher, R., et al.: Recursive deep models for semantic compositionality over a sentiment treebank. In: Proceedings of the Conference on Empirical Methods in Natural Language Processing (EMNLP), vol. 1631, p. 1642 (2013)
25. Stefanescu, D., Banjade, R., Rus, V.: Latent semantic analysis models on wikipedia and tasa. In: Proceedings of the Language Resources and Evaluation Conference, pp. 1417–1422 (2014)
26. Turian, J., Ratinov, L., Bengio, Y.: Word representations: a simple and general method for semi-supervised learning. In: Proceedings of the 48th Annual Meeting of the Association for Computational Linguistics, pp. 384–394. Association for Computational Linguistics (2010)
27. Yu, M., Dredze, M.: Improving lexical embeddings with semantic knowledge. In: Association for Computational Linguistics (ACL), pp. 545–550 (2014)

Strategies to Select Examples for Active Learning with Conditional Random Fields

Vincent Claveau$^{(\boxtimes)}$ and Ewa Kijak

IRISA - CNRS - Univ. of Rennes 1, Campus de Beaulieu, Rennes, France
{vincent.claveau,ewa.kijak}@irisa.fr

Abstract. Nowadays, many NLP problems are tackled as supervised machine learning tasks. Consequently, the cost of the expertise needed to annotate the examples is a widespread issue. Active learning offers a framework to that issue, allowing to control the annotation cost while maximizing the classifier performance, but it relies on the key step of choosing which example will be proposed to the expert. In this paper, we examine and propose such selection strategies in the specific case of Conditional Random Fields (CRF) which are largely used in NLP. On the one hand, we propose a simple method to correct a bias of some state-of-the-art selection techniques. On the other hand, we detail an original approach to select the examples, based on the respect of proportions in the datasets. These contributions are validated over a large range of experiments implying several datasets and tasks, including named entity recognition, chunking, phonetization, word sense disambiguation.

Keywords: CRF · Conditional random fields · Active learning
Semi-supervised learning · Statistical test of proportion

1 Introduction

Many NLP tasks rely on supervised machine learning. Among the commonly used techniques, Conditional Random Fields (CRF) exhibit excellent performance for tasks related to the sequences annotation (tagging, named entity recognition and information extraction, transliteration...). However, as with all supervised approaches, the cost of the sequence annotation needed to train the models is an important criterion to consider. For simple problems, such as labeling parts-of-speech, some studies show that this cost is relatively low [7], but most of the problems mentioned above rather require a very large number of annotations (see Sect. 5.2).

To reduce, or at least control, this cost, semi-supervised approaches exploit, in addition to annotated examples, non-annotated examples that are more readily available. Among these approaches, Active Learning allows the expert to annotate additional examples iteratively, thereby controlling the compromise between annotation cost vs. performance of the classifier. Thus, a classifier can be learned or improved at each iteration, and can be used to guide the selection

© Springer Nature Switzerland AG 2018
A. Gelbukh (Ed.): CICLing 2017, LNCS 10761, pp. 30–43, 2018.
https://doi.org/10.1007/978-3-319-77113-7_3

of future examples to annotate. In this article, we are interested in this active learning process, and more specifically in the issue of the selection of examples which are provided to the expert, in the particular case of CRF.

Many methods of selection, either generic to any machine learning algorithm or specific to the CRF (Sect. 2) have already been developed. In this article, we show that some very conventional methods of the state of the art comprise a bias tending to favor the choice of long examples, that is examples that are expensive to annotate. The first contribution of the paper is to propose a simple technique to remove this bias (Sect. 3). Another contribution is to propose an original selection technique, relying on the data representations used by the CRF, and based on a criterion balancing the proportions of the attributes in the datasets (Sect. 4). These different proposals are experimentally evaluated on several datasets and traditional tasks of CRF (Sect. 5).

2   Context and Related Work

2.1   Basic Notions

Conditional Random Fields [9] are undirected graphical models that represent the probability distribution of annotation y on observations x. They are widely used in NLP thanks to their ability to take into account the sequential aspect and rich descriptions of text sequences. They have been successfully used in many tasks casted as annotation problems, and have become standard tools for information extraction, named entity recognition, tagging, etc. [4,17,18,26, inter alia]. In such cases, x is a sequence of letters or words and y the corresponding sequence of labels. In this context, the conditional probability $P(y|x)$ is defined through a weighted sum of so-called feature functions f_j:

$$P(y|x,\theta) = \frac{1}{Z_\lambda(x)} \exp\left(\sum_j \sum_t \lambda_j f_j(x, y_t, y_{t-1}, t)\right)$$

where $Z_\lambda(x)$ is a normalization factor and θ is the vector of λ_j weights. The feature functions are often binary, returning 1 when a certain combination of labels and observations attributes is satisfied, 0 otherwise. They are applied to each position t of the sequence and the weight λ_j reflects their importance to determine the label. It is important to note that in practice the vector x is not considered as a whole, but only some combinations of attributes on observations around the position t in x are considered. These combinations are user-defined, usually indirectly through a set of patterns $\{\text{Pat}_i\}$. They are applied at each position t of each sequence x ($\text{Pat}_i(x,t)$), and with the information of the labels (y_{t-1} and y_t), they define all the possible feature functions.

The learning step for a CRF consists in estimating the weights λ_j from data with known labels. The weights are those that maximize the model log-likelihood on the training (labeled) sequences, for instance with quasi-Newton type algorithms such as L-BFGS [20]. Once learned, applying the CRF model to the new data consists in finding, for a sequence of observations x, the most probable sequence of labels (denoted y^* in the rest of this article), for example with a Viterbi algorithm.

2.2   Semi-supervised Learning and Active Learning

Semi-supervised learning consists in using annotated data (noted $\mathcal{T}$ hereafter) and non-annotated data ($\mathcal{N}$). Its purpose is to reduce the number of annotations and therefore the cost of the annotation, and/or to yield the best classifier performance for a given annotation cost. Different semi-supervised learning approaches have been explored in the context of CRF. Several studies use unlabeled data directly in training the model by modifying the expression of entropy. This change makes the objective function non-concave and therefore requires to adapt the learning process. Another family of approaches consists in adapting the learning and decoding procedures of CRF so that they are able to handle some other knowledge about the sequences rather than completely annotated sequences. For example, this knowledge may be partial annotation of the sequences (labels are known only for a few words [19]). It can also be a priori knowledge on the distribution of labels knowing certain attributes [12]. Although this is not strictly semi-supervised learning, let us mention the work using close techniques exploiting non-annotated data to improve learning on annotated data. For instance, [13] and [6] propose to cluster non-annotated data to build new feature –in this case, word classes– then used to better describe the (labeled) data. In this vein, it is also worth mentioning the work of [2] and those of [23]. They exploit the proximity of an annotated sequence with other sequences to bias the estimation of the CRF parameters. Although the framework of these studies is different from the work presented in this paper, they nonetheless share the idea of exploiting similarity between sequences seen as sets of features.

In this paper, the specific semi-supervised learning framework considered is known as active learning. Its principle is that supervision is carried out by an expert (or oracle) iteratively and interactively [22]. This is often set out in an algorithm whose main steps are as follows:

(1) infer a classifier from $\mathcal{T}$;
(2) apply the classifier to $\mathcal{N}$;
(3) select examples from $\mathcal{N}$;
(4) make an expert label these examples and add them to $\mathcal{T}$;
(5) go to step 1.

This process is repeated until a stopping criterion is reached (e.g. maximal cost of annotation, minimum classifier performance, or $\mathcal{N}$ is empty).

The crux of these active learning algorithms is step 3, that is the selection of examples to be labeled by the expert. One wants to choose the most beneficial examples for learning, in order to get the best classification performance. This selection problem is often based on the results of the current classifier (Step 2). Much work has been proposed in this regard, particularly in the field of NLP

[14] where these labeling problems are common. Regardless of the classifiers used, several families of selection strategies were proposed. The most common one is the uncertainty-based selection: the results from Step 2 are used to select examples for which the current classifier is less confident (see Sect. 3). A known drawback of this approach is that, at the beginning of the process, when there are few examples annotated, the classifier uncertainty measurements are unreliable. Another very common selection strategy is the selection by committee. Its principle is to learn not one but several classifiers in Step 1, then apply them to $\mathcal{N}$, and finally select examples on which they disagree the most. This approach is often implemented by techniques such as *bagging* and/or *boosting* [1], or by learning different classifiers from different representations of the data [16]. Beside the important computational cost generated by these multiple learning, these techniques also suffer from the same problem as uncertainty-based selection: classifiers are unreliable in the early rounds of iteration when $|\mathcal{T}|$ is small. Another family of selection techniques relies on the expected change in the model caused by adding new examples. The principle here is to select the sample that would impact most the model, assuming that this impact would result in improved performance. The underlying intuition is that the examples chosen in $\mathcal{N}$ will cover cases that are not covered by the examples of $\mathcal{T}$. Practical implementation of this approach heavily depends on the classifier used. [21] proposed several variants of this approach for CRF; only one, named *Information Density*, gave some positive results. It works by selecting the most different sequence in $\mathcal{N}$ with respect to those of $\mathcal{T}$. To assess this difference, the authors represent the sequences by a vector representing the combination of the sequence attributes, as captured by the feature functions. Since the labels of the sequences of $\mathcal{N}$ are unknown, only the features concerning x are considered. The most dissimilar sequence is simply defined as the one having the smallest average cosine with the sequences of $\mathcal{T}$.

This latter approach is close to those presented in this article: we also make use of sequence representation as sets of attributes, although the criteria we propose is more efficient than [21]'s one (Sect. 4). Furthermore, the evaluation method used in their study does not properly account for the annotation effort at each iteration: the authors evaluate performance based on the number of labeled sequences, without considering that some can be much longer than others. For our part, a more realistic setting is adopted: the annotation effort is measured in terms of annotated words (or sequence elements), which has implications for selection strategies tested by these authors (next section).

2.3   Experimental Context

In the remainder of this article, we will validate our proposals for sequence selection on different tasks for which the CRF are conventionally used. We briefly describe these tasks and data below; for details, the interested reader can refer to the provided references.

We use the dataset of the entity recognition task named the ESTER campaign [8]. It contains 55,000 breath groups from transcripts of radio broadcasts in

French; the named entities are annotated into 8 classes (person, place, time...). The CoNLL2002 dataset was proposed for the named entity recognition task in Dutch proposed at CoNLL 2002 [24]. It contains 4 different entity types; 14,000 sequences (sentences) are used in the experiments reported in the following section. The CoNLL2000 dataset contains English newspapers annotated with chunks [25], totaling about 11,000 sentences and 4 classes (3 types of chunks and a label 'other'). We also experiment with the Sense Disambiguation dataset from Senseval-2 [5]: disambiguation of *hard, line, serve, interest*, each of their senses being represented by a different label in about 16,000 sentences. A somewhat different task is the phonetic transcription of isolated words in English provided by Nettalk dataset. The goal is to transcribe these words in a specific phonetic alphabet. This task is seen as a letter-by-letter annotation task. It has 18,000 words and 52 different labels corresponding to the phonetic alphabet. A preliminary step of data was to align words with their phonetic transcription (and thus to introduce the appropriate symbols 'empty' when needed).

The data are described with usual attributes and patterns for these tasks, with the parts-of-speech, lemmas, capital presence/absence, etc., and the BIO annotation scheme is adopted when necessary (ESTER, CONLL2002, CONLL2000). Nine tenths is used for training (set $\mathcal{T}$ and $\mathcal{N}$) and the remaining tenth is used for performance evaluation. In most cases, the performance measure used is the word accuracy (rate of correctly labeled words), except for the phonetization task, which is evaluated by the sequence accuracy rate (a word must be completely and correctly phonetized). This evaluation is performed at each iteration and related to the annotation effort i.e. the number of words (or symbols) to which the expert added a label.

The CRF implementation used is WAPITI [10], with its default settings unless stated otherwise. It should be noted that tests with other settings (optimization algorithms, normalization ...), not reported in the article, do not change the conclusions presented.

3   Uncertainty-Based Selection

As we have seen, a common solution for the selection of examples to annotate at each iteration is to propose to the oracle those for which the classifier learned at the previous iteration is less certain. With CRF, this means choosing the sequence x by looking at the probabilities $P(y|x; \theta)$.

3.1   Minimal Confidence and Sequence Entropy

Among the different ways to proceed, [21] shows that two strategies in this family perform well in most cases: (i) the selection with minimal confidence, and (ii) selection from sequence entropy. The first simply consists in choosing in $\mathcal{N}$ the (automatically labeled) sequence whose probability is minimal with the current model: $x = \mathrm{argmin}_{x \in \mathcal{N}} P(y^*|x, \theta)$. The entropy method selects the sequence x

with the greatest entropy over all the possible labels y of this sequence:

$$x = \underset{x \in \mathcal{N}}{\operatorname{argmax}} \left(-\sum_y P(y|x, \theta) \log P(y|x, \theta) \right)$$

3.2 Length Bias

One of the problems of these state-of-the-art approaches is that they tend to choose the longest sequences, as they often have lower probabilities than short sequences. However, the annotation cost is proportional to the sequence length. If one seeks to maximize performance for a minimal cost annotation, it is then potentially an undesirable behavior. To illustrate this, we report in Table 1 correlation between the sequence lengths in the ESTER dataset and their probabilities given by two models respectively trained on 20 and 10,000 randomly chosen sequences.

Table 1. Correlation (Pearson r, Spearman ρ, Kendall τ) with their p-value between the sequence lengths and their probabilities according to two models respectively trained on 20 and 10,000 sequences.

Size of training set	Pearson r	(p-value)	Spearman ρ	(p-value)	Kendall τ	(p-value)
20 seq	-0.52	$(<10^{10})$	-0.59	$(<10^{10})$	-0.44	$(<10^{10})$
10,000 seq	-0.47	$(<10^{10})$	-0.56	$(<10^{10})$	-0.40	$(<10^{10})$

The length bias can be observed in both cases: in average, the sequence probability given by a CRF model is correlated to its length. This is particularly more pronounced when the model is trained on few sequences, which is precisely characteristic of the first iterations of active learning. Thus, this selection criterion is particularly unsuited at the beginning of active learning. Conversely, a simple normalization of the probabilities by the length of the sequences tends to favor very short sequences which does not provide enough useful information for learning.

3.3 Normalization

Based on the above findings, it seems important to normalize with respect to the sequence length. We propose a local, adaptive method of normalization based on the average probability of sequences for a given length. For this, we propose a method of normalization inspired by the Parzen window estimation method [15,27]. The underlying idea is that for a fixed sequence length (plus or minus ϵ), the normalized probability scores should be distributed uniformly between 0 and 1. For a sequence x of $\mathcal{N}$ of length l, we estimate the average $\hat{\mu}_l$ and standard deviation $\hat{\sigma}_l$ probabilities on all sequences of $\mathcal{N}$ of length $l \pm \epsilon$, i.e. the set $\{P(y'^*|x') \mid x' \in \mathcal{N}, |x'| = |x| \pm \epsilon\}$. These values are estimated at each iteration, and then used to center and reduce the probabilities used in the previous selection strategies. For example, the selection by minimal confidence is now:

$$x = \underset{x \in \mathcal{N}}{\operatorname{argmin}} \left(\frac{P(y^*|x, \theta) - \hat{\mu}_l}{\hat{\sigma}_l} \right)$$

For each considered close length, it should modify the probability dispersion for sequences of this length, and thus cancel the bias of sequence length previously observed. In practice, in the experiments reported in Sect. 5, same length sequences are not found using a fixed ϵ but by neighborhood: $\hat{\mu}_l$ is calculated over a fixed number of sequences whose lengths are closest to the one considered. This k-nearest-neighbor approach can better handle cases of *outlier* sequences with very different lengths for which a neighborhood defined with a small ϵ would not cover any other sequence.

4   Representativity of Feature Functions

The main proposal of this article is to consider that the distribution of attributes, such as captured by the feature functions, can guide the selection of examples to be annotated during an active learning iteration. To support this intuition, we first study how these attributes are distributed in terms of frequency and in terms of use in the models (Subsect. 4.1). Based on these considerations, Subsect. 4.2 proposes an original method to select sequences to annotate.

4.1   Preliminary Study

The feature functions encode the relationship between the description of sequences and labels, as expressed by the patterns $\{\text{Pat}_i\}$. It is interesting to observe their frequencies in the data, in order to see which ones among them are actually used for the prediction. CRF are known to produce large models in the sense that many parts of the data, as seen through the feature functions, are kept in the model [3, 28, for elements of discussion].

In order to study which functions are actually used in the model for the prediction, we first calculate the distribution of the occurrences of all possible feature functions f_j on ESTER data:

$$occ(f_j) = |\{f_j(x^{(m)}, y_{t-1}^{(m)}, y_t^{(m)}, t) = 1 | \forall \text{ example } m, \forall \text{ position } t\}|$$

We then extract from a model trained on the data the feature functions whose weight $|\lambda_j| > 0$. Among the learning settings for CRF, L1 or L2 normalization greatly influences the number of feature functions with non-zero weight. So, a model with a standard L1 and another with a normalization *elastic-net* (mixing equally L1 and L2) are trained on the whole ESTER dataset (full supervision). Figure 1 reports these three distributions.

As expected, we observe that these three distributions are very similar except for the rarest feature functions, especially with the L1 model. Most combinations

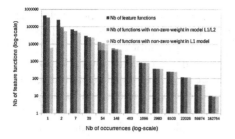

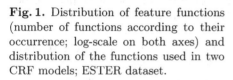

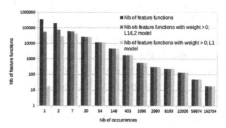

Fig. 1. Distribution of feature functions (number of functions according to their occurrence; log-scale on both axes) and distribution of the functions used in two CRF models; ESTER dataset.

Fig. 2. Distribution of feature functions without label information (number of functions according to their occurrence number; log-scale on both axes) and distribution of the functions used in two CRF models; ESTER dataset.

of attributes/labels from the data therefore appear useful (i.e. their weight $|\lambda_j| > 0$) for predictions in our two models. It means that the CRF models exploit a vast majority of attributes/labels combinations present in the data, in proportion to their frequency in the data: the fact that combinations are very common or rare does not intervene (except for the rarest configurations with L1 model). Thus, to build a smaller training set leading to models with similar characteristics, it seems important to offer the maximum variety of combinations accordingly with these proportions, i.e. respecting the distribution of attribute/label combinations of the whole dataset. This result is not specific to the ESTER dataset: the same distributions are observed for every tested dataset (see Sect. 2.3).

In our semi-supervised case, most of the data are not annotated. It is therefore important to check whether these earlier findings are still true without considering the labels. We therefore examine the distribution of feature functions regardless of labels, i.e. only by looking at the attributes concerning x in $\{f_j\}$. These incomplete feature functions (without label information) are noted f_j^*. Formally, we count in the data:

$$occ(f_j^*) = |\{f_j(x^{(m)}, y_1, y_2, t) = 1 | \forall \text{ example } m, \forall \text{ position } t, \forall \text{ labels } y_1, \ y_2\}|.$$

Figure 2 thus illustrates again the occurrences of feature functions, but regardless of the label. The same trends as before can be observed. These experiments suggest the importance of a varied and representative training set of all combinations of attributes (with no information on the label) defined by the feature functions.

4.2   Test of Proportion

We build on the previous observation to propose a new selection strategy. At each iteration of the active learning algorithm, we want the training set which is the most representative of the whole dataset. In other words, we want the sequence distribution, as seen by CRF via feature functions, to be as close as possible to those of $\mathcal{T} \cup \mathcal{N}$. As before, each sequence is seen as the set of feature functions that can be generated from it, not including labels.

To select the sequence x to add to the training set at each iteration (once annotated by the oracle), we need to evaluate how the resulting training set $\mathcal{T} \cup \{x\}$ compares with the whole data at our disposal (annotated or not, i.e. $\mathcal{T} \cup \mathcal{N}$). For each feature function, we propose to simply examine whether the proportion of this function observed in the sample $\mathcal{T} \cup \{x\}$ is comparable to that of the sample $\mathcal{T} \cup \mathcal{N}$. These two samples are not independent, but can be considered as such when $|\mathcal{N}| \gg |\mathcal{T}|$, which is ensured in the first iterations of active learning.

More specifically, we perform a statistical test of proportion between the two samples $\mathcal{T} \cup \{x\}$ and $\mathcal{T} \cup \mathcal{N}$, respectively denoted 1 and 2, with size n_1 and n_2. Let $\hat{p}_1^j = r_1^j/n_1$ be the estimator of the proportion of occurrences of a given feature function f_j appearing r_1^j times in sample 1, and $\hat{p}_2^j = r_2^j/n_2$ be the one for sample 2. We can then calculate the z-score:

$$z_{j,x} = \frac{\hat{p}_1^j(f_j) - \hat{p}_2^j(f_j)}{\sqrt{\hat{p}^j \times (1 - \hat{p}^j) \times (1/n_1 + 1/n_2)}} \qquad \text{with} \qquad \hat{p}^j = \frac{r_1^j + r_2^j}{n_1 + n_2}$$

The z-score follows a standard normal distribution, allowing us to calculate the probability $P(z_{j,x})$ to observe such a difference in proportion between the two samples. A high probability intuitively means that sample 1 contains a proportion of the feature function f_j comparable to that of sample 2.

It is necessary to combine these probabilities for all feature functions. In order to do so, we make a simplifying assumption by considering that the observations of feature functions are independent. Although this assumption is invalid in most cases, it allows us to propose a simple estimate of the overall probability of the sample x as the product of $P(z_{j,x})$ for every feature function f_j. Finally, the choice of the sequence to add to the training set is the one maximizing this probability: $x^* = \text{argmax}_{x \in \mathcal{N}} \prod_j P(z_{j,x})$.

5   Experiments

In this section, we compare experimentally the different selection strategies for active learning previously discussed. The experimental framework is detailed below, and learning curves are presented in Subsect. 5.2.

5.1   Settings

Several selection strategies are experimented: on the one hand, for comparison purposes, we implemented state-of-the-art strategies, namely, selection by minimal confidence, entropy and information density. We also added a simple baseline in which the sequences are selected at random. On the other hand, we tested the normalization process for the minimal confidence selection (cf. Sect. 3.3) and the approach based on proportion (cf. Sect. 4). We do not report results based on selection by committee as they yield lower results than the previous ones in almost every case [21].

All these methods are tested under the same conditions (CRF parameters, patterns...). For initialization, a sequence is randomly chosen to serve as the first example (the same for all selection methods). At each iteration, a single example is selected to be annotated by the oracle and the classifier is re-trained on all annotated data (therefore, this is not an update of the previous CRF model).

5.2   Results

Figures 3, 4, 5, 6 and 7 give the learning curves on our different datasets. The performance of the classifiers learned at each iteration is expressed in function of the cost of accumulated annotation of the set $calT$ (i.e. total number of words or symbols seen, according to the task). In the figures, the cost is reported on a logarithmic scale, so one can appreciate the different cases (few annotations vs. many annotations). Several observations stand in. First, these curves have very different appearance from a dataset to another. This is explained by the characteristics of tasks and data, implying that some are more readily feasible with good performance with few annotations (CoNLL2000) or not (CoNLL2002). For all datasets except Nettalk, differences, especially when the annotation cost is small, are sensitive. Regarding Nettalk, it is more difficult to bring out a selection method better than the other. This can certainly be explained by the difficulty of the task and, more precisely, by the huge number of possible labels. Indeed, there are a very large number of possible attributes/labels configurations; therefore, in all cases, it requires an extremely large number of examples to cover all these configurations.

Second, we observe that the three strategies from literature offer an average performance sometimes not far from the *random* strategy. Strategies by minimal confidence and entropy are even sometimes well below *random* (SenseEval-2), obviously penalized by their biases discussed in Sect. 3. This is important to

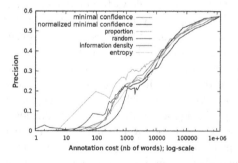

Fig. 3. Learning curve (precision rate vs. annotation cost expressed in words); ESTER dataset; log-scale

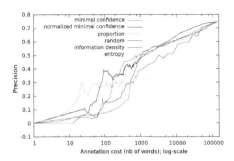
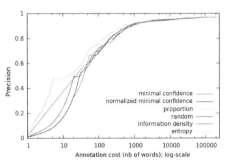

Fig. 4. Learning curve (precision rate vs. annotation cost expressed in words); CoNLL2002 dataset; log-scale

Fig. 5. Learning curve (precision rate vs. annotation cost expressed in words); CoNLL2000 dataset; log-scale

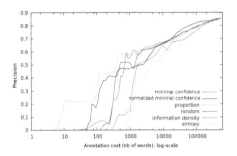
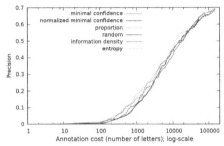

Fig. 6. Learning curve (precision rate vs. annotation cost expressed in words); SensEval-2 dataset; log-scale

Fig. 7. Learning curve (precision rate in terms of correctly phonetized vs. annotation cost expressed in letters); Nettalk dataset; log-scale

note; it is often overshadowed by evaluations taking into account the number of sequences, as we have already pointed in the work of [21].

Third, our normalization approach, applied to the minimal confidence strategy, gives satisfying results since it allows to get better or similar results to the non normalized version. It especially performs best when the number of annotation is important (ESTER, CoNLL2002, Senseval-2) even if the logarithmic scale in the figures hides a little this long domination.

Finally, our selection proposal based on proportion tests obtains very good results overall. It behaves generally better than other selection techniques, including *information density*, from which it is conceptually close. It may be noted that our strategy brings a significant gain when dealing with few annotations. This outcome is explained by the fact that the method does not rely on the predictions, unreliable at this stage, of the current classifier. However, this gain is less or even absent compared to other methods when the amount of annotated data becomes very important. This shows the limits of our approach, which does not exploit any information from the classifier, but it also allows to devise joint strategies in which the classification information would also be used when a minimal number of annotations is reached.

6   Conclusive Remarks

At a time when most NLP problems are tackled as supervised learning tasks, the cost of annotations by expert is a significant problem. Active learning provides a framework to control this cost while maximizing, hopefully, the classifier performance. As we have seen, it is in fact largely dependent on the example selection strategy implemented. In this article, we looked at some of these strategies and we have demonstrated a bias lowering their annotation cost/performance ratio. The normalization that we have proposed can solve this problem in a very simple manner while providing a significant performance gain. And when the annotation costs are limited, our strategy based on an original criterion of proportionality, appears the most advantageous on the several NLP taks examined. Of course, these gains are only appreciable in a real semi-supervised context in which one wants to get the best performance from a few annotated data; when a large amount of data is available, all the strategies tends to give similar results.

Many variations, improvements and research avenues can be explored. Among them, we would try to take into account the dependence between feature functions. In our current proposal, they are considered to be independent for simplification purpose, which is never the case in practice. These dependencies may even be very important because the patterns used to build these feature functions often exploit several times to the same elements (lemma of the current word, PoS the current word ...), and that these elements are themselves in a dependency relationship. This can strongly impact the estimate of the overall proportion probabilities, and ultimately distort the choice of the best example.

Another promising approach is to mix these different selection techniques to combine their benefits. They can obviously be simply merged (vote, product of scores or ranks ...), but it seems more interesting to aim more complex combinations, which could be achieved with *learning to rank* approaches [11].

Finally, in our current framework, the selected sequences are fully annotated. It would be interesting to study the case of partial annotations, under the same constraints to optimize the cost/performance ratio, taking inspiration for example from [19].

Acknowledgments. This work was partly funded by a French government support granted to the CominLabs LabEx managed by the ANR in *Investing for the Future* program under reference ANR-10-LABX-07-01.

References

1. Abe, N., Mamitsuka, H.: Query learning strategies using boosting and bagging. In: Proceedings of the Fifteenth International Conference on Machine Learning. Morgan Kaufmann Publishers Inc, Madison (1998)
2. Ando, R.K., Zhang, T.: A high-performance semi-supervised learning method for text chunking. In: Proceedings of the 43rd Annual Meeting on Association for Computational Linguistics. ACL 2005, pp. 1–9. Association for Computational Linguistics, Stroudsburg (2005). https://doi.org/10.3115/1219840.1219841
3. Chen, S.: Performance prediction for exponential language models. In: Proceedings of Annual Conference of the North American Chapter of the Association for Computational Linguistics, pp. 450–458, June 2009
4. Constant, M., Tellier, I., Duchier, D., Dupont, Y., Sigogne, A., Billot, S.: Intégrer des connaissances linguistiques dans un CRF: application à l'apprentissage d'un segmenteur-étiqueteur du français. In: Traitement Automatique du Langage Naturel (TALN'11), Montpellier, France (2011)
5. Edmonds, P., Cotton, S.: Senseval-2: Overview. In: Proceedings of SENSEVAL-2 Second International Workshop on Evaluating Word Sense Disambiguation Systems, pp. 1–5. Association for Computational Linguistics (2001). http://aclweb.org/anthology/S01-1001
6. Freitag, D.: Trained named entity recognition using distributional clusters. In: Proceedings of the Conference EMNLP (2004)
7. Garrette, D., Baldridge, J.: Learning a part-of-speech tagger from two hours of annotation, pp. 138–147, June 2013. http://www.cs.utexas.edu/users/ai-lab/?garrette:naacl13
8. Gravier, G., Bonastre, J.F., Geoffrois, E., Galliano, S., Tait, K.M., Choukri, K.: ESTER, une campagne d'évaluation des systèmes d'indexation automatique. In: Actes des Journées d'Étude sur la Parole, JEP, Atelier ESTER2 (2005)
9. Lafferty, J., McCallum, A., Pereira, F.: Conditional random fields: probabilistic models for segmenting and labeling sequence data. In: International Conference on Machine Learning (ICML) (2001)
10. Lavergne, T., Cappé, O., Yvon, F.: Practical very large scale CRFs. In: Proceedings the 48th Annual Meeting of the Association for Computational Linguistics (ACL), pp. 504–513. Association for Computational Linguistics, July 2010. http://www.aclweb.org/anthology/P10-1052
11. Liu, T.Y.: Learning to rank for information retrieval. Found. Trends Inf. Retrieval **3**(3), 225–331 (2009)
12. Mann, G.S., McCallum, A.: Generalized expectation criteria for semi-supervised learning of conditional random fields. In: Proceedings of ACL-08: HLT, Colombus, Ohio, USA, pp. 870–878 (2008)
13. Miller, S., Guinness, J., Zamanian, A.: Name tagging with word clusters and discriminative training. In: Procedings of the Conference ACL (2004)
14. Olsson, F.: A literature survey of active machine learning in the context of natural language processing. Technical report. Swedish Institute of Computer Science, Swedish Institute of Computer Science (2009)
15. Parzen, E.: On estimation of a probability density function and mode. Ann. Math. Stat. **33**, 1065–1076 (1962)
16. Pierce, D., Cardie, C.: Limitations of co-training for natural language learning from large datasets. In: Proceedings of the 2001 Conference on Empirical Methods in Natural Language Processing (EMNLP 2001), Pittsburgh, Pennsylvania, USA (2001)

17. Pranjal, A., Delip, R., Balaraman, R.: Part Of speech Tagging and Chunking with HMM and CRF. In: Proceedings of NLP Association of India (NLPAI) Machine Learning Contest (2006)
18. Raymond, C., Fayolle, J.: Reconnaissance robuste d'entités nommées sur de la parole transcrite automatiquement. In: Actes de la conférence Traitement Automatique des Langues Naturelles. Montréal, Canada (2010)
19. Salakhutdinov, R., Roweis, S., Ghahramani, Z.: Optimization with EM and Expectation-Conjugate-Gradient. In: Proceedings of the conference ICML (2003)
20. Schraudolph, N.N., Yu, J., Günter, S.: A stochastic quasi-Newton method for online convex optimization. In: Proceedings of 11th International Conference on Artificial Intelligence and Statistics. Workshop and Conference Proceedings, vol. 2, San Juan, Puerto Rico, pp. 436–443 (2007)
21. Settles, B., Craven, M.: An analysis of active learning strategies for sequence labeling tasks. In: Proceedings of the Conference on Empirical Methods in Natural Language Processing (EMNLP), pp. 1069–1078. ACL Press (2008)
22. Settles, B.: Active learning literature survey. Computer Sciences Technical report 1648, University of Wisconsin-Madison (2010)
23. Smith, N., Eisner, J.: Contrastive estimation: Training log-linear models on unlabeled data. In: Proceedings of ACL (2005)
24. Tjong Kim Sang, E.F.: Introduction to the CoNLL-2002 shared task: language-independent named entity recognition. In: Proceedings of CoNLL-2002, Taipei, Taiwan, pp. 155–158 (2002)
25. Tjong Kim Sang, E.F., Buchholz, S.: Introduction to the conll-2000 shared task: Chunking. In: Cardie, C., Daelemans, W., Nedellec, C., Tjong Kim Sang, E. (eds.) Proceedings of CoNLL-2000 and LLL-2000, Lisbon, Portugal, pp. 127–132 (2000)
26. Wang, T., Li, J., Diao, Q., Wei Hu, Y.Z., Dulong, C.: Semantic event detection using conditional random fields. In: IEEE Conference on Computer Vision and Pattern Recognition Workshop (CVPRW 2006) (2006)
27. Wasserman, L.: All of Statistics: A Concise Course in Statistical Inference. Springer Texts in Statistics. Springer, New York (2005). https://doi.org/10.1007/978-0-387-21736-9
28. Zhou, H., Hastie, T.: Regularization and variable selection via the elastic net, pp. 301–320 (2005)

Label-Dependencies Aware Recurrent
Neural Networks

Yoann Dupont[(✉)], Marco Dinarelli, and Isabelle Tellier

LaTTiCe (UMR 8094), CNRS, ENS Paris, Université Sorbonne Nouvelle - Paris 3
PSL Research University, USPC (Université Sorbonne Paris Cité),
1 rue Maurice Arnoux, 92120 Montrouge, France
`yoa.dupont@gmail.com, marco.dinarelli@ens.fr,`
`isabelle.tellier@univ-paris3.fr`

Abstract. In the last few years, Recurrent Neural Networks (RNNs) have proved effective on several NLP tasks. Despite such great success, their ability to model *sequence labeling* is still limited. This lead research toward solutions where RNNs are combined with models which already proved effective in this domain, such as CRFs. In this work we propose a solution far simpler but very effective: an evolution of the simple Jordan RNN, where labels are reinjected as input into the network, and converted into embeddings, in the same way as words. We compare this RNN variant to all the other RNN models, Elman and Jordan RNN, LSTM and GRU, on two well-known tasks of Spoken Language Understanding (SLU). Thanks to label embeddings and their combination at the hidden layer, the proposed variant, which uses more parameters than Elman and Jordan RNNs, but far fewer than LSTM and GRU, is more effective than other RNNs, but also outperforms sophisticated CRF models.

1 Introduction

In the last few years Recurrent Neural Networks (RNNs) [1–3] have proved very effective in several Natural Language Processing (NLP) tasks such as Part-of-Speech tagging (POS tagging), chunking, Named Entity Recognition (NER), Spoken Language Understanding (SLU), machine translation and even more [4–10]. These models are particularly effective thanks to their recurrent architecture, which allows neural models to *keep in memory* past information and re-use it at the current processing step.

In the literature of RNNs applied to NLP, several architectures have been proposed. At first Elman and Jordan RNNs, introduced in [1, 2], and known also as simple RNNs, have been adapted to NLP. The difference between these two models is in the type of connection giving the recurrent character to these two architectures: in the Elman RNN the recursion is a loop at the hidden layer, while in the Jordan RNN it relies the output layer to the hidden layer. This last recursion allows to use at the current step labels predicted for previous positions in a sequence.

ⓒ Springer Nature Switzerland AG 2018
A. Gelbukh (Ed.): CICLing 2017, LNCS 10761, pp. 44–66, 2018.
https://doi.org/10.1007/978-3-319-77113-7_4

These two recurrent models have shown limitations in learning relatively long contexts [11]. In order to overcome this limitation the RNNs known as *Long Short-Term Memory* (LSTM) have been proposed [3]. Recently, a simplified and, apparently, more effective variant of LSTM has been proposed, using *Gated Recurrent Units* and thus named *GRU* [12].

Despite outstanding performances on several NLP tasks, RNNs have not been explicitly adapted to integrate effectively label-depency information in sequence labelling tasks. Their sequence labelling decisions are based on intrinsecally local functions (e.g. the softmax). In order to overcome this limitation, sophisticated hybrid *RNN+CRF* models have been proposed [13–15], where the traditional output layer is replaced by a CRF neural layer. These models reach state-of-the-art performances, their evaluation however is not clear. In particular it is not clear if performances derive from the model itself, or thanks to particular experimental conditions. In [15] for example, the best result on POS tagging on the Penn Treebank corpus is an accuracy of 97.55, which is reached using word embeddings trained using *GloVe* [16], on huge amount of unlabeled data. The model of [15] without pre-trained embeddings reaches an accuracy of 96.9, which doesn't seem that outstanding if we consider that a CRF model dating from 2010, trained from scratch, without using any external resource, reaches an accuracy of 97.3 on the same data [17]. We achieved the same result on the same data with a CRF model trained from scratch using the incremental procedure described in [18]. Moreover, the first version of the network proposed in this paper, but using a sigmoid activation function and only the L_2 regularization, with a slightly different data preprocessing, achieves an accuracy on the Penn Treebank of 96.9 [19].

The intution behind this paper is that embeddings allow a fine and effective modelling not only of words, but also of labels and label dependencies, which are crucial in some tasks of sequence labelling. In this paper we propose, as alternative to *RNN+CRF* models, a variant of RNN allowing this more effective modelling. Surprisingly, a simple modification to the RNN architecture results in a very effective model: in our variant of RNN the recurrent connection connects the output layer to the input layer and, since the first layer is just a *look-up* table mapping discrete items into embeddings, labels predicted at the output layer are mapped into embeddings the same way as words. Label embeddings and word embeddings are combined at the hidden layer, allowing to learn relations between these two types of information, which are used to predict the label at current position in a sequence. Our intuition is that using several label embeddings as context, a RNN is able to model correctly label-dependencies, the same way as more sophisticated models explicitly designed for sequence labelling like CRFs [20].

This paper is a straight follow-up of [21]. Contributions with respect to that work are as follows:

(i) An analisys of performances of *forward*, *backward* and bidirectional models. (ii) The use of *ReLU* hidden layer and *dropout* regularization [22] at the hidden and embedding layers for improved regularized models.

(iii) The integration of a character-level convolution layer. (iv) An in-depth evaluation, showing the effect of different components and of different information level on the performance. (v) A straightforward comparison of the proposed variant of RNN to Elman, Jordan, LSTM and GRU RNNs, showing that the new variant is at least as effective as the best RNN models, such as LSTM and GRU. Our variant is even more effective when taking label-dependencies into account is crucial in the task, proving that our intuition is correct.

An high level schema of simple RNNs and of the variant proposed in this paper is shown in Fig. 1, where w is the input word, y is the label, E, H, O and R are the model parameters, which will be discussed in the following sections.

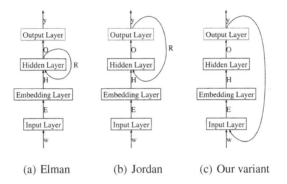

(a) Elman   (b) Jordan   (c) Our variant

Fig. 1. High level schema of simple RNNs (Elman and Jordan) and the variant proposed in this paper.

Since evaluations on tasks like POS tagging on the Penn Treebank are basically reaching perfection (state-of-the-art is at 97.55 accuracy), any new model would probably provide little or no improvement. Also, performances on this type of tasks seem to have reached a *plateau*, as models achieving 97.2 accuracy or even better, were already published starting from 2003 [23,24]. We propose instead to evaluate all the models on two different and widely used tasks of Spoken Language Understanding [25], which provide more variate evaluation settings: ATIS [26] and MEDIA [27].

ATIS is a relatively simple task and doesn't require a sophisticated modelling of label dependencies. This task allows to evaluate models in similar settings as tasks like POS tagging or Named Entity Recognition as defined in the *CoNLL Shared Task* 2003, both widely used as benchmarks in NLP papers. MEDIA is a very challanging task, where the ability of models to keep label dependencies into account is crucial to obtain good results.

Results show that our new variant is as effective as the best RNN models on a simple task like ATIS, providing the advantage of being much simpler. On the MEDIA task however, our variant outperforms all the other RNNs by a large margin, and even sophisticated CRF models, providing the best absolute result ever achieved on this task.

The paper is organized as follows: In the next section we describe the RNNs used in the literature for NLP, starting from existing models to arrive at describing the new variant we propose. In the Sect. 3 we present the corpora used for evaluation, the experimental settings and the results obtained in several experimental conditions. We draw some conclusions in Sect. 4.

2 Recurrent Neural Networks (RNNs)

In this section we describe the most popular RNNs used for NLP, such as Elman and Jordan RNNs [1,2], and the most sophisticated RNNs like LSTM and GRU [3,12]. We also describe training and inference procedures, and the RNN variant we propose.

2.1 Elman and Jordan RNNs

Elman and Jordan RNNs are defined as follows:

$$\mathbf{h_t}^{\text{Elman}} = \Phi(R\,\mathbf{h}_{t-1}^{\text{Elman}} + H\,\mathbf{I_t}) \tag{1}$$

$$\mathbf{h_t}^{\text{Jordan}} = \Phi(R\,\mathbf{y_{t-1}} + H\,\mathbf{I_t}) \tag{2}$$

The difference between these two models is in the way of computing hidden activities, while the output is computed in the same way:

$$\mathbf{y_t} = softmax(O\,\mathbf{h_t^*}) \tag{3}$$

h_t^* and y_t are respectively the hidden and output layer's activities[1], Φ is an activation function, H, O and R are the parameters at the hidden, output and recurrent layer, respectively (biases are omitted to keep equations lighter). h_{t-1}^{Elman} is the hidden layer activity computed at previous time step and used as context in the Elman RNN, while y_{t-1} is the previous predicted labels, used as context in the Jordan RNN. I_t is the input, which is often the concatenation of word embeddings in a fixed window d_w (for winDow of Words) around the current word w_t to be labelled. We define as $E(w_i)$ the embedding of any word w_i. I_t is then defined as:

$$\mathbf{I_t} = [E_w(w_{t-d_w})...E_w(w_t)...E_w(w_{t+d_w})] \tag{4}$$

where $[\ ]$ is the concatenation of vectors (or matrices in the following sections). The *softmax* function, given a set S of m numerical values v_i, associated to discrete elements $i \in [1, m]$, computes the probability associated to each element as:

$$\forall i \in [1, m]\ \ p(i) = \frac{e^{v_i}}{\sum_{j=1}^{m} e^{v_j}}$$

This function allows to compute the probability associated to each label and choose as predicted label the one with the highest probability.

---

[1] h_* means the hidden layer of any model, as the output layer is computed in the same way for all networks described in this paper.

2.2 *Long Short-Term Memory* (LSTM) RNNs

While LSTM is often used as the name of the whole network, it just defines a different way of computing the hidden layer activities. LSTMs use *gate* units to control how past and present information affect the network's internal state, and a *cell* to store past information that is going to be used as context at the current processing step. *Forget, input* gates and *cell state* are computed as:

$$f_t = \Phi(W_f h_{t-1} + U_f I_t) \tag{5}$$
$$i_t = \Phi(W_i h_{t-1} + U_i I_t) \tag{6}$$
$$\hat{c}_t = \Gamma(W_c h_{t-1} + U_c I_t) \tag{7}$$

Γ is used to indicate a different activation function from Φ^2. $\hat{c}_t$ is actually an intermediate value used to update the *cell state* value as follows:

$$c_t = f_t \odot c_{t-1} + i_t \odot \hat{c}_t \tag{8}$$

$\odot$ is the element-wise multiplication. Once these quantities have been computed, the *output* gate is computed and used to control the hidden layer activities at the current time step t:

$$o_t = \Phi(W_o h_{t-1} + U_o I_t) \tag{9}$$
$$h_t^{\text{LSTM}} = o_t \odot \Phi(c_t) \tag{10}$$

Once again (and in the remainder of the paper), biases are omitted to keep equations lighter. As we can see, each gate and the cell state have their own parameter matrices W and U, used for the linear transformation of the previous hidden state (h_{t-1}) and the current input (I_t). The evolution of the LSTM layer named GRU (*Gated Recurrent Units*) [12], combines together *forget* and *input* gates, and the previous hidden layer with the *cell state*:

$$z_t = \Phi(W_z h_{t-1} + U_z I_t) \tag{11}$$
$$r_t = \Phi(W_r h_{t-1} + U_r I_t) \tag{12}$$
$$\hat{h}_t = \Gamma(W(r_t \odot h_{t-1}) + U I_t) \tag{13}$$
$$h_t^{\text{GRU}} = (1 - z_t) \odot h_{t-1} + z_t \odot \hat{h}_t \tag{14}$$

GRU is thus a simplification of LSTM, it uses less units and it has less parameters to learn.

2.3 LD-RNN : Label-Dependencies Aware Recurrent Neural Networks

The variant of RNN that we propose in this paper can be thought of as having a recurrent connection from the output to the input layer. Note that from a different perspective, this variant can just be seen as a Feed-Forward Neural

---

[2] In the literature Φ and Γ are the *sigmoid* and *tanh*, respectively.

Network (FFNN) using previous predicted labels as input. Since Jordan RNN has the same architecture, the only difference being that in contrast to Jordan models we embed labels, we still prefer talking about recurrent network. This simple modification to the architecture of the network has important consequences on the model.

The reason motivating this modification is that we want embeddings for labels and use them the same way as word embeddings. Like we mentioned in the introduction, the first layer is a look-up table mapping discrete, or *one-hot*[3], representations into distributional representations.

Such representations can encode very fine syntactic and semantic properties, as it has already been proved by *word2vec* [28] or *GloVe* [16]. We want similar properties to be learnt also for labels, so that to encode in label embeddings the label dependencies needed for sequence labelling tasks. In this paper we learn label embeddings from the sequences of labels associated to word sentences in annotated data. But this procedure could be applied also when structured label information is available. We could thus exploit syntactic parse trees, structured named entities or entity relations for learning sophisticated label embeddings.

The idea of using label embeddings has been introduced in [29] for dependency parsing, resulting in a very effective parser. In this paper we go ahead with respect to [29] by using several label embeddings as context to predict the label at current position in a sequence. Also we pre-train label embeddings like it is usually done for words. As consequence, we learn first generic dependencies between labels without their interactions with words. Such interactions are then integrated and refined during the learning phase of the sequence labelling task. For this ability to learn label-dependencies, we name our variant *LD-RNN*, standing for *Label Dependencies aware RNN*.

Using the same formalism as before, we define E_w the matrix for word embeddings, while E_l is the matrix for label embeddings. The word-level input to our RNN is I_t as for the other RNNs, while the label-level input is:

$$\mathbf{L_t} = [E_l(y_{t-d_l+1})\ E_l(y_{t-d_l+2})\dots E_l(y_{t-1})] \tag{15}$$

which is the concatenation of vectors representing the d_l previous predicted labels (d_l stands (for win**D**ow of **L**abels)). The hidden layer activities of our RNN variant are computed as:

$$\mathbf{h_t}^{\text{LD-RNN}} = \Phi(H\ [\mathbf{I_t L_t}]) \tag{16}$$

We note that we could rewrite the equation above as $\Phi(H_w \mathbf{I_t} + H_l \mathbf{L_t})$ with a similar formalism as before, the two equations are equivalent if we define $H = [H_w H_l]$.

Thanks to the use of label embeddings and their combination at the hidden layer, our *LD-RNN* variant learns very effectively label dependencies. Since the

---

[3] The one-hot representation of a token represented by an index i in a dictionary, is a vector v of the same size as the dictionary and assigned zero everywhere, except at position i where it is 1.

other RNNs in general don't use explicetly the label information as context, they can predict incoherent label sequences. As we already mentioned, this limitation lead research toward hybrid *RNN+CRF* models [13–15].

Another consequence of the modification introduced in our RNN variant is an improved robustness to prediction mistakes. Since we use several label embeddings as context (see L_t above), once the model has learnt label embeddings, in the test phase it is unlikely that several prediction mistakes occur in the same context. Even in that case, thanks to properties encoded in the embeddings, mistaken labels have similar representations to correct labels, allowing the model to possibly predict correct labels. Reusing an example from [30]: if *Paris* is replaced by *Rome* in a text, this has no impact on several NLP tasks, as they are both proper nouns in POS tagging, localization in Named Entity Recognition etc. Using label embeddings provides the *LD-RNN* variant with the same robustness on the label side.

While the traditional Jordan RNN uses also previous labels as past information, it has not the same robustness because of the poor label representation used in adaptations of this model to NLP tasks. In Jordan RNNs used for NLP like [8–10], labels are represented either with the probability distribution computed by the *softmax*, or with the *one-hot* representation computed from the probability distribution.

In the latter case it is clear that a prediction mistake can have a bad impact in the context, as the only value being 1 in the *one-hot* representation would be in the wrong position. Instead, using the probability distribution may seem a kind of *fazzy* representation over several labels, but we have found empirically that the probability is very sharp and picked on one or just few labels. In any case this representation doesn't provide the desired robustness that can be achieved with label embeddings.

From another point of view, we can interpret the computation of the hidden activities in a Jordan RNN as using label embeddings. In the Eq. 2, the multiplication Ry_{t-1}, since y_{t-1} is a sparse vector, can be interpreted as the selection of an embedding from R.

Even with this interpretation there is a substantial difference between a Jordan RNN and our variant. In the Jordan RNN, once the label embedding has been computed with Ry_{t-1}, the result is not involved in the linear transformation applied by the matrix H, which is only applied to the word-level input I_t. The result of this multiplication is added to Ry_{t-1} and then the activation function is applied.

In our variant in contrast, labels are first mapped into embeddings with $E[y_i]$[4]. Word and label inputs I_t and L_t are then both transformed by multiplying by H, which is correctly dimensioned to apply the linear transformation on both inputs. In our variant thus, two different label transformations are always applied: (i) the conversion from sparse to embedding representation; (ii) the linear transformation by multiplying label embeddings by H.

---

[4] In our case, y_i is explicitely converted from probability distribution to *one-hot* representation.

2.4   Learning and Inference

We learn the *LD-RNN* variant like all the other RNNs, by minimizing the cross-entropy between the expected label l_t and the predicted label y_t at position t in the sequence, plus a L_2 regularization term:

$$C = -l_t \odot log(\mathbf{y_t}) + \frac{\lambda}{2} |\Theta|^2 \qquad (17)$$

λ is a hyper-parameter to be tuned, Θ is a short notation for E_w, E_l, H, O. l_t is the *one-hot* representation of the expected label. Since y_t above is the probability distribution over the label set, we can see the output of the network as the probability $P(i|\mathbf{I_t}, \mathbf{L_t})$ $\forall i \in [1, m]$, where $\mathbf{I_t}$ and $\mathbf{L_t}$ are the input of the network (words and labels), i is the index of one of the labels defined in the targeted task.

We can thus associate to the *LD-RNN* model the following decision function:

$$argmax_{i \in [1,m]} P(i|\mathbf{I_t}, \mathbf{L_t}) \qquad (18)$$

We note that this is still a local decision function, as the probability of each label is normalized at each position of a sequence. Despite this, the use of label-embeddings $\mathbf{L_t}$ as context allows the *LD-RNN* to effectively model label dependencies. Since the other RNNs like Elman and LSTM don't use the label information in their context, their decision function can be defined as:

$$argmax_{i \in [1,m]} P(i|\mathbf{I_t}) \qquad (19)$$

which can lead to incoherent predicted label sequences.

We use the traditional back-propagation algorithm with momentum to learn our networks [31]. Given the recurrent nature of the networks, the Back-Propagation Through Time (BPTT) is often used [32]. This algorithm consists in unfolding the RNN for N previous steps, N being a parameter to choose, and using thus the N previous inputs and hidden states to update the model's parameters. The traditional back-propagation algorithm is then applied. This is equivalent to learn a feed-froward network of depth N. The BPTT algorithm is supposed to allow the network to learn arbitrary long contexts. However [5] has shown that RNNs for language modelling learn best with only $N = 5$ previous steps. This can be due to the fact that, at least in NLP, a longer context does not lead necessarily to better performances, as a longer context is also more noisy.

Since the BPTT algorithm is quite expensive, [9] chose to explicitly use the contextual information provided by the recurrent connection, and to use the traditional back-propagation algorithm, apparently without performance loss.

In this paper we use the same strategy. When the contextual information is used explicitly in a Jordan RNN, the hidden layer state is computed as follows:

$$\mathbf{h_t} = \Phi(R[\mathbf{y_{t-d_1+1}} \, \mathbf{y_{t-d_1+2}} \cdots \mathbf{y_{t-1}}] + H \, \mathbf{I_t}) \qquad (20)$$

A similar modification can be applied also to Elman, LSTM and GRU RNNs to keep into account explicitly the previous hidden states. To our knowledge however, these networks are effectively learnt using only one previous hidden state [13–15].

From explanations above we can say that using explicit wide context of words and labels like we do in LD-RNN, can be seen as an approximation of the BPTT algorithm.

2.5 Toward More Sophisticated Networks: Character-Level Convolution

Even if word embeddings provide a very fine encoding of word features, several works such like [13–15,33] have shown that more effective models can be obtained using a convolution layer over characters of words. Character-level information is indeed very useful to allow a model generalizing over rare inflected surface forms and even out-of-vocabulary words in the test phase. Word embeddings are in fact much less effective in such cases. The convolution over word characters provide also the advantage of being very general: it can be applied in the same way to different languages, allowing to re-use the same system on different languages and tasks.

In this paper we focus on a convolution layer similar to the one used in [7] for words. For any word w of length $|w|$, we define $E_{ch}(w, i)$ the embedding of the character i of the word w. We define W_{ch} the matrix of parameters for the linear transformation applied by the convolution (once again we omit the associated bias). We compute a convolution of window size $2d_c + 1$ over characters of a word w as follows:

- $\forall i \in [1, |w|]\ Conv_i = W_{ch}[E_{ch}(w, i - d_c); \dots E_{ch}(w, i); \dots E_{ch}(w, i + d_c)]$
- $Conv_{ch} = [Conv_1 \dots Conv_{|w|}]$
- $Char_w = Max(Conv_{ch})$

the Max function is the so-called max-pooling [7]. While it is not strictly necessary mapping characters into embeddings, it would be probably less interesting applying the convolution on discrete representations. The matrix $Conv_{ch}$ is made of the concatenation of vectors returned from the application of the linear transformation W_{ch}. Its size is thus $|C| \times |w|$, where $|C|$ is the size of the convolution layer. The max-pooling computes the maxima over the word-length direction, thus the final output $Char_w$ has size $|C|$, which is independent from the word length. $Char_w$ can be interpreted as a distributional representation of the word w encoding the information at w's character level. This is a complementary information with respect to word embeddings, which encode inter-word information, and provide the model with an information similar to what is provided by discrete lexical features like word prefixes, suffixes, capitalization information etc., plus information about morphologically correct words of a given language.

2.6 RNN Complexities

The improved modelling of label dependencies in our *LD-RNN* variant is achieved at the cost of more parameters with respect to the simple RNN models.

However the number of parameters is still much less than sophisticated networks like LSTM. In this section we provide a comparison of RNNs complexity in terms of the number of parameters.

We introduce the following symbols: $|H|$ and $|O|$ are the size of the hidden and output layers, respectively. The size of the output layer is the number of labels; N is the embedding size, in *LD-RNN* we use the same size for word and label embeddings; d_w is the window size used for context words; and d_l is the number of label embeddings we use as context in *LD-RNN*. We analyze the hidden layer of all netwroks, and the embedding layer for *LD-RNN*. The other layers are exactly the same for all the networks described in this paper.

For Elman and Jordan RNNs, the hidden layer as the following number of parameters, respectively:

$$\{|H| * |H|\}_R + \{|H| * (2d_w + 1)N\}_{H \text{Elman}}$$
$$\{|O| * |H|\}_R + \{|H| * (2d_w + 1)N\}_{H \text{Jordan}}$$

Subscrits indicate from which matrix the parameters come. The factor $(2d_w + 1)N$ comes from the $(2d_w + 1)$ words used as input context and then mapped into embeddings. The factor $|O| * |H|$ in Jordan RNN is due to the fact that the matrix R connects output and hidden layers.

In *LD-RNN* we have:

$$\{|O| * N\}_{E_l} + \{((2d_w + 1 + d_l)N) * |H|\}_{H \text{LD-RNN}}$$

The factor $|O| * N$ is due to the use of the matrix E_l containing $|O|$ label embeddings of size N. Since in this paper we chose $N = |H|$ and $|O| < |H|$, and since in LD-RNN we don't use any matrix R on the recurrent connection, the fact of using label embeddings doesn't increase the number of parameters of the *LD-RNN* variant.

The hidden layer of *LD-RNN* however is dimensioned to connect all the word and label embeddings to all the hidden neurons. As consequence in the matrix H we have $d_l N$ more parameters than in the matrix H of Elman and Jordan RNNs.

In LSTM and GRU RNNs we have two extra matrices W and U for each gate and for the *cell state*, used to connect the previous hidden layer and the current input, respectively. These two matrices contain thus $|H| * |H|$ and $(2wd + 1)N * |H|$ parameters, respectively.

Using the same notation and the same settings as above, in the hidden layer of LSTM and GRU we have the following number of parameters:

$$\{4(|H| * |H| + |U| * (2d_w + 1)N)\}_{H \text{LSTM}}$$
$$\{3(|H| * |H| + |U| * (2d_w + 1)N)\}_{H \text{GRU}}$$

The 3 for GRU reflects the fact that this network uses only 2 gates and a *cell state*. It should be pointed out, however, that while we have been testing LSTM and GRU with a word window for a matter of fair comparison[5], these layers

---

[5] Indeed we observed better performances when using a word window with respect to when using a single word.

are applied on the current word and the previous hidden layer only, without the need of a word window. This is because this layer learns automatically how to use previous word information. In such case the complexity of the LSTM layer reduces to $\{4(|H|*|H|+|U|*N)\}_{H^{\text{LSTM}}}$. If we choose $|U| = |H|$, such complexity is comparable to that of *LD-RNN* in terms of number of parameters (slightly less actually). The LSTM is still more complex however because the hidden layer computation requires 4 gates and the cell state ($\hat{c}_t$) computations (each involving 2 matrix multiplications), the update of the new cell state c_t (involving also 2 matrix multiplications), and only after the hidden state can be computed. *LD-RNN*'s hidden state, in contrast, requires only matrix rows selection and concatenation to compute I_t and L_t, which are very efficient operations, and then the hidden state can already be computed.

As consequence, while the variant of RNN we propose in this paper is more complex than simple RNNs, LSTM and GRU RNNs are by far the most complex networks.

2.7   *Forward, Backward* and Bidirectional Networks

The RNNs introduced in this paper are proposed as forward, backward and bidirectional models [34]. The forward model is what has been described so far. The architecture of the backward model is exactly the same, the only difference is that the backward model processes data from the end to the begin of sequences. Labels and hidden layers computed by the backward model can thus be used as future context in a bidirectional model.

Bidirectional models are described in details in [34]. In this paper we utilize the version using separate forward and backward models. The final output is computed as the geometric mean of the output of the two individual models, that is:

$$y_t = \sqrt{y_t^f \odot y_t^b}$$

where y_t^f and y_t^b are the output of the forward and backward models, respectively.

In the developpement phase of our systems, we noticed no difference in terms of performance between the two types of bidirectional models described in [34]. We chose thus the version described above, since it allows to initialize all the parameters with the forward and backward models previously trained. As consequence the bidirectional model is very close to a very good optimum since the first learning iteration, and very few iterations are needed to learn the final model.

3   Evaluation

3.1   Corpora for Spoken Language Understanding

We evaluated our models on two tasks of Spoken Language Understanding (SLU) [25]:

The ATIS corpus (*Air Travel Information System*) [26] was collected for building a spoken dialog system able to provide flight information in the United States.

ATIS is a simple task dating from 1993. Training data are made of 4978 sentences chosen among dependency-free sentences in the ATIS-2 and ATIS-3 corpora. The test set is made of 893 sentences taken from the ATIS-3 NOV93 and DEC94 data. Since there are not official developpement data, we taken a part of the training set for this purpose. The word and label dictionaries contain 1117 and 85 items, respectively. We use the version of the corpus published in [35], where some word classes are available, such as city names, airport names, time expressions etc. These classes can be used as features to improve the generalization of the model on rare or unseen words. More details about this corpus can be found in [26].

An example of utterance transcription taken from this corpus is *"I want all the flights from Boston to Philadelphia today"*. The words *Boston, Philadelphia* and *today* in the transcription are associated to the concepts *DEPARTURE.CITY, ARRIVAL.CITY* and *DEPARTURE.DATE*, respectively. All the other words don't belong to any concept, they are associated to the void concept named *O* (for Outside). This example show the simplicity of this task: the annotation is sparse, only 3 words of the transcription are associated to a non-void concept; there is no segmentation problem, as each concept is associate to one word. Because of these two characteristics, the ATIS task is similar on the one hand to a POS tagging task, where there is no segmentation of labels over multiple words; on the other hand it is similar to a linear Named Entity Recognition task, where the annotation is sparse.

We are aware of the existence of two version of the ATIS corpus: the official version published starting from [35], and the version associated to the tutorial of *deep learning* made available by the authors of [9][6]. This last version has been modified, some proper nouns have been re-segmented (for example the token *New-York* has been replaced by two tokens *New York*), and a preprocessing has been applied to reduce the word dictionary (numbers have been converted into the conventional token *DIGIT*, and singletons of the training data, as well as out-of-vocabulary words of the developpement and test data, have been converted into the token *UNK*). Following the tutorial of [9] we have been able to download the second version of the ATIS corpus. However in this version word classes that are available in the first version are not given. We ran some experiments with these data, using only words as input. The results we obtained are comparable with those published in [36], in part from same authors of [9]. However without word classes we cannot fairly compare with works that are using them. In this paper we thus compare only with published works that used the official version of ATIS.

The French corpus MEDIA [27] was collected to create and evaluate spoken dialogue systems providing touristic information about hotels in France. This corpus is made of 1250 dialogs collected with *Wizard-of-OZ* approach. The

---

[6] Available at http://deeplearning.net/tutorial/rnnslu.html.

dialogs have been manually transcribed and annotated following a rich concept ontology. Simple semantic components can be combined to create complex semantic structures.[7] The rich semantic annotation is a source of difficulties, but also the annotation of coreference phenomena. Some words cannot be correctly annotated without knowing a relatively long context, often going beyond a single dialog turn. For example in the utterance transcription *"Yes, the one which price is less than 50 Euros per night"*, *the one* is a mention of an hotel previously introduced in the dialog. Statistics on the corpus MEDIA are shown in Table 2.

The task resulting from the corpus MEDIA can be modelled as a sequence labelling task by chunking the concepts over several words using the traditional *BIO* notation [37].

Thanks to the characteristics of these two corpora, together with their relatively small size which allows training models in a reasonable time, these two tasks provide ideal settings for the evaluation of models for sequence labelling. A comparative example of annotation, showing also the word classes available for the two tasks and mentioned above, is shown in the Table 1.

Table 1. An example of annotated utterance transcription taken from MEDIA (left) and ATIS (right). The translation in French is *"Yes, the one which price is less than 50 Euros per night"*

MEDIA			ATIS		
Words	Classes	Labels	Words	Classes	Labels
Oui	-	Answer-B	i'd	-	O
l'	-	BDObject-B	like	-	O
hotel	-	BDObject-I	to	-	O
le	-	Object-B	fly	-	O
prix	-	Object-I	Delta	airline	airline-name
à	-	Comp.-payment-B	between	-	O
moins	relative	Comp.-payment-I	Boston	city	fromloc.city-name
cinquante	tens	Paym.-amount-B	and	-	O
cinq	units	Paym.-amount-I	Chicago	city	toloc.city-name
euros	currency	Paym.-currency-B			

---

[7] For example the component *localization* can be combined with other components like `city`, `relative-distance`, `generic-relative-location`, `street` etc.

Table 2. Statistic of the corpus MEDIA

	Training		Dev.		Test	
# Sentences	12,908		1,259		3,005	
	words	concepts	words	concepts	words	concepts
# mots	94,466	43,078	10,849	4,705	25,606	11,383
# vocab.	2,210	99	838	66	1,276	78
# OOV%	–	–	1.33	0.02	1.39	0.04

3.2   Settings

The RNN variant *LD-RNN* has been implemented in *Octave*[8] using *OpenBLAS* for low-level computations.[9]

LD-RNN models are trained with the following procedure:

- Neural Network Language Models (NNLM), like the one described in [38], are trained for words and labels to generate the embeddings (separately).
- Forward and backward models are trained using the word and label embeddings trained at previous step.
- The bidirectional model is trained using as starting point the forward and backward models trained at previous step.

We ran also some experiments using embeddings trained with *word2vec* [28]. The results obtained are not significantly different from those obtained following the procedure described above. This outcome is similar to the one obtained in [10]. Since the tasks addressed in this paper are made of small data, we believe that any embedding is equally effective. In particular tools like *word2vec* are designed to work on relatively big amount of data. Results obtained with *word2vec* embeddings will not be described in the following sections.

We roughly tuned the number of learning epochs for each model on the developpement data of the addressed tasks: 30 epochs are used to traine word embeddings, 20 for label embeddings, 30 for the forward and backward models, 8 for the bidirectional model (the optimum of this model is often reached at the first epoch on the ATIS task, between the 3rd and the 5th epoch on MEDIA). At the end of the training phase, we keep the model giving the best prediction accuracy on the developpement data. We stop training the model is the accuracy is not improved for 5 consecutive epochs (also known as *Early stopping* strategy [31]).

We initialize all the weights with the "so called" *Xavier initialization* [31], theoretically motivated in [39] as keeping the standard deviation of the weights

---

[8] https://www.gnu.org/software/octave/; Our code is described at http://www.marcodinarelli.it/software.php and available upon request.
[9] http://www.openblas.net; This library allows a speed-up of roughly 330× on a single matrix-matrix multiplication using 16 cores. This is very attractive with respect to the speed-up of 380× that can be reached with a GPU, keeping into account that both Octave and OpenBLAS are available for free.

during the training phase when using *ReLU*, which is the hidden layer we chose for our variant of RNN.

We also tuned some of the hyper-parameters on the developpement data: we found out that the best initial learning rate is 0.5, this is linearly decreased with a value computed as the ratio between the initial learning rate and the number of epochs (*Learing Rate decay*). We combine *dropout* and L_2 regularization [31], the best value for the dropout probability is 0.5 at the hidden layer, 0.2 at the embedding layer on ATIS, 0.15 on MEDIA. The best coefficient (λ) for the L_2 regularization is 0.01 for all the models, expcet for the bidirectional model where the best is $3e^{-4}$.

We ran also some experiments for optimizing the size of the different layers. In order to minimize the time and the number of experiments, this optimization has been based on the result provided by the forward model on the two tasks, and using only words and labels as input (without word classes and character convolution, which were optimized separately). The best size for the embeddings and the hidden layer is 200 for both tasks. The best size for the character convolution layer is 50 on ATIS, 80 on MEDIA. In both cases, the best size for the convolution window is 1, meaning that characters are used individually as input to the convolution. A window of size 3 (one character on the left, one on the right, plus the current character) gives roughly the same results, we thus prefer the simpler model. With a window of size 5, results starts to slightly deteriorate.

We also optimized the size of the word and label context used in the *LD-RNN* variant. On ATIS the best word context size is 11 (5 on the lest, 5 on the right plus the current word), the best label context size is 5. On MEDIA the best sizes are 7 and 5 respectively. These values are the same found in [10] and comparable to those of [36].

The best parameters found in this phase has been used to obtain *baseline* models. The goal was to understand the behavior of the models with the different level of information used: the word classes available for the tasks, and the character level convolution. Some parameters needed to be re-tuned, as we will describe later on.

Concerning training and testing time of our models, the overall time to train and test *forward*, *backward* and bidirectionnel models, using only words and classes as input, is roughly 1 h 10 min on MEDIA, 40 min on ATIS. These times go to 2 h for MEDIA and 2 h 10 min for ATIS, using also word classes and character convolution as input. All these times are measured on a *Intel Xeon E5-2620* at 2.1 *GHz*, using 16 cores.

3.3   Results

All the results shown in this section are averages over 6 runs. Embeddings were learnt once for all experiments.

Incremental Results with Different Level of Information. In this section we describe results obtained with incremental levels of information given as input to the models: (i) Only words (previous labels are always given as input), indicated with *Words* in the tables; (ii) words and classes *Words+Classes*; (iii) words and character convolution *Words+CC*; (iv) All possible inputs *Words+Classes+CC*.

The results obtained on the ATIS task are shown in the Table 3, results on MEDIA are in Table 4.

Table 3. Results in terms of F1 measure on ATIS, using different level of information as input.

Model	F1 measure		
	Forward	*Backward*	Bidirectional
LD-RNN Words	94.23%	94.30%	94.45%
LD-RNN Words+CC	94.56%	94.69%	94.79%
LD-RNN Words+Classes	95.31%	95.42%	95.53%
LD-RNN Words+Classes+CC	**95.55%**	**95.45%**	**95.65%**

Table 4. Results in terms of F1 measure on MEDIA, using different level of information as input.

Model	F1 measure		
	Forward	*Backward*	Bidirectional
LD-RNN Words	85.39%	86.54%	87.05%
LD-RNN Words+CC	85.41%	86.48%	86.98%
LD-RNN Words+Classes	**85.46%**	86.59%	87.16%
LD-RNN Words+Classes+CC	85.38%	**86.79%**	**87.22%**

Results in these tables show that models have a similar behavior on the two tasks. In particular on ATIS, adding the different level of information results improve progressively and the best performance is obtained integrating words, labels and character convolution, though some of the improvements do not seem statistically significant, taking into account the small size of this corpus.

This observation is confirmed by results obtained on MEDIA, where adding the character level convolution leads to a slight degradation of performances. In order to understand the reason of this behavior we analysed the training phase on the two tasks. We found out that the main problem was an hidden layer saturation: with the number of hidden neurons chosen in the preliminar optimization phase using only words (and labels), the hidden layer was not able to model the whole information richness provided by all the inputs at the same time. We ran thus some experiments using a larger hidden layer with size 256, which gave the results shown in the two table with the model *LD-RNN Words+Classes+CC*. For lack of time we did not further optimized the size of the hidden layer.

Beyond all of that, results shown in the Table 3 and 4 are very competitive, as we will discuss in the next section.

Comparison with the State-of-the-Art. In this section we compare our results with the best results found in the literature. In order to be fair, the comparison is made using the same input information: words and classes. In the tables we use *E-RNN* for Elman RNN, *J-RNN* for Jordan RNN, *I-RNN* for the improved RNN proposed by [40].[10]

In order to give an idea of how our RNN variant compares to *LSTM+CRF* models like the one of [15], we ran an experiment on the Penn Treebank [41]. With a similar data pre-processing, exactly the same data split, using a *sigmoid* activation function, and using only words as input, the *LD-RNN* variant achieves an accuracy of 96.83. This is comaparable to the 96.9 achieved by the *LSTM+CRF* model of [15] without pre-trained embeddings.[11]

Results on the ATIS task are shown in Table 5. On this task we compare to results published in [40,42].

The results in the Table 5 show that all models obtain a good performance on this task, always higher than 94.5 $F1$. This confirm what we anticipated in the previous section concerning how easy is this task.

The GRU RNNs of [42] and our variant *LD-RNN* obtain equivalent results (95.53), which is slightly better than all the other models, in particular with the bidirectional models. This is a good outcome, as our variant of RNN obtains the same result as GRU while using much less parameters (see Sect. 2.6 for RNNs complexity). Indeed LSTM and GRU are considered very effective models for learning very long contexts. The way they are used in [42] allows to learn long contexts on the input side (words), they are not adapted however to learn also long label contexts, which is what we do in this paper with our variant. The fact that the best word context on this task is made of 11 words, show that this is the most important information to obtain good results on this task. It is thus not surprising that the GRU RNN achieves such good performance.

Comparing these our results on the ATIS task with those published in [40] with a Jordan RNN, which uses the same label context as our models, we can conclude that the advantage in the variant *LD-RNN* is given by the use of label embeddings and their combination at the hidden layer.

This conclusion is more evident if we compare results obtained with RNNs using label embeddings with the other RNNs on the MEDIA task. This comparison is shown in Table 6. As we mentioned in the Sect. 3.1, this task is very challenging for several reason, but in the context of this paper we focus on the label dependencies that we claim we can effectively model with our RNN variant.

In this context we note that a traditional Jordan RNN, the *J-RNN* of [40], which is the only traditional model to explicitly use previous label

---

[10] This is a publication in French, but results in the tables are easy to understand and directly comparable to our results.

[11] We did not run further experiments because without a GPU, experiments on the Penn Treebank are still quite expensive.

Table 5. Comparison of our results on the ATIS task with the literature, in terms of F1 measure.

Model	F1 measure		
	Forward	*Backward*	Bidirectional
[42] LSTM	95.12%	–	95.23%
[42] GRU	**95.43%**	–	**95.53%**
[40] E-RNN	94.73%	93.61%	94.71%
[40] J-RNN	94.94%	94.80%	94.89%
[40] I-RNN	95.21%	94.64%	94.75%
LD-RNN Words+Classes	95.31%	95.42%	**95.53%**

Table 6. Comparison of our results on the MEDIA task with the literature, in terms of F1 measure.

Model	F1 measure		
	Forward	*Backward*	Bidirectional
[10] CRF	86.00%		
[10] E-RNN	81.94%	–	–
[10] J-RNN	83.25%	–	–
[42] LSTM	81.54%	–	83.07%
[42] GRU	83.18%	–	83.63%
[40] E-RNN	82.64%	82.61%	83.13%
[40] J-RNN	83.06%	83.74%	84.29%
[40] I-RNN	84.91%	86.28%	86.71%
LD-RNN Words+Classes	**85.46%**	**86.59%**	**87.16%**

information as context, is more effective than the other traditional models, including LSTM and GRU (84.29 F1 with J-RNN, 83.63 with GRU, second best model among traditional RNNs). We note also that on MEDIA, CRFs, which are models specifically designed for sequence labelling, are by far more effective than the traditional RNNs (86.00 F1 with the CRF of [10]).

The only models outperforming CRFs on the MEDIA task are the *I-RNN* model of [40] and our *LD-RNN* variant, both using label embeddings.

Even if results on MEDIA discussed so far are very competitive, this task has been designed for Spoken Language Understanding (SLU) [25]. In SLU the goal is to extract a correct semantic representation of a sentence, allowing a correct interpretation of the user will by the spoken dialog system. While the F1 measure is strongly correlated with SLU evaluation metrics, the evaluation measure used most often in the literature is the *Concept Error Rate* (CER).

Table 7. Results on the MEDIA task in terms of *Concept Error Rate* (CER), compared with the best results published so far on this task.

Model	CER
[43] CRF	11.7%
[44] CRF	11.5%
[45] CRF	10.6%
LD-RNN Words	10.73% (10.63)
LD-RNN Words+Classes	10.52% (10.15)
LD-RNN Words+Classes+CC	**10.41% (10.09)**

CER is defined exactly in the same way as *Word Error rate* in automatic speech recognition, where words are replaced by concepts.[12]

In order to place our results on an absolute ranking among models designed for the MEDIA task, we propose a comparison in terms of CER to the best models published in the literature, namely [43–45]. This comparison is shown in Table 7.

The best individual models published by [45], [44] and [43] are CRFs, achieving a CER of 10.6, 11.5 and 11.7, respectively. These models use both word and classes, and a rich set of lexical features such like word prefixes, suffixes, word capitalisation information etc. We note that the large gap between these CRF models is due to the fact that the CRF of [45] is trained with an improved margin criterion, similar to the large margin principle of *SVM* [46,47]. We note also that comparing significance tests published in [43], a difference of 0.1 in CER is already statistically significant. Since results in this paper are higher, we hypothesize than even smaller gains are significant.

Our best *LD-RNN* model achieve a CER of 10.41. To the best of our knowledge this is the best CER obtained on the MEDIA task with an individual model. Moreover, instead of taking the mean of CER of several experiments, following a strategy similar to [8], one can run several experiments and keep the model obtaining the best CER on the developpement data of the target task. Results obtained using this strategy are shown in Table 7 between parenthesis. The best result obtained by our *LD-RNN* is a CER of 10.09, the best absolute result on this task so far, even better than the *ROVER* model [48] used in [45], which combines 6 individual models, including the indivudual CRF model achieving 10.6 CER.

3.4   Results Discussion

In order to understand the high performances of the *LD-RNN* variant on the MEDIA task, we made some simple analyses on the model output, comparing

---

[12] The errors made by the system are classified as Insertions (I), Deletions (D) and Substitutions (S). The sum of these errors is divided by the number of concepts in the reference annotation (R): $CER = \frac{I+D+S}{R}$.

them to the output of a Jordan RNN trained with our own system in the same conditions as *LD-RNN* models. The main difference between these two models is the general tendency of the Jordan RNN to split a single concept into two or more concepts, mainly for concepts instantiated by long surface forms, such like *command-tache*. This concept is used to mean the general user will in a dialog turn (e.g. *Hotel reservation, Price information* etc.). The Jordan RNN often split this concept into several concepts by introducing a void label, associated to a stopword. This is due to the limitation of this model to take relatively long label context into account, even if it is the only traditional RNN using explicitly previous labels as context information.

Surprisingly, *LD-RNN* never makes this mistake and in general never makes segmentation errors (concerning the BIO formalism). This can be due to two reasons. The first is that label embeddings learns similar representations for semantically similar labels. This allows the model to correctly predict start-of-concept (B) even if the target word has been seen in the training set only as continutation-of-concept (I), or viceversa, as the two labels acquire very similar representations. The second reason, which is not in mutual exclusion with the first, is that the model factorizes information acquired on similar words seen associated to start-of-concept labels. Thus if a word has not been seen associated to start-of-concept labels, but similar words do, the model is still able to provide the correct annotation. This second reason is what made neural networks popular for learning word embeddings in earlier publications [38]. In any case, in our experience, we never observed such precise behavior even with CRF models tuned for the MEDIA task. For this reason we believe *LD-RNN* deserves the name *Label Dependencies aware* RNN.

Still *LD-RNN* makes mistakes, which means that once a label annotation starts for a target word, even if the label is not the correct one, the same label is kept even if the following words provide evidence that the correct label is another one. *LD-RNN* tends to be coherent with previous labeling decisions. This behavior is due to the use of a local decision function which definitely relies heavily on the label embedding context, but it doesn't prevent the model from being very effective. Interestingly, this behavior suggests that *LD-RNN* could still benefit from a CRF neural layer like those used in [13–15]. We leave this as future work.

4   Conclusion

In this paper we proposed a new variant of RNN for sequence labelling using a wide context of label embeddings in addition to the word context to predict the next label in a sequence. We motivated our variant as being more effective at modelling label dependencies. Results on two Spoken Language Understanding tasks show that (i) on a simple task like ATIS our variant achieves the same performance as much more complex models such as LSTM and GRU, which are claimed the most effective RNNs; (ii) on the MEDIA task, where modelling label dependencies is crucial, our variant outperforms by a large margin all the other

RNNs, including LSTM and GRU. When compared to the best models of the literature in terms of Concept Error Rate (CER), our RNN variant results to be more effective, achieving a state-of-the-art CER of 10.09.

Acknowledgements. This work has been partially funded by the French ANR project Democrat ANR-15-CE38-0008.

References

1. Jordan, M.I.: Serial order: A parallel, distributed processing approach. In: Elman, J.L., Rumelhart, D.E. (eds.) Advances in Connectionist Theory: Speech. Erlbaum, Hillsdale, NJ (1989)
2. Elman, J.L.: Finding structure in time. Cogn. Sci. **14**, 179–211 (1990)
3. Hochreiter, S., Schmidhuber, J.: Long short-term memory. Neural Comput. **9**, 1735–1780 (1997)
4. Mikolov, T., Karafiát, M., Burget, L., Cernocký, J., Khudanpur, S.: Recurrent neural network based language model. In: 11th Annual Conference of the International Speech Communication Association, Makuhari, Chiba, Japan, pp. 1045–1048, 26–30 September 2010
5. Mikolov, T., Kombrink, S., Burget, L., Cernocky, J., Khudanpur, S.: Extensions of recurrent neural network language model. In: ICASSP, pp. 5528–5531. IEEE (2011)
6. Collobert, R., Weston, J.: A unified architecture for natural language processing: Deep neural networks with multitask learning. In: Proceedings of the 25th International Conference on Machine Learning, ICML 2008, pp. 160–167. ACM, New York (2008)
7. Collobert, R., Weston, J., Bottou, L., Karlen, M., Kavukcuoglu, K., Kuksa, P.: Natural language processing (almost) from scratch. J. Mach. Learn. Res. **12**, 2493–2537 (2011)
8. Yao, K., Zweig, G., Hwang, M.Y., Shi, Y., Yu, D.: Recurrent neural networks for language understanding. In: Interspeech (2013)
9. Mesnil, G., He, X., Deng, L., Bengio, Y.: Investigation of recurrent-neural-network architectures and learning methods for spoken language understanding. In: Interspeech 2013 (2013)
10. Vukotic, V., Raymond, C., Gravier, G.: Is it time to switch to word embedding and recurrent neural networks for spoken language understanding? In: InterSpeech, Dresde, Germany (2015)
11. Bengio, Y., Simard, P., Frasconi, P.: Learning long-term dependencies with gradient descent is difficult. Trans. Neur. Netw. **5**, 157–166 (1994)
12. Cho, K., van Merrienboer, B., Gülçehre, Ç., Bougares, F., Schwenk, H., Bengio, Y.: Learning phrase representations using RNN encoder-decoder for statistical machine translation. CoRR abs/1406.1078 (2014)
13. Huang, Z., Xu, W., Yu, K.: Bidirectional lstm-crf models for sequence tagging. arXiv preprint arXiv:1508.01991 (2015)
14. Lample, G., Ballesteros, M., Subramanian, S., Kawakami, K., Dyer, C.: Neural architectures for named entity recognition. arXiv preprint arXiv:1603.01360 (2016)
15. Ma, X., Hovy, E.: End-to-end sequence labeling via bi-directional LSTM-CNNs-CRF. In: Proceedings of the 54th Annual Meeting of the Association for Computational Linguistics, ACL 2016 (2016)

16. Pennington, J., Socher, R., Manning, C.D.: Glove: Global vectors for word representation. In: Empirical Methods in Natural Language Processing (EMNLP), pp. 1532–1543 (2014)
17. Lavergne, T., Cappé, O., Yvon, F.: Practical very large scale CRFs. In: Proceedings the 48th Annual Meeting of the Association for Computational Linguistics (ACL), pp. 504–513. Association for Computational Linguistics (2010)
18. Dinarelli, M., Rosset, S.: Models cascade for tree-structured named entity detection. In: Proceedings of International Joint Conference of Natural Language Processing (IJCNLP), Chiang Mai, Thailand (2011)
19. Dinarelli, M., Tellier, I.: Improving recurrent neural networks for sequence labelling. CoRR abs/1606.02555 (2016)
20. Lafferty, J., McCallum, A., Pereira, F.: Conditional random fields: Probabilistic models for segmenting and labeling sequence data. In: Proceedings of the Eighteenth International Conference on Machine Learning (ICML), Williamstown, MA, USA, pp. 282–289 (2001)
21. Dinarelli, M., Tellier, I.: New recurrent neural network variants for sequence labeling. In: Gelbukh, A. (ed.) CICLing 2016. LNCS, vol. 9623, pp. 155–173. Springer, Cham (2018). https://doi.org/10.1007/978-3-319-75477-2_10
22. Srivastava, N., Hinton, G., Krizhevsky, A., Sutskever, I., Salakhutdinov, R.: Dropout: a simple way to prevent neural networks from overfitting. J. Mach. Learn. Res. **15**, 1929–1958 (2014)
23. Toutanova, K., Klein, D., Manning, C.D., Singer, Y.: Feature-rich part-of-speech tagging with a cyclic dependency network. In: Proceedings of the 2003 Conference of the North American Chapter of the Association for Computational Linguistics on Human Language Technology, NAACL 2003, pp. 173–180. Association for Computational Linguistics, Morristown (2003)
24. Shen, L., Satta, G., Joshi, A.: Guided learning for bidirectional sequence classification. In: Proceedings of the 45th Annual Meeting of the Association of Computational Linguistics, pp. 760–767. Association for Computational Linguistics, Prague (2007)
25. De Mori, R., Bechet, F., Hakkani-Tur, D., McTear, M., Riccardi, G., Tur, G.: Spoken language understanding: a survey. IEEE Sig. Process. Mag. **25**, 50–58 (2008)
26. Dahl, D.A., et al.: Expanding the scope of the ATIS task: The ATIS-3 corpus. In: Proceedings of the Workshop on Human Language Technology, HLT 1994, pp. 43–48. Association for Computational Linguistics, Stroudsburg (1994)
27. Bonneau-Maynard, H., et al.: Results of the French EVALDA-MEDIA evaluation campaign for literal understanding. In: LREC, Genoa, Italy, pp. 2054–2059 (2006)
28. Mikolov, T., Chen, K., Corrado, G., Dean, J.: Efficient estimation of word representations in vector space. CoRR abs/1301.3781 (2013)
29. Chen, D., Manning, C.: A fast and accurate dependency parser using neural networks. In: Proceedings of the 2014 Conference on Empirical Methods in Natural Language Processing (EMNLP), pp. 740–750. Association for Computational Linguistics, Doha (2014)
30. Mikolov, T., Yih, W., Zweig, G.: Linguistic regularities in continuous space word representations. In: Human Language Technologies: Conference of the North American Chapter of the Association of Computational Linguistics, pp. 746–751 (2013)
31. Bengio, Y.: Practical recommendations for gradient-based training of deep architectures. CoRR abs/1206.5533 (2012)
32. Werbos, P.: Backpropagation through time: what does it do and how to do it. Proc. IEEE **78**, 1550–1560 (1990)

33. Chiu, J.P.C., Nichols, E.: Named entity recognition with bidirectional LSTM-CNNs. CoRR abs/1511.08308 (2015)
34. Schuster, M., Paliwal, K.: Bidirectional recurrent neural networks. Trans. Sig. Proc. **45**, 2673–2681 (1997)
35. Raymond, C., Riccardi, G.: Generative and discriminative algorithms for spoken language understanding. In: Proceedings of the International Conference of the Speech Communication Assosiation (Interspeech), Antwerp, Belgium, pp. 1605–1608 (2007)
36. Mesnil, G., et al.: Using recurrent neural networks for slot filling in spoken language understanding. IEEE/ACM Trans. Audio Speech. Lang. Process. (2015)
37. Ramshaw, L., Marcus, M.: Text chunking using transformation-based learning. In: Proceedings of the 3rd Workshop on Very Large Corpora, Cambridge, MA, USA, pp. 84–94 (1995)
38. Bengio, Y., Ducharme, R., Vincent, P., Jauvin, C.: A neural probabilistic language model. J. Mach. Learn. Res. **3**, 1137–1155 (2003)
39. He, K., Zhang, X., Ren, S., Sun, J.: Delving deep into rectifiers: surpassing human-level performance on imagenet classification. In: 2015 IEEE International Conference on Computer Vision, ICCV 2015, Santiago, Chile, pp. 1026–1034, 7–13 December 2015
40. Dinarelli, M., Tellier, I.: Etude des reseaux de neurones recurrents pour etiquetage de sequences. In: Actes de la 23eme conférence sur le Traitement Automatique des Langues Naturelles, Paris, France, Association pour le Traitement Automatique des Langues (2016)
41. Marcus, M.P., Santorini, B., Marcinkiewicz, M.A.: Building a large annotated corpus of english: the penn treebank. Comput. Linguist. **19**, 313–330 (1993)
42. Vukotic, V., Raymond, C., Gravier, G.: A step beyond local observations with a dialog aware bidirectional GRU network for Spoken Language Understanding. In: Interspeech, San Francisco, United States (2016)
43. Dinarelli, M., Moschitti, A., Riccardi, G.: Discriminative reranking for spoken language understanding. IEEE Trans. Audio Speech Lang. Process. (TASLP) **20**, 526–539 (2011)
44. Dinarelli, M., Rosset, S.: Hypotheses selection criteria in a reranking framework for spoken language understanding. In: Conference of Empirical Methods for Natural Language Processing, Edinburgh, U.K., pp. 1104–1115 (2011)
45. Hahn, S., et al.: Comparing stochastic approaches to spoken language understanding in multiple languages. IEEE Trans. Audio Speech Lang. Process. (TASLP) **99** (2010)
46. Herbrich, R., Graepel, T., Obermayer, K.: Large Margin Rank Boundaries for Ordinal Regression. MIT Press (2000)
47. Hahn, S., Lehnen, P., Heigold, G., Ney, H.: Optimizing CRFs for SLU tasks in various languages using modified training criteria. In: Proceedings of the International Conference of the Speech Communication Assosiation (Interspeech), Brighton, U.K. (2009)
48. Fiscus, J.G.: A post-processing system to yield reduced word error rates: recogniser output voting error reduction (ROVER). In: Proceedings 1997 IEEE Workshop on Automatic Speech Recognition and Understanding (ASRU), Santa Barbara, CA, pp. 347–352 (1997)

Universal Computational Formalisms and Developer Environment for Rule-Based NLP

Svetlana Sheremetyeva$^{(\boxtimes)}$

Department of Linguistics and Translation, South Ural State University,
76, Lenin pr., Chelyabinsk 454080, Russia
sheremetevaso@susu.ru, lanaconsult@mail.dk

Abstract. The paper explores the issues of universal computational formalisms and reusable developer environment as applied to rule-based NLP. It suggests a portable grammar framework and modular NLP architecture that by combining certain modules can be reused for different unilingual and multilingual applications through a universal developer environment. The developer environment includes a lexicon shell with flexible settings to define, among others, tag descriptions, entry structures, depth of knowledge, and a number of compilers with universal rule-writing formalisms. The formalisms and compilers described have been successfully used (in different combinations and with different depth of analysis) in a number of unilingual and multilingual applications that involved English, Danish, French and Russian.

Keywords: Rule-based NLP · Reusability · Computational formalism
Compilers

1 Introduction

After more than a decade of the dominance of the statistical paradigm in NLP, a new wave of R&D has reverted to the primacy of rule-based approaches. This is particularly true for processing tasks where highly inflecting languages are involved, for which, though certain attempts are made to substitute costly rule-based procedures with purely statistical methods, hidden costs are recognized associated with the use of pure statistics [11]. However, high quality NLP demands rich knowledge resources (world models, grammar rules and lexicons), which are often handcrafted from scratch for every new application, language or language pair.

The idea to reduce development and maintenance costs, by sharing and reusing processing methods and knowledge has been in focus of researchers' attention for many years. Certain attempts have been made to develop universal tagsets [2, 7], portable cross-linguistic knowledge [6, 16], and reusable rule-based components [9]. A formalism for simultaneous rule-based morphosyntactic tagging and partial parsing is suggested in [10]. Universal computational formalisms have been already explored in early works on portability [3, 8], which outline the following major principles: (a) universal computational formalisms are to be based on grammars that in a uniform

© Springer Nature Switzerland AG 2018
A. Gelbukh (Ed.): CICLing 2017, LNCS 10761, pp. 67–78, 2018.
https://doi.org/10.1007/978-3-319-77113-7_5

way deal with atomic informational structures and then manipulate these structures by means of a few well-defined operations, which build new more complex structures; (b) within the frame of these formalisms both the atomic and complex structures, are to be application oriented and motivated by processing considerations; (c) a computational formalism is to be well-defined, which means that its semantics (not the semantics of the language described by the formalism) is also well defined.

To be used in practice, computational formalisms, apart from being natural language frameworks, should be available for a developer through developer tools for knowledge acquisition, code validation, navigation, test suite management, etc. Most of such tools (see, e.g., [1]) require programming qualification and are primarily developed for programmers. However, the contribution of linguistic knowledge in developing rule-based applications cannot be cannot be but appreciated, though the needs of linguists, not so experienced in programming, are often neglected.

In this work we attempt to cover this gap and suggest universal computational formalisms and developer environment addressing both, programmers, and, primarily, linguists without extensive programming training. In what follows, we first present the overall framework of a line of rule-based applications, sharing linguistic and programming resources and then describe the main modules of the developer environment. The work in summarized in Conclusions.

2   Overall Framework

2.1   Grammar

In an attempt to introduce robustness into the grammar itself, rather than adjusting a parser algorithm, we combine the formalisms of context free lexicalized Phrase Structure Grammar (PSG) and Dependency Grammar (DG). The PSG component consists of a subset of regular PSG rewriting rules. However, this subset includes neither the basic PSG rule "S = NP + VP", nor any rules for rewriting VP.

The PSG grammar component covers only those sentence constituents that are not clause predicates (be it a main clause or a subordinate/relative clause). It is the basis for a chunking procedure and does not give any description of syntactic dependencies. The PSG component is specified over *a space of supertags* [14] augmented with local information, such as lexical preference and some of rhetorical knowledge, - the knowledge about text segments, anchored to tabulations, commas and periods.

The DG grammar component is a strongly lexicalized case-role grammar specified *over the space of phrases (NP, PP, etc.)* and a residue of tagged "ungrammatical" words, i.e., words that do not satisfy any of the rules of the PSG component. All syntactic and semantic knowledge within this grammar is anchored to one type of lexemes, namely *predicates*.

The grammar assigns a final parse (a universal content representation) to a sentence as shown in Fig. 1, where *label* is a unique identifier of the elementary predicate-argument structure (by convention, marked by the number of its predicate as it appears in the sentence, *predicate-class* is a label of an ontological concept, *predicate* is a string corresponding to a predicate from the lexicon, *status* is a semantic status of a case-role, such as place, instrument, etc., and *value* is a string which fills a case-role. *Supertag* is a tag, which conveys morphological, syntactic and semantic features as specified in the lexicon. *Word* and *phrase* are a word and phrase (NPs, PPs, etc.) in a standard understanding. This representation is universal in that, in case of multilingual applications, e.g. machine translation, the format of predicate-argument structures stays invariant after transfer to a TL.

> text::={ template){template}*
> template::={label predicate-class predicate ((case-role)(case-role)*}
> case-role::= (status value)
> value::= phrase{(phrase(word supertag)*)}*

Fig. 1. A universal format of content representation, invariant between languages.

2.2 Processing Steps and Applications Architecture

We here present a number of rule-based modules that in different configurations can be used to solve different unilingual and multilingual NLP tasks. This architecture resulted from our multi-year research, in the course of which formalisms and programs first created for authoring of patent claims in English were further ported and updated to develop a family of other multilingual scientific- and technical information-related applications, see, e.g., [5, 14, 15], to name just a few.

The reuse of earlier developed modules was to a great extent possible due to the universal grammar formalism (see Sect. 2.1) and elaborate developer environment for rule and lexicon acquisition.

An umbrella configuration of our processing modules shown in Fig. 2 covers the traditional top level procedures of RBMT (analysis, transfer and generation), while in particular applications only selected modules can be used (e.g., in case of unilingual authoring or summarization the Transfer module is skipped, and the Analyzer is pipelined directly into the Generator). All modules are compatible and can provide different depth of processing, the grammar formalism being the same. Every top level procedure includes a number of application-specific sub procedures. In our computational formalism the basic analysis scenario consists of the following sequence of procedures: *Tokenization, Tagging, Chunking and Shallow semantic analysis*. *Tokenization* can be tuned to detect generally used and more specific features and to flag them with different types of "border" tags, thus significantly augmenting the feature space for disambiguation rules. *Tagging* includes assigning tags by lexicon look up and tag disambiguation according to disambiguation rules. The specificity of the tagging procedure is that it does not require any lemmatization due to the amount of knowledge stored in the lexicon, where for all lexemes their paradigms are explicitly listed [13]. *Chucking* is performed by a bottom-up heuristic parser with a recursive pattern matching technique. It identifies and classifies text constituents as typed phrases. *Shallow semantic analysis* determines *semantic dependency relations* between the classified text chunks and predicates.

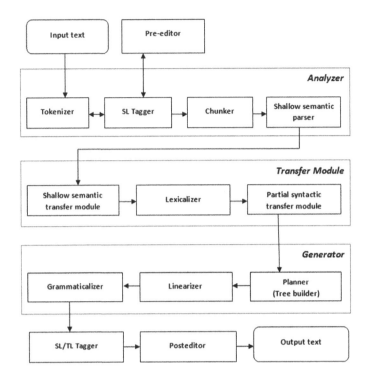

Fig. 2. An umbrella configuration of the rule-based NLP architecture.

For every identified phrase-chunk its governing predicate is detected and, then, the case-role status for every chunked phrase is determined. The final parse is a shallow semantic representation in the form of predicate/argument structures filled with SL text strings (see Fig. 1). Transfer is applied in case of multilingual applications and is a combination of interlingual transfer, lexicalization and syntactic transfer. The interlingual transfer substitutes a SL predicate-argument structure with a TL predicate-argument structure and links the latter with the TL knowledge in the lexicon. Lexicalization, called *Base transfer* substitutes every SL word in a predicate-argument structure with the base form of its cross-language equivalent by lexicon look-up. The predicate-argument format is kept unchanged (invariant). The *Syntactic transfer* is responsible for substituting lexicalized TL fillers resulting from the lexicon look up with well formed TL chunks, which is in fact translation of SL chunks-case-role fillers into a target language. *Generation* module linearizes predicate-argument structures filled with case-role strings into well-formed sentences. In case of a multilingual application a "real" translation procedure is thus reduced to the phrase level which, though not without problems, is still much simpler than machine translation that involves a full syntactic analysis and generation of possibly very complex sentences.

Every procedure relies on the system static knowledge and on the dynamic knowledge collected by the previous processing procedures. The knowledge handling and reuse is maintained by the universal developer environment.

3 Developer Environment

3.1 General Characteristic

The developer environment, we describe here, can be used to acquire and handle knowledge for multiple languages, domains. It includes a lexicon shell with flexible settings, several rule-acquisition compilers, where a linguist can write rules in a very simple formal language, and rule-control interfaces. The compilers provide for a computer environment, which can be used by a linguist not very experienced in programming. The multilingual lexicon developer tool with flexible settings is described in detail in [13]. The program shell of the lexicon allows defining entry structures, tagsets, contains knowledge that is directly used for text generation.

The lexicon program permits porting entry structures, tags and knowledge between languages and applications. The lexicon knowledge is directly pipelined to the rule acquisition compilers. Any changes made in the lexicon, e.g., tagsets, instantaneously propagate to the compilers and are displayed in the compiler interfaces. Developer environment is multilingual and every compiler is linked to a lexicon in a corresponding language.

Rules for different languages are accessed from a single program startup window. It makes it possible for the developer to freely navigate between knowledge bases for different applications and languages and to easily reuse appropriate amounts of linguistic knowledge, which proves to be quite possible, especially, in highly restricted domains, like, e.g., patent claims [14].

All compilers are equipped with front-end interfaces with a lot of effort saving functionalities, - it is possible on a mouse click to automatically get rule templates, transfer tag notations from the displayed list of tags to the curser position and check the consistency of tags and rule syntax. The compilers thus provide for an easy way to experiment with knowledge by simply copying the rules from one compiler to another and then, using the « check » and « control » functionalities update the language-specific rules. Figure 3 shows a startup window to access sets of language-dependent rules for tokenization, tag disambiguation, preediting, chunking, defining shallow semantic dependencies, predicate template tree builder, case-roles syntactic transfer, linearization, postediting.

Fig. 3. Compilers' startup window

3.2   Generic Features of Rule Formalisms

All rules are completely or partially formulated over the strings of tag variables and/or word variables. Though particular tagsets for different languages and applications can differ, what only matters in the universal computational formalism is a special top level classification of tags into *single* or *multiple*, *fine* or *coarse* tags. A *single tag* is a one tag symbol as assigned to an unambiguous wordform.

A *single tag* can code any range of linguistic information specified by developers, from simple POS classes [7] to what is included in "supertags" [4, 13–15] or morphosyntactic descriptions MSD [12], etc. Besides, we use the so called border tags to mark the start and end of a certain text segment.

A *multi-tag* is a string of *several single tags* assigned to one ambiguous wordform after, say, lexicon look up.

A *fine tag* is a tag assigned to one wordform after lexicon look up. A *fine tag* can be a *single-tag or multi-tag*.

A *coarse tag* is a tag that codes a *group of single tags* with the same morphosyntactic behavior. For example, if a tagset includes separate tags for different verbforms depending upon voice, person, tense, number etc., then a coarse tag can correspond, for example, to all verb forms or to the groups of verb tags in passive or active voice, correspondingly.

The rules for a very particular application, evidently, instead of generic tags should contain language-specific tags that should either be declared in a particular compiler or pipelined from lexicons. But the rule formalism is still the same. Every compiler program consists of two parts – a declaration part (optional) and a rule part. In the declaration part a developer can set variables, like lists of specific words or new valid tags to be used in the rules. The formalisms for writing rules are language independent, quite simple, though well-defined semantically, and, as said above, are formulated in terms of generic types of tags (see next section).

All compiler descriptions share the declaration part and rule format, - the IF-THEN-ELSE-ENDIF structure, where the IF block (rule conditions) can contain simple or complex conditions. Complex conditions can be formulated with the use of the binary logic operators AND, OR, and the operator NOT. The difference between particular rule formalisms used by every compiler program lies in the rule simple conditions and actions that are specific for every particular processing procedure. Below, we illustrate the developer environment and rule formalisms with the detailed descriptions of the tag disambiguation compiler and syntactic transfer compiler.

3.3   Tag Disambiguation Compiler

The input to the disambiguation compiler is the output of the first tagging procedure that is "welded in" the tagger. The input consists of border tags, as specified in the tokenization compiler, and text strings tagged with single or multiple fine tags assigned by the lexicon look up. The (condition) right-hand side of the tag disambiguating rules uses context information in terms of tags with attributes or words within a 5-word window with the tag/word in question in the middle. Figure 4 shows the compiler interface for writing rules. The left panes display tagset and help to format rules.

Below, for our fellow linguists, we in detail comment the disambiguation rules formalism, where the following notions are used.

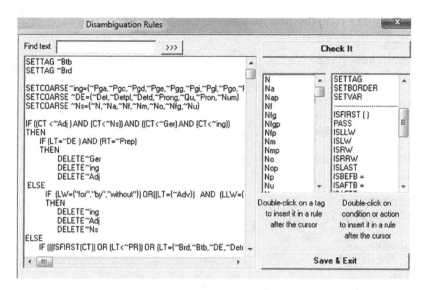

Fig. 4. A screenshot of the compiler interface for tag disambiguation rules (Here and in other exemples language-dependent tags are those specified in the patent-related application lexicon for English, see, e.g., (Sheremetyeva, 2004, 2007)).

CW means a current word, LW is a word to the left of CW, RW is a word to the right of CW, LLW is a word to the left of the left of CW, RRW means a word to the right of the right of CW. Everyone of these variable names can be used either on their own, or specified by some attributes, for example, as follows:

CW = {"means"}, LW = {"comprising", "having", "including"} or RRW = ListA, where listA is declared as SETVAR ListA = {"comprising", "having", "including"}.

ISLW, ISRW, ISLLW, ISRRW are conditions specifying that in the analyzed string there are words to the left, right, left-of-left, right-of-right of the current word;

CT means a current tag, LT means a tag to the left of CT, RT is a tag to the right of CT, LLT is a tag to the left of the left of CT, RRT is a tag to the right of the right of CT. Similar to the word variables the tag variables can be used either on their own, or specified by the tag attributes (parameters), for example, as follows: RT = { ~Brd, ~Btb, ~Conj, ~DE, ~Pg}, LT = { ~Adv}, or CT < ~Ns meaning that a current tag CT (multiple or single) includes a single tag component that is listed in a set of values of the coarse tag ~Ns. This coarse tag should be declared in the compiler, and in our example ~Ns is declared with the tag values for nouns of different semantic classes, see Fig. 4. The rule displayed in the screenshot in Fig. 4 disambiguates a multiple tag that could be assigned, e.g., to such a word as "opening": {opening} ~Adj~N~P~Ger, which means that this word can be an adjective, singular noun, present participle active or gerund. ISLAST, ISFIRST denote conditions, specifying that CT is the last/first in the segment. A processing module (the tagger in this case) can make several passes through he rules and the condition PASS <compare operation> <number> specifies the ordinal number of the pass, during which a certain part of the rules should be applied.

The compiler program description in the EBNF language

```
<program> :=
   [<declare-part>]
    <rule-part>

<declare-part> :=
[SETVAR <variable-name> = {"<string>" [, "<string>"
[,...]] } | SETTAG <tag>]
    . . .
[SETCOARSE <coarse-tag> = {<single-tag> [,<single-tag>
[,...]]}]

<rule-part> :=
   [<if-block>]
    . . .
<if-block> :=
   IF <condition>
   THEN
      <if-block> | <action-block>
   [ELSE
      <if-block> | <action-block>]
   ENDIF

<condition>  := <simple condition> | ( <condition> ) <bi-
nary logic operator> ( <condition> ) | NOT ( <condition>)

<simple condition> :=
   <tag variable> = {<multi-tag> [,<multi-tag> [,...]]}|
   ISFIRST (<tag variable>) | ISLAST|
   PASS <compare operation> <number> |
   <word variable> = {"<string>" [,"<string>" [,...]]} |
   <word variable> = <variable-name> |
   <tag variable> < <sinlge-tag> |
   ISLW | ISRW | ISLLW | ISRRW |

<compare operation> := = | < | >

<binary logic operator> := AND | OR |

<tag variable> := CT | LT | RT | LLT | RRT |

<word variable> := CW | LW | RW | LLW | RRW |

<action-block> :=
   [<action>]
    . . .
   <action> :=
   CT = <multi-tag> |
   DELETE <single-tag>

<multi-tag> := <single-tag>[<multi-tag>]
<single-tag> := ~<tag symbols>
```

3.4   Syntactic Transfer Rule Compliler

This compiler is used in our machine translation applications to acquire rules for syntactic restructuring and agreement in the strings of words that fill the case-role slots in the final parse structure (see Fig. 1) after TL lexicalization. In other words, the input to this compiler is separate strings of words in TL in base forms; the output is the correctly translated TL phrases filling particular case-roles in the TL predicate-argument structures.

At this stage, the feature space to formulate rule conditions includes the knowledge about SL and TL equivalents as specified in the bilingual lexicon and the knowledge produced by the previous processing steps that, in turn, includes

(a) the base forms of TL words with their base form tags,
(b) semantic classes of the predicates (cross-linguistic invariants) governing case-roles filled with the strings to be translated,
(c) case-role types (cross-linguistic invariants), to which the strings belong,
(d) phrase types (e.g., NP), to which the strings belong (in case processing included the phrase chunking stage[1]), and
(e) tag histories. The tag history is the knowledge about the "old" SL disambiguated tag of the SL word of its TL equivalent. The tags can only be single fine or coarse tags as multiple tags are disambiguated by this time.

All coarse tags, both for TL and SL should be necessarily declared in the compiler like, e.g., SETCOARSE $\sim$ NounInstr = { $\sim$ Nfi, $\sim$ Ni, $\sim$ Nni, $\sim$ Detdi, $\sim$ Npersi, $\sim$ Nai}. The TL fine tags are automatically taken from the lexicon, while the SL tags that are to be used in tag history should be declared in the compiler as SET-TAG $\sim$ Nameg. It is also necessary to declare lists of TL words that can be used as attributes (or parameters): SETVAR ListDifferent = {"different", "various", "similar"}.

The syntactic transfer rules are formulated in terms of tag templates composed of the strings of fine and/or coarse tags. The condition part contains "raw" TL tag templates as produced by lexicalization; the action part contains a template corresponding to a well-formed TL phrase. A "raw" tag template can be converted into a well-formed template by reordering, deleting, changing or inserting new tags associated with certain words. To make it possible the tags can be conditioned by the following attributes or parameters (Fig. 5):

(a) the number of their possible repetitions in the tag template, for which the Kleene + or * are used,
(b) their "parent" old SL tag: $\sim$ Gerund{ $\sim$ oldNoun},
(c) inclusion of the words from a certain list: Adj{ListDifferent},
(d) exclusion of the words, from a certain list: $\sim$ Adj(- ListDifferent} and (e) ordinal numbers in the output template in case a "raw" template contains several coinciding tags. This is done to apply appropriate changes to the relevant template component, in case the raw template contains, e.g. several nouns.

---

[1] It might not always be the case as there are applications that skip phrase chunking and work directly on lexicon tags.

Fig. 5. A screenshot of the compiler interface with the rules for machine translation for patent claims and scientific and technical papers from Russian into English.

In the EBNF description of the compiler the following notations are used: T = ~ Tag ~ Tag2... is a "raw" template; ~ Tag can have parameters (a) – (e) listed above;

PATTERN is an output pattern, indicating all changes.

PREDSC means semantic class of the predicate as specified in the lexicon. CRTYPE is case-role type SUBJ, DOBJ, IOBJ, PLC, MANN, ... PAR are case-role notations as specified in the system lexicon, meaning, "subject", "direct object", "indirect object", "place", "manner", ... "parameter", correspondingly.

The compiler program description in the EBNF language

```
<program> :=
 [<declare-part>]   // as in tagging;
 <rule-part>        // as in tagging but simple condition;
 ...
 <simple-condition>:=
   T = ~Tag[ {<tag-parameters>} ] ...  |
   PREDSC = {"sem-class", ... } |
   CRTYPE = {<cr-type> , ...}   |
   CURRENT = {~Border-tag=Phrase tag | ~CASEROLE, ...}

   <cr-type> := SUBJ | DOBJ | IOBJ | PLC | MANN | MEAN |
PURP | COND | TIME | SRC | DEST | PAR

   <tag-parameters> := [ + | * | <Number>]; [[-]ListName];
[[-]OldTagInLanguageFrom]

   <Number> := 1 | 2 | 3 | ...

<action>:=

   PATTERN <tags>[+]...

<tags> :=

   ~Tag [(<Number>)] |
   ~Tag (<word or words>) |
   ~NewTag (~Tag[(<number>)]])
```

4   Conclusions

The paper presented a set of portable computational formalisms for rule-based tagging, analysis, transfer and generation that allow migrating from one rule-based application to another within one language or cross-linguistically. All formalisms are implemented in processing modules and developer environment that includes a lexicon shell with flexible settings and a number of compilers for writing rules in language independent formalisms. The developer environment can be used by linguists without advanced programming skills and allows acquiring or editing lexical resources, specifying tags, writing processing rules, and control the correctness of processing. The formalisms and compilers described have been successfully used (in different combinations and with different depth of analysis) in a number of unilingual and multilingual applications in English, Danish, French and Russian.

References

1. Camilleri, J.J.: An IDE for the grammatical framework. In: Proceedings of a Workshop on Free/Open-Source Rule-Based Machine Translation, Gothenburg, Sweden, pp. 1–12 (2012)
2. Feldman, A., Jirka, H., Brew, Ch.: A cross-language approach to rapid creation of new morpho-syntactically annotated resources. In: Proceedings of LREC 2006 (2006)
3. Joshi, A.K.: Unification and some new grammatical formalisms. In: Proceedings of the Workshop on Theoretical Issues in Natural Language Processing TINLAP 1987, pp. 45–50. Las Cruces, New Mexico. USA (1987)
4. Joshi, A.K, Srinivas, B.: Disambiguation of super parts of speech (or Supertags): almost parsing (1994). http://acl.ldc.upenn.edu/C/C94/C94-1024.pdf
5. Neumann, C.: Generating patent claims in english from a Japanese-only interface. In: Proceedings of the Workshop on Patent Translation in conjunction with the MT Summit X. Phluket, Thailand, September 2005
6. Paul, M.: Translation knowledge recycling for related languages. In: Proceedings of MT Summit VIII. Santiago de Compostela, Galicia, Spain, 18–22 September 2001
7. Petrov, S., Das, D., McDonald, R.: A universal part-of-speech tagset. In: Proceedings of the Conference on Language Resources and Evaluation (LREC 2012). Istanbul, Turkey (2012)
8. Pereira, F.C.N., Shieber, S.M.: The semantics of grammar formalisms seen as computer languages. In: Proceedings of the Tenth International Conference on Computational Linguistics. Stanford, CA, USA (1984)
9. Pinkham, J., Corston-Oliver, M., Smets, M., Pettenaro, M.: Rapid assembly of a large-scale French-English MT system. In: Proceedings of MT Summit VIII. Santiago de Compostela, Galicia. Spain, 18–22 September 2001
10. Przepiórkowski, A.: A preliminary formalism for simultaneous rule-based tagging and partial parsing data structures for linguistic resources and applications. In: Rehm, G., Witt, A., Lemnitzer, L. (eds.) Tübingen:Gunter Narr Verlag, pp. 81–90 (2007)
11. Sharoff, S., Nivre, J.: The proper place of men and machines in language technology processing Russian without any linguistic knowledge. In: Proceedings of the International Conference on Computational Linguistics "Dialogue"- 2011. Russian State Humanitarian University, Moscow (2011)
12. Sharoff, S., Kopotev, M., Erjavec, T., Feldman, A., Divjak, D.: Designing and evaluating a Russian tagset. In: Proceedings of the Sixth Language Resources and Evaluation Conference, LREC 2008, Marrakech (2008)
13. Sheremetyeva, S.: Application adaptive electronic dictionary with intelligent interface. In: Proceedings of the Workshop on Enhancing and Using Electronic Dictionaries in Conjunction with the 20th International Conference on Computational Linguistics. COLING 2004, Geneva, Switzerland, 23–28 August 2004
14. Sheremetyeva, S.: On portability of resources for quick ramp-up of multilingual MT for patent claims. In: Proceedings of the workshop on Patent Translation in conjunction with MT Summit XI, Copenhagen, Denmark, 10–14 September 2007
15. Sheremetyeva, S.: On teaching technical writing with an authoring tool. In: Proceedings of the 10th International Technology, Education and Development Conference INTED 2016, Valencia, Spain, 7–9 March 2016
16. Takeda, K.: Portable knowledge sources for machine translation. In: Proceedings of the 15th International Conference on Computational Linguistics, Kyoto, Japan, 5–9 August 1994

Morphology and Text Segmentation

Several Ways to Use the Lingwarium.org Online MT Collaborative Platform to Develop Rich Morphological Analyzers

Vincent Berment[1(✉)], Christian Boitet[2(✉)],
Jean-Philippe Guilbaud[3(✉)], and Jurgita Kapočiūtė-Dzikienė[4(✉)]

[1] INaLCO, 65 rue des Grands Moulins, 75013 Paris, France
Vincent.Berment@inalco.fr
[2] GETALP, LIG, UGA, 700 avenue Centrale, 38041 Grenoble, France
Christian.Boitet@imag.fr
[3] GETALP, LIG, CNRS, 700 avenue Centrale, 38041 Grenoble, France
Jean-Philippe.Guilbaud@imag.fr
[4] Vytautas Magnus University, Kaunas, Lithuania
Jurgita.K.Dz@gmail.com

Abstract. We will demonstrate several morphological analyzers of languages for which morphological analysis is very difficult, and/or that are under-resourced. It will cover at least French, German, Khmer, Lao, Lithuanian, Portuguese, Quechua, Spanish and Russian. These morphological analyzers all run on the collaborative platform lingwarium.org that supports the ARIANE-H lingware development environment. Some will also be presented as stand-alone Windows applications.

1 Introduction

The online platform lingwarium.org was opened in July 2016. It provides a means for geographically scattered groups of language experts to develop new machine translation systems collaboratively, especially for under-resourced languages. The main linguistic programming toolkit is ARIANE-H, the version of ARIANE-G5 recently produced by Vincent Berment [1]. Lingwarium.org also offers other tools such as MOTOR, dedicated to the word segmentation of texts in languages using an unsegmented writing system, such as many Asian languages (Burmese, Khmer, Lao, Thai...). It also contains some programs used to speed up the development process.

The present paper details the different approaches used under lingwarium.org to develop rich morphological analyzers (as first steps of MT systems), using the ARIANE-H toolkit and some other tools. The demonstration will include morphological analysers for several languages, including the ones detailed in this paper: French, German, Khmer, Lao, Lithuanian, Portuguese, Quechua, Spanish and Russian.

© Springer Nature Switzerland AG 2018
A. Gelbukh (Ed.): CICLing 2017, LNCS 10761, pp. 81–86, 2018.
https://doi.org/10.1007/978-3-319-77113-7_6

2 Word Segmentation

The MOTOR word segmenter relies on the minimum matching algorithm that computes the segmentation of a text which contains the smallest possible number of words. In case it finds several solutions, it outputs the first one.

To run its algorithm, MOTOR only needs a list of words (word forms) for the language to be treated. MOTOR is currently used operationally in analysers for Burmese (27,493 words), Khmer (85,655 words), Lao (50,078 words), Thai (20,574 words) and old Tibetan (26,730 words). We also tested it with Japanese for a limited corpus (see below, the "Little Prince" project).

3 Tokenization, Stemming and POS Tagging

Another important operation in morphological analysers is to compute a lemma for each word of the texts to be analysed. In LINGWARIUM, this task is handled by writing inflectional and compositional rules in ATEF. ATEF is the SLLP (specialized language for linguistic programming) of the ARIANE framework used for writing morphological analysers.

Though this language is quite easy to use, a number of tools have been developed to simplify the task of the lexicographers. These tools can generate ATEF code from simple tables, typically Excel sheets or database tables, or from other frameworks such as NooJ.

For **Lithuanian**, we took all the distinct words (word forms) from an extract of the "*Corpus of the Contemporary Lithuanian Language*", created at Vytautas Magnus University [2]. The corpus extract we used contains about 1 million running words and covers different domains: fiction texts, newspaper texts, legislative texts, parliamentary transcripts, etc.

These word forms have been associated with their lemmas and grouped into 17 parts-of-speech: nouns (16,321 distinct lemmas); adjectives (4,937); adverbs (2,017); numerals (78); several verb forms differing in their inflection as verbs (11,831), participles (11,831), half participles (11,751), adverbial participles (lith. padalyviai) (11,831), adverbial participles1 (lith. būdiniai) (11,751); pronouns (43); particles (117); interjections (59); onomatopoeias (40); conjunctions (62); prepositions (73); abbreviations (109); and acronyms (156).

For each lemma, stable and unstable parts (changing due to inflection) are indicated. Where possible, word forms have been annotated with values of several attributes: polarity (positive, negative), degree of comparison (comparative, superlative), reflexivity (non-pronominal, pronominal), gender (masculine, feminine, neuter), number (singular, plural), and case (nominative, genitive, dative, accusative, instrumental, locative, vocative). The same morphological information has also been associated with the appropriate list of affixes (suffixes and endings) that vary to produce inflected forms.

The data have then been compiled automatically into a lexical database that can be used directly to produce the "lingware files" that make up the Lithuanian analyser in ARIANE-H. Basically, this database was obtained by transforming:

Table 1. Extract from the dictionary table

Id	Lemma	Morphological information	Paradigm
1	abatinis	FSAdjP	ADJ001
2	abdominalinis	FSAdjP	ADJ001
3	abejingas	FSAdjP	ADJ002
4	abejotinas	FSAdjP	ADJ002
5	abiotinis	FSAdjP	ADJ001
6	abipusis	FSAdjP	ADJ001
7	abonentinis	FSAdjP	ADJ001
8	abraomiškas	FSAdjP	ADJ002
9	abrazinis	FSAdjP	ADJ001
10	absoliutus	FSAdjP	ADJ004
...	...	...	...

- lemmas and morphological information into a *dictionary table* (see Table 1) containing lemmas and associated morphological information (expressed using so-called $ATEF$ *formats* that are simple property lists, or *decorations* in ARIANE terminology),
- endings and their associated morphological information into a *paradigm table* (see Table 2).

Table 2. Extract from the paradigm table

Id	Ending	Morphological information	Paradigm	Nb Char[a]
1	is	FAD1MSNN	ADJ001	2
2	io	FAD1MSNG	ADJ001	2
3	iam	FAD1MSND	ADJ001	2
4	į	FAD1MSNA	ADJ001	2
5	iu	FAD1MSNI	ADJ001	2
6	iame	FAD1MSNL	ADJ001	2
7	iam	FAD1MSNL	ADJ001	2
8	i	FAD1MSNV	ADJ001	2
9	iai	FAD1MPNN	ADJ001	2
10	ių	FAD1MPNG	ADJ001	2
...	...	...	...	...

[a]The "Nb Char" column contains the number of characters that have to be removed from the end of the lemma to build the radical that will be put in the ATEF dictionaries.

Here are several examples of how this is made for several other analysers (examples given for the inflectional analysis).

For **French**, we transformed two tables of a database built by Sylviane Chappuy, which contained (1) a list of words with their morphological paradigms, and (2) the paradigms themselves (the endings for each existing person, gender, number, tense...). This morphological analyser has been developed in the Traouiéro ANR project [3].

For **Russian**, we started from the NooJ lexical data built by Vincent Bénet [4], which contains Zaliznyak's dictionary.

The ATEF "variables" file DVM + DVS was derived from the _properties.def file. For example:

NooJ : *"A_Forme = fc | fl | adv;"* → ATEF: *"A_Forme: = (fc, fl, adv).".*

The ATEF radicals file was derived from the NooJ dictionary file. For example:

NooJ : *"багреный,A + FLX = новый"* → ATEF: *"багрен ==P1 (A,багреный).",*

where багрен is the radical obtained by removing a number of characters corresponding to the highest <B1> in the новый paradigm, P1 is the *morphological format* (it triggers the analysis rules) corresponding to the новый paradigm, A is the *syntactic format* (the combination of P1 and A contains the lexical information of the NooJ entry) and багреный is the *lexical unit* or LU. In many analysers for MT, the LU is a derivational class, but in this analyser, it is simply the lemma[1].

The grammar rules (GRAM component) and the endings dictionary are derived from the NooJ paradigms file _russe-morph.nof. The other ATEF files — the morphological formats file FTM (these formats trigger the rules) and the syntactical formats file FTS (which contain the lexical information) — are also derived from the NooJ dictionaries.

For **Quechua**, we started from the lexical data built by Maximiliano Duran. For many years, Duran compiled a bilingual dictionary between the Ayacho dialect, an agglutinative and under-resourced language, and French. We derived the radical file from this data, and the other ATEF files were written manually from the information detailed in his PhD thesis: parts of speech, suffixes... [5].

For **German**, Jean-Philippe Guilbaud directly writes in ATEF [6]. In June 2016, his analyser contained 18,219 verbs, 142,321 nouns and 21,747 adjectives, totalising 182,725 different lemmas.

For **Portuguese**, Paltonio Daun Fraga has also written the system directly in ATEF . From the Portuguese system, he derived a **Spanish** analyser in a very short period of time (less than six months) that even outperforms the Portuguese one.

For several Southeast Asian languages, a group of language experts scattered in many places around the world joined their efforts to develop a set of small but consistent analysers for Khmer, Lao, Myanmar, Thai, Tibetan and Vietnamese. The linguistic scope of this project is limited to the text of Saint Exupéry's "Little Prince". Going beyond this reduced perimeter, Vincent Berment and Guillaume de Malézieux are developing morphological analysers for **Lao** and **Khmer** with broader coverage.

---

[1] The RUS-FRA MT system built in the 70's by N. Nédobejkine in Ariane-G5 contains a very good MA for Russian, where the LUs are indeed derivational families. Its 13000 LUs correspond to about 40,000 lemmas, themselves corresponding to about 400,000 different accented word forms.

4 Named Entity Extraction

It is easy, with the ATEF language, to describe exhaustively all closed classes. By cons, if the affixes dictionaries may contain the full list of endings, prefixes and suffixes of the concerned language (grammatical morph[eme]s), the lexemes of the language constitute an unbound set, hence the lexical dictionaries can never be exhaustive. The "unknown word problem" is a recurring unavoidable phenomenon.

To handle it, we use the possibility offered by ATEF to write a whole subgrammar to handle unknown words. That subgrammar is triggered by the obligatory MODINC morphological format, and must contain at least a special rule, MOTINC, that is guaranteed to produce at least one result (it unconditionnally produces as LU value the input form itself and stops). When the analysis of a form fails, ATEF restarts it in a special configuration, as if the empty string had been segmented as a prefix, and would be associated with the MODINC morphological format (and hence all rules callable by it) in the dictionaries.

The MODINC subgrammar can be very simple (containing then only the MOTINC rule), or it can implement an elaborate strategy, for example to handle some classes of proper nouns, acronyms, neologisms, etc. For example, a verbal neologism such as "lispified" (transformed into LISP) can be assessed to be the participle past of an unknown verb "lispify", thanks to a normal Markov rewriting method that produces the hypothetical lemma with a few extra ATEF rules and a dictionary of special affixes obtained by a systematic transformation of the subset of normal affixes which are supposed to intervene in the inflectional morphology of unknown words [7].

5 Chunking, Parsing, and Coreference Resolution for Disambiguation

In order to process separate particles (such as the particle "*an*" in the German verb "*ankommen*") and also to disambiguate to some extent the output of the lemmatizer, we can use a sequence of two specialized modules after the ATEF phase: a first module written in EXPANS and a second one written in ROBRA[2].

The EXPANS module contains a dictionary whose entries are the base verbs accepting separable particles (e.g. "*kommen*"). For each such base verb, the dictionary provides a tree containing as many leaves as there are possible combinations of "particle + base verb" (e.g. "*an*" + "*kommen*"). Each leaf is actually a decorated structure containing the new value of lexical unit corresponding to the combination (e.g. "*ankommen*") together with a tactical variable used for coding the particle.

Then, the next (ROBRA) module executes a grammar that looks for the separable particles in the sentence and compares them with the expected values of particles for the processed verb tree. When the correct candidate is found, the others leaves are removed from the tree. This disambiguation process, able to recognize compound

---

[2] EXPANS and ROBRA are specialized languages of ARIANE, just as ATEF.

words and verbs with separate particles, is implemented by Jean-Philippe Guilbaud in his German morphological analyzer (AMALD).

6 Access Through an API

LEXTOH. Ying ZHANG has developed LEXTOH, a middleware to call morphological analysis web services, and then normalize, merge and filter the results.

7 Conclusion

Reusing software and relying on a community help make the efforts for developing new morphological analysers more efficient. Beyond the most advanced analysers presented in this paper, several prototypes are currently being developed for Ngazidja (the Comorian dialect of Gran Comoros), Swahili, Somali, and Breton. The "Little Prince" project is another approach to help language experts developing new systems, especially for the under-resourced languages on which we are focusing.

References

1. Berment, V., Boitet, C.: Heloise — An Ariane-G5 compatible environment for developing expert MT systems online. In: Proceedings of the 24th International Conference on Computational Linguistics (COLING), Mumbai, 9 p (2012)
2. Marcinkevičienė, R.: Tekstynų lingvistika: teorija ir praktika [Corpus Linguistics: Theory and Practice]. Darbai ir Dienos **24**, 7–64 (2000)
3. Chappuy, S., Guilbaud, J.-P., Berment, V.: T7o — Lemmatiseur du français FR4 en ATEF avec 100 000 lemmes. L4.2.b, deliverable L234.1, Traouiero ANR project, 8 p, 15 June 2011
4. Bénet, V.: Conception et réalisation de ressources lexicales et grammaticales pour le russe. Semaine NOOJ Inalco, 40 p, 31 January 2012
5. Duran, M.: Dictionnaire electronique de verbes français-quechua pour le TAL. Thèse de doctorat en linguistique, 286 p (2016). (should be defended before the COLING 2016 conference)
6. Guilbaud, J.-P., Boitet, C., Berment, V.: Un analyseur morphologique étendu de l'allemand traitant les formes verbales à particule séparée. TALN-RÉCITAL 2013, Les Sables d'Olonne, 9 p, 17–21 June 2013
7. Guilbaud, J.-P., Boitet, C.: Comment rendre une morphologie robuste du français encore plus robuste en traitant finement les mots inconnus avec les données disponibles. TALN 1997, Grenoble, 12 p, 12–13 June 1997

A Trie-structured Bayesian Model for Unsupervised Morphological Segmentation

Murathan Kurfalı[1(✉)], Ahmet Üstün[1], and Burcu Can[2]

[1] Cognitive Science Department, Informatics Institute Middle East Technical University (ODTÜ), Ankara 06800, Turkey
{kurfali,ustun.ahmet}@metu.edu.tr
[2] Department of Computer Engineering, Hacettepe University Beytepe, Ankara 06800, Turkey
burcucan@cs.hacettepe.edu.tr

Abstract. In this paper, we introduce a trie-structured Bayesian model for unsupervised morphological segmentation. We adopt prior information from different sources in the model. We use neural word embeddings to discover words that are morphologically derived from each other and thereby that are semantically similar. We use letter successor variety counts obtained from tries that are built by neural word embeddings. Our results show that using different information sources such as neural word embeddings and letter successor variety as prior information improves morphological segmentation in a Bayesian model. Our model outperforms other unsupervised morphological segmentation models on Turkish and gives promising results on English and German for scarce resources.

Keywords: Unsupervised learning · Morphology
Morphological segmentation · Bayesian learning

1 Introduction

Morphological segmentation is the task of segmenting words into their meaningful units called *morphemes*. For example, the word *transformations* is split into *trans*, *form*, *ation*, and *s*. This process serves mainly as a preprocessing task in many natural language processing (NLP) applications such as information retrieval, machine translation, question answering, etc. This process is essential because sparsity becomes crucial in those NLP applications due to morphological generation that produces various word forms from a single root. It is infeasible to build a dictionary that involves all possible word forms in a language in order to use in an NLP application. Hankamer [14] suggests that the number of possible word forms in an agglutinative language such as Turkish is infinite. Therefore, instead of building a model based on word forms, morphological segmentation is applied to reduce the sparsity principally in any NLP application.

© Springer Nature Switzerland AG 2018
A. Gelbukh (Ed.): CICLing 2017, LNCS 10761, pp. 87–98, 2018.
https://doi.org/10.1007/978-3-319-77113-7_7

Various features have been used for morphological segmentation. Many approaches use orthographic features. However, morphology is tightly connected with syntax and semantics. Syntactic and semantic features have also been used for the segmentation task.

Features are normally used in Bayesian models in the form of a prior distribution. For example, [7] utilize frequency and length information of morphemes as prior information, which provide some orthographic features.

In this paper, we aggregate prior information from different sources in morphological segmentation within a Bayesian framework. We use orthographic features such as letter successor variety (LSV) counts obtained from tries, semantic information obtained from the neural word embeddings [17] to measure the semantic relatedness between substrings of a word, and we use the presence information of a stem in a dataset after its suffixes are stripped off assuming a concatenative morphology. Our results show that combining prior information from different sources give promising results in unsupervised morphological segmentation.

In this study, we learn tries based on semantic and orthographic features. Therefore, the output of our model is not only segmentation, but also tries that are composed of semantically and morphologically related words.

The paper is organized as follows: Sect. 2 presents the previous work on unsupervised morphological segmentation, Sect. 3 defines the mathematical model, Sect. 5 describes the inference algorithm to learn the mathematical model, Sect. 7 presents the experimental results, and finally Sect. 8 concludes the paper with a discussion and potential future work.

2   Related Work

Morphological segmentation, as one of the oldest fields in NLP, has been excessively studied. Deterministic methods are the oldest ones used in morphological segmentation. Harris [15] defines the distributional characteristics of letters in a word for the first time for unsupervised morphological segmentation. LSV model is named after Harris, which defines the morpheme boundaries based on letter successor counts. If words are inserted into a trie, branches correspond to potential morpheme boundaries. An example is given in Fig. 1. In the example, *re-* is a potential prefix, and *-s*, *-ed* and *-ing* are potential suffixes in the trie due to branching that emerges before those morphemes. LSV model has been applied in various works [1–3,11,13]. In our study, we also use a LSV-inspired prior information, but this time in a Bayesian framework.

Stochastic methods have also been extensively used in unsupervised morphological segmentation. Morfessor is the name of the family of a group of unsupervised morphological segmentation systems which are all stochastic [8–10]. Non-parametric Bayesian models have also been applied in morphological segmentation [5,12,19].

Neural-inspired features are used in the recent studies. Narasimhan et al. [18] use semantic similarity obtained from neural word embeddings by word2vec [17].

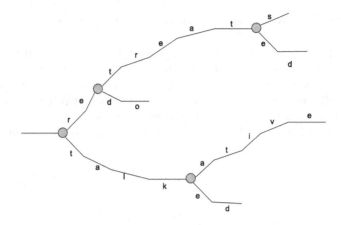

Fig. 1. Potential morpheme boundaries on a trie used in LSV model [4]

Narasimhan et al. [18] adopt the semantic similarity as a feature in a log-linear model. Soricut and Och [20] use word embeddings to learn morphological rules in an unsupervised setting.

In this work, we are both inspired by the oldest works and the recent works in terms of various features used. Thus our model has inspirations from the LSV methods, stochastic methods, and neural-based models in a combined framework.

3   Building Neural Word Embedding-Based Tries

Our model is based on neural word embedding-based tries that are built by using two different methods:

3.1   Tries Structured from the Same Stem

These tries contain semantically related (morphologically derived or inflected from each other) words having the same stem. In order to find the stem of a given word in the training set, we used the algorithm which is introduced in [21]. In the algorithm, all potential prefixes of a word are extracted. For example, *fe, fea, fear, fearf, fearfu, fearful, fearfull, fearfully* are the prefixes of *fearfully*. The rightmost segmentation point where the cosine similarity between the word and the first prefix (from the right of the word; i.e. *fearful*) is higher than a manually set threshold[1] gives the first valid prefix, which refers to the first segmentation point.

Other segmentation points are found by repeating the process towards the head of the word by checking the cosine similarity between the just detected valid prefix and the subsequent prefix to the left of the word. The final detected

---

[1] We assign 0.25 as the threshold following [21].

prefix with the leftmost segmentation point in the word becomes the stem of the word.

Among the nearest 50 neighbors of the stem which are obtained from word2vec [17], the ones that begin with the same stem are inserted to the same trie. This process is repeated for each word that is inserted on the trie recursively until all the words that are semantically similar which share the same stem (detected by using the same algorithm described above) are covered. An example trie that is built with the words having the same stem is given in Fig. 2.

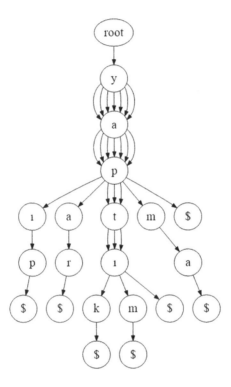

Fig. 2. Visualization of a trie portion that includes the word forms derived from the same stem. *yapıp, yapar, yaptık, yaptım, yapma* are inflected forms of the stem *yap* (means *to do*). The number of edges refers to the number of words in the corpus flowing in that direction on the trie. $ denotes the end of the word.

3.2   Tries Based on Semantic Relatedness

Semantically related 50 words are retrieved for each word in the training set by using word2vec [17]. For each word, a trie is built and 50 similar words are inserted on the word's trie. Eventually, a trie that consists of 51 words is created for each word in the training set. A portion of a trie that involves semantically related words is given in Fig. 3.

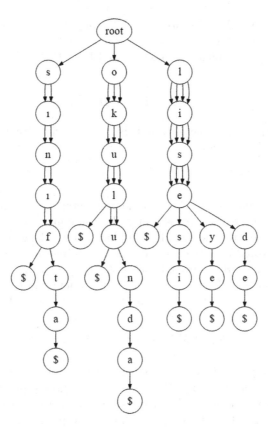

Fig. 3. Visualization of a trie portion built by using semantic relatedness. The trie consists of the stems *sınıf*, *okul*, *lise* (means *high school*, *class*, *school*) and affixed forms of these stems. The number of edges refers to the number of words in the corpus flowing in that direction on the trie. $ denotes the end of the word.

4   Bayesian Model Definition

We define a Bayesian model in order to find the morpheme boundaries on the tries:

$$p(Model|Corpus) \propto p(Corpus|Model)p(Model) \tag{1}$$

where *Corpus* is a list of raw words and *Model* denotes the segmentation of the corpus. The *Model* that maximizes the given posterior probability will be searched for the segmentation task. We apply a unigram model for the likelihood:

$$p(Corpus|Model) = \prod_{i}^{|W|} p(w_i = (m_{i1} + m_{i2} + \cdots + m_{it_i}|Model)$$

$$= \prod_{i}^{|W|} \prod_{j=1}^{t_i} p(m_{ij}|Model) \tag{2}$$

where w_i is the ith word in $Corpus = \{w_1, \cdots, w_{|W|}\}$, m_{ij} is the jth morpheme in w_i, t_i is the number of morphemes in word w_i, and $|W|$ is the number of words in the corpus. Here, morphemes are generated by a Dirichlet Process (DP) as follows:

$$m_{ij} \propto DP(\alpha, H) \qquad (3)$$

with the concentration parameter α and the base distribution H that is formed with a geometric distribution:

$$H(m_{ij}) = \gamma^{|m_{ij}|+1} \qquad (4)$$

Here, $|m_{ij}|$ is the length of m_{ij} and γ is the parameter of the geometric distribution. We assume that each letter is uniformly distributed. Therefore, we assign $\gamma = 1/L$ where L denotes the size of the alphabet in the language. Shorter morphemes will be favored with the usage of length-inspired base distribution in the DP. From the Chinese Restaurant Process (CRP) perspective, each morpheme is generated proportionally to the number of morphemes of the same type that have already been generated (i.e. customers having the same dish):

$$p(m_{ij} = k | Model) = \frac{n_k + \alpha H(k)}{N + \alpha} \qquad (5)$$

This computes the probability of m_{ij} being of type k where k refers to a distinct morpheme (i.e. morpheme type). Here, n_k is the number of morphemes of type k and N is the total number of morpheme tokens in the model. We generate each morpheme regardless of its type, such as stem, prefix, or suffix.

As for the prior information, we model the morpheme boundaries:

$$p(Model) = \prod_{i}^{|W|} \prod_{j=1}^{t_i} p(b_{ij}) \qquad (6)$$

Here, b_{ij} refers to the jth morpheme boundary in $w_i = m_{i1} + m_{i2} + \cdots + m_{it_i}$ where $w_i = \{b_{i1}, b_{i2}, \cdots, b_{it_i}\}$.

The probability of each b_{ij} is decomposed in terms of the number of branches leaving that node (when inserted on the trie), semantic similarity that is introduced between the two word forms that is split with b_{ij}, and the presence of the word form once the suffix is stripped off from the word:

$$p(b_{ij}) = p(b_{ij\,branch}) \cdot p(b_{ij\,semantics}) \cdot p(b_{ij\,presence})$$

where $p(b_{ij\,branch})$ denotes the probability of b_{ij} being a morpheme boundary based on the branches leaving the trie node, $p(b_{ij\,semantics})$ is based on the semantic similarity of the two word forms where b_{ij} separates the two forms, and $p(b_{ij\,presence})$ is estimated based on the word form whether it exists in the corpus once the suffix after b_{ij} is stripped off.

Based on the LSV, the branching on the tries corresponds to the potential morpheme boundaries. We model the branching with a Poisson distribution:

$$p(b_{ij\,branch}) = p(z_{ij} = l|\lambda) \tag{7}$$

$$= \frac{\lambda^l e^{-\lambda}}{l!} \tag{8}$$

where $z_{ij} = l$ denotes the number of branches leaving the node below b_{ij} and λ is the parameter of the Poisson distribution[2].

We use the cosine similarity (which is always between 0 and 1) between the neural word embeddings of the two word forms that are separated by b_{ij} for the semantic distribution:

$$p(b_{ij\,semantics}) = cos(x_{m_{i1}+\cdots+m_{ij}}, x_{m_{i1}+\cdots+m_{ij+1}}) \tag{9}$$

Here, $x_{m_{i1}+\cdots+m_{ij}}$ corresponds to the word vector of the word form $m_{i1}+\cdots+m_{ij}$ obtained from word2vec. It is the full word vector and not the compositional vector obtained from morpheme vectors.

As for the presence of the word form in the word list, we compute the likelihood of the word form $m_{i1} + \cdots + m_{ij}$:

$$p(b_{ij\,presence}) = \frac{f(m_{i1} + \cdots + m_{ij})}{\sum_{i=1}^{|Corpus|} f(w_i)} \tag{10}$$

where $f(m_{i1} + \cdots + m_{ij})$ denotes the frequency of the word form in the corpus.

5   Inference

We use Gibbs sampling [6] for the inference. In each iteration, a word is uniformly selected from any trie and removed from the corpus. A binary segmentation of the word is sampled from the given posterior distribution:

$$p(w_i = m_{i1} + m_{i2}|Corpus^{-w_i}, Model^{-w_i}, \alpha, \lambda, \gamma)$$
$$\propto p(m_{i1}|Model^{-w_i}, \alpha, \gamma)p(m_{i2}|Model^{-w_i}, \alpha, \gamma)p(b_{i1}) \tag{11}$$

Once a binary segmentation is sampled, another binary segmentation is sampled for m_{i1}. Therefore, a left-recursion is applied for the left part of the word. This is because of the cosine similarity that is computed between neural word embeddings of word forms and not suffixes by the original word2vec.

This process is repeated recursively until having at least 4 letters in the stem or having sampled the word itself from the posterior distribution (i.e. when the word is not segmented). An illustration is given in Fig. 4.

---

[2] In the experiments, we assign $\lambda = 4$.

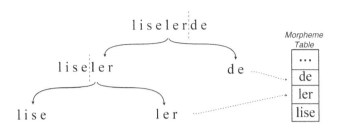

Fig. 4. The binary segmentation of the word *liselerde* (means *in the high schools*)

6 Segmentation

Once the model is learned, any unseen word can be segmented by using the learned model. Each word is split based on the maximum likelihood in the learned model:

$$\arg\max_{m_{i1},\cdots,m_{it_i}} p(w_i = m_{i1} + \cdots + m_{it_i}|Model, \alpha, \gamma) \tag{12}$$

For the segmentation, we apply two different strategies. In both methods, we select the segmentation with the maximum likelihood, however the set of possible segmentations for the given word differs. In the first method, we only consider the segmentations learned by the model. Since the same word can exist in multiple tries, a word may have more than one different segmentation. In the second method, we consider all possible segmentations of a word and choose the one with the maximum likelihood.

7 Experiments and Results

We did experiments on Turkish, English and German. For each language, we built two sets of tries based on the methods described in Sects. 3.1 and 3.2. We aggregated the publicly available training and development sets provided by Morpho Challenge 2010 [16] for English, Turkish and German for training. Although gold segmentations are provided in the datasets, we only used the raw words in training. Gold segmentations were only used for evaluation purposes.

We began with 1686 English words, 1760 Turkish words, and 1779 German words obtained from the aggregated sets. Once the tries have been built by recursively augmenting the tries by using word2vec [17], eventually we obtained 2560 English word types, 43884 Turkish word types, and 13747 German word types in the tries structured from similar stems (see Sect. 3.1). Additionally, we obtained 34594 English word types, 67292 Turkish word types, and 23875 German types in the tries that were built based on the semantic relatedness (see Sect. 3.2).

We used 200-dimensional word embeddings that were obtained by training word2vec [17] on 361 million word tokens and 725.000 word types in Turkish, 129 million word tokens and 218.000 word types in English, and 651 million word

Table 1. Size of the datasets used in the experiments. *m1* denotes the train set built by the first method (Sect. 3.1) and *m2* denotes the train set built by the second method (Sect. 3.2)

Language	Train-m1	Train-m2	Train word2vec	Test
Turkish	43884 types	67292 types	725K types	1760 types
			361M tokens	
English	2560 types	34594 types	218K types	1686 types
			129M tokens	
German	13747 types	23875 types	608K types	1779 types
			651M tokens	

tokens and 608.000 word types in German. The size of all datasets used in the experiments are given in Table 1.

We compared our model with Morfessor Baseline [8] (M-Baseline), Morfessor CatMap [9] (M-CatMAP) and MorphoChain System [18]. For that purpose, we trained these models on the same training sets. We obtained the frequency information from the full word lists provided by Morpho Challenge which was need by other systems. The evaluation was performed on the aggregated training and development sets of Morpho Challenge 2010 using the Morpho Challenge evaluation method [16]. All word pairs that have a common morpheme are extracted from the results and checked whether they really share a common morpheme in the gold standard data. One point is given for each correct pair. The Precision is the proportion of the collected points to the total number of words. Recall is computed analogously. This time all word pairs that share a common morpheme are extracted from the gold standard data and checked whether they have a common morpheme in the results. For each correct pair, one point is given. Finally, the Recall is the proportion of the collected points is to the total number of words.

The results are given in Tables 2 and 3 for tries that are composed of words structured from the same stem (see Sect. 3.1) and for tries that are based on semantic relatedness (see Sect. 3.2). According to the results, tries that contain semantically similar words achieve a better performance on morphological segmentation proving that semantically similar words also manifest similar syntactic and thus similar morphological features.

Our trie-structured model (TST) performs better than Morfessor Baseline [8], Morfessor CatMAP [9] and Morphological Chain [18] on Turkish with a F-measure of %44.16 on the tries based on semantic relatednesss. We obtained a F-measure of %39.89 for Turkish from the tries structured from the same stem, which is poorer than the other method. This shows that for morphologically rich languages, semantic relatedness plays a more important role in segmentation. That is because of the sparseness of the word forms in morphologically rich languages. Here we overcome the sparsity problem with semantic information that is used in semantically built tries.

Table 2. Results obtained from the tries based on semantic relatedness (see Sect. 3.2). TST denotes our trie-structured model.

	Precision (%)	Recall (%)	F-measure (%)
Turkish			
TST	58.27	**35.55**	**44.16**
M-CatMAP	77.78	22.91	35.40
M-Baseline	**84.39**	19.27	31.38
MorphoChain	69.45	18.29	28.95
English			
M-Baseline	64.82	**64.07**	**64.44**
TST	56.40	47.90	51.81
MorphoChain	**86.26**	25.95	39.90
M-CatMAP	76.37	19.23	30.72
German			
M-Baseline	**64.74**	30.10	**41.09**
TST	38.66	**38.57**	38.61
M-CatMAP	62.32	15.68	25.06
MorphoChain	56.39	13.72	22.07

Table 3. Results obtained from the tries structured from the same stem (see Sect. 3.1). TST denotes our trie-structured model.

	Precision (%)	Recall (%)	F-measure (%)
Turkish			
M-CatMAP	59.44	**33.41**	**42.78**
TST	58.85	30.17	39.89
M-Baseline	**74.09**	20.52	32.14
Morpho-Chain	72.28	25.77	38.00
English			
M-Baseline	**75.28**	**61.05**	**67.42**
TST	58.69	51.22	54.70
MorphoChain	91.74	30.39	45.66
M-CatMAP	90.20	5.86	11.00
German			
M-Baseline	59.65	29.47	**39.45**
TST	39.62	**35.28**	37.33
MorphoChain	**79.06**	16.36	27.11
M-CatMAP	55.96	16.41	25.38

Our TST model performs better on the tries structured from the same stem on English with a F-measure of %54.70 compared to the tries based on semantic relatedness, which has a F-measure of %51.81. Since English is not a morphologically rich language, obtaining the correct stem plays an important role in segmenting the word. Words usually do not have more than one suffix and therefore finding the stem is normally sufficient to do morphological segmentation in morphologically poor languages such as English.

Our German results are close to each other obtained from two types of tries. We obtain a F-measure of %38.61 from the tries based on semantic relatedness and it performs better than Morfessor CatMAP and Morphological Chain. The F-measure is %37.33 on German, which is obtained from the tries structured from the same stem.

The results also show that Morfessor CatMAP suffers from sparsity in small datasets (especially in English), whereas our trie-structured model learns also well in small datasets.

8   Conclusion and Future Work

We propose a Bayesian model that utilizes semantically built trie structures that are built by using neural word embeddings (i.e. obtained from word2vec [17]) for morphological segmentation in an unsupervised setting. The current study constitutes the first part of the on-going project which in the end aims to learn part-of-speech tags and morphological segmentation jointly. To this end, the fact that the tries having semantically related words achieves the best performance paves the way of using semantically similar words in learning syntactic features.

Moreover, considering the resource-scarce languages like Turkish, our trie-structured model shows a good performance on comparably smaller datasets. In comparison to other available systems, our model outperforms them in spite of the limited training data. This shows that the small size of data can be compensated to a certain extent with structured data, that is the main contribution of this paper.

Acknowledgments. This research is supported by the Scientific and Technological Research Council of Turkey (TUBITAK) with the project number EEEAG-115E464.

References

1. Bordag, S.: Unsupervised knowledge-free morpheme boundary detection. In: Proceedings of the RANLP 2005 (2005)
2. Bordag, S.: Two-step approach to unsupervised morpheme segmentation. In: Proceedings of 2nd Pascal Challenges Workshop, pp. 25–29 (2006)
3. Bordag, S.: Unsupervised and knowledge-free morpheme segmentation and analysis. In: Peters, C., et al. (eds.) CLEF 2007. LNCS, vol. 5152, pp. 881–891. Springer, Heidelberg (2008). https://doi.org/10.1007/978-3-540-85760-0_113
4. Can, B.: Statistical models for unsupervised learning of morphology and POS tagging. Ph.D. thesis, Department of Computer Science, The University of York (2011)

5. Can, B., Manandhar, S.: Probabilistic hierarchical clustering of morphological paradigms. In: Proceedings of the 13th Conference of the European Chapter of the Association for Computational Linguistics, EACL 2012, pp. 654–663. Association for Computational Linguistics (2012)

6. Casella, G., George, E.I.: Explaining the Gibbs sampler. Am. Stat. **46**(3), 167–174 (1992)

7. Creutz, M.: Unsupervised segmentation of words using prior distributions of morph length and frequency. In: Proceedings of the 41st Annual Meeting on Association for Computational Linguistics, pp. 280–287. Association for Computational Linguistics (2003)

8. Creutz, M., Lagus, K.: Unsupervised discovery of morphemes. In: Proceedings of the ACL-02 Workshop on Morphological and Phonological Learning, pp. 21–30. Association for Computational Linguistics (2002)

9. Creutz, M., Lagus, K.: Inducing the morphological lexicon of a natural language from unannotated text. In: Proceedings of the International and Interdisciplinary Conference on Adaptive Knowledge Representation and Reasoning (AKRR 2005), pp. 106–113 (2005)

10. Creutz, M., Lagus, K.: Unsupervised models for morpheme segmentation and morphology learning. ACM Trans. Speech Lang. Process. **4**, 1–34 (2007)

11. Déjean, H.: Morphemes as necessary concept for structures discovery from untagged corpora. In: Proceedings of the Joint Conferences on New Methods in Language Processing and Computational Natural Language Learning, pp. 295–298. Association for Computational Linguistics (1998)

12. Goldwater, S., Johnson, M., Griffiths, T.L.: Interpolating between types and tokens by estimating power-law generators. In: Weiss, Y., Schölkopf, B., Platt, J. (eds.) Advances in Neural Information Processing Systems, vol. 18, pp. 459–466. MIT Press, Cambridge (2006)

13. Hafer, M.A., Weiss, S.F.: Word segmentation by letter successor varieties. Inf. Storage Retriev. **10**(11–12), 371–385 (1974)

14. Hankamer, J.: Finite state morphology and left to right phonology. Proc. West Coast Conf. Formal Linguist. **5**, 41–52 (1986)

15. Harris, Z.S.: From phoneme to morpheme. Language **31**(2), 190–222 (1955)

16. Kurimo, M., Lagus, K., Virpioja, S., Turunen, V.: Morpho challenge 2010. http://research.ics.tkk.fi/events/morphochallenge2010/ (2011). Accessed 31 Jan 2017

17. Mikolov, T., Chen, K., Corrado, G., Dean, J.: Efficient estimation of word representations in vector space. CoRR abs/1301.3781 (2013). http://arxiv.org/abs/1301.3781

18. Narasimhan, K., Barzilay, R., Jaakkola, T.S.: An unsupervised method for uncovering morphological chains. Trans. Assoc. Comput. Linguist. **3**, 157–167 (2015)

19. Snyder, B., Barzilay, R.: Unsupervised multilingual learning for morphological segmentation. In: Proceedings of ACL-08: HLT, pp. 737–745. Association for Computational Linguistics, June 2008

20. Soricut, R., Och, F.: Unsupervised morphology induction using word embeddings. In: Proceedings of the Human Language Technologies: The 2015 Annual Conference of the North American Chapter of the ACL, pp. 1627–1637. Association for Computational Linguistics (2015)

21. Üstün, A., Can, B.: Unsupervised morphological segmentation using neural word embeddings. In: Král, P., Martín-Vide, C. (eds.) SLSP 2016. LNCS (LNAI), vol. 9918, pp. 43–53. Springer, Cham (2016). https://doi.org/10.1007/978-3-319-45925-7_4

Building Morphological Chains
for Agglutinative Languages

Serkan Ozen[1] and Burcu Can[2(✉)]

[1] Department of Computer Engineering, Middle East Technical University (ODTÜ),
Ankara 06800, Turkey
`serkan1ozen@gmail.com`
[2] Department of Computer Engineering, Hacettepe University Beytepe,
06800 Ankara, Turkey
`burcucan@cs.hacettepe.edu.tr`

Abstract. In this paper, we build morphological chains for agglutinative languages by using a log linear model for the morphological segmentation task. The model is based on the unsupervised morphological segmentation system called MorphoChains [1]. We extend MorphoChains log linear model by expanding the candidate space recursively to cover more split points for agglutinative languages such as Turkish, whereas in the original model candidates are generated by considering only binary segmentation of each word. The results show that we improve the state-of-art Turkish scores by 12% having a F-measure of 72% and we improve the English scores by 3% having a F-measure of 74%. Eventually, the system outperforms both MorphoChains and other well-known unsupervised morphological segmentation systems. The results indicate that candidate generation plays an important role in such an unsupervised log-linear model that is learned using contrastive estimation with negative samples.

Keywords: Unsupervised learning · Morphological segmentation
Morphology · Log-linear models · Contrastive estimation

1 Introduction

Unsupervised morphological segmentation has been one of the fundamental tasks in natural language processing. Segmentation of words is required normally as a pre-processing task in many natural language processing applications, such as machine translation, question answering, sentiment analysis and so on. One of the main reasons to perform morphological segmentation before applying any natural language processing task is the out-of-vocabulary (OOV) problem. The number of different word forms can be theoretically infinite in agglutinative languages [2].

Morphological analysis is also required for some natural language processing tasks. In a full morphological analysis, morphemes are tagged according to their

© Springer Nature Switzerland AG 2018
A. Gelbukh (Ed.): CICLing 2017, LNCS 10761, pp. 99–109, 2018.
https://doi.org/10.1007/978-3-319-77113-7_8

syntactic roles in addition to finding the morpheme boundaries. For example, in order to distinguish the word that is inflected with the negation suffix *ma* (or *me* depending in the vowel harmony) from another word that has a derivational suffix *ma* (or *me*) in Turkish requires a full morphological analysis. Here, we aim to perform morphological segmentation rather than a full morphological analysis. Thus we only aim to find the morpheme segmentation points of each word.

In this paper, we propose an improvement to the MorphoChains segmentation system [1] by extending the candidate space used in contrastive estimation, thereby covering also agglutinative languages for multiple split points. Normally, log-linear models are supervised. However, using contrastive estimation by shifting the probability mass from the unobserved data (and possibly that are impossible to observe in data) to observed data enables unsupervised learning. Unobserved data is generated with negative sampling using the observed data through some transformations on the observed data (such as transpose, deletion, insertion etc.).

In this paper, rather than extending the probability mass assigned for unobserved data, we target the probability mass assigned for the observed data. For that purpose, we generate more segmentation points (i.e. candidates) for each observed word to extend the observed space.

Unsupervised models seem to be a good alternative for discovering both orthographic and semantic features of words. We also adopt both orthographic features and semantic features in this paper as proposed in the original model.

We perform all experiments on publicly available Turkish, English and German datasets provided by Morpho Challenge 2010 [3]. The evaluation method will be the same with the one used in MorphoChains segmentation system [1].

The paper is organized as follows: Sect. 2 addresses the related work on unsupervised morphological segmentation, Sect. 3 describes the extended log-linear model, Sect. 4 explains the improvements performed on the original log-linear model, Sect. 5 presents the experiments and scores for English, Turkish and German along with a discussion over the scores, and finally Sect. 6 concludes the paper with the potential future work.

2   Related Work

Morphological segmentation is one of the oldest natural language processing tasks that has been excessively studied.

The oldest works have been usually based on deterministic methods. One of the earliest works is *Linguistica* that is proposed by Goldsmith [4]. The model is based on Minimum Description Length (MDL) principle, which is deterministic. *Linguistica* employs morphological structures called signatures in order to represent words. Signatures reflect the internal structure of words. Words with similar morphological structure reside in the same signature. For example, { *order*, *walk*}-{*ing*, *s*} make a signature that covers words such as *walking, ordering, walks, orders*, and {*paper*, *pen*}-{*s*} make another signature that covers *papers, pens*.

Probabilistic methods have also been used in unsupervised morphological segmentation. Creutz and Lagus [5] introduce another well-known unsupervised morphological segmentation Morfessor Baseline, the first member of the Morfessor family. One of the versions is based on MDL principle and the other one is based on Maximum Likelihood (ML) estimate. In another member of the same family, Creutz and Lagus [6], suggest using priors by converting the model into a Maximum a Posteriori model, thereby introducing another member of the same family, called Morfessor Categories MAP (Maximum A-posterior). Morfessor has been one of the main reference segmentation systems to compare with most of the unsupervised segmentation systems. In this paper, we also compare our extended model with Morfessor Baseline and Morfessor CatMAP.

Non-parametric Bayesian methods have also been used in segmentation task. Goldwater et al. [7] present a framework that generates power-laws by using word frequencies. Pitman-Yor Process [8] (the two parameter extension of a Dirichlet Process) is used as a stochastic process in their framework. Snyder and Barzilay [9] use Dirichlet Process, the simplified version of the Pitman-Yor Process, to induce morpheme boundaries on a bilingual aligned corpus simultaneously by finding the cross-lingual morpheme relations. Lee et al. [10] address the connection between syntax and morphology in a statistical model. Syntactic knowledge is incorporated in their morphological segmentation system. Their results show that using syntactic information helps in morphological segmentation.

Some of the systems not only attempt to perform morphological segmentation, but also aim to learn hidden structures behind words. Chan [11] applies Latent Dirichlet Allocation (LDA) to learn morphological paradigms as latent classes. The model assumes that correct segmentations of words are known but morphological paradigms are to be learned. Chan discovers that the final morphological paradigms can be matched with syntactic tags (such as noun, verb etc.). Can and Manandhar [12] obtain syntactic categories from a context distributional clustering algorithm [13] and learn paradigms by using the pairs of syntactic categories that have common stems.

Similar to MorphoChains system, log-linear models have also been utilized in morphological segmentation. Poon et al. [14] suggest using bi-gram morpheme contexts in a log linear model similar to the current study in this paper. In addition to morpheme contexts, Minimum Description Length-inspired (MDL) prior information is also used in their model to keep the lexicon and corpus size small.

In the recent years, deep neural networks are used for learning morphology. Cao and Rei [15] propose a model where word embeddings and segmentation are learned simultaneously. Soricut and Och [16] learn the morphological transformations between words using a high dimensional vector space (i.e. word-embedding space).

In this paper, 200-dimensional neural word embeddings obtained from word2vec [17] are also used to capture semantic similarities between words that are derived from each other.

3   Model

3.1   Model Definition

In this paper, we extend the MorphoChains [1] segmentation system where each word and its morphological roots are represented as a chain structure. For example, {*walking, walk*} and {*undoable, doable, do*} make morphological chains. In the morphological chain, each word appears in a parent-child relation. Here, *walk* is the parent of *walking*; *doable* is the parent of *undoable*, and *do* is the parent of *doable*.

In the MorphoChains system, a log-linear model is used to extract the chain structure in an unannotated corpus. The model has a feature vector ϕ: W × Z → R^d and a corresponding weight vector $\theta \in R^d$, where W denotes words and Z denotes candidates. A candidate is a potential parent set of a word. For example, the word *doors* has the following candidates: (*door, suffix*), (*doo*, suffix), (*do*, suffix), (*rs*, prefix), (*ors*,prefix), (*oors*, prefix). Every word and candidate pair has a feature vector associated with it.

Probability of a word-candidate pair (w, z) is modeled as:

$$P(w, z) = e^{\theta \cdot \phi(w,z)} \tag{1}$$

where w is a word and z is a candidate of w. Thus, the conditional probability of a candidate given its word is computed by:

$$P(z|w) = \frac{e^{\theta \cdot \phi(w,z)}}{\sum_{z' \in C(w)} e^{\theta \cdot \phi(w,z)}} \tag{2}$$

where $C(w)$ corresponds to the candidates of w.

The log-linear model proposed in the original paper uses features and their weights in order to learn the underlying segmentation of words. These features are described in the following section.

3.2   Features

Features play a key role in a log-linear model as they represent both orthographic and semantic properties of word-candidate pairs. The features in the model are as follows:

Semantic Similarity is applied by the cosine similarity of a word-parent pair. The cosine similarity is computed by using the word embeddings obtained from word2vec [17]. The paper indicates that morphologically related word-parent pairs tend to have high cosine similarity. For example, *(fly, flying)* pair will have higher cosine similarity when compared to *(flyi, flying)* and this will favor *fly* to be the parent of *flying* rather than having *flyi* as the parent.

Affixes are automatically generated as a list of most frequent affixes in the corpus. In order to build the affix list, each word having a higher frequency than a manually set threshold is analyzed through its potential suffixes and prefixes.

All potential suffixes are added into the affix list. If another word in the corpus having the same suffix is met, then the frequency of the suffix is incremented.

Affix correlation shows how related two affixes are in terms of the rate of their common stems. For example *(ing, ed)* suffix pair is expected to have a high correlation since many verbs in English can take both of the suffixes. If two affixes share common stems, then they are called neighbor suffixes. Again the same pair *(ing, ed)* are called neighbors since they share many common stems. For example, regarding the word *(laughing)* and the suffix *(-ing)*, since another word *(laughed)* exists in the corpus, the parent-candidate pair *(laugh, laughing)* gets a feature stating that *(-ing)* is most probably a suffix which in turn favors *(laugh)* to be a strong candidate for *(laughing)*.

Presence in the wordlist represents whether the parent is seen in the corpus or not. This provides a bias on the likelihood of a parent to be a valid word. This feature assumes that the language is concatenative.

Transformation features are used for stem changes during affixation. There are three types of transformation features, namely *repeat, delete, modify*. For example, *(running, run)* word-candidate pair has a repeat feature set to 1 due to the repetition of *n* at the end of the word, and *(deleting, delete)* pair has a delete feature set to 1 due to the deletion of the letter *e* at the end of the word.

Stop features help to identify whether a parent is the root or not. One of the key features to handle this is the highest cosine similarity between a word and its parents. For example, for the word *flying*, *fly* is more likely to be the base word than *fl* because cosine similarity between *flying* and *fly* is higher than the cosine similarity between *fl* and *flying*.

4   Improvements to the Model

The model is learned in an unsupervised setting my maximizing the likelihood of observed words in a given corpus. The likelihood of the model for a given unannotated word list D is given as follows in the original paper:

$$L(\theta, D) = \prod_{w^* \in D} P(w^*) \tag{3}$$

$$= \prod_{w^* \in D} \sum_{z \in C(w*)} P(w^*, z) \tag{4}$$

$$= \prod_{w^* \in D} \left[\frac{\sum_{z \in C(w*)} e^{\theta \cdot \phi(w^*, z)}}{\sum_{w \in \Sigma*} \sum_{z \in C(w)} e^{\theta \cdot \phi(w, z)}} \right] \tag{5}$$

where $\Sigma*$ denotes the alphabet, which is problematic to calculate for all possible unobserved data for a given language. Contrastive estimation is used to apply negative sampling and replace the normalization term with the neighbors of each word. This process creates a large space of unobserved data from which the probability mass will be shifted to observed space and therefore the likelihood will be normalized through the unobserved data.

Function *RCG (word, candidateList)*

    Data: word

    Result: candidateList

    for *i=word.length-1;i ≥word.length-4 and i ≥0; i=i-1* **do**

        parent←word.substring(0, i);

        if *2 * parent.length ≥word.length and word.length >2* **then**

            candidateList ← parent;

            AddSuffixFeature(parent);

            RCG(parent, candidateList);

        end

        parent←word.substring(i,word.length);

        if *2 * parent.length ≥word.length* **then**

            candidateList ← parent;

            AddPrefixFeature(parent);

        end

    end

Algorithm 1. RCG (Recursive candidate generation) algorithm

Here, we have noticed that although candidates play an important role in the model, they are generated by only binary segmentation of each word. For example, the Turkish word *kitap+lar+dan* (from the books) will never have the suffix *lar* in any of its candidates. This holds true for any word with more than one suffix. With this intuition, we aim to increase the candidate space generated from each word by including all possible segmentations of each word, therefore introducing the suffixes in the middle of words as candidates as well.

In our approach, in order to generate all possible candidates of a word, each binary segmentation of the word is proposed as a candidate. For each candidate stem obtained from the binary segmentation, candidate segmentations are generated again with a binary segmentation. Therefore, a left-recursion is applied for each word in all levels of the binary segmentation in order to generate candidates. In other words, the process is repeated recursively for each candidate stem in each iteration.

We restrict some candidate generations with some heuristics. Each candidate has a maximum suffix length of 4. The recursion continues as long as the base word's length is greater than 2. Another heuristic in the original model is that twice of parent's length must be greater than or equal to the twice of the child word's length. For example, candidates of the word *cars* will be *car, ca, rs, ars*. Words *(c, s)* are detained from being candidates since they do not meet any of the heuristics.

An example is given in Fig. 1. The candidates of the word *kitapçılar* (means *bookshops*) is generated recursively. As can be seen in the figure, base word has at most 4 child nodes that corresponds to the first level candidates[1] (i.e. having a suffix with maximum 4 letters).

---

[1] We chose 4 because of the fact that the longest suffix in Turkish language is 4, e.g. *-iyor*.

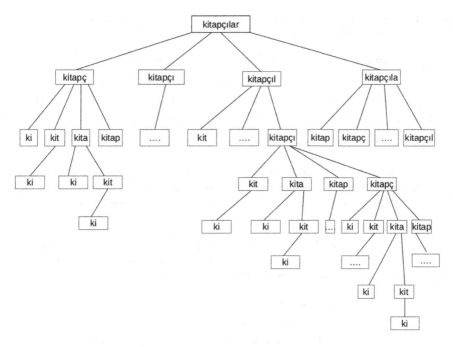

Fig. 1. Recursive candidate generation of a Turkish word *kitapçılar* (means *bookshops*). For the simplicity, only the stems (left part of the binary segmentation) are shown in the segmentation tree.

This recursive generation creates a large candidate space that also enlarges the observed space. The recursive procedure is given in Algorithm 1.

The model is learned by optimizing the feature weights according to the model likelihood given in Eq. 3. Gradient-descent algorithm is used for the optimization similar to the original model. Once the model is learned, the prediction is performed for a novel word through the optimized weights for each feature, where again a recursive segmentation is applied.

5   Results

We use the publicly available datasets provided by Morpho Challenge [3] for both training and testing. The training sets contain 878K words, 617K words, and 2M for English, Turkish, and German respectively. The test sets contain 2200 words, 2500 words, and 785 words for English, Turkish, and German respectively that are also obtained from Morpho Challenge gold standard datasets by aggregating the gold sets in Morpho Challenge 2005–2010.

We also use large datasets for training the neural word embedding model, word2vec [17] in order to build the neural word embeddings for the semantic similarity feature. All neural word embeddings are 200-dimensional. The corpora size is given in Table 1.

Table 1. Corpora size. MC: Morpho-challange. MC-05:10 Aggregated test data from Morpho Challange 2005–2010

Language	Train	Test	WordVectors
Turkish	MC-2010 (617K)	MC-05:10 (2534)	Vectors-Gencor (361M)
English	MC-2010 (878K)	MC-05:10 (2218)	Wikipedia-Normalized (129M)
German	MC-2010 (2M)	MC-2010 (785)	Manually collected (651M)

Experiments and evaluation are held as in the original paper. Segmentation points in the results are compared to those given in gold segmentation data, and Precision, Recall and F-1 measure values are calculated accordingly.

Table 2. Comparison of MorphoChains-R (with recursive candidate generation) with MorphoChains-O (the original MorphoChains system), Lee Segmenter, and other Morfessor members for Turkish, English and German

Language	Method	Precision	Recall	F-1
Turkish	MorphoChains-R	0.70	0.74	**0.72**
	MorphoChains-O	0.49	0.76	0.60
	Morfessor-CatMAP	0.52	0.60	0.56
	Morfessor-Baseline	0.82	0.36	0.50
	Lee Segmenter	0.78	0.35	0.48
English	MorphoChains-R	0.88	0.64	**0.74**
	MorphoChains-O	0.67	0.79	0.71
	Morfessor-Baseline	0.74	0.62	0.67
	Lee Segmenter	0.82	0.52	0.64
	Morfessor-CatMAP	0.67	0.58	0.62
German	Morfessor-Baseline	0.55	0.54	**0.54**
	MorphoChains-O	0.33	0.49	0.38
	MorphoChains-R	0.21	0.33	0.25

We compare our Turkish and English results with MorphoChains-O [1] (original MorphoChains system), Morfessor Baseline [5], Morfessor CatMAP [6] and Lee Segmenter [10]. All models are trained on the same train and test sets. Recursive candidate generation notably improves the scores with 12% on Turkish with a final F-measure of 72%, whereas the original MorphoChains system has a F-measure of 60%. The same improvement also applies in English, having a F-measure of 74% with 3% improvement compared to the original MorphoChains system which has a F-measure of 71%.

We compare the recursive MorphoChains system with the original MorphoChains system and Morfessor Baseline on also German. Morfessor Baseline outperforms two other models with a F-measure of 54%. The German results are better in the original model with a F-measure of 38%, whereas the recursive model gives 25%. This is possibly because of the morphological structure of the German language. German is not an agglutinative language and the left-recursion applied in the candidate generation will generate more erroneous candidate suffixes. This is also because of the common compounds in German language. All results for English, Turkish and German are given in Table 2.

Results suggest that enlarging the candidate space will also enlarge the neighborhood size. Since contrastive estimation performs better on larger datasets, enlarging the size of the candidate space improves the precision scores because of the improved sub-word counts. For example, for the word *kitapçılar*, the recursive candidate space will contain many valid candidates which in turn will help learning correct weights for correct candidates.

Table 3. Example to correct and incorrect segmentations in the original (MorphoChains-O) and the recursive (MorphoChains-R) MorphoChains system

Language	MorphoChains-O (Incorrect)	MorphoChains-R (Correct)
Turkish	s-ön-d-ü-rme-ye	sön-dür-me-ye
	cerrah-lara	cerrah-lar-a
	b-a-ğ-lan-ma-mIz	bağ-lan-ma-mIz
	öğr-en-me-si-dir	öğren-me-si-dir
	kı-ş-k-ı-r-t-ma-lar-In	kışkırt-ma-lar-ın
English	sid-e-s-wipes	sides-wipe-s
	mediterranea-n	mediterranean
	lef-t-'s	left-'s
	to-t-ed	tot-ed
	pelle-t	pellet
	piz-za	pizza

Some examples to correct and incorrect segmentations are given in Table 3. In the original MorphoChains system, words are prone to be oversegmented, especially in Turkish. In the recursive MorphoChains system, more words are segmented correctly by overcoming the oversegmentation problem in the original model.

All this information can let us claim that increasing the candidate space in log-linear models improves the segmentation results especially in agglutinative languages such as Turkish. Enlarging the unobserved word space has been studied before via negative sampling. However, enlarging the observed space has not been studied before to our knowledge. In this paper, we show how it affects to enlarge the observed space in such a log-linear model. The results show that its affect is noticeably high.

6   Conclusion and Future Work

In this paper, we extend the unsupervised morphological segmentation system called MorphoChains [1]. We adopt the original log-linear model that uses contrastive estimation with negative sampling and aim to enlarge the observed space from which probability mass will be shifted.

We enlarge the observed candidate space by generating candidates recursively, whereas in the original model candidates are generated through binary segmentations of each word. Therefore, for each word the number of candidates is equal to the number of letters in each word. The recursion provides generating candidates that extract the suffixes in the middle of the word and this increases the probability assigned to these suffixes. However, in the original model only the probability of suffixes at the end of the words are increased with their occurrence counts in the corpus.

We aim to try different optimization algorithms in the original log-linear model as a future goal. We believe that using a better optimization technique will also improve the results further.

Acknowledgments. This research is supported by the Scientific and Technological Research Council of Turkey (TUBITAK) with the project number EEEAG-115E464 and we are grateful to TUBITAK for their financial support.

References

1. Narasimhan, K., Barzilay, R., Jaakkola, T.: An unsupervised method for uncovering morphological chains. Trans. Assoc. Comput. Linguist. **3**, 157–167 (2015)
2. Hankamer, J.: Finite state morphology and left to right phonology. Proc. West Coast Conf. Formal Linguist. **5**, 41–52 (1986)
3. Kurimo, M., Lagus, K., Virpioja, S., Turunen, V.: Morpho challenge 2010. http:// research.ics.tkk.fi/events/morphochallenge2010/ (2011). Accessed 1 Jan 2017
4. Goldsmith, J.: Unsupervised learning of the morphology of a natural language. Comput. Linguist. **27**(2), 153–198 (2001)
5. Creutz, M., Lagus, K.: Unsupervised discovery of morphemes. In: Proceedings of the ACL-02 Workshop on Morphological and Phonological Learning, pp. 21–30. Association for Computational Linguistics (2002)
6. Creutz, M., Lagus, K.: Inducing the morphological lexicon of a natural language from unannotated text. In: Proceedings of the International and Interdisciplinary Conference on Adaptive Knowledge Representation and Reasoning (AKRR 2005, pp. 106–113 (2005)
7. Goldwater, S., Johnson, M., Griffiths, T.L.: Interpolating between types and tokens by estimating power-law generators. In: Advances in Neural Information Processing Systems, vol. 18, pp. 459–466. MIT Press (2006)
8. Ishwaran, H., James, L.F.: Generalized weighted Chinese restaurant processes for species sampling mixture models. Statistica Sinica **13**, 1211–1235 (2003)
9. Snyder, B., Barzilay, R.: Unsupervised multilingual learning for morphological segmentation. In: ACL, pp. 737–745 (2008)

10. Lee, Y.K., Haghighi, A., Barzilay, R.: Modeling syntactic context improves morphological segmentation. In: Proceedings of the Fifteenth Conference on Computational Natural Language Learning, pp. 1–9. Association for Computational Linguistics (2011)
11. Chan, E.: Learning probabilistic paradigms for morphology in a latent class model. In: Proceedings of the Eighth Meeting of the ACL Special Interest Group on Computational Phonology and Morphology, pp. 69–78. Association for Computational Linguistics (2006)
12. Can, B., Manandhar, S.: Clustering morphological paradigms using syntactic categories. In: Peters, C. (ed.) CLEF 2009. LNCS, vol. 6241, pp. 641–648. Springer, Heidelberg (2010). https://doi.org/10.1007/978-3-642-15754-7_77
13. Clark, A.: Inducing syntactic categories by context distribution clustering. In: Proceedings of the 2nd Workshop on Learning Language in Logic and the 4th Conference on Computational Natural Language Learning, ConLL 2000, vol. 7, pp. 91–94. Stroudsburg, PA, USA. Association for Computational Linguistics (2000)
14. Poon, H., Cherry, C., Toutanova, K.: Unsupervised morphological segmentation with log-linear models. In: Proceedings of Human Language Technologies: The 2009 Annual Conference of the North American Chapter of the Association for Computational Linguistics, pp. 209–217. Association for Computational Linguistics (2009)
15. Cao, K., Rei, M.: A joint model for word embedding and word morphology. CoRR abs/1606.02601 (2016)
16. Soricut, R., Och, F.: Unsupervised morphology induction using word embedddings. In: Human Language Technologies: The 2015 Annual Conference of the North American Chapter of the ACL, pp. 1627–1637. Association for Computational Linguistics (2015)
17. Mikolov, T., Chen, K., Corrado, G., Dean, J.: Efficient estimation of word representations in vector space. CoRR abs/1301.3781 (2013)

Joint PoS Tagging and Stemming for Agglutinative Languages

Necva Bölücü[(✉)] and Burcu Can[(✉)]

Department of Computer Engineering, Hacettepe University,
Beytepe, 06800 Ankara, Turkey
{necva,burcu}@cs.hacettepe.edu.tr

Abstract. The number of word forms in agglutinative languages is theoretically infinite and this variety in word forms introduces sparsity in many natural language processing tasks. Part-of-speech tagging (PoS tagging) is one of these tasks that often suffers from sparsity. In this paper, we present an unsupervised Bayesian model using Hidden Markov Models (HMMs) for joint PoS tagging and stemming for agglutinative languages. We use stemming to reduce sparsity in PoS tagging. Two tasks are jointly performed to provide a mutual benefit in both tasks. Our results show that joint POS tagging and stemming improves PoS tagging scores. We present results for Turkish and Finnish as agglutinative languages and English as a morphologically poor language.

Keywords: Unsupervised learning
Part-of-Speech tagging (PoS tagging) · Stemming
Bayesian learning · Hidden Markov Models (HMMs)

1 Introduction

Part-of-speech (PoS) tagging is one of the essential tasks in many natural language processing (NLP) applications, such as machine translation, sentiment analysis, question answering etc. The task is especially crucial for the disambiguation of a word. For example, the word *saw* can correspond to either a noun or a verb. The meaning is ambiguous unless its syntactic category is known. Once its syntactic category is assigned a noun, it becomes clear that the word corresponds to the tool, *saw*.

Agglutinative languages introduce the sparsity problem in NLP tasks due to their rich morphology. Hankamer [12] claims that the number of various word forms in an agglutinative language like Turkish is theoretically infinite. The sparsity emerges with out-of-vocabulary (OOV) problem and is often a bottleneck in PoS tagging. Therefore, PoS tagging in agglutinative languages becomes even more challenging compared to other languages with a poorer morphology.

In this paper, we tackle the sparsity problem by combining PoS tagging with stemming in the same framework by reducing the number of distinct word forms to distinct stem types. Stemming is the process of finding the stem of the word by

© Springer Nature Switzerland AG 2018
A. Gelbukh (Ed.): CICLing 2017, LNCS 10761, pp. 110–122, 2018.
https://doi.org/10.1007/978-3-319-77113-7_9

removing its suffixes. In stemming, normally inflectional suffixes are stripped off, whereas the derivational suffixes are kept because the stem refers to a different word type (i.e. lemma). For example, the stem of *bookings* is *booking* since -*s* is an inflectional suffix, whereas -*ing* is a derivational suffix. Moreover, *booking* exists in dictionary as a word itself.

Many PoS tagging models ignore the morphological structure of the agglutinative languages. In this paper, we present an unsupervised model for PoS tagging that jointly finds stems and PoS tags. We propose different approaches to the same model, where all of them learn the tags and stems from a given raw text in a fully unsupervised setting. Different approaches show that using stems rather than words in learning PoS tagging improves PoS tagging performance, which also helps in learning stems cooperatively. Our model is based on a Bayesian hidden Markov Model (HMM) with a second order Markov chain for the tag transitions. We test with different emission types and the results show that emitting stems rather than words improves PoS tagging accuracy.

The paper is organized as follows: Sect. 2 addresses the related work on unsupervised POS tagging and stemming, Sect. 3 describes our Bayesian HMM model and the different settings of the same Bayesian model applied for joint learning of PoS tags and stems, Sect. 4 explains the inference algorithm to learn the model, Sect. 5 presents the experimental results obtained from different datasets for English, Turkish and Finnish languages along with a discussion on the results, and finally Sect. 6 concludes the paper with the future goals.

2   Related Work

2.1   PoS Tagging

Various methods have been applied for PoS tagging. Some of them have seen PoS tagging as a clustering/classification problem. Brown et al. [4] introduce a class-based n-gram model that learns either syntactic or semantic classes of words depending on the adopted language model; Schütze [28] classifies the vector representation of words using neural networks to learn syntactic categories; Clark [6] proposes a probabilistic context distributional clustering to cluster words occurring in similar contexts, thereby having similar syntactic features. Bienmann [2] introduces a graph clustering algorithm as a PoS tagger. The graph based tagger involves two stages: In the first stage, words are clustered based on their contextual statistics; in the second stage, less frequent words are clustered using their similarity scores.

Some other approaches have tackled PoS tagging as a sequential learning problem. For that purpose, hidden Markov models (HMMs) are commonly used for PoS tagging. HMM-based PoS tagging models go back to Merialdo [18]. Merialdo uses a trigram HMM model with maximum likelihood (ML) estimation. Trigrams'n'Tags (TnT) [3] is another statistical PoS tagger that uses a second order Markov Model with also maximum likelihood estimation.

Johnson [13] compares the estimators used in HMM PoS taggers. He discovers that Expectation-Maximization (EM) is not good at estimation in HMM-based PoS taggers. Gao et al. [7] also compare different Bayesian estimators for HMM PoS taggers. Gao et al. state that Gibbs sampler performs better on small datasets with few tags, but variational Bayes performs better on larger datasets.

Bayesian methods have also been used in PoS tagging. Goldwater and Griffiths [10] adopt Bayesian learning in HMMs. HMM parameters are modeled as a Multinomial-Dirichlet distribution. In this paper, we also use their model as a baseline to our joint model.

Van Gael et al. [30] and Synder et al. [29] introduce infinite HMMs that are non-parametric Bayesian where the number of states is not set and the states can grow with data.

2.2   Stemming

Stemmers are mainly based on three approaches: rule-based, statistical, and hybrid.

Rule based stemmers, as the name implies, extract the base forms by using manually defined rules. The oldest stemmers are rule-based [14, 25, 26].

One of the earliest statistical stemmers is developed by Xu and Croft [31]. Their method makes use of co-occurences of words to deal with words grouped in equivalence classes that are built by aggressive stemming. Mayfield and McNamee [16] propose a language independent n-gram stemmer. In their approach, stems are induced using n-gram letter statistics obtained from a corpus. Melucci et al. [17] implement a HMM-based stemmer using ML estimate to select the most likely stem and suffix based on the substring frequencies obtained from a corpus.

Linguistica [9], although being an unsupervised morphological segmentation system, is also used as a stemmer. The method is based on Minimum Description Length (MDL) model that aims to minimize the size of the lexicon by segmenting words into its segments.

GRAph-based Stemmer (GRAS) is introduced by Paik et al. [21] that groups words to find suffix pairs. These suffix pairs are used to build an undirected graph. Another unsupervised stemmer of the same authors [22] use co-occurence statistics of words to find the common prefixes.

HPS [5] is one of the recent unsupervised stemmers that exploits lexical and semantic information to prepare large-scale training data in the first state, and use a maximum entropy classifier in the second stage by using the training data obtained from the first stage.

Hybrid stemmers involve different methods in a single model. Popat et al. [24] propose a hybrid stemmer for Gujarati that combines statistical and rule-based models. MAULIK [19] introduce another hybrid stemmer that combines the rule-based stemmers. A word is searched in the lexicon. If not found, suffix stripping rules are used to detect the stem. Adam et al. [1] apply PoS tagging and then use a rule-based stemmer to strip off the suffix from the word based on its tag in a pipelineframework.

3   Model and Algorithm

We define a joint PoS tagger and stemmer that extends the fully Bayesian PoS tagger by Goldwater and Griffiths [10]. By joining PoS tagging and stemming, we aim to reduce the sparsity in PoS tagging for agglutinative languages while also improving the stemming accuracy using the tag information.

3.1   Word-Based Bayesian HMM Model

The word-based Bayesian HMM model (Goldwater and Griffiths [10]) for PoS tagging is defined as follows (see Fig. 1):

$$t_i|t_{i-1} = t, \tau^{(t,t')} \propto Mult(\tau^{(t,t')}) \tag{1}$$

$$w_i|t_i = t, \omega^{(t)} \propto Mult(\omega^{(t)}) \tag{2}$$

$$\tau^{(t,t')}|\alpha \propto Dirichlet(\alpha) \tag{3}$$

$$\omega^{(t)}|\beta \propto Dirichlet(\beta) \tag{4}$$

where w_i denotes the ith word in the corpus and t_i is its tag. $Mult(\omega^t)$ is the emission distribution in the form of a Multinomial distribution with parameters $\omega^{(t)}$ that are generated by $Dirichlet(\beta)$ with hyperparameters β. Analogously, $Mult(\tau^{(t,t')})$ is the transition distribution with parameters $\tau^{(t,t')}$ that are generated by $Dirichlet(\alpha)$ with hyperpameters α.

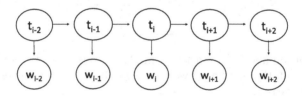

Fig. 1. Word-based HMM Model for PoS tagging

The conditional distribution of t_i under this model is:

$$P(t_i|t_{-i}, w, \alpha, \beta) = \frac{n_{(t_i,w_i)}+\beta}{n_{t_i}+W_{t_i}\beta} \cdot \frac{n_{(t_{i-1},t_i)}+\alpha}{n_{t_{i-1}}+T\alpha} \tag{5}$$

$$\cdot \frac{n_{(t_i,t_{i+1})}+I(t_{i-1}=t_i=t_{i+1})+\alpha}{n_{t_i}+I(t_{i-1}=t_i)+T\alpha}$$

where W_{t_i} is the number of word types in the corpus, T is the size of the tag set, n_{t_i} is the number of words tagged with t_i, $n_{(t_{i-1},t_i)}$ is the frequency of tag bigram $<t_{i-1}, t_i>$. $I(.)$ is a function that gives 1 if its argument is true, and otherwise 0.

3.2   Stem-Based Bayesian HMM Model

We extend the basic HMM model for PoS tagging introduced by Goldwater and Griffiths [10] by replacing the word emissions with stem emissions in order to reduce the emission sparsity, thereby mitigating the size of the out-of-vocabulary words. Therefore, we obtain a joint PoS tagger and stemmer with this model.

The stem-based model is defined as follows (see Fig. 2):

$$t_i|t_{i-1} = t, \tau^{(t,t')} \propto Mult(\tau^{(t,t')}) \tag{6}$$

$$s_i|t_i = t, \omega^{(t)} \propto Mult(\omega^{(t)}) \tag{7}$$

$$\tau^{(t,t')}|\alpha \propto Dirichlet(\alpha) \tag{8}$$

$$\omega^{(t)}|\beta \propto Dirichlet(\beta) \tag{9}$$

Here, t_i and s_i are the ith tag and stem, where $w_i = s_i + m_i$, m_i being the suffix of w_i.

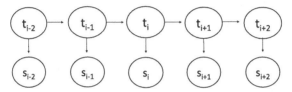

Fig. 2. Stem-based HMM Model for PoS Tagging

Under this model, the conditional distribution of t_i becomes as follows:

$$P(t_i|t_{-i}, s_i, \alpha, \beta) = \frac{n_{(t_i,s_i)}+\beta}{n_{t_i}+S_t\beta} \cdot \frac{n_{(t_{i-1},t_i)}+\alpha}{n_{t_{i-1}}+T\alpha} \tag{10}$$

$$\cdot \frac{n_{(t_i,t_{i+1})}+I(t_{i-1}=t_i=t_{i+1})+\alpha}{n_{t_i}+I(t_{i-1}=t_i)+T\alpha}$$

where S_t is the number of stem types in the corpus. When compared to the word-based model, the number of word types reduces to stem types. Therefore, sparsity also decreases.

3.3   Stem/Suffix-Based Bayesian HMM Model

Words belonging to the same syntactic category take also similar suffixes. For example, words ending with *ly* are usually adverbs, whereas words ending with *ness* are usually nouns. We include suffixes in the emissions in addition to the stems (see Fig. 3):

$$t_i|t_{i-1} = t, \tau^{(t,t')} \propto Mult(\tau^{(t,t')}) \tag{11}$$

$$s_i|t_i = t, \omega^{(t)} \propto Mult(\omega^{(t)}) \tag{12}$$

$$m_i|t_i = t, \psi^{(t)} \propto Mult(\psi^{(t)}) \tag{13}$$

$$\tau^{(t,t')}|\alpha \propto Dirichlet(\alpha) \tag{14}$$

$$\omega^t|\beta \propto Dirichlet(\beta) \tag{15}$$

$$\psi^{(t)}|\gamma \propto Dirichlet(\gamma) \tag{16}$$

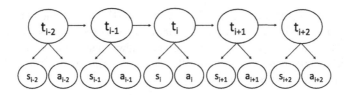

Fig. 3. Stem/suffix-based HMM Model for PoS tagging

where m_i is the suffix of $w_i = s_i + m_i$ which is generated by $Mult(\psi^{(t)})$ with parameters drawn from $Dirichlet(\gamma)$ with hyperparameters γ.

The new conditional distribution of t_i becomes:

$$P(t_i|t_{-i}, s_i, m_i, \alpha, \beta, \gamma) = \frac{n_{(t_i, s_i)} + \beta}{n_{t_i} + S_t\beta} \cdot \frac{n_{(t_i, m_i)} + \gamma}{n_{t_i} + M_t\gamma} \cdot \frac{n_{(t_{i-1}, t_i)} + \alpha}{n_{t_{i-1}} + T\alpha}$$
$$\cdot \frac{n_{(t_i, t_{i+1}) + I(t_{i-1} = t_i = t_{i+1})} + \alpha}{n_{t_i} + I(t_{i-1} = t_i) + T\alpha} \quad (17)$$

where M_t is the number of suffix types in the corpus.

4   Inference

We use Gibbs sampling [8] for the inference of the model. **t** are drawn from the posterior distribution $P(\mathbf{t}|\mathbf{w}, \alpha, \beta) \propto P(\mathbf{w}|\mathbf{t}, \beta)P(\mathbf{t}|\alpha)$ in the word-based Bayesian HMM model, $P(\mathbf{t}|\mathbf{s}, \alpha, \beta) \propto P(\mathbf{s}|\mathbf{t}, \beta)P(\mathbf{t}|\alpha)$ in the stem-based model, and $P(\mathbf{t}|\mathbf{s}, \mathbf{m}, \alpha, \beta, \gamma) \propto P(\mathbf{s}|\mathbf{t}, \beta)P(\mathbf{m}|\mathbf{t}, \gamma)P(\mathbf{t}|\alpha)$ in the stem/suffix-based model.

In the word-based Bayesian HMM model, all tags are randomly initialized at the beginning of the inference. Then each word's tag is sampled from the model's posterior distribution given in Eq. 5. This process is repeated until the system converges.

In the stem-based and stem/suffix-based model, all tags are randomly initialized and all words are split into two segments randomly. In each iteration of the algorithm, a tag and a stem are sampled for each word from the posterior distribution given in Eqs. 10 and 17 respectively.

5   Experiments and Results

Data: We used three datasets for the experiments and evaluation:

- *Turkish:* METU-Sabancı Turkish Treebank [20] that consists of 53751 word tokens.
- *English:* The first 12K and 24K words from the WSJ Penn Treebank [15].
- *Finnish:* The first 12k and 24k words from FinnTreeBank corpus that is a revised version of the original FTB1[1].

---

[1] Available at http://www.ling.helsinki.fi/kielikteknologia/tutkimus/treebank/ sources/.

There are 41 tags in both Penn Treebank and METU-Sabancı Turkish Treebank, and 12 tags in FinnTreeBank. We mapped the three tagsets to the Universal tagset [23] that involves 12 categories. The new tagsets for Turkish, English and Finnish are given in Tables 1 and 2. Therefore, the size of the tagset is 12 in all experiments for three languages.

Table 1. The mapping of the Universal tagset to the Penn Treebank tagset and the FinnTreeBank tagset

Universal tagset	Penn TreeBank tagset	FinnTreeBank tagset
VERB	VBP, VBD, VBG, VBN, VB, VBZ, MD	V
PRON	WP, PRP, PRP, WP	Pron
PUNCT	("), (,), -LRB-, -NONE-, -RRB-, (.), (:), ("), $	Punct
PRT	RP, TO	Pcle
DET	WDT, EX, PDT, DT	Det
NOUN	NN, NNP, NNPS, NNS	N
ADV	RB, RBR, WRB, RBS	Adv
ADJ	JJ, JJS	A
UNKNOWN	FW, UH	Symb, Foreign, Interj
ADP	IN	Adp
NUM	CD	Num
CONJ	CC	C

We ran each model with four settings of parameters. In the first setting, we assigned $\alpha = 0.001$, $\beta = 0.1$, and $\gamma = 0.001$ (indicated as setting 1 in the tables); in the second setting, $\alpha = 0.003$, $\beta = 1$, and $\gamma = 0.003$ (indicated as setting 2 in the tables); in the third setting, $\alpha = 0.001$, $\beta = 0.1$, and $\gamma = 0.001$ (indicated as setting 3 in the tables); and in the fourth setting we assigned $\alpha = 0.003$, $\beta = 1$, and $\gamma = 0.003$ (indicated as setting 4 in the tables).

The stemming results are evaluated based on the accuracy measure. We compare our stemming results obtained from the stem-based Bayesian HMM (Bayesian S-HMM) and stem/suffix-based HMM (Bayesian SM-HMM) with HPS [5] and FlatCat [11]. The results for Turkish and Finnish obtained from the Metu-Sabancı Turkish Treebank and FinnTreeBank respectively are given in Table 3. Although our stemming results are far behind the results of the HPS algorithm for Turkish, our Finnish results are on a par with HPS and Morfessor FlatCat. The results show that using suffixes does not help in stemming. Using stem emissions alone gives the best accuracy for stemming in the joint task.

Since the English stems are not covered in Penn TreeBank, we were not able to evaluate the English stemming results.

Examples to correct and incorrect stems in all languages are given in Table 4. The results show that our joint model can find the common endings, such as *s*, *ed, ted, er, d, e, ing* in English. However, since we do not exploit any semantic features in the model, words such as *filter* can be stemmed as *filt+er*. This is also

Table 2. The mapping of the Universal tagset to the Metu-Sabancı Turkish Treebank tagset

Universal tagset	Metu-Sabancı Turkish Treebank tagset
Noun	Noun_Pron, Noun_Ins, Noun_Nom, Noun_Verb, Noun_Loc, Noun_Acc, Noun_Abl, Noun_Gen, Noun_Dat, Noun_Adj, Noun_Num, Noun_Pnon, Noun_Postp, Noun_Equ
Adj	Adj_Noun, Adj_Verb, Adj, Adj_Pron, Adj_Postp, Adj_Num
Adv	Adv_Verb, Adv_Adj, Adv_Noun, Adv
Conj	Conj
Det	Det
Interj	Interj
Ques	Ques
Verb	Verb, Negp, Verb_Noun, Verb_Postp, Verb_Adj, Verb_Adv, Verb_Verb
Postp	Postp
Num	Num
Pron	Pron, Pron_Noun
Punc	Punc

Table 3. Stemming results for Turkish and Finnish based on four parameter settings

	Accuracy (%)			
	Model	Metu	Finn 12K	Finn 24K
	HPS[a] [5]	**53.79**	**28.19**	27.04
	Morfessor FlatCat[b] [11]	52.06	24.47	25.93
1	Bayesian S-HMM	46.21	23.24	22.32
	Bayesian SM-HMM	34.97	26.69	26.40
2	Bayesian S-HMM	46.39	23.49	22.14
	Bayesian SM-HMM	34.82	27.95	**27.38**
3	Bayesian S-HMM	46.57	23.28	22.03
	Bayesian SM-HMM	34.97	27.45	26.74
4	Bayesian S-HMM	46.46	18.62	22.57
	Bayesian SM-HMM	32.13	24.78	24.40

[a]HPS: http://liks.fav.zcu.cz/HPS/
[b]Morfessor FlatCat: https://github.com/aalto-speech/flatcat

Table 4. Examples to correct and incorrect stems

Turkish		English		Finnish	
Correct	Incorrect	Correct	Incorrect	Correct	Incorrect
siz-lere	öğre-ncilere	year+s	chairma+n	niska+an	sai+si
dur-du	jandar+malığına	york-bas+ed	repor+ted	suomenmaa+#	tänn+e
gör-düğünüz	rastla+dığı	talk+ing	filt+er	pappila+ssa	tul+ee
göz-leri	iznin+e	the+#	sai+d	piste+ttä	kotii+n
abone+#	geti+riliyor	inform+ation	institut+e	valinta+nsa	oll+a

Table 5. POS tagging evaluation results for Turkish

	Model	Metu	
		Many-to-1	VI
1	Bayesian HMM	57.58	10.58
	Bayesian S-HMM	55.34	10.57
	Bayesian SM-HMM	56.17	**10.47**
2	Bayesian HMM	56.56	11.01
	Bayesian S-HMM	**57.70**	10.48
	Bayesian SM-HMM	55.30	10.60
3	Bayesian HMM	55.89	10.70
	Bayesian S-HMM	54.99	10.59
	Bayesian SM-HMM	57.08	**10.47**
4	Bayesian HMM	56.64	10.94
	Bayesian S-HMM	57.01	10.48
	Bayesian SM-HMM	55.99	10.58
	Brown Clustering[a] [4]	54.91	10.83

[a]Brown Clustering: http://www.cs.berkeley.edu/~pliang/software/brown-cluster-1.2.zip(Percy Liang)

one of the main problems in morphological segmentation models that rely only on the orthographic features. Our stemming results are promising, but it shows that it is not sufficient to reduce the sparsity based on the common segments and it requires more features.

We evaluate PoS tagging results with many-to-one accuracy and variation of information (VI) measure [27]. Turkish, English and Finnish results are given in Tables 5, 6, and 7 respectively. The overall results show that using stems rather than words leads to better results in three languages. Therefore, the Bayesian S-HMM model outperforms other two models in three languages in general. Although English has got a poor morphology when compared to Turkish and Finnish, the Bayesian S-HMM model still outperforms other two models. Using suffixes also does not help in PoS tagging and its scores are generally behind the

Table 6. PoS tagging evaluation results for English

	Model	Penn 12k		Penn 24k	
		Many-to-1	VI	Many-to-1	VI
1	Bayesian HMM	49.10	7.80	52.34	7.94
	Bayesian S-HMM	50.84	**7.54**	49.76	7.92
	Bayesian SM-HMM	51.04	7.86	51.65	8.12
2	Bayesian HMM	41.88	8.46	47.13	8.52
	Bayesian S-HMM	**54.92**	7.58	54.52	7.86
	Bayesian SM-HMM	48.76	7.92	51.93	8.13
3	Bayesian HMM	50.70	7.75	46.74	8.15
	Bayesian S-HMM	50.57	7.71	52.05	7.97
	Bayesian SM-HMM	53.06	7.64	51.39	7.95
4	Bayesian HMM	45.15	8.33	45.65	8.55
	Bayesian S-HMM	52.67	7.69	**55.32**	**7.67**
	Bayesian SM-HMM	53.15	7.67	51.41	8.17
	Brown Clustering[a] [4]	53.78	7.58	54.11	7.78

[a]Brown Clustering: http://www.cs.berkeley.edu/~pliang/
software/brown-cluster-1.2.zip(Percy Liang)

Table 7. PoS tagging evaluation results for Finnish

	Model	FinnTreeBank 12k		FinnTreeBank 24k	
		Many-to-1	VI	Many-to-1	VI
1	Bayesian HMM	42.43	10.65	44.96	11.43
	Bayesian S-HMM	47.29	10.39	46.83	11.16
	Bayesian SM-HMM	48.84	10.27	48.45	11.09
2	Bayesian HMM	42.94	10.74	44.50	11.56
	Bayesian S-HMM	**51.15**	10.34	51.27	11.10
	Bayesian SM-HMM	48.66	10.32	49.04	11.12
3	Bayesian HMM	42.57	10.61	46.47	11.28
	Bayesian S-HMM	45.91	10.36	51.58	11.07
	Bayesian SM-HMM	49.40	**10.25**	48.43	11.08
4	Bayesian HMM	42.96	10.75	44.27	11.58
	Bayesian S-HMM	50.47	10.38	**51.52**	**11.02**
	Bayesian SM-HMM	49.02	10.31	48.43	11.14
	Brown Clustering[a] [4]	44.33	10.64	47.95	11.33

[a]Brown Clustering: http://www.cs.berkeley.edu/~pliang/
software/brown-cluster-1.2.zip(Percy Liang)

Bayesian S-HMM model. However, in some parameter settings Bayesian SM-HMM model outperforms other two Bayesian models.

The overall PoS tagging results show that our stem-based and stem/suffix-based Bayesian models outperform both Brown Clustering [4] and word-based Bayesian HMM model [10] for three languages according to both many-to-one measure and VI measure.

6   Conclusion and Future Work

In this paper, we extend the Bayesian HMM model [10] for joint learning of PoS tags and stems in a fully unsupervised framework. Our model reduces the sparsity by using stems and suffixes instead of words in a HMM model. The results show that using stems and suffixes rather than words outperforms a simple word-based Bayesian HMM model for especially agglutinative languages such as Turkish and Finnish. Although English has got a poor morphology, the English PoS tagging results are also better when the stems are used instead of words.

Although our Turkish stemming results are far behind the other compared models, our Finnish stemming results are on par with other models.

We aim to use other features (such as semantic features) in our model to capture the semantic similarity between the stems and their derived forms, which is left as a future work.

Our model does not deal with irregular word forms. We also leave this as a future work.

Acknowledgments. This research is supported by the Scientific and Technological Research Council of Turkey (TUBITAK) with the project number EEEAG-115E464.

References

1. Adam, G., Asimakis, K., Bouras, C., Poulopoulos, V.: An efficient mechanism for stemming and tagging: the case of Greek language. In: Setchi, R., Jordanov, I., Howlett, R.J., Jain, L.C. (eds.) KES 2010. LNCS (LNAI), vol. 6278, pp. 389–397. Springer, Heidelberg (2010). https://doi.org/10.1007/978-3-642-15393-8_44

2. Biemann, C.: Unsupervised part-of-speech tagging employing efficient graph clustering. In: Proceedings of the 21st International Conference on Computational Linguistics and 44th Annual Meeting of the Association for Computational Linguistics: Student Research Workshop, pp. 7–12. Association for Computational Linguistics (2006)

3. Brants, T.: TNT: a statistical part-of-speech tagger. In: Proceedings of Sixth Conference on Applied Natural Language Processing, pp. 224–231. Association for Computational Linguistics (2000)

4. Brown, P.F., deSouza, P.V., Mercer, R.L., Pietra, V.J.D., Lai, J.C.: Class-based n-gram models of natural language. Comput. Linguist. **18**(4), 467–479 (1992)

5. Brychcín, T., Konopík, M.: HPS: high precision stemmer. Inf. Process. Manage. **51**(1), 68–91 (2015)

6. Clark, A.: Inducing syntactic categories by context distribution clustering. In: Proceedings of the 2nd Workshop on Learning Language in Logic and the 4th Conference on Computational Natural Language Learning-Volume 7, pp. 91–94. Association for Computational Linguistics (2000)
7. Gao, J., Johnson, M.: A comparison of Bayesian estimators for unsupervised hidden Markov model PoS taggers. In: Proceedings of the Conference on Empirical Methods in Natural Language Processing, pp. 344–352. Association for Computational Linguistics (2008)
8. Geman, S., Geman, D.: Stochastic relaxation, Gibbs distributions, and the Bayesian restoration of images. IEEE Trans. Pattern Anal. Mach. Intell. **6**, 721–741 (1984)
9. Goldsmith, J.: Unsupervised learning of the morphology of a natural language. Comput. Linguist. **27**(2), 153–198 (2001)
10. Goldwater, S., Griffiths, T.: A fully Bayesian approach to unsupervised part-of-speech tagging. In: Annual Meeting-Association for Computational Linguistics, vol. 45, p. 744. Citeseer (2007)
11. Grönroos, S.A., Virpioja, S., Smit, P., Kurimo, M.: Morfessor FlatCat: an HMM-based method for unsupervised and semi-supervised learning of morphology. In: COLING, pp. 1177–1185 (2014)
12. Hankamer, J.: Finite state morphology and left to right phonology. In: Proceedings of the West Coast Conference on Formal Linguistics, vol. 5, pp. 41–52 (1986)
13. Johnson, M.: Why doesn't EM find good HMM PoS-taggers? In: EMNLP-CoNLL, pp. 296–305 (2007)
14. Lovins, J.B.: Development of a stemming algorithm. Mech. Transl. Comput. Linguist. **11**, 22 (1968)
15. Marcus, M.P., Marcinkiewicz, M.A., Santorini, B.: Building a large annotated corpus of English: the penn treebank. Comput. Linguist. **19**(2), 313–330 (1993)
16. Mayfield, J., McNamee, P.: Single n-gram stemming. In: Proceedings of the 26th Annual International ACM SIGIR Conference on Research and Development in Informaion Retrieval, pp. 415–416. ACM (2003)
17. Melucci, M., Orio, N.: A novel method for stemmer generation based on hidden Markov models. In: Proceedings of the Twelfth International Conference on Information and Knowledge Management, pp. 131–138. ACM (2003)
18. Merialdo, B.: Tagging English text with a probabilistic model. Computat. Linguist. **20**(2), 155–171 (1994)
19. Mishra, U., Prakash, C.: MAULIK: an effective stemmer for Hindi language. Int. J. Comput. Sci. Eng. **4**(5), 711 (2012)
20. Oflazer, K., Say, B., Hakkani-Tür, D.Z., Tür, G.: Building a Turkish treebank. In: Abeillé, A. (ed.) Treebanks. Text, Speech and Language Technology, vol. 20, pp. 261–277. Springer, Dordrecht (2003). https://doi.org/10.1007/978-94-010-0201-1_15
21. Paik, J.H., Mitra, M., Parui, S.K., Järvelin, K.: GRAS: an effective and efficient stemming algorithm for information retrieval. ACM Trans. Inf. Syst. (TOIS) **29**(4), 19 (2011)
22. Paik, J.H., Pal, D., Parui, S.K.: A novel corpus-based stemming algorithm using co-occurrence statistics. In: Proceedings of the 34th International ACM SIGIR Conference on Research and Development in Information Retrieval, pp. 863–872. ACM (2011)
23. Petrov, S., Das, D., McDonald, R.: A universal part-of-speech tagset. arXiv preprint arXiv:1104.2086 (2011)
24. Popat, P.P.K., Bhattacharyya, P.: Hybrid stemmer for Gujarati. In: Proceedings of the 23rd International Conference on Computational Linguistics, pp. 51–55 (2010)

25. Porter, M.F.: An algorithm for suffix stripping. Program **14**(3), 130–137 (1980)
26. Porter, M.F.: Snowball: a language for stemming algorithms (2001)
27. Rosenberg, A., Hirschberg, J.: V-measure: a conditional entropy-based external cluster evaluation measure. In: Proceedings of the 2007 Joint Conference on Empirical Methods in Natural Language Processing and Computational Natural Language Learning, vol. 7, pp. 410–420 (2007)
28. Schütze, H.: Part-of-speech induction from scratch. In: Proceedings of the 31st Annual Meeting on Association for Computational Linguistics, pp. 251–258. Association for Computational Linguistics (1993)
29. Snyder, B., Naseem, T., Eisenstein, J., Barzilay, R.: Unsupervised multilingual learning for PoS tagging. In: Proceedings of the Conference on Empirical Methods in Natural Language Processing, pp. 1041–1050. Association for Computational Linguistics (2008)
30. Van Gael, J., Vlachos, A., Ghahramani, Z.: The infinite HMM for unsupervised PoS tagging. In: Proceedings of the 2009 Conference on Empirical Methods in Natural Language Processing, vol. 2, pp. 678–687. Association for Computational Linguistics (2009)
31. Xu, J., Croft, W.B.: Corpus-based stemming using cooccurrence of word variants. ACM Trans. Inf. Syst. (TOIS) **16**(1), 61–81 (1998)

Hungarian Particle Verbs
in a Corpus-Driven Approach

Ágnes Kalivoda(⊠)

Faculty of Humanities and Social Sciences, Pázmány Péter Catholic University,
Budapest, Hungary
kalivoda.agnes@itk.ppke.hu

Abstract. The meaning and the argument structure of particle verbs
are determined by the combination of a verb and a particle. In Hungarian,
verbal particles (preverbs) can occupy various positions in the sentence:
they can be preverbal (detached from the verb), immediately preverbal
or postverbal. The syntax of these particles is discussed in a wide range of
theoretical literature. This paper presents a performance-based analysis,
using corpus-driven method to reveal the distribution patterns of verbal
particles in more than 21.5 million sentences. In order to obtain these
data, it was needed to improve the POS-tagging of verbal particles and
to develop a semi-automatic method to decide which verb the detached
particle belongs to. The distribution patterns give better insight into the
phonological and pragmatic factors that may determine the position of
verbal particles.

Keywords: Particle verbs · Verbal particles · Hungarian syntax
Distribution patterns · Corpus-driven approach

1 Introduction

Hungarian particle verbs have a property which raises difficulties in computational linguistics: the verbal particle (in other terms: preverb, verbal prefix)
can occupy various places in the sentence. Its canonical position is immediately
before the verb, in this case it is written together with the verb as a single token.
In some syntactic structures, however, the particle can precede or follow the verb
with one or more words inserted between them.

On the one hand, Hungarian verbal particles have a common feature with
Slavic prefixes: they usually – but not exclusively – express an aspectual change
by rendering the predicate telic. On the other hand, they are similar to Germanic
particles: they can be separated from the verb [15].

In the case of Hungarian, there aren't any strict syntactic constraints to the
extent of the maximal distance. This might lead to problems, especially in the
automatic parsing process of a sentence containing a verb and a detached verbal
particle. A promising framework is VFRAME which can parse the Hungarian
verbal complex by using a relatively small look-ahead window, see [9].

© Springer Nature Switzerland AG 2018
A. Gelbukh (Ed.): CICLing 2017, LNCS 10761, pp. 123–133, 2018.
https://doi.org/10.1007/978-3-319-77113-7_10

The following examples shed light on the possible positions of the particle and its verb (marked with boldface):

(1)   a. **Kimegyünk** a   kertbe.
         out_go+Pl1 the garden+ILL

         'We are going out to the garden.'   immediately preverbal

     b. **Ki** is   **megyünk** a   kertbe.
         out also **go+Pl1**   the garden+ILL

         'We are going out to the garden as well.'   detached preverbal

     c. Nem **megyünk** ma   **ki** a   kertbe.
         not **go+Pl1**   today **out** the garden+ILL

         'We are not going out to the garden today.'   (detached) postverbal

The aim of this paper is to answer the following questions: How far can a particle be placed from its verb in Hungarian? Which factors determine whether a particle should stay close to its verb or can be moved into a remote position?

The syntax of Hungarian particle verbs is a frequently discussed problem in the theoretical literature. This paper presents a performance-based analysis, focusing on the distribution patterns of verbal particles, which is a whole new approach in this topic.

The paper is structured as follows: Sect. 2 describes the methodology of the research, including some technical and theoretical problems. Sections 3 and 4 present the obtained results of preverbal and postverbal particles, respectively. Finally, I conclude my results in Sect. 5.

2 Methodology

A corpus-driven method (see [16] for details of the concept) was used to study particle verbs. My statements reflect the evidence provided by the HUNGARIAN GIGAWORD CORPUS, version 2.0.3 (hereafter called HGC) [13]. This corpus contains 785 million tokens – 987 million with punctuation marks – from various domains having the following registers: journalism, literature, science, personal, official and – transcribed – spoken.

The distribution of verbal particles is measured by defining the finite verb as 0 position and counting the other positions compared to this. For example, both *megy* 'goes' and *be+megy* 'goes in', lit. 'in+goes' are in 0 position, while *Menj be!* 'Go in!' has a particle in +1 position, *Be se menj!* 'Don't even go in!' has a particle in −2 position (with one intervening word between the particle and its verb). To sum up, immediately preverbal particles get zero position, while detached preverbal ones are found in an interval less than zero and detached postverbal ones in an interval greater than zero.

Verbal particles get often erroneous annotation in HGC, as a lot of them have homographs, e.g. *ki* can be a verbal particle ('out') and an interrogative or relative pronoun ('who') as well. It was needed to improve the POS-tagging of such particles in order to minimalize the errors in the final results. Furthermore, an automatic method was required to decide whether a detached particle belongs to the finite verb or to an infinitive or a participle – the latter ones are not included in this research. The following subsections describe the solutions.

2.1   Improving the POS-Tagging of Verbal Particles

Hungarian particle verbs get erroneous annotation if **(a)** they have homographs in Hungarian, **(b)** they have homographs in another language which is represented in HGC, **(c)** there is an abbreviation or acronym written exactly like the particle. It is also possible that a verbal particle can be affected by two or three error types, as shown on Fig. 1.

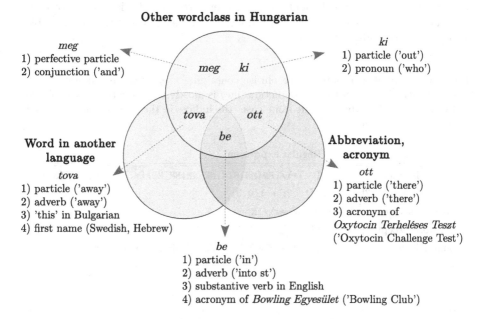

Other wordclass in Hungarian

meg
1) perfective particle
2) conjunction ('and')

ki
1) particle ('out')
2) pronoun ('who')

Word in another language

tova
1) particle ('away')
2) adverb ('away')
3) 'this' in Bulgarian
4) first name (Swedish, Hebrew)

Abbreviation, acronym

ott
1) particle ('there')
2) adverb ('there')
3) acronym of
Oxytocin Terheléses Teszt
('Oxytocin Challenge Test')

be
1) particle ('in')
2) adverb ('into st')
3) substantive verb in English
4) acronym of *Bowling Egyesület* ('Bowling Club')

Fig. 1. Five verbal particles which often get wrong POS-tags, ordered in three sets corresponding to possible error types. See [10] for more examples.

The correction of the POS-tags is executed by a script written in Python. It creates a dictionary of potential verbal particles. These words are the keys, and their values are their possible error types, represented as a list of numbers from 1 to 3. While the script is processing a sentence, it looks for the potential verbal particle and checks if it is erroneously annotated as a particle. This is done by regular expressions which are sensitive to the context of the given word. Thus, the disambiguation method is rule-based. In the current state of research, the script works with 98 rules. 7 of them are based on the ideas of Makrai [11] and used here with slight modifications. Table 1 presents the detailed description of an effective rule concerning the word *meg*.

The current functionality of the script is limited to one direction: it can filter words which are wrongly annotated as a verbal particle, but it can not find verbal particles which have wrong POS-tags. In this direction, it was possible to improve POS-tagging of verbal particles by 69.6% (f-score).

The script was tested on a 5000 sentence-long minicorpus extracted from HGC. In every sentence, there was a word annotated as a verbal particle, and the amount of these words reflected their actual frequency in the whole HGC. For example, *el* 'away' gives 19.1% of the verbal particles in HGC, therefore 956 of 5000 sentences contained the word *el*.

The rule-based method achieved 88.2% precision and 57.5% recall, resulting in an f-score of 69.6%. The two main causes of the relatively low recall are the following: (1) the context of the problematic word can not be used if it contains other wrongly annotated elements too, (2) there are a lot of misspelled words or complete sentences where accents are missing. My method is currently not able to cope with this kind of errors.

Table 1. A rule which matches the phrase containing *meg* if *meg* is annotated as a verbal particle, although it is a conjunction: *meg* is mostly a conjunction if it is placed between two nominals bearing the same case (the indices in the pseudocode show that the cases are identical).

Pseudocode	Regular expression
nominal+case$_i$	`[^ ]+\/(FN|MN|SZN|MN_NM|NM_MN)[+.][^ ]*([^ ]{3})`
0-3 word except .?!	`([^ .?!]+ ){0,3}`
meg (as a verbal particle)	`meg\/meg\/IK`
0-3 word except .?!	`([^ .?!]+ ){0,3}`
nominal+case$_i$	`[^ ]+\/(FN|MN|SZN|MN_NM|NM_MN)[+.][^ ]*\2`

2.2 Using a List of Combinations as Lexical Resource

I extracted every particle+verb combination from HGC and created a list of more than 27.000 words.[1] Only combinations occurring at least 5 times were added to the list. This list is used to make an automatic decision whether a finite verb and a detached particle could belong together in a sentence or not. My research is based on sentences that contain a particle+verb combination validated by this list. The disadvantage of the method is that rare combinations and some neologisms are excluded.

---

[1] Verbs having a modal or a causative suffix are counted as separate types because they have different lemmata in HGC, e.g. *csinál* 'to do st', *csinálhat* 'to be able/allowed to do st' and *csináltat* 'to make sy do st'.

3   Results of Preverbal Particles

3.1   The Distribution of Preverbal Particles

The data extracted from the HGC corpus show that the dominant position among preverbal particles is the zero position (when the particle and the verb are written together). This covers 99.26% of the preverbal cases.

The left periphery of the finite verb has a strict syntax, the verbal particle can not move far away. The maximal left-position turned out to be −4, meaning that there were no more than three intervening words between the particle and the verb. This can be considered a rather rare phenomenon (with a 6 out of 13.817.958 ratio).

Table 2. Distribution of preverbal particles.

FIN	−4	−3	−2	0
SUM	6	91	102768	13715093
%	-	-	0.74	99.26

−1 position is left out of Table 2, as it contains verbal particles only if the particle belongs to an other verbal class (e.g. an infinitive) or the finite particle verb is misspelled. There is only one case when the verbal particle is in −1 position and does in fact belong to the finite verb: when two particles of opposite meaning are concatenated, e.g. *fel-alá járkál* 'to pace up and down', lit. 'to up-down pace'. This type of verbal particles was not covered in this research.

3.2   Sentence Types Having Particle-Verb Order

In neutral sentences, verbal particles are mostly in 0 position.[2] These sentences are characterized by neutral intonation and the additional meaning arising from the structure is minimal [1]. There are also some non-neutral sentence types having particle-verb order, e.g. yes-no questions (these cover approx. 6.4% of 0 position) and focussed sentences where the verb is in focus.

A stylistically marked type of imperative sentences – which are non-neutral – has also particle-verb order. These imperative structures often have strong emotional excess (see examples 2a–b). This type has altogether 4547 examples, thus it covers approx. 0.03% of sentences in 0 position.

(2)   a. **Elhallgass**   végre!
          away_keep_silent+Imp+Sg2   at_last
          'Shup up at last!'

---

[2] There are, however, neutral sentences with verb-particle order, e.g. presenting sentences [14] in Csángó dialects and some rare examples from old Hungarian [8]. These are not represented in HGC.

  b. **Eltakarodjatok**   a   szemem   elől!
     away_pack_off+Imp+Pl2 the eye+Ps+Sg1 before
     'Get out of my sight!'

In the case of −2 position, only five words can be inserted between the detached verbal particle and the verbs, these are: *is* 'also' (49.4%), *sem* (10.12%) or *se* (5.46%) 'neither/either', *nem* (10.85%) or *ne* (24.17%) 'not'. *Nem* is typical for negative, *ne* for prohibitive sentences.

  In the case of −3 and −4 positions, the verbal particle functions as contrastive topic [5], which is a rare phenomenon in Hungarian. It is questionable if these are verbal particles or rather adverbs (example 3 allows both interpretations).

  (3) De **haza**   is   csak látogatni **megyek** ezentúl.
       but **to_home** also only visit+Inf **go+Sg1** from_now_on
       'But from now on, I will go home only to visit.'

4   Results of Postverbal Particles

4.1   The Distribution of Postverbal Particles

In Hungarian, the order of the sentence's constituents after the finite verb is not fixed [6]. Thus, it should be expected that the verbal particle can be in any position behind the verb. My measurements showed that the postverbal particles can indeed be located in a wider scope than the preverbal ones, but they are in +1 or +2 positions in 99.9% of the cases (see Table 3).

Table 3. Distribution of postverbal particles.

FIN	+1	+2	+3	+4	+5	+6	+7	+8	+9	+10	+11
SUM	7527308	163993	5126	1193	267	101	27	5	3	2	1
%	97.78	2.13	0.07	0.02	-	-	-	-	-	-	-

There are only 11 examples (out of 7.698.015) where the verbal particle is more than six words away. The most extreme sentence was the following, containing a verbal particle in the +11 position:

  (4) 27 gyereket   **vitt**   egy feltehetően részeg buszsofőr
       27 children+ACC **take+Past+Sg3** a   presumably drunk bus_driver
       Szentesen   még csütörtökön   egy sportrendezvény után **vissza**
       Szentes+SUP still Thursday+SUP a   sports_event   after **back**
       az iskolába.
       the school+ILL
       'On Thursday a presumably drunk bus driver brought 27 children back to school after a sports event in Szentes.'

There is an interesting tendency of how the different particles are distributed in the postverbal positions. There is a group of longer particles which have not completely grammaticalized and can still function as adverbs. This group can be placed further away from the verb, while short, grammaticalized particles prefer to follow the verb immediately.

4.2   Sentence Types Having Verb-Particle Order

Sentences with verb-particle order are – except of some rare structures mentioned in Footnote 2 – non-neutral. Alberti's detailed description of non-neutral sentences [1] was used to create eight categories which are more or less recognizable using automatic method. These are:

1. *Sentences containing a focussed constituent.* These are difficult to identify automatically, since almost any kind of word can be focussed, regardless to its wordclass. In this case, the focussed word gets the privileged preverbal position of the verbal particle, moving it away into a postverbal position. The focus could not be detected with help of pattern matching. So it was checked if regular expressions could match the other seven sentence types first, than the rest of the sentences was ordered into this category. In the case of focus, manual check would be inevitable.
2. *Sentential negations* (including also embedded negatives). These can be covered well with regular expressions.
3. *Imperative sentences* (including also embedded imperatives). These are easy to detect, since the imperative is morphologically marked in Hungarian.
4. *Optative sentences* (mostly expressing irreal wishes). These can be found if a verb in conditional – which is marked morphologically – and the word *csak* ('only') co-exist in the sentence and they are relatively close to each other.
5. *Prohibitions.* These are in the intersection of imperative and negative sentences, but listed here as a different category, since they can be covered perfectly with regular expressions.
6. *Wh-questions and exclamatives.* These are linguistically two distinct categories – having characteristic intonation patterns –, but look alike in written corpora. Punctuation marks (question and exclamation mark, respectively) may help distinguishing them, but only in case they are not embedded in another sentence. Example 5a–b illustrates the problem:

(5)   a.   Milyen későn **kelt fel?**
           how   late   **woke+Sg3 up**
           'How late did he/she wake up?'

       b.   Milyen későn **kelt fel!**
           how   late   **woke+Sg3 up**
           'How late he/she woke up!'

7. *Sentences with progressive aspect.* These sentences – where the only reason of moving the verbal particle behind the verb is that it has a progressive reading, expressing that something happens right now, at the moment – occur sparsely in HGC.

8. *Existential sentences.* In these sentences, the fact that something occurred at least once is being emphasized [4].

Table 4. Sentence types having verb-particle order. The table presents their distribution in HGC, according to the positions taken by the verbal particle. Columns % and *SUM* show the frequency of the given types, taking the whole postverbal corpus as 100%. We can see which positions are preferred by the different sentence types, given in percent.

Sentence type	+1	+2	+3	+4	+5	+6	+7	%	SUM
Focussed sentence	99.87	0.11	0.02	0.01	0	0	0	72.0648	5547562
Negative	99.85	0.11	0.03	0.01	0	0	0	14.2314	1095539
Imperative	99.82	0.13	0.03	0.01	0	0	0	10.3561	797209
Optative	99.79	0.15	0.04	0.01	0	0	0	1.6316	125605
Prohibitive	99.75	0.18	0.04	0.01	0	0	0	1.0885	83794
Wh-question/exclamative	99.43	0.22	0.26	0.05	0.02	0.02	0.01	0.6253	48136
Progressive	98.45	1.55	0	0	0	0	0	0.0021	154
Existential	68.75	25	0	6.25	0	0	0	0.0002	16

An interesting tendency can be discovered by wh-questions and exclamatives (see Table 4). They have detached verbal particles also in distant positions. This is caused by discourse particles and interjections, frequently used in these sentences. By existential sentences, the relatively high percents in +2 and +4 position are due to the almost irrelevant, low frequency of this type (only 16 examples were found in HGC).

4.3   Determining Factors of the Verb-Particle Distance

99.9% of postverbal particles follow the finite verb immediately or with one preceding word (see Table 3). The main reason for this high percent is that the argument structure of the verb can vary depending on the verbal particle. If the postverbal particle is far behind the finite verb, calculating the meaning of the whole sentence becomes more complicated for the listener/reader.

However, there are still 0.1% of postverbal particles that need an explanation. It can be assumed that two factors determine the possible distance of postverbal particles: (**a**) the opposition of written and spoken – edited and unedited – text, (**b**) phonological constraints.

Opposition of Written and Spoken Text. Basically, spoken texts are spontaneous and unplanned, compared to written ones. Thus, they are less precise syntactically [7]. It can be expected that a lot of sentences having postverbal particle in remote positions are coming from live speech. Using the metadata found in HGC, it was quite simple to detect whether a given sentence was originally said or written.

I put the metadata of text types into two categories. *Written:* press, personal (mostly blogs and forum comments), science, literature (prose and poetry, separately), official (mostly law). *Spoken:* radio broadcasts and parliamentary speeches. Figure 2 presents the percentage of written and spoken texts in seven positions of the postverbal particles.

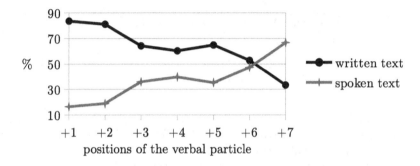

Fig. 2. Proportion of written and spoken text in seven positions, given in percents.

In order to understand the results, we have to see that the proportion of spoken texts is very low in HGC. Radio broadcasts form 5.4% of the corpus. Assuming that half of the official texts is spoken, the final amount of spoken data is no more than 9.4%.

The distribution of text types in +1 position does not differ substantially from the entire HGC: there are significantly more written texts here (83.55%) than spoken (16.45%). The proportion of spoken data, however, keeps growing as we look to more distant positions. In the case of +7 position, two thirds of the sentences come from spoken text.

Phonological Constraints. As mentioned in Sect. 4.1, verbal particles which are relatively long – consisting of two or more syllables – can be found in marginal positions as well, while short ones are never placed too far from the verb. In this case, the most extreme position is +5. Therefore, it can be assumed that phonology plays an important role in the distribution of postverbal particles.

In the fourth volume of *Deutsche Syntax* ('German Syntax'), Otto Behaghel presented five cross-language principles related to the position of constituents in a sentence, these are known as Behaghel's Laws [2]. The fourth of them is called the *Gesetz der wachsenden Glieder* ('Law of Increasing Terms'). According to this principle, the shorter constituent prefers to precede the longer one, if there is no syntactic rule that could prevent it.[3]

According to É. Kiss [6], this principle holds in the case of the right periphery – phrases placed behind the finite verb – in Hungarian. This tendency can indeed be quantified.

I measured the length of the first three words behind the finite verb in the sub-corpora of +1, +2 and +3 positions, and counted the average of length-values in these three positions, separately. The obtained results are summarized in Table 5.

Table 5. Average length of the first three word after the finite verb in the sub-corpora if +1, +2 and +3 positions of the postverbal particles.

Sub-corpus	Average length of words			Frequency
+1	2.8	4.1	4.9	97.78
+2	3.0	3.2	4.5	2.13
+3	3.3	5.2	4.1	0.07

As shown in Table 5, the Law of Increasing Terms holds in the case of sentences having a postverbal particle in +1 or +2 position. It fails, however, by +3 positioned postverbal particles, where the particle (with an average length of 4.1 characters) is preceded by a longer word (5.2 characters). It must be noted that only 0.07% of the sentences are affected by this.

To sum up, the order of constituents on the right periphery is mostly influenced by their phonological weight. The averaged data of +1 and +2 positions show the tendency predicted by the Law of Increasing Terms. In case of positions greater than +2, this tendency does not hold.

5   Conclusion

The corpus-driven research of Hungarian particle verbs has proven to be useful for theoretical and computational linguistics as well. The theoretical novelty comes from the data of postverbal particles. We saw that verbal particles move away in spontaneous speech more often than in edited texts. By examining the phonological effects, the Law of Increasing Terms became quantifiable.

Context-based rules were applied to decide whether a homograph word is a verbal particle or not. By following this method, it was possible to improve the POS-tagging of verbal particles by 69.6%.

---

[3] This tendency was recognised also by Pāṇini, a Sanskrit grammarian living in the 4th century. Therefore, it is sometimes referred to as Pāṇini's Law (see [3,12]). He observed that in Sanskrit compound words, typically the shorter component is placed before the longer.

For further research, the distribution of detached verbal particles could be measured in other Hungarian verbal structures (e.g. infinitives and participles). A challenging problem is to decide which verb the verbal particle belongs to, especially if it can form a valid combination with more than one verbs in the sentence.

References

1. Alberti, G.: A magyar mondattan elmélete és gyakorlata generatív megközelítésben III. A háttérelmélet. Pécsi Tudományegyetem, Pécs (2006)
2. Behaghel, O.: Deutsche Syntax IV. Carl Winters, Heidelberg (1932)
3. Cooper, W.E., Ross, J.R.: World order. In: Grossman, R.E., San, L.J., Vance, T.J. (eds.) Papers from the Parasession on Functionalism, pp. 63–111. Chicago linguistic society, Illinois, Chicago (1975)
4. É. Kiss, K.: Az egyszerű mondat szerkezete. In: Kiefer, F. (ed.) Strukturális magyar nyelvtan I., Mondattan, pp. 74–160. Akadémiai Kiadó, Budapest (1992)
5. É. Kiss, K.: Mondattan. In: Kiss Katalin, É., Kiefer, F., Siptár, P. (eds.) Új magyar nyelvtan, pp. 15–186. Osiris Kiadó, Budapest (2003)
6. É. Kiss, K.: Az ige utáni szabad szórend magyarázata. Nyelvtudományi Közlemények **104**, 124–152 (2007)
7. Goody, J.: The Interface Between the Written and the Oral. Cambridge University Press, Cambridge (1987)
8. Hegedűs, V.: Particle-verb order in Old Hungarian and complex predicates. In: 14th Diachronic Generative Syntax Conference (DiGS14), University of Lisbon, Lisbon (2012). Handout
9. Indig, B., Vadász, N.: Windows in human parsing - how far can a preverb go? In: Tadić, M., Bekavac, B. (eds.) Tenth International Conference on Natural Language Processing (HrTAL2016) 2016, Dubrovnik, Croatia, 29–30 September 2016, Proceedings. Springer, Cham (2016)
10. Kalivoda, Á.: A magyar igei komplexumok vizsgálata. MA thesis. Budapest, Pázmány Péter Katolikus Egyetem, Bölcsészet- és Társadalomtudományi Kar (2016). https://github.com/kagnes/hungarian_verbal_complex
11. Makrai, M.: Többértelműségek magyar mondatok számítógépes elemzésében - a "meg" szó szófajának vizsgálata gyakoriságokkal (2007). Term paper. http://hlt.bme.hu/media/pdf/makrai07_temalabor_meg.pdf
12. O'Connor, M.P.: Hebrew Verse Structure, vol. 1. Eisenbrauns, Winona Lake (1978)
13. Oravecz, C., Váradi, T., Sass, B.: The Hungarian Gigaword Corpus. In: Calzolari, N., et al. (eds.) Proceedings of the 9th International Conference on Language Resources and Evaluation, 26–31 May 2014, pp. 1719–1723. ELRA, Reykjavik, Iceland (2014)
14. Peredy, M.: Az ige-igekötő sorrend a Jókai-kódexben. In: Bakró-Nagy, M., Forgács, T. (eds.) A nyelvtörténeti kutatások újabb eredményei VI, pp. 181–197. Szegedi Tudományegyetem, Magyar Nyelvészeti Tanszék, Szeged (2011)
15. Surányi, B.: Verbal particles inside and outside vP. Acta Linguistica Hungarica **56**(2–3), 201–249 (2009)
16. Tognini-Bonelli, E.: Corpus Linguistics at Work. John Benjamins Publishing Company, Amsterdam (2001)

HANS: A Service-Oriented Framework for Chinese Language Processing

Lung-Hao Lee[1($\boxtimes$)], Kuei-Ching Lee[2], and Yuen-Hsien Tseng[1]

[1] Graduate Institute of Library and Information Studies,
National Taiwan Normal University,
No. 162, Sec. 1, Heping East Road, Taipei 10610, Taiwan
{lhlee,samtseng}@ntnu.edu.tw
[2] China Development Lab, IBM,
No. 13, Sanchong Road, Taipei 11501, Taiwan
jkclee@tw.ibm.com

Abstract. A service-oriented architecture called as HANS is proposed to facilitate Chinese natural language processing. This unified framework seamlessly integrates fundamental NLP tasks including word segmentation, part-of-speech tagging, named entity recognition, chunking, paring, and semantic role labeling to enhance Chinese language processing functionality. A basic Chinese word segmentation task is used to illustrate the function of the proposed architecture. to demonstrate the effects. Evaluated benchmarks are taken from the SIGHAN 2005 bakeoff and the NLPCC 2016 shared task. We implement publicly released toolkits including Stanford CoreNLP, FudanNLP and CKIP as services in our HANS framework for performance comparison. Experimental results confirm the feasibility of the proposed architecture. Findings are also discussed to point to potential future developments.

Keywords: Service-oriented architecture · Chinese word segmentation
Chinese natural language processing

1 Introduction

Chinese natural language processing has been extensively studied, including word segmentation, part-of-speech tagging, named entity recognition, chunking, syntactic/dependency parsing and other tasks [1]. The most fundamental task for Chinese language processing is to segment a Chinese sentence into words, as written Chinese does not feature word delimiters, such as blank spaces, and a Chinese word can consist of single or multiple characters. The key challenges for Chinese word segmentation include segmentation ambiguity and unknown word identification in Chinese text. Many approaches have been proposed to deal with these problems. For examples, a corpus-based learning method was used to derive sets of syntactic rules, which were then applied to distinguish monosyllabic words from monosyllabic morphemes [2]. In addition to statistical information, morphological, syntactic, semantic, word type-specific knowledge is used to identify unknown words [3]. A study proposes a pragmatic mathematical framework to segment known words and detect unknown

© Springer Nature Switzerland AG 2018
A. Gelbukh (Ed.): CICLing 2017, LNCS 10761, pp. 134–142, 2018.
https://doi.org/10.1007/978-3-319-77113-7_11

words [4]. Conditional random fields are used to detect new words [5]. Discriminative pruning rules for language models are designed for Chinese word segmentation [6]. Punctuation is considered as implicit clues for word segmentation [7]. A unified character-based tagging framework is used to segment Chinese texts into words [8]. Romanized pinyin is used with search engines to deal with problems stemming from segmentation ambiguity, unknown word detection, and stop words [9]. Large scale unlabeled data is used to represent Chinese characters and enhance word segmentation through deep learning techniques [10]. A dual-decomposition algorithm is proposed to combine the strength of word-based and character-based segmentation schemes for joint inference [11]. The max-margin tensor neural network is used to model complicated interactions between tags and contextual characters for Chinese word segmentation [12].

Benchmark data exist for evaluating Chinese Word Segmentation (CWS) performance. The key landmarks are bake-offs, competition-based evaluations organizing by Special Interest Group for Chinese Language Processing (SIGHAN) of the Association for Computational Linguistics (ACL). The first CWS bakeoff was held in 2003 [13], with subsequent events in 2005 [14], 2006 [15], and 2008 [16]. Another series of CWS evaluations are shared tasks organized by the conference on Natural Language Processing and Chinese Computing (NLPCC). Recently, NLPCC 2015 [17] and 2016 [18] organized CWS shared tasks for micro-blog texts.

Although different methods and several toolkits for Chinese language processing have been proposed and released publicly, their advantages remain unclear regarding relative suitability for various text genres, such as news stories, novels, and social media texts, even in terms of linguistic similarities/differences using different character systems in traditional or simplified Chinese. One possible reason is a lack of large-scale comparisons based on the same benchmarks. Another is the difficulty of implementation due to different usage interfaces in different programming languages. These observations motivate us to develop a unified framework for processing Chinese text, using a word segmentation task to demonstrate the feasibility of our architecture and to compare system performance based on selected benchmark data and toolkits.

The rest of this paper is organized as follows. Section 2 provides a detailed description of the HANS framework for Chinese language processing. Section 3 presents the experimental results for performance evaluation. Section 4 discusses related findings. Conclusions along with future research directions are finally presented in Sect. 5.

2 HANS Framework for Chinese Language Processing

We adopt a Service-Oriented Architecture (SOA) to develop our proposed framework for Chinese language processing, HANS (pronounced like 漢思 'han si', which can be translated as "Mandarin thinking"). Figure 1 shows the overview of the designed architecture. Users can submit Chinese texts to the broker server according to the predefined message formats. The broker server plays a message exchange role for communication. Submitted messages are parsed to extract Chinese texts and send them to specific services for further processing. All results returned by involved services will

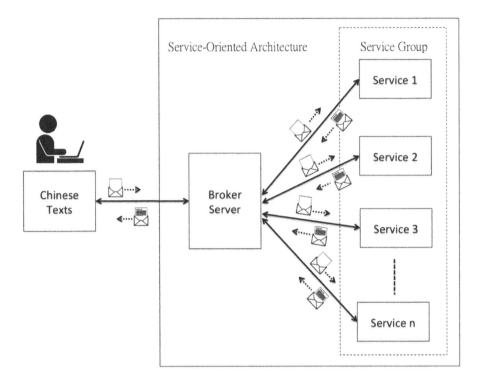

Fig. 1. System architecture of the proposed HANS framework.

be aggregated together to generate a response to the user. The flexibility of SOA allows developers to incrementally use any later-developed approaches to enhance system effectiveness without affecting running services.

The submission message, denoted as an envelope with rightwards arrow in Fig. 1, should contain the following items.

- Content: this item contains Chinese texts that are targets for processing. The default encoding is UTF-8.
- Target Service: the service name to be used to process submitted texts, such as CKIP, Stanford CoreNLP, FudanNLP, and so on.
- Task Code: the NLP task to be applied; to name a few, "cws" denotes Chinese word segmentation; "post" represents part-of-speech tagging; "ner" expresses named entity recognition.
- Query Queue: the queue name in the broker server used to receive users' submitted messages, after which the broker server will parse received messages for further processing.
- Reply to: this is web socket (IP address and port number) of the broker server. The services will return results to this address.
- User ID: the unique identifier of a user who submits the messages.
- Timestamp: the timestamp used to record submission time. This is an optional item according to the submitting user's preferences.

We also define the response message, indicated as an envelope with a leftwards arrow in Fig. 1, which should include items as follows.

- Service Name: the service name used for the processed submissions and replied messages.
- Task Code: the task code should be identical to that of a specific task.
- Return: the results from a specific service successfully processing the submitted text. The format is the result and the service name, separated by a tab. If the service failed to process the submission, this item will contain associated error information.
- Answer Queue: the queue name in the broker server, which is stored to await services' response messages. If the submitting user does not acquire his/her results in time, the messages will be abandoned automatically.
- Timestamp: the timestamp used to note the time of message receipt. This is also an optional item.

The user submits an example sentence "唐納川普宣誓就職成為美國第45任美國總統" ("Donald Trump sworn in as the 45th President of the United States of America") to the broker server. Services A, B, C, and D are designated to process this sentence. The returned results may be as follows:

- 唐納 川普 宣誓 就職 成為 美國 第45 任 總統 <tab> Service A
- 唐納川 普宣誓 就職 成為 美國 第45 任 總統 <tab> Service B
- 唐納 川普 宣誓 就職 成為 美國 第 45 任 總統 <tab> Service C
- Service D: out of service.

Services A, B, and C returned different segmented results, while Service D is suspended at submission time.

The proposed framework is implemented using Apache ActiveMQ [19], a messaging server with full Java Message Service (JMS). JMS aims to provide a public API for Message-Oriented Middleware (MOM), which is a category of software for communication in an asynchronous, loosely-coupled, reliable, scalable, and secure manner among distributed applications.

3 Experiments on Chinese Word Segmentation

Experimental benchmark data are taken from the SIGHAN bakeoff 2005 and the NLPCC 2016 shared task. Table 1 presents statistical results of test sets, where #Sents, #Words, and #Types respectively denote the number of sentences, the number of words, and the number of word types. We briefly describe the corpora as follows.

- SIGHAN 2005 Bakeoff [14]: This is the most complete and representative benchmark. The training, testing, and gold-standard data sets, as well as the scoring script, are available for research use. Four corpora and accompanying segmentation guidelines are adopted from the following organizations: Academia Sinica (AS), City University of Hong Kong (CityU), Beijing University (PKU), and Microsoft Research (MSR). All texts are selected and extracted from formal news. Among

Table 1. Statistics of the test sets

Source	Data	#Sents	#Words	#Types
SIGHAN 2005 Bakeoff	AS	14,429	122,564	18,871
	CityU	1,492	40,959	9,007
	PKU	1,944	210,687	13,159
	MSR	3,985	210,550	12,933
NLPCC 2016	Weibo	8,592	187,877	27,804

these data, AS and CityU use Traditional Chinese, while PKU and MSR use simplified Chinese.

- NLPCC 2016 Shared Task [18]: This corpus consists of informal texts taken from Sina Weibo instant message chats. The character system is simplified Chinese.

The performance of three publicly released toolkits is compared. Each toolkit corresponds to a service in Fig. 1.

- CKIP [20]: this segmentation system uses a hybrid method of mixed heuristic and statistical rules to solve segmentation ambiguities. One implementation is available online at http://ckipsvr.iis.sinica.edu.tw/.
- FudanNLP [21]: this open source toolkit uses statistics-based and rule-based methods to deal with Chinese NLP tasks, such as word segmentation, part-of-speech tagging, named entity recognition, dependency parsing, and so on. FudanNLP is distributed under license LGPL 3.0, and is available online at https://github.com/FudanNLP/fnlp
- Stanford CoreNLP [22]: a suite of core NLP tools, including part-of-speech tagger, named entity recognizer, parser, co-reference resolution system, sentiment analysis, bootstrapped pattern learning, and open information extraction tools. In addition to English, it also provides packaged models for Chinese. Stanford CoreNLP is licensed under the GNU General Public License version 3 or later. It can be obtained from http://stanfordnlp.github.io/CoreNLP/. We used version 3.6 for experiments.

For performance evaluation, we use three traditional metrics: precision, recall and F1 score. The scoring script released by the SIGHAN 2005 bakeoff was used to measure performance.

Tables 2 and 3 respectively show the SIGHAN 2005 bakeoff results for two traditional Chinese datasets. Stanford slightly performed better than Fudan on AS data, while the latter was clearly preferable for CityU data. CKIP achieved the best results for both data sets for all metrics.

Tables 4 and 5 respectively show the effects on two simplified Chinese corpora from the SIGHAN 2005 bakeoff. The performance differences among these three toolkits were not obvious. However, Stanford performed best for all three metrics.

Table 6 shows the results on Weibo data from the NLPCC 2016 shared task. CKIP and Stanford performed similarly, but not as well as Fudan.

Table 2. SIGHAN 2005 bakeoff results for AS data.

AS data	Precision	Recall	F1 score
CKIP	0.932	0.945	0.938
Fudan	0.875	0.856	0.865
Stanford	0.869	0.871	0.870

Table 3. SIGHAN 2005 bakeoff results for CityU data.

CityU data	Precision	Recall	F1 score
CKIP	0.863	0.886	0.875
Fudan	0.867	0.854	0.861
Stanford	0.824	0.831	0.827

Table 4. SIGHAN 2005 bakeoff results for MSR data.

MSR data	Precision	Recall	F1 score
CKIP	0.795	0.851	0.822
Fudan	0.816	0.844	0.830
Stanford	0.817	0.853	0.835

Table 5. SIGHAN 2005 bakeoff results for PKU data.

PKU data	Precision	Recall	F1 score
CKIP	0.870	0.884	0.877
Fudan	0.891	0.877	0.884
Stanford	0.900	0.894	0.897

Table 6. NLPCC 2016 results for Weibo data

Weibo data	Precision	Recall	F1 score
CKIP	0.856	0.895	0.875
Fudan	0.899	0.900	0.900
Stanford	0.864	0.879	0.872

These experimental results indicate the CKIP segmentation system is a good choice to process formal text in traditional Chinese, while Stanford's CoreNLP toolkit provides advantages for simplified Chinese, and FudanNLP performs better for informal texts such as those from micro-blogs.

4 Discussion

A word is a minimal unit used to represent a semantic meaning in a sentence. Word segmentation systems are trained using discrete data sets and different word segmentation standards may result in performance differences, even when using similar machine learning methods. Performance comparisons in this study only used publicly released toolkits with pre-trained models. Without retraining using the corresponding data source, performance may be worse than the competition entries. While incorporating suitable training data may improve performance, it is often impractical because available annotated data is insufficient to optimize system performance. The performance shown above represent an estimate of the effectiveness of the systems incorporated in our HANS framework.

In fact, flexibility is the most representative characteristic of our HANS framework. HANS can easily integrate multiple methods by service implementation for practical use or performance evaluation. In terms of system management, a logging service can be added to record operation details. An authentication service can be used to manage users and permissions. All services can also be integrated with the broker sever at the same site to reduce network latency. Moreover, an interface provided by a web sever can reside between the client side and the broker site to provide relatively convenient Chinese manipulation.

Distribution reliability is also an advantage of this framework. The same method can be implemented as more than one stored service for fault tolerance. If a service fails at submission, the back-up service can run automatically to process the submission. In addition, a distributed method can be easily adopted to enhance processing efficiency.

5 Conclusions and Future Work

This study proposes the HANS framework based on a service-oriented architecture for Chinese language processing. Experimental results on Chinese word segmentation tasks are used to compare various toolkits. The major contributions of this work are twofold: (1) demonstrating the feasibility of a unified framework to facilitate Chinese language processing; and (2) using the proposed framework to evaluate existing toolkits based on benchmark data. Our HANS framework is scalable to incorporate reliable services and, most importantly, allows for easy implementation when dealing with Chinese NLP tasks by way of a simple message transmission protocol.

In addition to comparing Chinese word segmentation methods to understand their effects and suitability, future work will focus on new solutions for enhancing word segmentation and other NLP tasks.

Acknowledgments. This study was partially supported by the Ministry of Science and Technology, under the grant MOST 105-2221-E-003-020-MY2 and the "Aim for the Top University Project" and "Center of Language Technology for Chinese" of National Taiwan Normal University, sponsored by the Ministry of Education, Taiwan.

References

1. Wong, K.-F., Li, W., Xu, R., Zhang, Z.: Introduction to Chinese natural language processing. Synth. Lect. Hum. Lang. Technol. **2**, 1–148 (2009)
2. Chen, K.-J., Bai, M.-H.: Unknown word detection for Chinese by a corpus-based learning method. Int. J. Comput. Linguist. Chin. Lang. Process. **3**(1), 27–44 (1998)
3. Chen, K.-J., Ma, W.-Y.: Unknown word extraction for Chinese documents. In: 19th International Conference on Computational Linguistics, pp. 169–175. ACL Anthology (2002)
4. Gao, J., Li, M., Wu, A., Huang, C.-N.: Chinese word segmentation and named entity recognition: a pragmatic approach. Comput. Linguist. **31**(4), 531–574 (2005)
5. Peng, F., Feng, F., MaCallum, A.: Chinese segmentation and new word detection using conditional random fields. In: 20th International Conference on Computational Linguistics, pp. 562–568. ACL Anthology (2004)
6. Li, J., Wang, H., Ren, D., Li, G.: Discriminative pruning of language models for Chinese word segmentation. In: 44th Annual Meeting of the Association for Computational Linguistics, pp. 1001–1008. ACL Anthology (2006)
7. Li, Z., Sun, M.: Punctuation as implicit annotations for Chinese word segmentation. Comput. Linguist. **35**(4), 505–512 (2009)
8. Zhao, H., Huang, C.-N., Li, M., Lu, B.-L.: A unified character-based tagging framework for Chinese word segmentation. ACM Trans. Asian Lang. Inf. Process. **9**(2) (2010). Article 5
9. Wang, F.L., Yang, C.C.: Mining web data for Chinese segmentation. J. Am. Soc. Inf. Sci. Technol. **58**(12), 1820–1837 (2007)
10. Zheng, X., Chen, H., Xu, T.: Deep learning for Chinese word segmentation and POS tagging. In: 2013 Conference on Empirical Methods in Natural Language Processing, pp. 647–657. ACL Anthology (2013)
11. Wang, M., Voigt, R., Manning, C.D.: Two knives cut better than one: Chinese word segmentation with dual decomposition. In: 52nd Annual Meeting of the Association for Computational Linguistics, pp. 193–198. ACL Anthology (2014)
12. Pei, W., Ge, T., Chang, B.: Max-margin tensor neural network for Chinese word segmentation. In: 52nd Annual Meeting of the Association for Computational Linguistics, pp. 293–303. ACL Anthology (2014)
13. Sproat, R., Emerson, T.: The first international Chinese word segmentation bakeoff. In: 2nd SIGHAN Workshop on Chinese Language Processing. ACL Anthology (2003)
14. Emerson, T.: The second international Chinese word segmentation bakeoff. In: 4th SIGHAN Workshop on Chinese Language Processing, pp. 123–133. ACL Anthology (2005)
15. Levow, G.-A.: The third international Chinese language processing bakeoff: word segmentation and named entity recognition. In: 5th SIGHAN Workshop on Chinese Language Processing, pp. 108–117. ACL Anthology (2006)
16. Jin, G., Chen, X.: The fourth international Chinese language processing bakeoff: Chinese word segmentation, named entity recognition and Chinese POS tagging. In: 6th SIGHAN Workshop on Chinese Language Processing, pp. 69–81. ACL Anthology (2008)
17. Qiu, X., Qian, P., Yin, L., Wu, S., Huang, X.: Overview of the NLPCC 2015 shared task: Chinese word segmentation and POS tagging for micro-blog texts. In: Li, J., Ji, H., Zhao, D., Feng, Y. (eds.) NLPCC 2015. LNCS (LNAI), vol. 9362, pp. 541–549. Springer, Cham (2015). https://doi.org/10.1007/978-3-319-25207-0_50

18. Qiu, X., Qian, P., Shi, Z.: Overview of the NLPCC-ICCPOL 2016 shared task: Chinese word segmentation for micro-blog texts. In: Lin, C.-Y., Xue, N., Zhao, D., Huang, X., Feng, Y. (eds.) ICCPOL/NLPCC -2016. LNCS (LNAI), vol. 10102, pp. 901–906. Springer, Cham (2016). https://doi.org/10.1007/978-3-319-50496-4_84
19. ActiveMQ. http://activemq.apache.org
20. Ma, W.-Y., Chen, K.-J.: Design of CKIP Chinese word segmentation system. Int. J. Asian Lang. Process. **14**(3), 235–249 (2004)
21. Qiu, X., Zhang, Q., Huang, X.: FudanNLP: a toolkit for Chinese natural language processing. In: 51st Annual Meeting of the Association for Computational Linguistics, pp. 49–54. ACL Anthology (2013)
22. Manning, C.D., Surdeanu, M., Bauer, J., Finkel, J., Bethard, S.J., McClosky, D.: The stanford CoreNLP natural language processing toolkit. In: 52nd Annual Meeting of the Association for Computational Linguistics: System Demonstrations, pp. 55–60. ACL Anthology (2014)

Syntax and Parsing

Learning to Rank for Coordination Detection

Xun Wang[1][(✉)], Rumeng Li[2], Hiroyuki Shindo[2], Katsuhito Sudoh[1],
and Masaaki Nagata[1]

[1] NTT Communication Science Laboratories, Kyoto, Japan
{wang.xun,sudoh.katsuhito,nagata.masaaki}@lab.ntt.co.jp
[2] Nara Institute of Science and Technology, Nara, Japan
alicerumeng@gmail.com, shindo@is.naist.jp

Abstract. Coordinations refer to phrases such as "A and/but/or/... B". The detection of coordinations remains a major problem due to the complexity of their components. Existing work normally classified the training data into two categories: correct and incorrect. This often caused the problem of data imbalance which inevitably damaged performances of the models they used. We propose to fully exploit the differences between training data by formulating the detection of coordinations as a ranking problem to remedy this problem. We develop a novel model based on the long short-term memory network. Experiments on Penn Treebank and Genia verified the effectiveness of the proposed model.

1 Introduction

Coordinations widely exist in languages for their great expressive capacity. Below we show a simple example *Sentence* 1 in which two words "dogs" and "cats" are coordinated as one phrase to serve as the object. But more often is the case that long and complex phrases are coordinated to express sophisticated intentions of speakers as shown in *Sentence* 2. Researchers have found that almost no components in a sentence cannot be coordinated and this often results in complex structures with complicated coordinations, as is demonstrated by *Sentence* 2. The underlined part in *Sentence* 2 is a coordination which contains not only a very long conjunct but also other coordinations in one of its conjuncts (nestified coordinations).

Sentence 1: I like *(dogs)* and *(cats)*.
Sentence 2: The hurricane left hundreds of thousands without access to their homes or jobs, has (separated people from relatives,) and (inflicted both ***physical and mental*** distress on those who suffered through *the storm and its aftermath*).

Parsing sentences with coordinations is a challenging task. We tested on *Sentence* 2 using a state-of-the-art syntactic parser, Enju [20]. Figure 1 shows the result. As can be seen, though it detects the existence of coordinations and

A. Gelbukh (Ed.): CICLing 2017, LNCS 10761, pp. 145–157, 2018.
https://doi.org/10.1007/978-3-319-77113-7_12

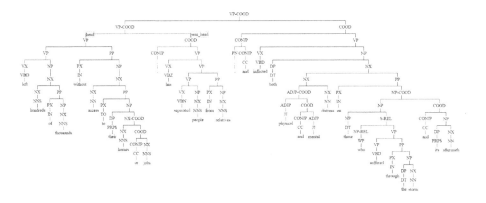

Fig. 1. Outcome of the Enju Parser

correctly determines the scopes of two simple coordinations, Enju fails when comes to the long and complex one.

This is not just an isolated example. When tested on the Wall Street Journal (WSJ) part of Penn Treebank, Enju, as one of the best parsers available, reports recalls of about 60% for sentences containing coordinations [27], which is much lower compared with about 90% for ordinary sentences. On the other hand, sentences containing coordinations occupy a large proportion in corpora. In the WSJ part of Penn Treebank, about 70% sentences contain at least one coordination. For corpora of some peculiar domains, such as Genia [10] for bio-text mining, the proportion is even higher [27]. The great importance of coordinations and the relative poor performance of existing parsers in this aspect lead to intensive study on the task of coordination detection.

Coordination detection involves identifying the components of coordinations: the coordinator(s) and the conjuncts. Coordinators are relatively easy to detect for they are limited in number and we can enumerate every one. The difficulty lies in deciding the scopes of conjuncts. [27] propose a classification-based method for detecting conjunct scopes. They define a simple grammar to generate all the possible structures for coordinations and then train a perceptron model for selecting the correct one based on various features. Other work follows this guideline and obtains sustaining improvements by introducing new models and features [6,7].

These works without exception classify the training data into two categories. Take *Sentence* 1 for example, there are three possible conjunct pairs: { "I like dogs", "cats"}, { "like dogs", "cats"}, and { "dogs", "cats"}. They are classified into two classes, the wrong and the correct. Apparently the correct class contains only one sample which is { "dogs", "cats"}, and the wrong class contains the other two. For this example, the size of the wrong class is twice that of the correct class. For some coordinations, there may exist dozens of candidates and still only one of them is correct. The imbalanced training data inevitably damages the performances of classification models.

To remedy this problem, we propose to exploit the difference among examples inside the wrong class: { *"like dogs"*, *"cats"*} and { *"I like dogs"*, *"cat"*}. We notice that although both of them are wrong, the latter is more wrong than the former because the former is closer to the correct one when measured by edit distance. Previous classification-based works generally fail to fully explore the training data regarding this point.

In this work, we rank all the candidates according to their edit distances with correct answers and then train *a ranker* rather than *a classifier* from the ranked lists. Hence we avoid the drawbacks of classification models and fully take advantage of the training data. Moreover, the recent booming neural network models enable us to use semantics and syntax more efficiently than traditional models and features. We use a long short-term memory (LSTM) network to calculate representations for candidates and then employ a bilinear transformation for ranking.[1] Experiments on Penn Treebank and Genia prove the effectiveness of the proposed method and report satisfactory results on two tasks, general coordination scope detection and NP coordination detection.

The main contributions of this paper include not only improvements in the task of coordination detection but also the idea of using ranking methods to take advantages of imbalanced data towards better performances for classification tasks.

The remaining of the paper is organized as follows: Sect. 2 introduces the related work. Section 3 presents the proposed model. Section 4 describes the experimental results and our analysis. Section 5 concludes this paper.

2   Related Work

2.1   Coordination Detection

Coordination has been intensively studied by theoretical and computational linguists for a long time.

Some early work includes grammatical analysis [25] and rule based methods [11,23]. Machine learning methods are later developed for they require less human labour and usually produce better results. Some work regards coordination detection as a generative problem. [8] presented a generative lexicalized parser which considered the symmetry of part-of-speech tags and phrase categories of conjuncts. [9] further developed it by considering parallelism and selectional preferences. Another research guideline is to formulate coordination detection as a discriminative problem. Influential work includes that of [2,3,7,27]. [2,3] used a linear-chain CRF model to select the correct answer from many candidates. [7,27] employed the perceptron learning. [6] further developed a model based on dual composition and utilized rich semantic features to find the best candidate.

---

[1] Other neural network models can also be employed to learn representations. Here we choose this one for its effectiveness and simplicity.

The development of coordination detection benefits other tasks, such as syntactic parsing, a lot. [8] improved the performance of Bikel-Collins Parser [1] by considering coordination structures. [32] also mentioned the improvement for dependency parsing via coordination structure analysis.

Some other related work includes that of [4, 21, 22] which we do not elaborate due to the space limitation. As we have stated, none of them considers exploring the training data using ranking methods. Therefore we introduce our model to remedy previous ones towards a better coordination detection method.

The proposed model uses neural network models to learn representations for candidate conjuncts. Learning representations for long text (sentence/paragraph/document) is an important problem which draws much attention. In this work, we adopt the long short-term memory network (LSTM) which is regarded as an improvement of the traditional Recurrent neural networks (RNN).

3   Coordination Detection by Ranking

As stated, a coordination includes coordinators and conjuncts. Coordinators are easy to resolve as we can enumerate all of them. In this paper, as is the same with previous work [6, 7, 27], we focus on the detection of conjuncts: to identify the scopes of the left and right conjunct for each coordinator in sentences.

3.1   Problem Formulation

Available corpora usually annotate coordinations in syntactic trees. We need to extract the correct coordinations and produce all possible candidates from the annotated data set for the training and testing.

There are several special kinds of coordinations to which we need pay peculiar attention. As we know, a coordination contains n coordinators (n is a positive integer and $n \geq 1$), and $n + 1$ conjuncts. Most coordination contains only one coordinator and two conjuncts. These coordinations, however, contain two or even more coordinators, such as "A, B and C" (here the comma is also considered as a coordinator) or "(A and B) and C". Such coordinations are dealt with using different strategies.

Nestification. Structures like "((A and B) and C)" contains two related coordinations, "A and B", and "(A and B) and C". "(A and B)" is a conjunct of "(A and B) and C". Thus their left conjuncts should share the same left boundary and the right boundary of the right conjunct of "A and B" should be the right boundary of the left conjunct of "(A and B) and C". But here we consider them as different and independent coordinations for simplicity.

Multiple Conjuncts. For the "A, B and C" style coordinations, we regard "A, B" as one conjunct, and treat them as "(A, B) and C" for simplification.

To construct data for training and testing, we start from generating all the candidates. For each coordinator co in $\{w_{l_n}, w_{l_{(n-1)}}, ...w_{l_1}, w_{l_0}, co, w_{r_0}, w_{r_1}, ..., w_{r_n}\}$, we collect its left neighbours $\{w_{l_n}, w_{l_{(n-1)}}, ...w_{l_1}, w_{l_0}\}$ and right neighbours $\{w_{r_0}, w_{r_1}, ..., w_{r_n}\}$. The right part $\{w_{r_0}, w_{r_1}, ..., w_{r_n}\}$ ends where it has to end, e.g., a period, a comma, a semi-comma or other punctuations. The left part $\{w_{l_n}, w_{l_{(n-1)}}, ...w_{l_1}, w_{l_0}\}$ ends at w_{l_0} and starts where a sentence or clause starts. Note the right part cannot start after a comma, because we need to deal with the "(A, B) and C" style coordinations which contains a comma inside the conjunct.

The candidate set C which contains all the possible conjunct pairs is as follows:

$(L_i, R_j) \in C$

$$\begin{cases} L_i = \{w_{l_i}, w_{l_{i-1}}, ..., w_{l_0}\}; i \le l_n \\ R_j = \{w_{r_0}, w_{r_1}, ..., w_{r_j}\}; j \le r_n \end{cases}$$

C is a Cartesian product of all the left candidate conjuncts and all the right candidate conjuncts.

In the work of [6,7,27], a simple grammar is proposed for generating all the possible trees and then a model is learnt to find the tree containing the correct coordination(s). The problem is that the simple grammars they used produce hundreds of trees which makes the following step of identification inefficient. In this work it is not a serious problem for most examples have small candidate sets. But still we remove some extremely long coordinations in the following experiments for efficiency.

We then rank all the items in C for the training. The top one is the correct one and the remaining are ranked according to their edit distances with the correct one. We employ the following function to map candidate pairs to $\mathbb{R}$, $f : C \to \mathbb{R}$:

$$\begin{cases} t = max(0, 1 - \frac{edit\_distance(\{L_c, R_c\}, \{L_i, R_j\})}{\|\{L_c, R_c\}\|}) \\ f((L_i, R_j)) = exp(2t) \end{cases}$$

$\{L_c, R_c\}$ is the correct conjunct pair. $\|\{L_c, R_c\}\|$ is the number of words in the correct conjunct pair. f is a function which takes t as input. It is also plausible to use other functions. But we have the following benefits from this exponentiation function. Since $\frac{df}{dt} = 2exp(2t)$, it can be known that when t becomes large, f is more sensitive to the change of t, which means the penalty will increase more rapidly when candidates are more similar to the correct one.

3.2   Learning to Rank

Using the data generated above, $\{L_i, R_j\}$ and their scores, we are able to train a ranker. In the testing, for each coordinator, we calculate scores for all the candidates using the learnt ranker and then choose the one with the highest score as the output.

Figure 2 shows the structure of the proposed model. As stated, each internal node represents a LSTM memory cell except the one on the top, which represents a bilinear transformation. Take {"*I like dogs*", "*cats*"} as an example. We obtain the representations of candidate conjuncts, "I like dogs" and "cats" using LSTM separately. Their representations are then fed to the bilinear transformation node which outputs the final score.

The LSTM node conducts the following operations:

$$\begin{cases} i_t = \sigma(W_i x_t + U_i h_{t-1} + b_i) \\ \widetilde{C_t} = tanh(W_c x_t + U_c h_{t-1} + b_c) \\ f_t = \sigma(W_f x_t + U_f h_{t-1} + b_f) \\ C_t = i_t \widetilde{C_t} + f_t C_{t-1} \\ out_t = \sigma(W_o x_t + U_o h_{t-1} + V_o C_t + b_o) \\ h_t = out_t * tanh(C_t) \end{cases} \quad (1)$$

Here W_*, U_*, V_* are weight matrices. b_* are bias vectors. x_t is the input and $output_t$ is the output at time t. f_*, h_*, C_* are some internal states.

The bilinear function takes as input out_{L_i} and out_{R_j} and outputs the score:

$$\begin{aligned} Score(out_{L_i}, out_{R_j}) = \\ f(out_{L_i} W_0 + out_{R_j} W_1 + out_{L_i} W_2 out_{R_j} + b) \end{aligned} \quad (2)$$

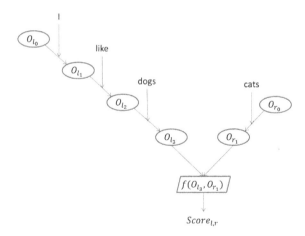

Fig. 2. An Example for Coordination Detection

We minimize the loss over parameters: $Loss = \sum_{L_i, R_j} max(0, \gamma - (f((L_i, R_j)) - Score(out_{L_i}, out_{R_j})))$. γ is the margin. It is also plausible to use other loss functions to optimize the top N of the ranked list [31].

4   Experiments

We held experiments on two tasks to investigate the performance of the proposed model. One is the general coordination detection and the other is the NP coordination detection.

4.1   Datasets

In this work, we have the WSJ part of Penn Treebank [17] which consists of news articles on finance and economy. We extract all the sentences with coordinations and keep those which contain the three most frequent coordinators: "and[2]", "but", and "or". They occupy about 90% of all the coordinations. And among them, coordinations with "and" occupies the largest proportion which is about 80% as is shown in Table 1.

Table 1. Distribution of coordinators in WSJ

Coordinator	No	Proportion
Total	18612	100
and	15088	81.0
or	2543	13.7
but	981	5.3

Also we have the Genia data set which is a semantically annotated corpus for bio-textmining [10]. Genia is composed of abstracts of medical research papers and is also widely used in previous research. As is the case with WSJ, "and" occupies the largest proportion.

4.2   Settings

Word representations come from GloVe [24]. The dimension of word vectors, the learning rate and the margin are to be decided using the development data set of WSJ in the following section. We use *chainer*[3], a flexible framework of neural networks, for the implementation of the proposed model and use the stochastic gradient descent for the optimization.

4.3   Task I: General Coordination Detection

Following the settings of previous work, we split the WSJ data into 3 parts: WSJ2-21 as the training data, WSJ22 as the development data and WSJ 23 as

Table 2. Distribution of Coordinations in WSJ

	Total	NP	VP	ADJP	S	SBAR	PP	UCP	ADVP	QP	Others
Total	18528	11497	3382	934	1245	194	380	480	153	202	61
WSJ_2_21	16896	10482	3083	841	1142	178	356	433	143	184	54
WSJ_22	704	429	142	42	44	5	13	19	4	5	1
WSJ_23	928	586	157	51	59	11	11	28	6	13	6

Table 3. Distribution of coordinations in Genia

	Total	NP	VP	ADJP	S	SBAR	PP	UCP	ADVP	QP	Others
Total	17093	11914	1842	988	530	146	787	219	96	14	557
Genia_1_1600	14264	9956	1520	810	448	124	658	192	74	8	474
Genia_1601_1800	1483	1098	160	33	45	8	77	4	14	0	44
Genia_1801_1999	1346	860	162	145	37	14	52	23	8	6	39

the test data. The coordinations are classified into different subtypes as is shown in Table 2.

Using the WSJ data, we firstly tune some parameters. When we are tuning one parameter, we keep the others fixed. According to overall recall reported on the development set, we decide the learning rate is 0.0005, the margin is 0.001 and the dimension of the word vector is 50. This setting is used for all the experiments conducted in this work. The development data sets in each experiment are used to decide the number of iterations (Table 3). We conduct two experiments using different data for this task.

Experiment on WSJ. This first experiment is conducted on WSJ. The number of iteration is set to be 7. We report our results on the top 1/2/5 candidates for each category on the test set in Table 4.

Table 4. Recall (top 1/2/5) on WSJ for general coordination detection (test data)

	Total	NP	VP	ADJP	S	SBAR	PP	UCP	ADVP	QP	Others
re@1	71.2	81.1	41.0	92.2	45.8	63.6	36.4	89.3	100	0	100
re@2	72.1	82.0	42.7	92.2	45.8	63.6	36.4	89.3	100	0	100
re@5	74.4	84.0	47.1	94.1	45.8	63.6	36.4	89.3	100	0	100

We compare our model with several previous state-of-the-art models in Table 5. T is a model using a dual composition method with alignment-based

---

[2] "&" which is usually regarded as a special form of "and" is excluded for it only appears in proper nouns and constitutes simple coordinations that are easy to identify.

[3] http://chainer.org/.

features extracted from HPSG parsing. *Enju* is a HPSG parser and the coordination detection results are extracted from the parse trees. Both T and *Enju* come from [6]. Recalls reported by T and *Enju* on the test data set are 70.6 and 69.0 respectively. And the proposed model obtains a recall of 71.2 as is shown in Table 4.

Table 5 shows the performances on each model on subtypes. Note that all the figures in Table 5 are based on the development data set, as previous works did not report their performances for each category on the test data set. Also the methods for data split and preprocessing are slightly different.

Table 5. Performance comparison on WSJ for general coordination detection (development data)

	Proposed		T		Enju
	Percentage(%)	Recall	Percentage(%)	Recall	Recall
Overall	100	73.4	100	71.6	68.1
NP	61.0	84.4	56.3	67.5	61.4
VP	20.2	41.5	18.4	79.8	78.8
ADJP	6.0	90.5	7.2	58.5	59.1
S	7.0	61.2	13.9	51.4	52.3
PP	1.8	38.5	1.9	64.5	59.1
Others	4.0	69.0	2.3	78.3	73.9

Experiment on Genia. Genia is another popular corpus for coordination detection, and is also used in many previous work. We also test our system on Genia and report the results in Table 6. We use the WSJ(2-21) and Genia(1-1600) for the training, Genia(1601-1800) for parameter tuning and Genia(1801-1900) for test. This is consistent with previous work [6]. Note that the annotation standards of WSJ and Genia are slightly different from each other. Here we select several major types which are not only shared by both corpora but also cover the majority of both corpora.

Table 6. Recall (top 1/2/5) on Genia for general coordination detection (test data)

	Total	NP	VP	ADJP	S	SBAR	PP	UCP	ADVP	QP	Others
re@1	69.2	79.1	35.2	78.0	43.2	42.9	34.6	65.2	87.5	66.7	41.0
re@2	70.7	79.5	37.0	82.1	48.6	50	36.5	73.9	87.5	66.7	41.0
re@5	73.8	80.4	40.1	91.0	59.4	50	44.2	95.6	1	100	41.0

Table 6 presents our results on the Genia test data set. With an overall recall of 69.2%, we outperform the previous work of T(67.8%) and *Enju* (65.5%) on the same data set.

Table 7. Performance comparison on Genia for general coordination detection (development data).

	Proposed		T		Enju	S	
	Percentage(%)	Recall	Percentage(%)	Recall	Recall	Percentage(%)	Recall
Test	100	69.8	100	67.7	63.3	100	61.5
NP	74.0	79.8	65.6	67.5	61.4	65.8	64.2
VP	10.8	34.3	11.2	79.8	78.8	12.7	54.2
ADJP	2.2	78.8	10.4	58.5	59.1	8.7	80.4
S	3.6	55.6	6.0	51.4	52.3	5.2	22.9
PP	5.2	25.6	6.0	64.5	59.1	4.6	59.9
Others	4.2	42.0	0.7	78.3	73.9	3.9	49.3

S is a chart-parsing method leveraging alignment-based features proposed by [7]. Note that S is also based on Genia, but uses a 5-fold cross validation. Table 7 compares their performances for subtypes on the Genia data set (on the development data set except S).

Table 8. Performance comparison on Genia for NP coordination detection

Sys	P	R	F
Proposed	75.3	80.1	77.6
S	61.7	57.9	59.7

As can be seen from the two experiments, the proposed model achieves the best overall recall for both WSJ and Genia, and achieves the best results on subtypes of NP, ADJP and S. Meanwhile, T is good at detecting VP. $Enju$ is good at detecting NP and VP, while S is good at detecting NP, VP and ADJP. These subtypes are not equally important. In corpora, NPs occupy more than 60%[4] and the proposed model shows a huge improvement on NP. This leads to an improvement in the overall performance.

4.4   Task II: NP Coordination Detection

NP coordinations occupy the largest proportion and we test the proposed model on the task of *NP coordination detection* as previous work did. This task is to detect NP coordinations and their scopes. We use Genia and divide coordinations into two categories, NP and Non-NP. We report the precision, recall and F1 using five-fold cross validation. S also [7] reported the micro-averaged results of five-fold cross validation on Genia. Table 8 shows the results. Though the settings and preprocessing are slightly different, we obtain improvements when tested on

---

[4] In Genia, the proportion is even higher.

the same data set. T [6] did not conduct this experiment. But from the result they reported in the general coordination detection task, which is 67.5 on NPs, we are able to conclude that the proposed model is among the best if not the best in detecting NP coordinations.

4.5 Analysis

Neural networks are more powerful at representing and calculating meaning than traditional models. It also becomes easy to encode features that are derived from parse trees [29]. Here we did not explore such features but will consider them in further work. We use LSTM to process information over extended time interval when using recurrent neural networks and is better at handling long distance relations than previous models. This is vital for coordination detection as this task heavily relies on long distance relations. In fact, LSTM has been proved useful in many other tasks for such a characteristic [26, 28].

We analyse the performances of the proposed model on the test data. Also we compare the proposed model with previous work on the development data because the previous work failed to provide details on performances on the test data.

In the analysis, we found that among all the coordinations, the proposed method achieves the best performance for NP.

One reason is that for NP coordinations, their conjuncts are not too long, thus allowing the recurrent LSTM model to learn relatively good representations for them. While conjuncts of coordinations of VP/S/PP are relatively too long for neural networks to deal with. For them, performances of the proposed model drop. We analyze the relation between performances and average lengths of coordinations of different categories. We found that generally for shorter coordinations, we obtain higher recalls. It is the same case with ADJP and UCP which are shorter than VP/S/PP. As we all know, neural networks learn very good distributed representations for words [18, 24, 30]. But for long text such as sentences or phrases, we are still faced with many difficulties [5, 12].

Long coordinations are hard to deal with. This is also consistent with the intuition that length has a positive correlation with the difficulty level of sentence analysis. For the remaining, such as ADVP, QP, SBAR, and Others, the data sets are too small to support any meaningful conclusions.

Besides, it also seems that distributed representations are better at capturing semantics for nouns than for other word classes. [19] show that for the word similarity task the performance on nouns is the best. [16] also presents a result which supports this conclusion. And in our work, we obtain better results for Nouns than others.

With the rapid development of neural network methods, more sophisticated models can be employed to obtain better sentence representations hence better results for the targeted task. The recursive neural network which uses parsing trees to generate text representations proves useful [13] provides a possible improvement. Besides, since coordination deals with relations between two pieces of text, it is possible to use conversation model to evaluate the quality of candidates [14,15]. Our future work will explore this point. This work verified the meaning of using ranking to explore imbalanced data to improve performances for classification tasks and is of help to future research.

5   Conclusion

We regard coordination detection as a ranking problem and use a neural network model leveraging recurrent LSTM to resolve it. Satisfactory results are obtained when experimented on Penn Treebank for both the coordination scope detection task and the NP coordination detection task. We also find that the proposed model which uses neural networks produces better results on short text and nouns.

Besides, our model does not require pre-defined grammars to generate candidate parse trees nor employ other complex features, hence it is easy to be applied to languages other than English. This is important as not all the languages enjoy plenty of ready-to-use tools as English does. In the future, we will further explore how to integrate our method with parsing to improve the performance of parsers.

References

1. Bikel, D.M.: Intricacies of collins' parsing model. Comput. Linguist. **30**(4), 479–511 (2004)
2. Buyko, E., Hahn, U.: Are morpho-syntactic features more predictive for the resolution of noun phrase coordination ambiguity than Lexico-semantic similarity scores? In: COLING, vol. 1, pp. 89–96. ACL (2008)
3. Buyko, E., Tomanek, K., Hahn, U.: Resolution of coordination ellipses in biological named entities using conditional random fields. In: PACLING, pp. 163–171 (2007)
4. Chantree, F., Kilgarriff, A., De Roeck, A., Willis, A.: Disambiguating coordinations using word distribution information. In: Proceedings of RANLP 2005 (2005)
5. Elman, J.L.: Distributed representations, simple recurrent networks, and grammatical structure. Mach. Learn. **7**(2–3), 195–225 (1991)
6. Hanamoto, A., Matsuzaki, T., Tsujii, J.: Coordination structure analysis using dual decomposition. In: EACL, pp. 430–438. ACL (2012)
7. Hara, K., Shimbo, M., Okuma, H., Matsumoto, Y.: Coordinate structure analysis with global structural constraints and alignment-based local features. In: ACL-AFNLP, vol. 2, pp. 967–975. ACL (2009)
8. Hogan, D.: Coordinate Noun Phrase Disambiguation in a Generative Parsing Model. ACL (2007)
9. Kawahara, D., Kurohashi, S.: Generative modeling of coordination by factoring parallelism and selectional preferences. In: IJCNLP, pp. 456–464 (2011)
10. Kim, J.D., Ohta, T., Tateisi, Y., Tsujii, J.: Genia corpurs-semantically annotated corpus for bio-textmining. Bioinformatics **19**(suppl 1), i180–i182 (2003)
11. Kurohashi, S., Nagao, M.: A syntactic analysis method of long Japanese sentences based on the detection of conjunctive structures. Comput. Linguist. **20**(4), 507–534 (1994)

12. Le, Q.V., Mikolov, T.: Distributed representations of sentences and documents. arXiv preprint arXiv:1405.4053 (2014)
13. Li, J., Luong, M.T., Jurafsky, D., Hovy, E.: When are tree structures necessary for deep learning of representations? arXiv preprint arXiv:1503.00185 (2015)
14. Li, J., Monroe, W., Jurafsky, D.: A simple, fast diverse decoding algorithm for neural generation. arXiv preprint arXiv:1611.08562 (2016)
15. Li, J., Monroe, W., Shi, T., Ritter, A., Jurafsky, D.: Adversarial learning for neural dialogue generation. arXiv preprint arXiv:1701.06547 (2017)
16. Lopyrev, K.: Learning distributed representations of phrases (2014)
17. Marcus, M.P., Marcinkiewicz, M.A., Santorini, B.: Building a large annotated corpus of English: the penn treebank. Comput. Linguist. **19**(2), 313–330 (1993)
18. Mikolov, T., Sutskever, I., Chen, K., Corrado, G.S., Dean, J.: Distributed representations of words and phrases and their compositionality. In: NIPS, pp. 3111–3119 (2013)
19. Mitchell, J., Lapata, M.: Composition in distributional models of semantics. Cogn. Sci. **34**(8), 1388–1429 (2010)
20. Miyao, Y., Tsujii, J.: Deep linguistic analysis for the accurate identification of predicate-argument relations. In: COLING, p. 1392. ACL (2004)
21. Nakov, P., Hearst, M.: Using the web as an implicit training set: application to structural ambiguity resolution. In: HLT-EMNLP, pp. 835–842. ACL (2005)
22. Nyblom, J., Kohonen, S., Haverinen, K., Salakoski, T., Ginter, F.: Predicting conjunct propagation and other extended stanford dependencies. In: Proceedings of the International Conference on Dependency Linguistics (Depling 2013), pp. 252–261 (2013)
23. Okumura, A., Muraki, K.: Symmetric pattern matching analysis for English coordinate structures. In: ANLP, pp. 41–46. ACL (1994)
24. Pennington, J., Socher, R., Manning, C.D.: GloVe: global vectors for word representation. EMNLP **14**, 1532–1543 (2014)
25. Schachter, P.: Constraints on coördination. Language, pp. 86–103 (1977)
26. Schmidhuber, J., Gers, F.A., Eck, D.: Learning nonregular languages: a comparison of simple recurrent networks and LSTM. Neural Comput. **14**(9), 2039–2041 (2002)
27. Shimbo, M., Hara, K.: A discriminative learning model for coordinate conjunctions. In: EMNLP-CoNLL, pp. 610–619. ACL (2007)
28. Sundermeyer, M., Schlüter, R., Ney, H.: LSTM neural networks for language modeling. In: INTERSPEECH, pp. 194–197 (2012)
29. Wang, X., Sudoh, K., Nagata, M.: Empty category detection with joint context-label embeddings. In: HLT-NAACL, pp. 263–271 (2015)
30. Wang, X., Sudoh, K., Nagata, M.: Enhanced word embeddings from a hierarchical neural language model. In: CIKM, pp. 1927–1930. ACM (2015)
31. Weston, J., Bengio, S., Usunier, N.: WSABIE: scaling up to large vocabulary image annotation. IJCAI **11**, 2764–2770 (2011)
32. Yoshimoto, A., Hara, K., Shimbo, M., Matsumoto, Y.: Coordination-aware dependency parsing (preliminary report). IWPT **2015**, 66 (2015)

Classifier Ensemble Approach to Dependency Parsing

Silpa Kanneganti[✉], Vandan Mujadia[✉], and Dipti M. Sharma[✉]

LTRC, International Institute of Information Technology, Hyderabad,
Hyderabad, India
{silpa.kanneganti,vandan.mujadia,dipti}@research.iiit.ac

Abstract. In this paper we propose a neural network based classifier voting approach to dependency parsing using multiple classifiers as component systems in an ensemble and a neural network algorithm as an oracle. We show significant improvements over the best component systems for both transition-based and graph-based dependency parsing. We also investigate different weighting schemes for voting among individual classifiers in the ensemble. All our experiments were conducted on Hindi and Telugu language data but the approach is language-independent.

1 Introduction

Transition-based and graph-based parsing models are two of the most dominant approaches in dependency parsing. Transition-based parsers learn a model conditioned on parse history to score transitions from one parser state to the next. It employs a greedy algorithm, by taking the highest-scoring transition out of every parser state until a complete dependency graph is achieved. Graph-based parsers, learn a model to score dependency graphs for a given sentence by factoring them into their component arcs and perform parsing by searching for the highest-scoring graph.

Both models have been used to achieve state-of-the-art accuracy on dependency parsing for a wide range of languages (Buchholz and Marsi 2006; Nivre et al. 2007). Nivre and McDonald 2008 proposes a method to integrate both the aforementioned models by letting the output of one guide the features for the other resulting in two stacked approaches, graph-based models guided by transition-based models and transition-based models guided by graph-based models. This approach is known to produce better accuracies than the isolated individual models.

While their approaches couldn't be more different from each other, both transition and graph-based models use single classifier based linear models to predict arcs or decisions for a given instance.

Recent studies suggest a classifier ensemble works better than a single classifier approach Dietterich 2000. If a learning algorithm can be viewed as searching a space H of hypotheses to identify the best suited one in them, below are few issues faced by a single classifier approaches.

© Springer Nature Switzerland AG 2018
A. Gelbukh (Ed.): CICLing 2017, LNCS 10761, pp. 158–169, 2018.
https://doi.org/10.1007/978-3-319-77113-7_13

1. Statistical problems arise when the amount of training data available is too small compared to the size of the hypothesis space. Without enough data, the learning algorithm may find many different hypotheses in H that all give the same accuracy on the training data.
2. Computationally many learning algorithms work by performing some form of local search that may get stuck in local optima. Even in cases where there is enough training data it may still be very difficult computationally for the learning algorithm to find the best hypothesis.
3. Representationally speaking in most applications of machine learning, the true function f(classifier or regression function) cannot be represented by any of the hypotheses in H.

Hence the space H needs to be an effective space of hypotheses searched by the learning algorithm for a given training data set. Ensemble methods have the promise of reducing these three key shortcomings of standard learning algorithms.

In this paper as an alternative to the aforementioned single classifier approaches, we propose an ensemble method with several learners (multiple linear models) whose individual predictions are combined into a voting mechanism. Instead of using a majority based voting approach, to choose the best prediction, we train a neural network algorithm to predict the best classifier model given a feature vector. All the classifier models in the ensemble learn on the same set of train instance to produce predictions which are then validated against the gold standards values. These validations are then used as to train a neural network algorithm to learn the best possible performing classifier model for a given instance.

The key idea is to combine a number of classifiers such that the resulting system achieves higher classification accuracy and efficiency than the original single classifier models. Diversified multiple classifiers trained by different classifier parameters over the same train data prove to be more efficient (Kittler et al. 1998). Hence, for each model we pick varied classifier models to get the best of each of them which helps reduce classifier related errors while confining to practical time constraints. Through experiments we show that our proposed ensemble model reports higher performances than the current state of the art single classifier models. All our experiments are performed on Hindi and Telugu language data.

2   Background and Related Work

Hindi is an Indo-Aryan language with richer morphology as compared to English. It exerts a relatively free word order with SOV being the default configuration. Due to the flexible word order, dependency representations are preferred over constituency for its syntactic analysis (Bharati and Sangal 1993).

Telugu, a morphologically,syntactically complex language, is highly inflectional and agglutinative. It is a nominative-accusative language, with SOV as its default word order where the verbs exhibit a rich inflectional morphology.

Hence it encodes various grammatical categories like tense, case, gender, number, person, negatives, imperatives etc. The dependency grammar formalism, used for both the languages is Computational Paninian Framework (CPG) (Begum et al. 2008; Bharati et al. 2009). The data set we use for both the languages is from the ICON10 parsing contest (Husain et al. 2010). For the purpose of this work we only deal with inter-chunk dependency trees.

Previous work on parser ensembling was based on models where, integration takes place at parsing time as well as at learning time, and requires at least three different base parsers. Hall et al. 2007 combines six transition-based parsers and is so far the best performing system. Nivre and McDonald 2008 integrates the two parser models by allowing the output of one define features for the other. Feature-based integration performed by McDonald et al. 2006 to substantial improvements in accuracy, lets a subset of the features for one model be derived from the output of a different model. In addition, feature-based integration has been used by Taskar et al. 2005, who train a discriminative word alignment model using features derived from the IBM models, and by Florian et al. 2004, who trained classifiers on auxiliary data to guide named entity classifiers. Collins 2000 perform perser re-ranking, where one parser produces a set of candidate parses and a second stage classifier chooses the most likely one. However, feature-based integration since is not explicitly constrained to any parse decisions that the guide model might make, is more efficient than parser re-ranking. Nakagawa 2007 and Hall 2007 try to add global features to overcome the limited feature scope of graph-based models. Titov and Henderson 2007a,b try to reduce error propagation, by performing beam search with globally normalized models for scoring transition sequences.

3   Proposed Model

In this section we describe in detail the proposed a neural network based classifier voting approach to dependency parsing.

3.1   Feature Representation

As explained in Sect. 3, all the models essentially learn a scoring function s : $X \rightarrow R$, where the domain X is different for the two models. While for the transition-based model, X is the set of possible configuration-transition pairs (c, t), for the graph-based model, X is the set of possible dependency arcs (i, j, l); But in both cases, the input is represented by a k-dimensional feature vector f : $X \rightarrow R \ k$. For transition based approach, we use the feature models described in Nivre et al. 2007; For the graph-based models, we use the feature vectors defined in Husain et al. 2010 and for integrated models we use the feature vectors from Nivre and McDonald 2008.

3.2   Ensemble Selection

Different kind of classifiers (i.e. independent, informed, diverse, etc) pick up different patterns in the data. Diversity, accuracy and run time are the three significant factors taken into account while picking the classifier ensemble. A necessary and sufficient condition for an ensemble of classifiers to be more accurate than any of its individual members is if the classifiers are accurate and diverse (Hansen and Salamon 1990. Two classifiers are diverse if they make different errors on new data points. In addition to this the selection and the number of classifiers used in the model are also confined by the total run time of the parser.

While the efficiency of the model keeps improving with the number of classifiers in the ensemble, keeping the aforementioned parameters in mind and various experiments, we picked 4 diverse classifiers each for both transition and graph based models without compromising on accuracy or run time of the parsers. While, Stochastic gradient descent and Linear ridge regression are common for both models, we use support vector machines (SVM) and Random Forest Tree classifier for transition-based models and Maximum Entropy and Decision Tree classifiers for graph-based models. We chose SVM for transition-based parsing for its proven high accuracy (Nivre et al. 2007), its theoretical guarantees to over-fitting, for higher dimension and non-linear data. We choose Maximum Entropy classifier for graph based parsing because of its efficiency with conditionally dependent data. We choose Decision Tree Classifier because they easily handle feature interactions. They are non-parametric, so one doesn't have to worry about outliers or whether the data is linearly separable. We picked random dom forest tree for its is non-parametric and easily handle feature interactions with less concern for outliers.

We picked linear ridge regression because features co-relation is not necessary, the coefficients of linear transformation are normal distributed and the model is easily interpretable to analyze outputs, stochastic gradient descent because it is very efficient in discriminative learning of linear classifiers under convex loss functions. It is successful in sparse and large-scale learning because it is easier to scale sparse data. It has proven to be very successful in NLP tasks where large number of unique features are possible. We picked Multi-layer perceptron as the neural network classifier because of its ability to give more well structure blocks of layers to derive useful patterns from input data and its remarkable capability at deriving meanings (complex patterns) from imprecise data which may be too complex for other learning techniques.

3.3   Voting/Ensemble Function

Given a feature vector, the output of each classifier is evaluated against the gold annotated output and corresponding binary values of 1 if the output is accurate and 0 if not is assigned to each vector. The resulting binary values are the mapped against their corresponding input feature vectors to create the training data for the Neural network. In order to handle redundancy and confusion, while creating the training data for the neural network, we ignore the instances where all the classifier predict the same value (0,1). This we believe creates a better learning model. We also added the corresponding accuracies of each of the classifiers as prior weights into the training instances for neural network. The neural nets are then trained on this data to produce the classifier method that works best for a given feature vector in the test data. Based on these predictions, we use the output of the respective model to predict the output.

Table 1 shows the example validation(nn_train data) data used to train the neural network model.

Table 1. Accuracies

	Classifier 1	Classifier 2	Classifier 3	Classifier 4	nn prediction
Feat 1	1	0	1	0	1
Feat 2	1	0	1	1	3
Feat 3	0	1	1	1	4
Feat 4	1	0	1	1	1

Given a set of training examples, $S = (f(x_1), y_1), (f(x_2), y_2)...., (f(x_n), y_n)$ where $f(x_i)$ is the feature vector and y_i their corresponding predictions, the proposed combined model C_n where n is the number of classifiers, produces a set of predictions $P_i = y_{i1}, y_{i2}, ...y_{in}$ for a given feature vector $f(x_i)$ and are used to train the neural network (MLP-Multi layer perception) which learns to assigns scores to each of the classifiers based on their performance on the validation data. Given an test instance $f(x_j)$, the resulting learning algorithm picks one of the classifiers in the ensemble c_n, as the possible model that works best for $f(x_j)$. The data is divided into 3 parts:

- Train data: Used to train the feature vectors of the classifier models.
- NN_train data: Used to train the neural network (MLP) model.
- Test data: Used to test the model

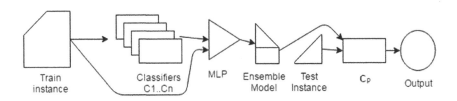

Fig. 1. Working of ensemble function C = Classifier, MLP = Multi Layer Perception

Figure 1 details the working of the proposed model, where the upper part shows the training of each of the classifiers in the ensemble on training instances and the predictions being passed to a Neural Network Model (MLP) as training data to help learn to decide the classifier model that works best for a given instance. The below part of the figure, shows the use of MLP to predict the best classifier (c_p) for a given test instance which in turn makes the final prediction.

We have used tensorflow software library (Abadi et al. 2015) for neural network implementation with 6 layered MLP with 4 hidden layers and they are consist of m,m/3,m/9,m/27 neurons respectively We have used RMSPropopti-mizer[1] from tensorflow to minimize our objective function.

4   Experiments and Results

In this section, we present an experimental evaluation of the all the four afore-mentioned models. We conduct our experiments on Hindi and Telugu language data provided by Husain et al. 2010 respectively. Due to data sparsity, 2 fold cross validation is done on the integrated models while 5 fold cross validation is done on transition and graph-based models. For transition-based and graph-based models,

- The 1452 annotated sentences in Telugu treebank data, are divided into 870 sentences of train data, 292 sentences of NN_train data and 290 of test data.
- Of the 19254 sentences, available in Hindi treebank data, 15404 sentences are used as train data, 3850 sentences are used as nn_train data and 3850 sentences as test data.
- For the integrated models, 400 sentences are used to train each of the graph-based and transition based models, 200 sentences each to train the respective neural network models and 252 sentences are used as test data.
- Of the 19254 sentences in HDTB data, 5000 sentences each are used to train the transition and graph based models, 3000 sentences each are used to train the neural network models and 3254 sentences are used as test data.

4.1   Transition Based Model

Transition based parsing systems use a model parameterized over transitions, such that every transition sequence from the designated initial configuration to some terminal configuration derives a valid dependency graph. The set of training instances for the learning problem are pairs (c, t) such that t is the correct transition out of c in the transition sequence that derives the correct dependency graph for some sentence x in the training set T. Each training instance (c, t) is represented by a feature vector f(c, t), where features are defined in terms of arbitrary properties of the configuration c, including the state of the stack, the input buffer and the partially built dependency graph.

Many features involve properties of the two target tokens that could be connected by an edge, the token on top of the stack and the first token in the input buffer. The full set of features used by the base model for Hindi is described in Husain et al. 2010. We use an implemented version of arc eager algorithm by

---

[1] http://www.cs.toronto.edu/~tijmen/csc321/slides/lecture_slides_lec6.pdf.

Bhat et al. 2016 and support vector machines as the base linear model to learn transition scores. In the proposed model, we use multiple discriminative methods to predict transitions over the same set of configurations. The validity vectors of resulting transitions are then fed to the neural network algorithm which inturn predicts the transitions of the test feature configurations. The data is divided into 3 parts, data to train the classifier models, data to train the neural network model, test data.

Table 2. Accuracies

Classifier Model	Hindi			Telugu		
	LAS	UAS	LS	LAS	UAS	LS
Support Vector Machines (Base model)	**78.29%**	**85.36%**	**80.02%**	**62.17%**	**70.35%**	**64.55%**
Random Forest Tree	77.58%	85.03%	79.71%	61.50%	69.71%	63.74%
Stochastic Gradient Descent	76.64%	84.56%	78.5%	61.71%	69.97%	63.23%
Linear Regression (RidgeCV)	76.14%	84.22%	78.63%	62.10%	70.25%	64.58%
Ensemble model	**79.04%**	86.16%	81.63%	**62.66%**	**71.03%**	**64.98%**

Table 2 shows the accuracies[2] of best accuracies for each of the classifiers and the ensemble model.

4.2   Graph Based Model

Graph-based dependency parsers parameterize a model over smaller substructures in order to search the space of valid dependency graphs and produce the most likely one. The simplest parameterization is the arc-factored model that defines a real-valued score function for arcs $s(i, j, l)$ and further defines the score of a dependency graph as the sum of the score of all the arcs it contains. The specific graph-based model studied in this work is that presented by McDonald et al. 2005, which factors scores over pairs of arcs (instead of just single arcs) and uses near exhaustive search for unlabeled parsing coupled with a separate classifier to label each arc. We use Maxent as the base line classifier with the settings suggested in Husain et al. 2010. Since the ensemble model is used only to predict the labels of predicted trees, there is no change the Unlabeled Attachment score (UAS) and hence the Labeled Attachment score (LAS) and Labeled Accuracy (LA) are the same.

Table 3 shows the best accuracies for each of the classifiers as well as the ensemble model. The **UAS** on graph based models is same for all the models. This can be attributed to the fact that these classifier models are used only to

---

[2] LAS: Labeled Accuracy Score; UAS: Unlabled Accuracy Score; LS: Labeled Accuracy score.

Table 3. Accuracies

Classifier model	Hindi			Telugu		
	LAS	UAS	LS	LAS	UAS	LS
Maxent (Base model)	**78.63%**	**85.94%**	**78.63%**	**62.34%**	**70.93%**	**62.34%**
Decision Tree	77.34%	85.94%	77.34%	61.90%	70.93%	61.90%
Stochastic Gradient Descent	77.05%	85.94%	77.05%	61.21%	70.93%	61.21%
Linear Regression (RidgeCV)	78.64%	85.94%	78.64%	62.82%	70.93%	62.82%
Ensemble Model	**79.42%**	**85.94%**	**79.42%**	**62.92%**	**70.93%**	**62.92%**

predict the labels while the trees predicted my the parser remain constant for all of them.

4.3   Guided Transition Based Model

Guided transition-based models follow the same perspective as transition-based models but for modified feature configurations. The basic training instances of a transition-based model are extended to add the features predicted by the graph-based parser for a given sentence. We therefore use the same classifier settings and approach we used for the transition-based parsing model discussed in Sect. 4.1. The data is divided into 5 parts. Data used to train the base model, the guide model, neural network models of the base and guide models as well as the test data run by the base model. The full set of features used by the guided transition-based model are described in Nivre and McDonald 2008.

Table 4. Accuracies

Classifier model	Hindi			Telugu		
	LAS	UAS	LS	LAS	UAS	LS
Support Vector Machines (Base model)	**71.09%**	**79.15%**	**73.3%**	**55.59%**	**63.81%**	**57.24%**
Random Forest Tree	70.46%	78.92%	72.71%	54.02%	62.71%	56.19%
Stochastic Gradient Descent	70.53%	78.02%	72.18%	54.71%	62.97%	56.23%
Linear Regression (RidgeCV)	71.21%	79.43%	73.65%	55.10%	63.25%	57.58%
Ensemble model	**72.14%**	**80.21%**	**74.13%**	**56.16%**	**64.03%**	**58.98%**

Table 4 shows the best accuracies for each of the classifiers as well as the ensemble model.

4.4   Guided Graph Based Model

Guided graph-based models follow the same perspective as graph-based models but for modified feature vectors. The basic training instances of a graph-based model are extended to add the features predicted by the transition-based parser for a given sentence. We therefore use the same classifier settings and approach we used for the graph-based parsing model discussed in Sect. 4.2. The data is used as discussed in Sect. 4.3. The full set of features used by the guided graph-based model are discussed in (Nivre and McDonald 2008). Since the ensemble model is used only to predict the labels of predicted trees, there is no change the Unlabeled Attachment score (UAS). Similar to the graph-based models, since the ensemble model is used only to predict the labels of already parsed trees, there is no change the Unlabeled Attachment score (UAS) and hence the Labeled Attachment score (LAS) and Labeled Accuracy (LA) are the same.

Table 5. Accuracies

Classifier model	Hindi			Telugu		
	LAS	UAS	LS	LAS	UAS	LS
Maxent(Base model)	**71.56%**	**79.86%**	**71.56%**	**55.83%**	**64.25%**	**55.83%**
Decision Tree	70.44%	79.86%	70.44%	54.71%	64.25%	54.71%
Stochastic Gradient Descent	70.26%	79.86%	70.26%	54.38%	64.25%	54.38%
Linear Regression (RidgeCV)	71.94%	79.86%	71.94%	56.13%	64.25%	56.13%
Ensemble model	**72.44%**	**79.86%**	**72.44%**	**56.48%**	**64.25%**	**56.48%**

Table 5 shows the best accuracies for each of the classifiers and the ensemble model. Similar to the graph based models in Sect. 4.2, given that the models are only predicting dependency labels, the tree structures between them remain the same. Hence **UAS** is common among all of them.

5 Observations

One of the main assumptions in using diverse classifiers while choosing the ensemble is to learn to handle the cases where output generated by those models may differ. The notion that agreement between the models, an indication of correctness might create confusion for the neural network to learn in this scenario worked in our favor. Experiments with training the neural network with only the instances where atleast 2 classifiers disagree proved to be more helpful than all the samples in case of Hindi data for all the models. Telugu data on the other hand shows improvements only for transition and graph based model. For integrated models, due to scarcity of already split data in Telugu, no training instance could be ignored.

In all the models, assigning prior weights to classifiers based on their individual performances has proven to get the best results for both the languages. We also observed that a random initial weight settings to classifier voting for neural networks performed surprisingly well, although not better than performance based weight assignment. It is interesting to note that although the baseline

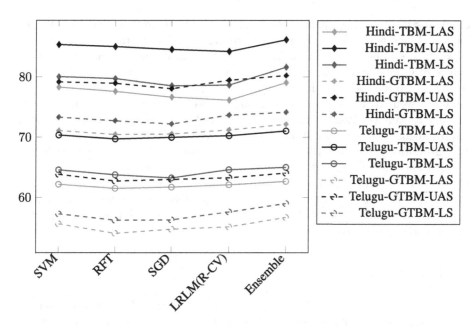

Fig. 2. accuracy graph, TBM = Transition based model; GTBM = Guided Transition based model, LAS = Label attachment score; UAS = Unlabeled attachment score; LS = Label accuracy score

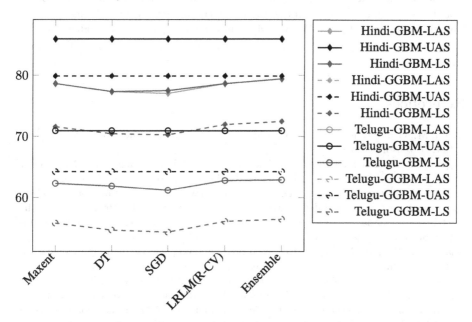

Fig. 3. accuracy graph, GBM = Graph based model; GGBM = Guided Graph based model, LAS = Label attachment score; UAS = Unlabeled attachment score; LS = Label accuracy score

classifier SVM is giving the best results compared to the rest in transition-based models, Liner regression model is performing better than the baseline model Maxent for graph-based models. Analysis on the results suggest that the performance of ensemble models are heavily dependent on the size of the data sets. In the integrated models a huge chunk of the data goes into training the guide model as well as the corresponding neural network models. Hence the drop in baseline as well as ensemble accuracies compared to the simple parsers. Also the improvement accuracies for Hindi data is higher than Telugu on all the models. We also noticed that performance enhancement for an ensemble comes with the first few classifiers combined peaking at a threshold even with increase in ensemble sizes. In conclusion, as a general technique with the right pick of classifier models and the weight assignment, ensemble models may produce larger gains in accuracy. Figures 2 and 3 show the plotted statistics of transition and graph based models respectively.

6 Conclusion

The idea of ensemble modeling in parsing is not new. While work on parser ensembling usually makes use of a voting strategy of some kind in order to derive a single prediction for an output, we deal with classifier ensembles within a parser. Our work differs from that of the parser ensembles in that integration is done during learning of the parse models. In addition to the simple parse models we also show our technique improves the integrated models described in Nivre and McDonald 2008. We also show how choosing diverse classifiers help in expanding to cover more hypotheses space resulting in reduced classifier based errors.

References

Abadi, M., et al.: TensorFlow: Large-scale machine learning on heterogeneous systems (2015). tensorflow.org

Begum, R., Husain, S., Dhwaj, A., Sharma, D.M., Bai, L., Sangal, R.: Dependency annotation scheme for Indian languages. In: IJCNLP, pp. 721–726. Citeseer

Bharati, A., Husain, S., Misra, D., Sangal, R.: Two stage constraint based hybrid approach to free word order language dependency parsing. In: Proceedings of the 11th International Conference on Parsing Technologies, pp. 77–80. Association for Computational Linguistics (2009)

Bharati, A., Sangal, R.: Parsing free word order languages in the paninian framework. In: Proceedings of the 31st annual meeting on Association for Computational Linguistics, pp. 105–111. Association for Computational Linguistics (1993)

Bhat, R.A., Bhat, I.A., Sharam, D.M.: Improving dependency parsing of Hindi and Urdu by modeling syntactically relevant phenomena. ACM Transactions on Asian and Low-Resource Language Information Processing (2016). (Under review)

Buchholz, S., Marsi, E.: Conll-x shared task on multilingual dependency parsing. In: Proceedings of the Tenth Conference on Computational Natural Language Learning, pp. 149–164. Association for Computational Linguistics (2006)

Collins, M.: Discriminative reranking for natural language parsing. In: Proceedings of the Seventeenth International Conference on Machine Learning, ICML 2000, pp. 175–182. Morgan Kaufmann Publishers Inc., San Francisco (2000)

Dietterich, T.G.: Ensemble methods in machine learning. In: Kittler, J., Roli, F. (eds.) MCS 2000. LNCS, vol. 1857, pp. 1–15. Springer, Heidelberg (2000). https://doi.org/10.1007/3-540-45014-9_1

Florian, R., Hassan, H., et al.: A statistical model for multilingual entity detection and tracking. Technical report, DTIC Document (2004)

Hall, J., Nivre, J., Nilsson, J.: A hybrid constituency-dependency parser for Swedish. In: Proceedings of NODALIDA, pp. 284–287 (2007)

Hall, K.: K-best spanning tree parsing. In: Annual Meeting-Association for Computational Linguistics, vol. 45, p. 392 (2007)

Hansen, L.K., Salamon, P.: Neural network ensembles. IEEE Trans. Pattern Anal. Mach. Intell. **12**, 993–1001 (1990)

Husain, S., Mannem, P., Ambati, B., Gadde, P.: The icon-2010 tools contest on Indian language dependency parsing. In: Proceedings of ICON-2010 Tools Contest on Indian Language Dependency Parsing, ICON, vol. 10, pp. 1–8 (2010)

Kittler, J., Hatef, M., Duin, R.P., Matas, J.: On combining classifiers. IEEE Trans. Pattern Anal. Mach. Intell. **20**(3), 226–239 (1998)

McDonald, R., Crammer, K., Pereira, F.: Online large-margin training of dependency parsers. In: Proceedings of the 43rd Annual Meeting on Association for Computational Linguistics, pp. 91–98. Association for Computational Linguistics (2005)

McDonald, R., Lerman, K., Pereira, F.: Multilingual dependency analysis with a two-stage discriminative parser. In: Proceedings of the Tenth Conference on Computational Natural Language Learning, pp. 216–220. Association for Computational Linguistics (2006)

Nakagawa, T.: Multilingual dependency parsing using global features. In: EMNLP-CoNLL, pp. 952–956. Citeseer (2007)

Nivre, J., et al.: MaltParser: a language-independent system for data-driven dependency parsing. Nat. Lang. Eng. **13**(02), 95–135 (2007)

Nivre, J., McDonald, R.T.: Integrating graph-based and transition-based dependency parsers. In: ACL, pp. 950–958 (2008)

Taskar, B., Lacoste-Julien, S., Klein, D.: A discriminative matching approach to word alignment. In: Proceedings of the conference on Human Language Technology and Empirical Methods in Natural Language Processing, pp. 73–80. Association for Computational Linguistics (2005)

Titov, I., Henderson, J.: Constituent parsing with incremental sigmoid belief networks. In: Annual Meeting-association For Computational Linguistics, vol. 45, p. 632 (2007a)

Titov, I., Henderson, J.: Fast and robust multilingual dependency parsing with a generative latent variable model. In: EMNLP-CoNLL, pp. 947–951 (2007b)

Evaluation and Enrichment of Stanford Parser Using an Arabic Property Grammar

Raja Bensalem Bahloul[1], Nesrine Kadri[1], Kais Haddar[1(⊠)],
and Philippe Blache[2]

[1] Multimedia InfoRmation Systems and Advanced Computing Laboratory,
Higher Institute of Computer Science and Multimedia, Sfax, Tunisia
raja_ben_salem@yahoo.com, knesrine@hotmail.fr,
kais.haddar@yahoo.fr
[2] Laboratoire Parole et Langage, Université de Provence,
Aix-en-Provence, France
pb@lpl.univ-aix.fr

Abstract. So far, the Stanford Arabic statistic parser is considered as the best parsing tool in terms of performance compared to other parsers. This performance is not stable and may vary depending on the given corpus. A more detailed method to evaluate this parser may help the users to address the causes of a performance loss. We propose, for this reason, to evaluate the Stanford Parser using the verification of the satisfaction of the syntactic constraints (called, properties) based on the analysis results of the corpus. We may obtain these properties from a reference Arabic property grammar. By the way, we enriched the simple representation of the parsing result with syntactic properties. This allows to explicit several implicit information that are the relations between syntactic units. Therefore, we had both a detailed method for the evaluation of parsers and a more syntactically informative representation for the analysis. We obtained widely detailed and encouraging results.

Keywords: Statistic parser · Property grammar · Evaluation · Enrichment
Arabic language

1 Introduction

The Stanford Arabic statistic parser [15] is considered as the best parsing tool in terms of performance compared with the other parsers. This performance is not stable and may vary depending on the given corpus. A more detailed method to evaluate this parser may help the users to addresses the causes of a performance loss. Verifying on the Parsed Corpus (PC) the satisfaction of the Property Grammar (GP) [6] constraints could be a powerful alternative. However, its development may face many difficulties. The technical ones concern the choice of computer platforms and the appropriate format of the PC. The linguistic difficulties consist of the unconformity between the GP and the PC tagsets and the choice of the evaluation scores.

According to this alternative, we propose, in the present paper, to evaluate the Stanford Parser using the verification of the satisfaction of the syntactic constraints

© Springer Nature Switzerland AG 2018
A. Gelbukh (Ed.): CICLing 2017, LNCS 10761, pp. 170–182, 2018.
https://doi.org/10.1007/978-3-319-77113-7_14

(called, properties) based on the parsing result (the PC). We may obtain these properties from a reference Arabic GP [4]. By the way, we enriched the simple representation of the parsing result with syntactic properties. This allows to explicit several implicit information that are the relations between syntactic units. Therefore, we had both a detailed method for the evaluation of parsers and a more syntactically informative representation for the analysis.

This paper is organized as follows: Sect. 2 is devoted to explain our work keywords and to present some related works. Section 3 describes our evaluation and enrichment approach. Section 4 presents the experimental results and discussions. Section 5 gives a conclusion and some perspectives.

2   Related Work

The Stanford parser, the Property Grammar (GP) and the parser evaluation are the main concepts that we use in our contribution. Let us define and give brief overviews about them separately:

2.1   The Stanford Parser

The Stanford parser is a statistical syntactic analyzer [15], which is adapted to work with many languages (e.g. English, Arabic, Chinese, German). For Arabic, it had trained its lexicalized PCFG model from the Penn Arabic treebank (ATB) [11]. In its new versions, the Stanford parser model adds a representation of lexical dependencies, whose preferences are combined with efficient exact inference, using an A* algorithm. This representation is a simple, uniform and quite accessible to non-linguists. It is a description of the grammatical relationships in the given sentence (e.g. the adjectival complement, the adverb modifier, the direct object).

The Stanford Parser is exploited in several NLP search domains. For head recognition, the work of [8] uses the Chinese Stanford parser, to train on the data of Tsinghua Chinese Treebank (TCT) into three models: the parsing model, the POS model, and the head recognition model. The sequences are labeled with a CRF tagging. They obtained an accuracy of 77% for head recognition and 94.82% for Pos-tagging. Seraji et al. developed the Persian corpus UPEC, an improved version of the Bijankhan corpus [14]. They also derived a treebank of 215 annotated sentences from UPEC with an annotation scheme based on Stanford typed dependencies and with an accuracy of 97.8%.

For Arabic, the language that interests us the most, we may find different application areas of the Stanford parser. Indeed, for the Author attribution problem, a classification issue, [2] generates in its training phase an extended PCFG Language Model (XPCFG) from training Arabic texts. First, the Stanford parser is run to obtain annotated texts, from which production rules are induced. The extended model is based on probabilities and scores calculated in terms of the production rules and the terminals/non-terminals. In the Arabic to English Machine Translation System of [1], the Stanford Parser is used to detect the constituent structures of the language source (Arabic). The Stanford dependencies are exploited to resolve some syntactic ambiguities, which reduces the ambiguity of choosing the word meanings. Even in the question-answering domain, the Stanford parser is used to analyze morphologically and syntactically each question [18]. This allows the extraction of the question portion that is a referring to the answer (the focus), the focus head, and the modifiers of the focus head. In the modeling, the Stanford Parser is used in the generation process of the use case models from Arabic user requirements. The Stanford parser analyzes the latter to obtain the categories and the phrasal structures that aid in finding the potential actors and the use cases [3].

2.2 The Property Grammar (GP)

The GP formalism [6] represents the linguistic information as local and decentralized constraints, so-called properties. These properties express the relations that may exist between the categories of the described syntactic structure. The syntactic properties in particular have six types: the linearity ($\prec$), the requirement ($\Rightarrow$), the exclusion ($\otimes$), the dependency ($\rightsquigarrow$), the uniqueness (Unic) and the obligation (Oblig).

Several works benefited from this formalism grammar such as the contribution [7] which gives a syntactic representation that enriches the French treebank FTB after inducing a property grammar. This representation is used then in a hybridization approach of a symbolic control and a probabilistic parsing. This approach is based on heuristics (weights) that are calculated on the occurrences of the satisfied properties. Bensalem et al. [4] developed also the same induction technique to generate an Arabic GP from the Arabic treebank ATB and then to enrich the ATB with syntactic properties thanks to a formal modeling method [5]. Before that, Duchier et al. presents a formal semantic definition of the GP [9]. They apply this definition to model GP parsing as a constraint satisfaction problem. In 2012, [10] proposes an extension from this model to process new property types. This extension transforms the syntactic relations on feature structures. By contrast, [13] built a GP based parser that produces the best constituent structure out of a given grammatical or ungrammatical sentence. However, Vanrullen et al. [17] formulates mathematically the GP formalism to control the parsing granularity.

2.3 The Parser Evaluation Tools

The choice of the appropriate parsing system depends on its accuracy, which is calculated by the evaluation tools. We present in this sub-section some freely available ones:

- Evalb[1] provides the precision, the recall, and the number of crossing brackets. It also reports the accuracy of the POS tagging. It calculates these scores separately on each sentence by comparing the gold standard analysis to the parses in the parser output.

---

[1] http://nlp.cs.nyu.edu/evalb/ (of Sekine, S. and Collins, M. in 2006)

- The metric GramRelEval[2] is based on the annotation scheme of the grammatical relations. It calculates also the precision, the recall and the F-measure for each grammatical relation and gives a confusion matrix of these relations.
- The package incr_tsdb [12] produces metrics the parameter settings such as the time and the memory consumption and the number of the successfully parsed sentences.

The evaluation method that we propose is original because it is not used to train new parsed texts but to evaluate the parsing result by verifying the satisfaction of the GP constraints different. The use of GP to evaluate parser is also an original idea.

3   Approach of the Enrichment and the Evaluation of the Stanford Parser

Our approach is based on a three-step method: the syntactic analysis of a plain Arabic corpus using the Stanford Parser, the enrichment of the parsing result with syntactic properties using an Arabic reference GP and the evaluation of the enriched parsing (Fig. 1).

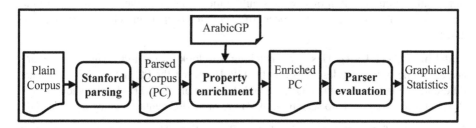

Fig. 1. Enrichment and evaluation of the Stanford Parser

3.1   Stanford Parsing

Before running the Stanford parser. The input corpus needs to be splitted and lexically annotated. In the following, three sub-steps should be applied:

1. **Splitting:** to segment the plain Arabic corpus into a sentence list.
2. **Morphological analysis:** to separate the agglutinated words to tokens and then to tag (to assign) the tokens the appropriate lexical categories (Part-Of-Speech tags).
3. **Syntactic analysis:** to generate the parse trees of the lexically annotated corpus.

3.2   Enrichment

This step is based on the enrichment method of the French treebank FTB [7]. It consists of four sub-steps shown in Fig. 2. Indeed, after converting the PC to an XML format,

---

[2] http://www.informatics.susx.ac.uk/re-search/nlp/carroll/greval.html (of Carroll, J. in 2006).

we run a loop task of the following sub-steps: matching each PC syntactic category to the GP one, verifying the GP properties and their integration in the PC to obtain the enriched one. They are more precisely explained below.

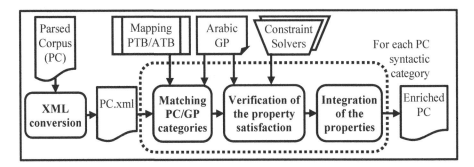

Fig. 2. Sub-steps of the parsed corpus enrichment with syntactic properties

XML Conversion. The PC needs to be converted to an xml format, as the data structure should be able to host the syntactic properties that will enrich the PC. The xml format allows the insertion of new tags unlike the actual parentheses format. For that, we made a conversion recursive process of encountered open and close parentheses to xml tags.

Matching PC/GP Categories. The matching is to look for each PC syntactic category and the property list that describes it in the GP. However, the PC and the GP tagsets are different; one is based on the PTB [16] and the other on the ATB [4]. We should use a PTB/ATB mapping table to search the ATB tag of the PC syntactic category in the GP.

Verification of the Satisfaction of the Properties. This is the most important step. For each PC syntactic category found in the GP, the GP properties that describe this category are verified on its phrasal structure. We used for that a set of methods to check the property satisfaction. Each method, so-called "constraint solver", verifies if a given PC phrasal structure tagged with a syntactic category respects a given GP property that describes this syntactic category. The solution produced by a solver is the result of this verification (satisfied or not). As these constraint solvers are the heart of the enrichment step, we have chosen to devote the next sub-section to introduce their descriptions.

Integration of the Verified Properties. This task adds to each phrasal structure in the PC the verification result of the properties that describe its syntactic category (either satisfied or not). New tags in the xml block of this phrasal structure will enrich the XML format of the PC. Therefore, the Stanford parser result and the GP are combined.

3.3   Descriptions of the Constraint Solvers

We ran these solvers in the step of the verification of the GP property satisfaction. We used for that the Arabic GP of [4]. In the following, we present the descriptions of these solvers in terms of set operations and we give some significant Arabic examples.

We note, first, L the ordered list of the constituents (grammatical categories of the PTB tagset) realized in the current phrasal structure of the PC. atb_tag(L) is the syntactic category XP that specifies L according to the ATB tagset. XP is described in the GP with different types of syntactic properties; c_i is a constituent of L at the position i; $L_{i..j}$ the sublist of L between the positions i and j; |S| is the cardinality of the set S. The GP syntactic properties can be unary or binary. The unary ones consists of constituent sets (constituency, obligation and uniqueness). Indeed the set C(XP) (resp. O(XP) and U (XP)) gathers the constituency (resp. obligation and uniqueness) properties of XP in the GP. The binary properties specify relations sets (linearity, requirement and exclusion). Each relation consists of two constituents x and y (x is in its left-hand side (lhs) and y is in its right-hand side (rhs)). We give the following Arabic sentence, parsed with Stanford to explain the application of our constraint solvers (Fig. 3):

```
<Category index="14" label="S">
 <Category index="14:0" label="VP">
  <Category index="14:0:0" label="VBD" value="انتهى" />
  <Category index="14:0:1" label="NP">
   <Category index="14:0:1:0" label="DTNN" value="اليوم" />
   <Category index="14:0:1:1" label="ADJ_NUM" value="الاول" />
  </Category>
  <Category index="14:0:2" label="PP">
   <Category index="14:0:2:0" label="IN" value="من" />
   <Category index="14:0:2:1" label="NP">
    <Category index="14:0:2:1:0" label="DTNNS" value="المفاوضات" />
    <Category index="14:0:2:1:1" label="DTJJ" value="السورية" />
   </Category>
  </Category>
 </Category>
 <Category index="14:1" label="PUNC" value="." />
</Category>
```

Fig. 3. Stanford parsing result of the sentence "انتهى اليوم الأول من المفاوضات السورية" (AnthY Alywm Al > wl mn AlmfAwDAt Alswryp/The first day of the Syrian negotiations is over)

The Solver of Constituency Properties

This solver verifies that all the categories realized in a given phrasal structure from the PC belongs to the constituent set of its GP syntactic category. In terms of set operations, the constituency solver can be seen as follows:

$$\textbf{if } (\forall c_i \in L \text{ \& } atb\_tag(L) = XP) \textbf{ then } atb\_tag(c_i) \in C(XP)$$

For each category c_i in the category structure L which is tagged with the ATB category XP, the ATB tag of c_i should belong to the XP constituent set C(XP). For example:

$L = DTNNS\ DTJJ; \mathbf{atb\_tag(L)} = NP; \mathbf{atb\_tag(c_1)} = DET + NOUN; \mathbf{atb\_tag(c_2)} = DET + ADJ;$
$C(NP) = \{NOUN, DET + ADJ, ADJ\_COMP, DET + NOUN, SBAR, \ldots\};$
$\mathbf{atb\_tag(c_1)} \in C(NP) \text{ and } \mathbf{atb\_tag(c_2)} \in C(NP)$

Thus, the ATB tags DET+NOUN and DET+ADJ of the constituents DTNNS and DTJJ of L are constituents in NP. The constituency properties of NP are satisfied.

The Solver of the Obligation Properties
This solver checks the presence in a given phrasal structure from the PC of one of the obligatory constituent set of its GP syntactic category. In terms of set operations, the obligation solver can be seen as follows:

$$\textbf{if } \textbf{L\_atb} = \{\textbf{atb\_tag}(c_i)|\forall c_i \in \textbf{L}\} \textbf{ then } |\textbf{O}(\textbf{XP}) \cap \textbf{L\_atb}| > 0$$

Indeed, the cardinality of the intersection between the set of obligatory constituents O(XP) and the set of the ATB tags of the categories in the structure L (L_atb) should be greater than zero. For example:

$L = $ DTNNS DTJJ; $\textbf{atb_tag}(L) = $ NP; $\textbf{atb_tag}(c_1) = $ DET + NOUN; $\textbf{atb_tag}(c_2) = $ DET + ADJ; $\textbf{L\_atb} = \{$ DET + NOUN, DET + ADJ$\}$ $O(NP) = \{$NOUN, DET + ADJ, ADJ_COMP, DET + NOUN, PRON, CV...$\}$; $|\textbf{L\_atb} \cap \textbf{O}(\textbf{NP})| = |\{$DET + NOUN, DET + ADJ$\}| = 2$

Thus, two constituents (DTNNS and DTJJ) of L are marked as obligatory constituents in the syntactic category NP, so the obligatory properties of NP are satisfied.

The Solver of the Uniqueness Properties
The satisfaction of this solver is reached when a given unique constituent of the specified syntactic category XP, is not repeated in the current PC phrasal structure L. This means that: $\forall c_i \in L, |\textbf{atb\_tag}(c_i) \cap U(XP)| \leq 1$

Indeed, the cardinality of the intersection of a unique constituent and the constituent list of this PC structure should not be greater than one. For example:

$L = $ VBD NP; $\textbf{atb_tag}(L) = $ VP; $\textbf{atb_tag}(c_1) = $ PV; $\textbf{atb_tag}(c_2) = $ NP; $\textbf{L\_atb} = \{$ PV, NP$\}$; $U(VP) = \{$NOUN.VN, ADJ.VN, CV, FRAG, PV_PASS,...$\}$; $|\textbf{atb\_tag}(c_1) \cap \textbf{U}(\textbf{VP})| = 0; |\textbf{atb\_tag}(c_2) \cap \textbf{U}(\textbf{VP})| = 0$

Thus, no unique category in U(VP) is found in L_atb, so no uniqueness property can be verified as satisfied or not in L. This uniqueness property is not integrated in the PC.

The Solver of the Linearity Properties
This solver checks when a constituent of the lhs of a given linearity property appears in the current PC structure, then no constituent of its rhs can precede it in this structure:
For p : $x \prec y$, if $x = $ atb_tag(c_i) & $y = $ atb_tag(c_j), then $c_i \in L_{1..k}$ & $c_j \in L_{k+1..|L|}$

Indeed, the linearity property $x \prec y$ is satisfied in the structure L if x is found in L in a position i ($i \in [1, k]$) lower than the position j of y in L ($j \in [k+1, |L|]$). For example,

> $L = VBD\,NP;\ \textbf{atb\_tag}(L) = VP;\ \textbf{atb\_tag}(c_1) = PV;\ \textbf{atb\_tag}(c_2) = NP;\ L\_atb = \{\ PV,\ NP\};$
> $\textbf{p}: PV \prec NP; PV \in \textbf{L}_{1.1}$ and $NP \in \textbf{L}_{2.2}$

Thus, the position of x = PV in L is equal to 1 and the position of y = NP is equal to 2, so PV appears before NP in L. Then, p is satisfied in L and will be inserted in the PC.

The Solver of the Requirement Properties

This solver checks when a constituent of the lhs of a given GP requirement property appears in the current PC phrasal structure, then the constituent of its rhs should too:

$$\textbf{For p}: \textbf{x} \Rightarrow \textbf{y, if x} = \textbf{atb\_tag}(c_i)\textbf{then y} \in \textbf{L\_atb}$$

Indeed, the requirement property x ⇒ y is satisfied in the structure L if when x appears in L, then y belongs also to L. For example:

> $L = DTNN\quad ADJ\_NUM;\quad \textbf{atb\_tag}(L) = NP;\quad \textbf{atb\_tag}(c_1) = DET + NOUN;\quad \textbf{atb\_tag}(c_2) =$
> $ADJ\_NUM; \textbf{L\_atb} = \{\ DET + NOUN,\ ADJ\_NUM\}; \textbf{p}: FRAG \Rightarrow NP;$
> $\forall c_i \in L, x \neq\ \text{atb\_tag}(c_i):\ FRAG \neq\ \text{atb\_tag}(DTNN)$ and $FRAG \neq \text{atb\_tag}(ADJ\_NUM)$

Thus, no category in L is equal to x = FRAG. The property p can not be verified as satisfied or violated in L, so this requirement property is not integrated in the PC.

The Solver of the Exclusion Properties

This solver checks when a constituent of the lhs of a given GP exclusion property appears in the current PC phrasal structure, then the constituent of its rhs should not and vice versa: **For p** : $\textbf{x} \otimes \textbf{y}$, **if x** = $\textbf{atb\_tag}(c_i)$ **then y** $\notin$ **L_atb**

Indeed, the exclusion property x ⊗ y is satisfied in the structure L if x appears in L, then y not belong to L. For example:

> $L = DTNN\quad ADJ\_NUM;\quad \textbf{atb\_tag}(L) = NP;\quad \textbf{atb\_tag}(c_1) = DET + NOUN;\quad \textbf{atb\_tag}(c_2) =$
> $ADJ\_NUM; \textbf{L\_atb} = \{\ DET + NOUN,\ ADJ\_NUM\}; \textbf{p}: DET + NOUN \otimes ADJ\_NUM;$
> $x = \text{atb\_tag}(c_1) = DET + NOUN;\ y =\ \text{atb\_tag}(c_2) = ADJ\_NUM$

Thus, the category x is found in L but y is also found in L. The exclusion property DET + NOUN ⊗ ADJ_NUM is then not satisfied. This property is integrated in the PC.

3.4 Evaluation

This step presents an evaluation tool of the Stanford parsing result. The scores of this evaluation are based on the results of the constraint solvers, presented in the previous sub-section. The enriched PC, which is normally the input of this evaluation, contains for each phrasal structure, the verified properties as satisfied or violated. The percentage of the satisfied/violated properties may give a more detailed evaluation level of a specific Stanford parsing result. We propose to calculate these in terms of the occurrence number of each satisfied property and the occurrence number of the described syntactic category. We may give this percentage per syntactic category, per property type or per property. We note $sn(p_c^t)$ the occurrence number when the property p $\in$ P (p of the type t $\in$ T describes the syntactic category c $\in$ C, P is the GP property set of t, C is the GP syntactic category set and T is the GP property type list), was satisfied. $vn(p_c^t)$ is the occurrence number when p is violated. $en(p_c^t)$ is the total occurrence number when p is verified $(sn(p_c^t) + vn(p_c^t))$. Now, we present the scores of our evaluation tool:

S_c^t (resp. V_c^t): the property satisfaction (resp. violation) percentage per property type

$$S_c^t = \frac{\sum_{p \in P} sn(p_c^t)}{\sum_{p \in P} en(p_c^t)} \qquad V_c^t = \frac{\sum_{p \in P} vn(p_c^t)}{\sum_{p \in P} en(p_c^t)} \tag{1}$$

S_c (resp. V_c): the property satisfaction (resp. violation) percentage per c:

$$S_c = \frac{\sum_{t \in T} \sum_{p \in P} sn(p_c^t)}{\sum_{t \in T} \sum_{p \in P} en(p_c^t)} \qquad V_c = \frac{\sum_{t \in T} \sum_{p \in P} vn(p_c^t)}{\sum_{t \in T} \sum_{p \in P} en(p_c^t)} \tag{2}$$

S^t (resp. V^t): the property satisfaction (resp. violation) percentage per type t:

$$S^t = \frac{\sum_{c \in C} \sum_{p \in P} sn(p_c^t)}{\sum_{c \in C} \sum_{p \in P} en(p_c^t)} \qquad V^t = \frac{\sum_{c \in C} \sum_{p \in P} vn(p_c^t)}{\sum_{c \in C} \sum_{p \in P} en(p_c^t)} \tag{3}$$

$sp_c^t(p)$ (resp. $vp_c^t(p)$): the satisfaction (resp. violation) distribution of the property p of the type t, that describes c:

$$sp_c^t(p) = \frac{sn(p_c^t)}{\sum_{p \in P} sn(p_c^t)} \qquad vp_c^t(p) = \frac{vn(p_c^t)}{\sum_{p \in P} vn(p_c^t)} \tag{4}$$

4 Experimentation and Evaluation

We evaluated the Stanford Parser (v3.6) on 100 Arabic sentences that are obtained from Arabic children stories[3]. The sentences contain in average 9 words. The parsing result gave us the syntactic trees of these sentences. We obtained 877 syntactic categories for 1047 lexical ones. The distribution of these categories is shown in Fig. 4:

As shown in Fig. 4, the Noun Phrases (NP) dominates all the syntactic categories with 384 occurrences. It is more frequent than the Verbal Phrases (VP) and the

---

[3] From the Arabic book "زهرة بابنج للعصفورة" (chamomile flower to the bird) of Talal Hassan: http://www.awu-dam.org/book/02/child02/105-t-h/105-t-h.zip.

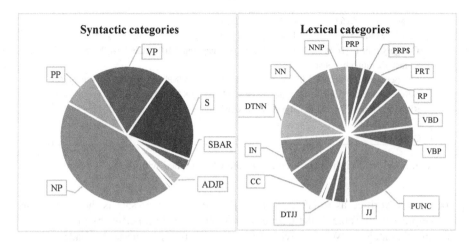

Fig. 4. Distribution of the syntactic and lexical categories in the corpus of Children Stories (CS)

sentences (S) because this corpus contains simple sentences for children. A simple sentence is tagged with one S that encapsulates a VP. This VP contains a verb and many NP. We justify this by the sum of the frequencies of the imperfect and perfective verbs (VBP and VBD) that is near to the frequency of the VP (162). The subordinate clause frequency (SBAR) is also low. The Children Stories (CS) are based also on punctuations. For this reason, the most frequent syntactic category is the punctuation (PUNC).

The Arabic GP is induced from the ATB (ATB2v1.3) that includes Ummah Arabic News. We varied the granularity of the GP categories to have a better PTB/ATB mapping. For example, the PTB tag "DT" describes the demonstrative pronouns. However, the PTB tag "PRP$" describes the possessive pronouns. The mapping of these tags with the ATB tagset assigns to the first ones the tag "PRON_DEM" and to the other ones the tag "PRON_POSS". For that, we should specify the feature "pronoun type" in the categories of the GP. We obtained finally the Arabic GP with the characteristics below (Table 1):

Table 1. Arabic GP characteristics (varied granularity version)

	Frequency	Occurrence number
Syntactic categories	21	189711
Lexical categories	39	184251
Properties	3081	6792698

We verified the satisfaction of the properties of the Arabic GP to evaluate the parsing result of the corpus of CS. By the way, we integrated these properties in the parsed corpus to obtain an enriched corpus. Figure 5 shows an extract of it:

In Fig. 5, the tag "characterization" encapsulates the properties of the GP that can be verified on the constituents "DTNN" and "DTJJ" of the category "NP". All these

```
<Category index="16:0:1:2" label="NP" pos-mapping="35">
  <Category index="16:0:1:2:0" label="DTNN" pos-mapping="29" value="الضبع" />
  <Category index="16:0:1:2:1" label="DTJJ" pos-mapping="10" value="العجوز" />
  <Caracterization nb_const="2" sat="+">
    <Property index="16:0:1:2:0" type="obligation" source="NP" target="DTNN" sat="+"/>
    <Property index="16:0:1:2:2" type="exclusion" source="DTNN" target="UH" sat="+"/>
    <Property index="16:0:1:2:3" type="exclusion" source="DTJJ" target="UH" sat="+" />
  </Caracterization>
</Category>
```

Fig. 5. Extract of the parsed corpus of the CS, enriched with the GP properties

properties are verified as satisfied in this NP. The parameter "sat" is then marked as "+". We present, in Table 2, the frequencies of all verified properties by property type and syntactic category. We abbreviated the property types in the table headers.

Table 2. Occurrence number of the verified properties on the parsed corpus of CS

Type	ADJP	ADVP	INTJ	NAC	NP	PP	VP	S	SBAR	UCP	SQ	SBARQ	WHADVP	WHNP	Σ
Const	25	7	1	5	384	74	162	184	26	1	1	1	3	3	877
Uniq	2	7	1	6	87	2	19	6	19	2	1	2	3	3	160
Oblig	39	7	1	5	538	148	390	410	21	3	1	3	3	3	**1572**
Lin	6	0	0	1	14	3	**166**	38	19	1	1	3	0	0	252
Req	0	0	0	1	3	2	8	1	**14**	1	2	1	0	0	33
Excl	51	7	0	6	1009	149	871	411	71	3	2	4	0	6	**2590**
Σ	123	28	3	24	**2035**	378	**1616**	**1050**	170	11	8	14	9	15	5484

According to Table 2, the half of the verified properties describes the most frequent category, which is the NP (2035/5484). However, we cannot observe this dominance for all types of properties. Most of the linearity properties describe the VPs (65%) and most of the requirement properties describe the SBARs (42%). The 14 linearity properties that are verified in the NP, corresponds to only 7 from 56 property forms and the 3 requirement properties corresponds to only 1 from 15 property forms. This is due to the simple and redundant constructions of the NP in the corpus of CS intended for children. The exclusion properties are the most verified in this corpus with the half of the total because the GP have an over generation for this property type. The obligation properties are also frequent (1572) and exceed the constituent occurrence number (877). Therefore, there are many constituents marked as obligatory. This proves the variety of the constructions of the ATB from which the GP is induced.

The properties are verified as satisfied or violated. For many categories and property types, all verified properties are satisfied in this corpus. For that, in Table 3 we propose to show the violated ones with their occurrence numbers (V) and the verified ones (E).

As shown in Table 3, most of the violated properties describes the VPs. The latter contain categories that require the presence of the interjection "UH". The presence of "UH" is required even for the UCP (Unlike Coordinated Phrase) and the SQ (Sentence Question) when there is a SBAR and respectively a PP (Prepositional Phrase).

Table 3. Occurrence numbers of the violated properties in the parsed corpus of CS

	ADJP			VP			SBAR			UCP			SQ			ΣV	ΣE
	Property	V	E	Property	V	E	Property	V	E	Property	V	E	Property	V	E		
Uniq	-	0	2	-	0	19	IN ∈ Unic(SBAR)	1	19	-	0	2	-	0	1		1	160
Lin	-	0	6	-	0	166	-		0	19	ADJP<SBAR	1	1	-	0	1	1	252
Req	-	0	0	NAC⇒UH 1 UCP⇒UH 1 VBN⇒UH 2 VN⇒UH 2		8	-		0	14	SBAR ⇒ UH	1	1	PP ⇒ UH	1	2	8	33
Excl	JJ⊗ADJ_NUM	2	51		0	871	-	0	71	-	0	3	-	0	2		2	2590
Σ	2		123	6		1616	1		170	2		11	1		8	12	5484	

Using the results of Tables 2 and 3, we can calculate the evaluation scores (1; 2; 3; 4) that we have proposed in the previous section. For example:

The percentages of satisfaction/violation of the requirement properties that describe VP are: $S_{VP}^{req} = \frac{\sum_{p \in P} sn(p_{VP}^{req})}{\sum_{p \in P} en(p_{VP}^{req})} = \frac{2}{8} = 25\%$ and $V_{VP}^{req} = \frac{\sum_{p \in P} vn(p_{VP}^{req})}{\sum_{p \in P} en(p_{VP}^{req})} = \frac{6}{8} = 75\%$ (score 1)

The percentages of the satisfaction/violation of all properties that describe VP are: $S_{VP} = \frac{\sum_{t \in T} \sum_{p \in P} sn(p_{VP}^{t})}{\sum_{t \in T} \sum_{p \in P} en(p_{VP}^{t})} = \frac{155}{161} = 99.6\%$ and $V_{VP} = \frac{\sum_{t \in T} \sum_{p \in P} vn(p_{VP}^{t})}{\sum_{t \in T} \sum_{p \in P} en(p_{VP}^{t})} = \frac{6}{161} = 0.4\%$

The percentage of satisfaction/violation of requirement properties of all categories: $S^{req} = \frac{\sum_{c \in C} \sum_{p \in P} sn(p_{c}^{req})}{\sum_{c \in C} \sum_{p \in P} en(p_{c}^{req})} = \frac{25}{33} = 75.76\%$ and $V^{req} = \frac{\sum_{c \in C} \sum_{p \in P} vn(p_{c}^{req})}{\sum_{c \in C} \sum_{p \in P} en(p_{c}^{req})} = \frac{8}{33} = 24.27\%$

The distribution of the satisfaction/violation of the property NAC ⇒ NP that describes VP are: $sp_{VP}^{req}(NAC \Rightarrow NP) = \frac{sn(NAC \Rightarrow NP)}{\sum_{p \in P} sn(p_{VP}^{req})} = \frac{1}{2} = 50\% = vp_{VP}^{req}(NAC \Rightarrow NP)$ (score 4).

These scores do not replace the evaluation tools that calculate the recall and the precision like Evalb and GramRelEval. They assume that the corpus is correctly parsed and give when the relations between the syntactic units are correct.

5   Conclusion et Perspectives

We have described in the present paper an evaluation method of the Stanford Arabic parser. This method is based on the verification of the satisfaction of syntactic constraints on the parsing result. We used the properties of the Arabic GP, an ATB-based grammar that describes different type constraints for each syntactic category. To verify the satisfaction of these properties, we developed for each property type a constraint solver. By the way, we enriched the simple representation of the parsing result with the verified properties to have relations between the syntactic units. As result, we had both a detailed method for the evaluation of Stanford parser and a more syntactically informative representation for the analysis. We had encouraging experimentation results.

As perspectives, we can enrich and evaluate other Arabic syntactic parsers. This needs to prepare a mapping table of the GP and the chosen parser tagsets. We can also compare any enriched parsed corpus with syntactic properties to the analysis of GP-based parsers. This comparison can be in terms of the number of described syntactic properties. We can go further by adding morphological and semantic properties to the GP. We can include these properties to ease the parsing process.

References

1. AbuShquier, M., Al-Howiti, K.M.: Fully automated arabic to english machine translation system: transfer-based approach of AE-TBMT. Int. J. Inf. Commun. Technol. (2015)
2. Abuhaiba, I.S., Eltibi, M.F.: Author attribution of arabic texts using extended PCFG language model. J. Intell. Syst. Appl. **6**, 27–39 (2016)
3. Arman, N., Jabbarin, J.: Generating use case models from arabic user requirements in a semi-automated approach using a NLP tool. J. Intell. Syst. (2014)
4. Bahloul, R.B., Elkarwi, M., Haddar, K., Blache, P.: Building an arabic linguistic resource from a Treebank: the case of property Grammar. In: Sojka, P., Horák, A., Kopeček, I., Pala, K. (eds.) TSD 2014. LNCS (LNAI), vol. 8655, pp. 240–246. Springer, Cham (2014). https://doi.org/10.1007/978-3-319-10816-2_30
5. Bensalem, R.B., Haddar, K., Blache, P.: A formal modeling method to enrich the arabic Treebank ATB with syntactic properties. In: Proceedings of KEOD (2015)
6. Blache, P.: Les Grammaires de Propriétés: Des contraintes pour le traitement automatique des langues naturelles. Hermès science publications (2001). 228 pages
7. Blache, P., Rauzy, S.: Hybridization and Treebank enrichment with constraint-based representations. In: Workshop on Advanced Treebanking (2013)
8. Cheng, Y., Sun, C., Liu, B., Lin, L.: CRF tagging for head recognition based on Stanford parser. In: CIPS-SIGHAN Joint Conference on Chinese Language Processing (2010)
9. Duchier, D., Prost, J.-P., Dao, T.-B.-H.: A model-theoretic framework for grammaticality judgements. In: Conference on Formal Grammar, Bordeaux, France (2009)
10. Duchier, D., Dao, T., Parmentier, Y.: Analyse Syntaxique par Contraintes pour les Grammaires de Propriétés à Traits. Journées Francophones de Programmation par Contraintes (2012)
11. Maamouri, M., Bies, A., Buckwalter, T., Mekki, W.: The Penn Arabic Treebank: Building a Large-Scale Annotated Arabic Corpus (2004)
12. Oepen, S., Carroll, J.: Parser engineering and performance profiling. J. Nat. Lang. Eng. **6**(1), 81–97 (2000)
13. Prost, J.-P.: Analyse relâchée à base de contraintes. In: TALN (Poster Session), Senlis (2009)
14. Seraji, M., Beata Megyesi, B., Nivre, J.: A basic language resource kit for persian. In: The international Conference on Language Resource Evaluation, pp. 2245–2252 (2012)
15. Green, S., Manning, C.D.: Better arabic parsing: baselines, evaluations, and analysis. In: International Conference on Computational Linguistics (COLING 2010) (2010)
16. Taylor, A., Marcus, M., Santorini, B.: The penn Treebank: an overview. In: Abeille, A. (ed.) Treebanks: the State of the Art in Syntactically Annotated Corpora. Kluwer (2003)
17. Vanrullen, T.: Analyse syntaxique à granularité variable. In: RECITAL (2004)
18. Waheeb, A., Babu, A.: Question analysis for arabic question answering systems. Int. J. Nat. Lang. Comput. (IJNLC) **5**(6) (2016)

Word Sense Disambiguation

SenseDependency-Rank: A Word Sense Disambiguation Method Based on Random Walks and Dependency Trees

Marco Antonio Sobrevilla-Cabezudo[(✉)], Arturo Oncevay-Marcos, and Andrés Melgar

Department of Engineering, Research Group on Pattern Recognition and Applied Artificial Intelligence, Pontificia Universidad Católica del Perú, Lima, Peru
{msobrevilla,arturo.oncevay,amelgar}@pucp.edu.pe

Abstract. Word Sense Disambiguation (WSD) is the field that seeks to determine the correct sense of a word in a given context. In this paper, we present a WSD method based on random walks over a dependency tree, whose nodes are word-senses from the WordNet. Besides, our method incorporates prior knowledge about the frequency of use of the word-senses. We observed that our results outperform several graph-based WSD methods in All-Word task of SensEval-2 and SensEval-3, including the baseline, where the nouns and verbs part-of-speech show the better improvement in their F-measure scores.

Keywords: Word Sense Disambiguation · Dependency tree
Random walks

1 Introduction

Ambiguity is one of the most relevant and difficult problems to outperform in the natural language processing field [11], and it must be faced at different linguistic levels. Specifically, Lexical Ambiguity (LA) is one of the most difficult problems to be solved in Semantics. This kind of ambiguity occurs when a word may express two or more senses in a dictionary but just one of these may be used in a given context.

Thereby, Word Sense Disambiguation (WSD) is the task that aims at determining the most appropriate sense of a word in its context using a specific sense-repository, trying to deal the ambiguity issues described before [1]. This represents a great challenge for the researchers, due to the relevance and usefulness of WSD in many other Natural Language Processing tasks, like machine translation, information extraction and question answering [14].

There are different approaches in order to achieve a proper working model for automatic WSD, and according to [4,9], the unsupervised WSD methods (knowledge-based) have become popular recently. A reason to explain this trend

© Springer Nature Switzerland AG 2018
A. Gelbukh (Ed.): CICLing 2017, LNCS 10761, pp. 185–194, 2018.
https://doi.org/10.1007/978-3-319-77113-7_15

could be related to the difficulty required in the creation of a high quality and large size sense-annotated corpus to train and test supervised algorithms.

Among unsupervised methods, graph-based methods have been widely used because they can take advantages of the WordNet (which contains relations between synsets, and a knowledge graph could be created). Most of them build a knowledge graph linking each word to all the words in the sentence, disregarding whether a specific word contributes to disambiguate another word, and then algorithms are applied over the graph to determine the correct sense of each word [2,13,16]. Only recently methods that create knowledge graphs based on dependency parsing have become popular [4,9].

In this study, we proposed a new graph-based method that builds a knowledge graph for a sentence, using the dependency relations obtained from a Dependency Tree, and a modified version of the PageRank algorithm, considering weights and prior probabilities, to rank the senses over the knowledge graph and obtain the best sense for each content word in a sentence.

The paper is organized as follows. First, we describe some related works to Graph-based Word Sense Disambiguation. Then, Sect. 3 introduces the proposed methodology, detailing the building of the knowledge graph and the modified PageRank. After that, the main results and discussion are shown in Sect. 4. Finally, we draw some conclusions and future works in Sect. 5.

2   Related Works

In first place, the work proposed in [13] tested the WSD problem extracting label annotation dependency patterns for the encode of a graph-based algorithm with random walks performing. In this way, the word senses and definitions (as labels) were obtained from the WordNet, while the word sense relations were derived using machine readable dictionaries in an unsupervised way. Also, for the relation extraction, a similarity metric was calculated using the common tokens between the definitions of the different senses. The achieved accuracy surpassed the best results from that time, in comparison with other methods using the same data sources in the SensEval-2 all-words dataset.

The paper presented in [16] proposed a generalization of their previous unsupervised graph-based method for word sense disambiguation. This study focused on two main elements: (1) the metric of word semantic similarity and (2) the use of graph centrality algorithm for encoding the sense relations. Regarding the first one, six metrics were evaluated (Leacock & Chodorow, Lesk, Wu & Palmer, Resnik, Lin, and Jiang & Conrath) using their WordNet-based implementations [15]. On the other side, several algorithms for graph centrality were developed in order to analyze the importance of the node regarding its relations with other nodes in the graph, and the algorithms used were: indegree, closeness, betweenness, and PageRank [3]. The results in the SensEval-2 dataset represented a significant improvement in the WSD task, indicating that the right combination of those two key elements could lead to promising results.

The work proposed in [2] focused on a large Lexical Knowledge Bases (LKB) using a random walks algorithm. Therefore, the main contribution in this work was the WSD algorithm based on the features described before. In that way, the LKB was built from WordNet and other resources related to the same, represented as graphs (the concepts of the words as the nodes and the relations between them as the edges, disregarding the type of relation). Besides, for the random walks process, the authors adapted the PageRank algorithm [3], considering the rank of a node as the random walk probability over the graph ending on that point. The complete method competed positively to other similar approaches in the state-of-the-art, in both English and Spanish datasets. In the case of the SensEval-2 all-words dataset, they surpass the results for Nouns, Verbs and Adverbs disambiguation, while using the SensEval-3 dataset the improvement was achieved in the Adjective category.

Finally, to the best of our knowledge, the most recent study is presented in [4], where the authors proposed a complete unsupervised WSD system setting based on two distinguishing elements. The first one is the sense dependency, which means that the sense of the words will only depend on the other word senses in the context (usually the sentence). Then, the second idea is the selective dependency, where the sense dependency will be filtered to a few selected words in the context. However, they also argued about considering the context as the complete sentence where the word belongs, so they decided to use a dependency parser in order to determine a context based on syntactic structure. For the model developed, they considered the join probability of senses of the words in the sentence using a Markov Random Field, modeling the WSD task as a Maximum A Posteriori (MAP) Inference Query. The linguistic resources used were the Princeton WordNet and the syntactic analyzer of the Stanford Parser and Link Parser. In their experimentation on SensEval-2 and SensEval-3 datasets, they outmatched all the previous works in the Noun and Verb individual categories, and in the general accuracy score for WSD.

3   Methodology

3.1   Initial Considerations

Unlike most graph-based methods, which explore algorithms over graphs created by relations between all word-senses or all WordNet graph, the proposed method tries to explore the PageRank algorithm over a graph generated by a dependency parser, relying on the idea that dependency relations give better support to word sense disambiguation. Also, this method incorporates a prior probability in the PageRank, in order to give priority to the first senses, considering that humans usually tends to use the first senses in their talks.

The pipeline for this Word Sense Disambiguation Method is shown in Fig. 1. As we may see, this method is divided into two steps: (1) the creation of knowledge graph and (2) the execution of the Pagerank over the built knowledge graph.

In the next subsections, there are more details about the developing of these steps.

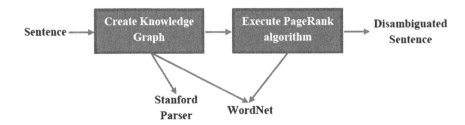

Fig. 1. Pipeline of the Word Sense Disambiguation Method

3.2   Building the Knowledge Graph

The first step in this method is the building of the Knowledge Graph of the target sentence. To achieve this goal, firstly, we used the Stanford Parser [5,12] to get the all dependencies between words in the sentence (via its Dependency Tree). For example, the output of the sentence *"The landlord had died."* was as follows:

det(landlord-2, The-1)
nsubj(died-4, landlord-2)
aux(died-4, had-3)
root(ROOT-0, died-4)

In this example, we may see the relation between "died" and "landlord" is *nsubj* (which means "nominal subject") and the relation between "landlord" and "the" is *det* (which means "determiner").

Secondly, we built the Knowledge Graph using all dependencies generated by the Dependency Parser. At this point, we perceived that all dependencies do not contribute to the disambiguation process, thus, we had to filter some dependency relations using some criteria. The filter criteria are described below:

- Parts-Of-Speech (POSs) of words linked by a Dependency Relation: To be considered, a dependency relation had to include words with content Part-of-Speech (POS) (i.e., Noun, Verb, Adjective and Adverb). These POS's contain a high semantic weight and they are useful to disambiguate the words. Relations that contained words with other POS-Tagging were excluded from the Knowledge Graph.
- Auxiliary Dependency: This kind of relation (aux) was excluded from the Knowledge Graph because *Auxiliary Verbs* it does not provide enough semantic weight in the disambiguation process.

An example of the filter process of dependency relations is presented in Fig. 2. In this Figure, we used the same example described at the begin of this subsection (this example will be used in all paper). As it may be seen, the dependency relations *aux* (died-4, had-3) and *det* (landlord-2, The-1) were excluded because *"The"* was identified as Determinant. In the case of the relation *aux* (died-4, had-3), this was excluded due to the auxiliary relation.

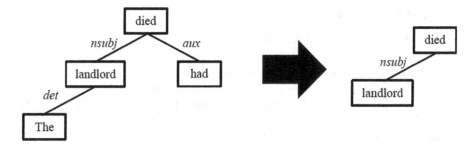

Fig. 2. Filter Process of Dependency Relations in the Dependency Tree

At this point, we have to note that explored relations were associated to senses instead words. Therefore, to finish the Knowledge Graph building, we got all senses of an element in a dependency relation and associated them to all senses of the second element. The senses used were represented by the synsets (synonym sets) of the WordNet 3.0 [7].

An example of how the synsets relations were represented will next be shown in Fig. 3. In this example, the relation *nsubj* (died-4, landlord-2) is presented, and, as it may be seen, each synset of the word *"die"* is related to all synsets of the word *"landlord"*.

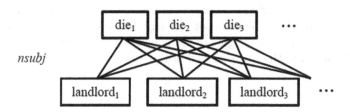

Fig. 3. Relation between the synsets in a dependency relation

Finally, we got the weights of all dependency relations in the knowledge graph. The weights were calculated using the Eq. 1, where S_1 and S_2 represent the synsets to be related, and length is the distance between S_1 and S_2 in WordNet-Pr. To calculate the distances between all synsets, we executed the Dijkstra algorithm for each synset in the WordNet graph. The WordNet graph used was composed by the WNet30 and gloss, which contains WordNet 3.0 synsets and relations, including manually disambiguated glosses. This graph has 117, 522 vertices and 525, 356 relations. We highlight that the built of the knowledge graph was undirected, therefore, the weight of an edge between a synset "a" and a synset "b" in a dependency relation was the same as the edge between the synset "b" and synset "a".

$$weight(S_1, S_2) = \frac{1}{length(S_1, S_2)} \tag{1}$$

3.3   Modified PageRank Algorithm

The final step in the method was the execution of the PageRank algorithm [3] over the graph built in the previous sub-step. Before explaining how we used this method, the Pagerank algorithm will be detailed:

Let "G" be a graph with "N" vertices $v_1, \ldots, v_N$ and d_i be the outdegree of node "i"; let "M" be a $N \times N$ transition probability matrix, where $M_{ji} = 1/d_i$ if a link from i to j exists, and zero otherwise. Then, the computation of the PageRank vector "Pr" over G equates to solve Eq. 2.

$$Pr = cMPr + (1 - c)v \tag{2}$$

In the equation, the first term of the sum models the voting scheme. The second term represents the probability of a surfer randomly jumping to any node, i.e., without following any paths on the graph. In this term, v is a $N \times 1$ vector whose elements are usually $1/N$ and "c" is called damping factor.

As mentioned in the previous paragraph, in the traditional PageRank, each element of the vector v has value of $1/N$, thus assigning equal probabilities to all nodes in the graph in case of random jumps.

A modification of this vector is proposed in [10]. In this work, the author mentioned that "v" can be non-uniform, and stronger probabilities to certain nodes were assigned, biasing the resulting PageRank vector to prefer these nodes. Also, in the case of a node received a higher prior probability, this would affect its neighbor nodes, receiving a higher rank too.

For our PageRank implementation, we modified the way to build the transition matrix to consider the weights in the knowledge graph. Specifically, we used the formula shown in the Eq. 3 instead of the formula $M_{ji} = 1/d_i$.

$$M_{ji} = \frac{w_{ij}}{\sum_z w_{iz}} \tag{3}$$

In this formula, w_{ij} represents the edge weight between the nodes "i" an "j" and the denominator calculates the sum of all edge weights whose source is the node "i".

Besides that modification, we used a non-uniform vector "v", in which each element contained the frequency shown by WordNet expressed in probabilities. This vector modification allowed to incorporate a bias into the algorithm to prefer the most frequent synsets, because this is the way like humans usually talk.

Finally, the damping factor considered in the experiments was 0.85 and the number of iterations was 30.

4   Experiments and Results

In order to evaluate our method, we conducted three evaluations. All evaluations were performed over the SensEval-2 [6] and SensEval-3 [17] English All-words WSD datasets.

The first evaluation was focused on determining the performance of our method. A point worth of note was that our method used the WordNet 3.0 and the data in SenSeval-2 and SensEval-3 was annotated using previous versions. Therefore, some of the words which need to be tagged have been removed from WordNet 3.0 (for example, "something" or "might"). For this reason, we executed twice the algorithm over the datasets: (1) considering only the words included in WordNet 3.0 and (2) considering all words, including words that have no corresponding sense in WordNet 3.0.

The F-measure of the two executions is presented in Table 1. The results of SensEval-2* and SensEval-3* represent the first execution, when only words included in WordNet 3.0 are considered, and the others results consider all words. As it may be seen, the results decreased when the words that have no corresponding sense in WordNet 3.0 were considered.

Table 1. SenseDependency-Rank in Senseval-2 and Senseval-3 considering modifications of WordNet

Method	All	N	V	Adj.	Adv.
Senseval-2*	65.32	71.95	45.99	63.92	79.47
Senseval-2	62.84	71.24	44.98	58.14	77.55
SensEval-3*	63.68	70.25	56.33	60.14	100
SensEval-3	62.54	69.86	55.48	56.58	92.31

The second evaluation was the performance comparison of our method with other graph-based WSD methods. In our case, the methods compared were the proposed in [13] (called Mih05), [16] (called Sinha07), [2] (called Agirre14) and [4] (called MRF-LP). Also, our method was compared with the baseline used in both SensEval datasets, which is the Most Frequent Sense method (MFS).

The results of all methods are presented in Tables 2 and 3 (expressed in terms of F-measure). As it may be seen, our method obtained the best performance, overcoming even the MFS method, that is usually hard to be outperformed.

One point to be highlighted is the fact that, in both datasets, the results for nouns and verbs were the best. This reinforces ideas such as that (1) dependency relations and their similarity measures provided a better contextual information to disambiguate Nouns and Verbs and (2) Noun and Verbs are useful to disambiguate among themselves and, principally, Nouns are good hints to disambiguate Verbs, like mentioned in [8].

In the case of adverbs and adjectives, the same cannot be said because the results were almost the worst. A hypothesis is that exploring other kind of similarity measures (more suitable) in the generated dependency trees could contributes to the improvement of performance.

Finally, the third evaluation was focused on comparing our method with other method that combines the MFS method in its execution. In the work presented in [2], the authors evaluated a combination between their original method

Table 2. Results in SensEval-2 all-words dataset

Method	All	N	V	Adj.	Adv.
MFS	60.1	71.2	39.0	61.1	75.4
Mih05	54.2	57.5	36.5	56.7	70.9
Sinha07	56.4	65.6	32.3	61.4	60.2
Agirre14	59.7	70.3	40.3	59.8	72.9
MRF-LP	60.5	66.9	42.7	**63.2**	**82.9**
SD-Rank	**62.84**	**71.24**	**44.98**	58.14	77.55

Table 3. Results in SensEval-3 all-words dataset

Method	All	N	V	Adj.	Adv.
MFS	62.3	69.3	53.6	63.7	92.9
Mih05	54.2	57.5	36.5	56.7	**100**
Sinha07	52.4	60.5	40.6	54.1	-
Agirre14	57.9	65.3	47.2	**63.6**	96.3
MRF-LP	58.6	65.8	50.1	59.9	87
SD-Rank	**62.54**	**69.86**	**55.48**	56.58	92.31

(Agirre14) and the MFS method in order to measure the possible improvement of their method. As it may be seen in Table 4, our method outperformed the results of Agirre14+MFS in SensEval-2 All-words datasets, but was outperformed by the same method in SensEval-3. Highlights of these results was the best performance obtained for Verbs in both datasets. This shows us that, as mentioned in the second evaluation, the dependency relations and similarity measures between Nouns and Verbs positively contribute to the disambiguation of Verbs.

Table 4. SenseDependency-Rank vs. Method proposed in [2]

Dataset	Method	All	N	V	Adj.	Adv.
SensEval-2	Agirre14+MFS	62.6	**73.3**	41.8	**64.2**	76.3
	SD-Rank	**62.84**	71.24	**44.98**	58.14	**77.55**
SensEval-3	Agirre14 + MFS	**63.0**	**70.7**	54.3	**65.5**	**96.3**
	SD-Rank	62.54	69.86	**55.48**	56.58	92.31

5   Conclusions and Future Works

In this paper, we proposed a new graph-based method that builds a knowledge graph for a sentence using the dependency relations obtained from a Dependency Tree and a modified version of PageRank algorithm (considering weights) to rank the senses over the knowledge graph and obtain the best sense for every content word in a sentence.

The results showed that our method outperformed previous approaches in English all-words datasets, including the MFS method. The content Part-of-Speech (POS) most benefited were Nouns and Verbs, but the Adjectives and Adverbs presented a decrease in the performance. A comparison with other method that incorporates MFS method was performed obtaining comparable results. In this evaluation, Verbs were the most benefited, showing that dependency relations and similarity measures with Nouns are useful in its disambiguation process.

As a future work, we are planning to explore other kind of similarity measures between synsets in different POS. In this work, we used the inverse of the length-path and this showed improvements on nouns and verbs disambiguation, however, this improvement was not obtained on adjectives and adverbs.

Finally, our algorithm can be ported to other languages, with two requirements: (1) having a WordNet and (2) having a Dependency Parser. The attempt is to research and apply this kind of method in less-resourced languages (with WordNet-like resources under development), and that would be a great challenge, because of the need to identify an enough and minimum size resource to work with.

Acknowledgments. For this study, the authors acknowledge the support of the "Programa Nacional de Innovación para la Competitividad y Productividad", Perú, under the contract 124-PNICP-PIAP-2015.

References

1. Agirre, E., Edmonds, P.: Word Sense Disambiguation: Algorithms and Applications, 1st edn. Springer, Heidelberg (2007). https://doi.org/10.1007/978-1-4020-4809-8
2. Agirre, E., López de Lacalle, O., Soroa, A.: Random walks for knowledge-based word sense disambiguation. Comput. Linguist. **40**(1), 57–84 (2014)
3. Brin, S., Page, L.: The anatomy of a large-scale hypertextual web search engine. In: Proceedings of the Seventh International Conference on World Wide Web 7, WWW7, pp. 107–117. Elsevier Science Publishers B. V., Amsterdam (1998)
4. Chaplot, D.S., Bhattacharyya, P., Paranjape, A.: Unsupervised word sense disambiguation using markov random field and dependency parser. In: Proceedings of the Twenty-Ninth AAAI Conference on Artificial Intelligence, January 25–30, 2015, Austin, Texas, USA, pp. 2217–2223 (2015)
5. Chen, D., Manning, C.: A Fast and Accurate Dependency Parser using Neural Networks. In: Proceedings of the 2014 Conference on Empirical Methods in Natural Language Processing (EMNLP), pp. 740–750. Association for Computational Linguistics, Doha, October 2014

6. Edmonds, P., Cotton, S.: Senseval-2: Overview. In: The Proceedings of the Second International Workshop on Evaluating Word Sense Disambiguation Systems, SENSEVAL 2001, pp. 1–5. Association for Computational Linguistics, Stroudsburg (2001)
7. Fellbaum, C. (ed.): WordNet An Electronic Lexical Database. The MIT Press, Cambridge (1998)
8. Fellbaum, C.: Wordnet and wordnets. In: Brown, K. (ed.) Encyclopedia of Language and Linguistics, pp. 665–670. Elsevier (2005). http://wordnet.princeton.edu/
9. Gao, N., Zuo, W., Dai, Y., Lv, W.: Word sense disambiguation using wordnet semantic knowledge. In: Wen, Z., Li, T. (eds.) Knowledge Engineering and Management. AISC, vol. 278, pp. 147–156. Springer, Heidelberg (2014). https://doi.org/10.1007/978-3-642-54930-4_15
10. Haveliwala, T.H.: Topic-sensitive pagerank. In: Proceedings of the 11th International Conference on World Wide Web, WWW 2002, pp. 517–526. ACM, New York (2002)
11. Jurafsky, D., Martin, J.H.: Speech and Language Processing, 2nd edn. Prentice-Hall Inc., Upper Saddle River (2009)
12. de Marneffe, M.C., MacCartney, B., Manning, C.D.: Generating typed dependency parses from phrase structure parses. In: Proceedings of International Conference on Language Resources and Evaluation. LREC, pp. 449–454 (2006)
13. Mihalcea, R.: Unsupervised large-vocabulary word sense disambiguation with graph-based algorithms for sequence data labeling. In: Proceedings of the Conference on Human Language Technology and Empirical Methods in Natural Language Processing, HLT 2005, pp. 411–418. Association for Computational Linguistics, Stroudsburg (2005)
14. Navigli, R.: Word sense disambiguation: A survey. ACM Comput. Surv. **41**(2), 10:1–10:69 (2009)
15. Pedersen, T., Patwardhan, S., Michelizzi, J.: Wordnet: Similarity: measuring the relatedness of concepts. In: Demonstration Papers at HLT-NAACL 2004, pp. 38–41. Association for Computational Linguistics (2004)
16. Sinha, R.S., Mihalcea, R.: Unsupervised graph-basedword sense disambiguation using measures of word semantic similarity. In: Proceedings of the First IEEE International Conference on Semantic Computing (ICSC 2007), 17–19 September 2007, Irvine, California, USA, pp. 363–369 (2007)
17. Snyder, B., Palmer, M.: The english all-words task. In: Mihalcea, R., Edmonds, P. (eds.) Senseval-3: Third International Workshop on the Evaluation of Systems for the Semantic Analysis of Text, pp. 41–43. Association for Computational Linguistics, Barcelona (2004)

Domain Adaptation for Word Sense Disambiguation Using Word Embeddings

Kanako Komiya[1]([⊠]), Shota Suzuki[1], Minoru Sasaki[1], Hiroyuki Shinnou[1], and Manabu Okumura[2]

[1] Ibaraki University, 4-12-1 Nakanarusawa, Hitachi-shi, Ibaraki 316-8511, Japan
{kanako.komiya.nlp,12t4042a,minoru.sasaki.01,
hiroyuki.shinnou.0828}@vc.ibaraki.ac.jp
[2] Tokyo Institute of Technology,
4259 Nagatuta, Midori-ku, Yokohama 226-8503, Japan
oku@pi.titech.ac.jp

Abstract. In this paper, we propose domain adaptation in word sense disambiguation (WSD) using word embeddings. The validity of the word embeddings from a huge corpus, e.g., Wikipedia, for WSD had already been shown, but their validity in a domain adaptation framework has not been discussed before. In addition, if they are valid, the difference in effects according to the domain of the corpora is still unknown. Therefore, we investigate the performances of domain adaptation in WSD using the word embeddings from the source, target, and general corpora and examine (1) whether the word embeddings are valid for domain adaptation of WSD and (2) if they are, the effects in accordance with the domain of the corpora. The experiments using Japanese corpora revealed that the accuracy of WSD was highest when we used the word embeddings obtained from the target corpus.

1 Introduction

Word embeddings [21–23] are the vector representations of the word meanings, which have compositionality. They are effective for word sense disambiguation (WSD) tasks [31] since they are low dimensional vectors and the sparseness of the meaning representations has been greatly alleviated through them.

On the other hand, domain adaptation involves adapting the classifier that has been trained from data in one domain (source domain) to data in another domain (target domain). This has been studied intensively including domain adaptation in WSD (see Sect. 2). In this paper, first, we propose using the word embeddings for domain adaptation of WSD (see Sect. 3). Since the domain adaptation suffers from data sparseness caused by a domain shift, we suppose that the word embeddings can also improve the performance of WSD in a domain adaptation framework.

However, one problem with the domain adaptation is the shift of the priors of the word senses in the texts [30], which means that the word meanings in one domain could be different from those in another domain.

© Springer Nature Switzerland AG 2018
A. Gelbukh (Ed.): CICLing 2017, LNCS 10761, pp. 195–206, 2018.
https://doi.org/10.1007/978-3-319-77113-7_16

Therefore, second, we investigate the performances of domain adaptation in WSD using the word embeddings obtained from the source, target, and general corpora and examine the effects in accordance with their domains. We investigated Japanese WSD. The experiments (see Sects. 4, 5, and 6) revealed that the accuracy of WSD was highest when we used the word embeddings obtained from the corpus in the target domain, which indicated that the domain of the corpus to generate the word embeddings affected the results more than the vocabulary size of the word embeddings. We lastly discuss the results (Sect. 7) and conclude this paper (Sect. 8).

2   Related Work

Much work using the word embeddings has recently been done. Sugawara et al. [31] investigated the context representation using word enbeddings for WSD. Taghipour and Ng [33] proposed semi-supervised WSD using the word embeddings and showed that the word embeddings of a general or specific domain improved the performance of WSD. Much work has improved distributed representation of word senses or concepts and used them for WSD ([7,28,35]) or other tasks ([34]).

On the other hand, many researchers have investigated domain adaptation within or outside the area of natural language processing. The domain adaptation problem can be categorized into three types depending on the information for learning, i.e., that in supervised, semi-supervised, and unsupervised approaches. According to Daumé [10], a classifier in a supervised approach is developed from a large amount of labeled source data and a small amount of labeled target data. A classifier in a semi-supervised approach is developed from a large amount of labeled source data, a small amount of labeled target data, and a large amount of unlabeled target data. Finally, a classifier is developed from a large amount of labeled source data and unlabeled target data in an unsupervised approach. We focused on the unsupervised domain adaptation in Japanese WSD in the research reported in this paper.

Escudero et al. demonstrated domain dependence of supervised WSD system [11]. Chan and Ng carried out the domain adaptations of WSD by estimating class priors using an EM algorithm, which were unsupervised domain adaptation and supervised domain adaptation using active learning [4,5]. Daumé augmented an input space and made triple length features that were general, source-specific, and target-specific for supervised domain adaptation [9]. Daumé et al. extended the earlier work to semi-supervised domain adaptation [10]. Agirre and de Lacalle applied singular value decomposition (SVD) to a matrix of unlabeled target data and a large amount of unlabeled source data, and trained a classifier with them [1,2]. Kunii and Shinnou proposed combined use of topic models on unsupervised domain adaptation in WSD [19]. Jiang and Zhai demonstrated that performance increased as examples were weighted when domain adaptation was applied [15]. Shinnou et al. reported active learning to remove source instances for domain adaptation in WSD [30]. Shinnou et al. also proposed learning under a covariate shift for domain adaptation in WSD [29]. Kouno et al. performed unsupervised domain adaptation in WSD using a stacked denoising autoencoder [18]. Izquierdo et al. [14] investigated Class-based WSD and it is useful for unsupervised domain adaptation.

The research by Blitzer et al. involved work on semi-supervised domain adaptation, where they calculated the weight of words around the pivot features (words that frequently appeared both in source and target data and behaved similarly in both) to model some words in one domain that behaved similarly in another [3]. They applied SVD to the matrix of the weights, generated a new feature space, and used the new features with the original features.

Komiya and Okumura determined an optimal method of domain adaptation using decision tree learning given a triple of the target word type of WSD, source data, and target data [16]. They discussed what features affected how the best method was determined. They also proposed determining the optimal method, i.e., the optimal training data set, for each instance using the degree of confidence for supervised domain adaptation in WSD [17]. There are researches of domain adaptation using deep learning [8,12]. Sun et al. [32] proposed CORAL, which minimizes domain shift easily.

The closest work to ours is the work by Sugawara et al. [31], which investigated the context representation using word enbeddings for WSD. We propose using the context representation using word enbeddings for domain adaptation of WSD and show that they can also improve the performance of WSD in a domain adaptation framework. In addition, we investigate the performances of domain adaptation in WSD using the word embeddings obtained from the source, target, and general corpora and examine the effects in accordance with their domains.

3 Domain Adaptation in WSD Using Word Embeddings

We carry out domain adaptation in WSD by concatenating the features of the word embeddings for surrounding words of the target word of WSD in addition to the baseline features in both the source and target data. We followed the work by Sugawara et al. [31] and used Context-Word-Embeddings, which was a concatenated vector of the real-valued vectors of the words in the context window [1]. As Sugawara et al. [31] described, if the window size is N and words appearing in the context window are $w_{-N}, ..., w_{-1}, w_{+1}, ..., w_{+N}$, this feature vector is a vector concatenating $v_{w_{-N}}, ..., v_{w_{-1}}, v_{w_{+1}}, ..., v_{w_{+N}}$, where v_w represents an embedding of word w. If the dimension of each word embedding is d, the size of this feature vector is $2 \times N \times d$. Table 1 and Fig. 1 show a simple example of Context-Word-Embeddings. Figure 1 shows the Context-Word-Embeddings of instance vector for the word "大きい (big)" in the context of the phrase "効果が大きいため、 (Because the effect is big,)" as shown in Table 1. In this example, N is two and d is three.

---

[1] Sugawara et al. [31] reported that Context-Word-Embeddings improved the result of WSD more than Average-Word-Embeddings, which was the average of vector representations of words in the context window.

Table 1. Simple example of Context-Word-Embeddings

Japanese word	Translation	word embedding
効果	effect	0.1 0.2 0.3
が	is	0.4 0.1 0.7
大きい	big	0.5 0.2 0.1
ため	because	-0.2 0.3 0.1
、	、	-0.7 0.8 0.3

0.1 0.2 0.3 0.4 0.1 0.7 -0.2 0.3 0.1 -0.7 0.8 0.3

Fig. 1. Example of the concept text

We investigated the following five cases depending on the domains of corpora to generate the word embeddings.

Add Target The case where the word embeddings obtained from the target data are used.

Add Source The case where the word embeddings obtained from the source data are used.

Add Wiki The case where the word embeddings obtained from a general huge corpus, i.e., Wikipedia, are used.

Add Target & Source The case where the word embeddings obtained from the source and those obtained from the target data are used together by concatenating them.

Add Target & Wiki The case where the word embeddings obtained from the target data and those obtained from the general huge corpus are used together by concatenating them.

Add Target Large The case where the word embeddings obtained from the large target data are used.

4   Experiment

We used word2vec[2] [21–23] to generate the word embeddings. The vector size, which is d in Sect. 3, and the window size to generate the word embeddings were set to 200 and five respectively. We used a skip-gram algorithm. The window size of each target word, which is N in Sect. 3, was set to two. We used default settings for other parameters.

Libsvm [6], which supports multi-class classification, was used as the classifier for WSD[3]. A linear kernel was used in accordance with the results obtained from preliminary experiments. Twenty features, which were the same as [16,17], were introduced as the baseline features to train the classifier.

---

[2] https://code.google.com/archive/p/word2vec/.

[3] We used the -b option of libsvm.

- Morphological features
  - Bag-of-words
  - Part-of-speech (POS)
  - Finer subcategory of POS
- Syntactic features
  - If the POS of a target word is a noun, the verb that the target word modifies is used.
  - If the POS of a target word is a verb, the case element of '{ヲ}' (wo, objective) for the verb is used.
- Semantic feature
  - Abstracted semantic class

Morphological features and a semantic feature were extracted from the surrounding words (two words to the right and left) of the target word and the target word itself. POS and the finer subcategory of POS could be obtained by using a morphological analyzer. We used Mecab[4] as a morphological analyzer, the Word List by Semantic Principles [24] for the abstracted semantic class of each word, and CaboCha[5] as a syntactic parser.

5 Data

Three labeled data were used for the experiments: (1) the sub-corpus of white papers in the Balanced Corpus of Contemporary Japanese (BCCWJ) [20] [6], (2) the sub-corpus of documents from a Q&A site on the WWW in BCCWJ, and (3) Real World Computing (RWC) text databases (newspaper articles) [13]. Domain adaptation was conducted in six directions in accordance with different source and target data. Word senses were annotated in these corpora in accordance with a Japanese dictionary, i.e., the Iwanami Kokugo Jiten [25]. It has three levels for sense IDs, and we used the middle-level sense in the experiments. Multi-sense words that appeared equal to or more than 50 times in all the data were selected as the target words in the experiment; 36 word types were used in the experiments. Table 2 lists the minimum, maximum, and average number of instances of each word type for each corpus. Table 3 summarizes the list of target word types. "No. of senses" in the first column is the number of the senses of each word type in the dictionary. For example, the word type "考える (think)" has two senses in the dictionary. Please note that there is no guarantee that all the senses in the dictionary will appear in the corpora.

In addition, we used seven types of unlabeled data to generate the word embeddings: (1) the sub-corpus of white papers in the BCCWJ, (2) the sub-corpus of documents from a Q&A site on the WWW in the BCCWJ, (3) RWC text databases, (4) large collections of white papers in BCCWJ, (5) large collections of documents from a Q&A site on the WWW in the BCCWJ, (6) large

---

[4] https://github.com/jordwest/mecab-docs-en.

[5] http://sourceforge.net/projects/cabocha/.

[6] SemEval-2010 Task: Japanese WSD [26] is included in this corpus.

Table 2. Minimum, maximum, and average number of instances of each word type for each corpus

Genre	Min	Max	Sum	Avg
White papers	60	8,691	76,889	2,136
Q & A site	158	1,7387	112,320	3,120
Newspaper	92	1,046	9,959	277

Table 3. List of target words types

No. of senses	Target words (in Japanese)	Sense example in English	No. of senses	Target words (in Japanese)	Sense example in English
2	考える	think	3	地方	area
	技術	technique		出る	go out
	経済	economy		入る	enter
	現在	present		開く	open
	子供	child		前	before
	自分	self	4	時間	time
	情報	information		進める	move forward
	高い	high		出す	emit
	場合	in case		乗る	ride
	ほか	other		計る	plan
3	言う	say		一つ	one
	今	now		持つ	have
	入れる	put in	5	良い	good
	大きい	big	6	合う・会う	meet
	関係	connection		見る	see
	聴く	hear	7	手	hand
	社会	society		上げる	raise
	進む	proceed	8	取る	take

collections of the newspaper articles of the Mainichi Shimbun from 1991 to 2005 including RWC text databases, and (7) dumped Japanese Wikipedia data (2015/10/02)[7]. (1) and (2) were obtained from original BCCWJ and (3) was created from the labeled data, by deleting the labels. The second three, i.e., (4), (5), and (6), are the larger corpora including the first three. Table 4 shows the number of word types of the word embeddings and its ratio to that of Wikipedia in accordance with each genre. The numbers in the table are different from the number of the word types of each corpus because word2vec generated the meaning representation vector if and only if a word appeared in the corpus equal to or more than five times [8].

---

[7] https://dumps.wikimedia.org/jawiki/.

[8] The ratios of the numbers of tokens for corpora (1)–(5) are less than two percent to that of Wikipedia respectively and the ratio of corpus (6) is about 46%.

Table 4. Number of word types each genre and its ratio to that of Wikipedia

No	Corpus	No. of Word Types	Ratio
(1)	White papers	13,336	1.24%
(2)	Q & A site	22,709	2.10%
(3)	Newspaper	11,685	1.08%
(4)	White papers Large	14,303	1.33%
(5)	Q & A site Large	37,893	3.51%
(6)	Newspaper Large	20,1421	18.67%
(7)	Wikipedia	1,078,930	100.00%

Table 5 shows the corpora for word embeddings according to the source and target corpora and the domains of corpus from which word embeddings were generated. For example, (5) large collections of documents from a Q&A site on the WWW in the BCCWJ was used for the domain adaptation from (1) the sub-corpus of white papers in the BCCWJ to (2) the sub-corpus of documents from a Q&A site on the WWW in the BCCWJ by *Add Target Large*. The average accuracies over six directions (Table 6) and the average accuracies over two directions according to the target corpora (Table 7) were evaluated in Sect. 6.

Table 5. Corpora for word embeddings according to source and target corpora and domains of corpus from which word embeddings were generated

Source → Target	(1) → (2)	(1) → (3)	(2) → (1)	(2) → (3)	(3) → (1)	(3) → (2)
Add Target	(2)	(3)	(1)	(3)	(1)	(2)
Add Source	(1)	(1)	(2)	(2)	(3)	(3)
Add Wiki	(7)	(7)	(7)	(7)	(7)	(7)
Add Target and Source	(1) + (2)	(1) + (3)	(1) + (2)	(2) + (3)	(1) + (3)	(2) + (3)
Add Target and Wiki	(2) + (7)	(3) + (7)	(1) + (7)	(3) + (7)	(1) + (7)	(2) + (7)
Add Target Large	(5)	(6)	(4)	(6)	(4)	(5)

6   Results

Table 6 lists the micro- and macro-averaged accuracies of WSD for the whole data set in accordance with the domains of corpus from which the word embeddings were generated, and Table 7 summarizes the micro- and macro-averaged accuracies of WSD in accordance with the genre of the test corpora and the domains of corpus from which the word embeddings were generated. Micro-average is the average over word tokens, and macro-average is that over word types. We tested Self, which is standard supervised learning with the whole target data by five-fold cross validation, assuming that fully annotated data were obtained and could be used for learning with word embedding features, MFS, which is the most frequent sense of the target corpus[9], and Baseline features, which is standard supervised learning without word embedding features, as references.

Self was an upper bound, and Baseline features were a baseline. The highest accuracies except for Self and MFS have been written in bold for each corpus in Tables 6 and 7. Plus and minus represent that the accuracy of the case is higher or lower than that of *Baseline features*. The asterisk means the difference between accuracies of each case and *Baseline features* is statistically significant according to a chi-square test. The level of significance in the test was 0.05.

Table 6. Summary of accuracies

Method	Macro Avg.	Micro Avg.
Baseline features	77.90%	79.79%
Add Target	78.35%+	79.92%+*
Add Source	73.48%−	75.98%−*
Add Wiki	78.60%+	79.73%−
Add Target & Source	78.13%+	79.77%−*
Add Target & Wiki	76.99%−	78.45%−*
Add Target Large	**78.64%+**	**80.02%+***
MFS	77.54%	82.05%
Self	89.88%	92.67%

Table 7. Average accuracies of WSD according to test corpora and domains of corpus from which word embeddings were genarated

Target data	Newspaper		Q & A site		White papers	
Method	Macro	Micro	Macro	Micro	Macro	Micro
Baseline features	78.84%	78.15%	74.01%	76.59%	80.86%	**84.70%**
Add Target	78.79%−	78.17%+	74.39%+	77.18%+*	81.87%+	84.16%−*
Add Source	73.13%−	78.21%+	73.93%−	75.94%−*	73.38%−	72.86%−*
Add Wiki	**79.19%+**	78.52%+	74.45%+	77.10%+*	82.15%+	83.73%−*
Add Target & Source	78.49%−	77.86%−	73.97%−	77.07%+*	81.93%+	83.96%−*
Add Target & Wiki	77.36%−	76.47%−*	73.66%−	76.28%−*	79.94%−	81.88%−*
Add Target Large	79.18%+	**78.69%+**	**74.59%+**	**77.30%+***	**82.15%+**	84.17%−*
MFS	73.20%	72.64%	77.71%	78.47%	81.72%	88.51%
Self	84.31%	85.23%	90.58%	89.84%	94.74%	97.76%

7   Discussion

First, Table 6 demonstrates that macro-averaged accuracy increased but the micro-averaged accuracy decreased when we used *Add Wiki*. Table 7 shows that

---

[9] Note that we cannot know the most frequent sense in the target corpus without the labeled target data and it is hard to beat [27].

the reason for the decline is the micro-averaged accuracy of white papers. We think this is because the senses in the corpus of white papers are very biased as shown in Table 7 and for these cases the word embedding features cannot improve the accuracy. Table 7 also shows that the best micro-averaged accuracy in white papers is *Baseline features*, and none of our proposed method, i.e., domain adaptation, could beat it. We think the reason is the same here.

Next, Table 6 shows that the best case is *Add Target Large* in both micro- and macro-averaged accuracies and the difference between the case and *Baseline features* in micro-averaged accuracies is statistically significant. The same table indicates that the second best case is *Add Target* in micro-averaged accuracy and the difference between the case and *Baseline features* in micro-averaged accuracies is also statistically significant. These results indicate that the word embedding features are effective in the domain adaptation framework and we should use the word embeddings obtained from the target corpus. In particular, the ratios of the vocabulary size of the word embeddings in the target corpora to that of Wikipedia are from one to three percent and those of the larger target corpora are from one to 19 percent according to Table 4. This indicates that *Add Target* or *Add Target Large* could improve the accuracies of WSD even though the number of word types of word enbeddings are much smaller than that of Wikipedia.

Table 6 shows the cases other than *Add Target* and *Add Target Large* could not improve both micro- and macro-averaged accuracies; one of them decreased in *Add Wiki* and *Add Target & Source* and both of them decreased in *Add Source* and *Add Target & Wiki*. We think that the accuracies decreased when *Add Source* is used because the sparseness of the meaning representations was not alleviated much in the target data. As we used the semantic classification codes as the meaning representations for the baseline features in our experiments unlike [31], the sparseness might not be alleviated as much as in their experiments. We think the second reason is that the meanings of the words in the source data were different from those in the target data. In addition, according to Table 7, all the accuracies of *Add Target & Wiki* are lower than those of *Baseline features* although those of *Add Target* and *Add Wiki* were higher, except for the micro-averaged accuracies in white papers, which indicates that the concatenation of the two good feature sets cannot always provide good results.

Finally, we will discuss why the word embedding features obtained from the target corpora are effective for domain adaptation. We think that the word embedding features are effective for WSD tasks since the sparseness of the meaning representations has been greatly alleviated through them. Our experiments show that the results of *Add Target Large* are always better than those of *Add Target*, which indicate that the word embeddings obtained from the larger corpora did improve the performance of WSD when their domains were the same. However, we cannot explain all the improvements of the results in a domain adaptation framework only through the alleviation of the sparseness. Table 8 shows the ratios of the unknown words of each case. It shows the ratio of the unknown words of *Add Target* (24.49%) and *Add Target Large* (7.88%) are greater than that of *Add Wiki* (2.31%) in the target data. This demonstrates that the improvement in the results comes from not only the alleviation of the sparseness, but also the domain of corpus from which the word embeddings were generated. In particular, the experimental results show that the domain of the corpus to generate the word embeddings affected the results more than the vocabulary size of the word embeddings when we compared *Add Target Large* with *Add Wiki*.

Table 8. Ratio of unknown words

Method	Ratio
Add Target	24.49%
Add Source	44.82%
Add Wiki	2.31%
Add Target & Source	19.45%
Add Target & Wiki	2.22%
Add Target Large	7.88%

8 Conclusion

We proposed using the word embedding features generated through word2vec and showed their validity for domain adaptation in Japanese WSD tasks. We investigated the performances of domain adaptation in WSD using various cases, i.e., *Add Target*, *Add Source*, *Add Wiki*, *Add Target & Source*, *Add Target & Wiki*, and *Add Target*, and examined the effects of the domains of the corpora to generate the word embeddings. Our experiments demonstrated that the word embeddings obtained from the target corpus improved the accuracies of WSD the most. They also revealed that improvement in the results of domain adaptation came from not only the alleviation of the sparseness, and showed that the domain of the corpus to generate the word embeddings affected the results more than the vocabulary size of the word embeddings.

Acknowledgment. This work was supported by JSPS KAKENHI Grant Number 15K16046.

References

1. Agirre, E., de Lacalle, O.L.: On robustness and domain adaptation using svd for word sense disambiguation. In: Proceedings of COLING 2008, pp. 17–24 (2008)
2. Agirre, E., de Lacalle, O.L.: Supervised domain adaption for WSD. In: Proceedings of EACL 2009, pp. 42–50 (2009)
3. Blitzer, J., McDonald, R., Pereira, F.: Domain adaptation with structural correspondence learning. In: Proceedings of EMNLP 2006, pp. 120–128 (2006)
4. Chan, Y.S., Ng, H.T.: Estimating class priors in domain adaptation for word sense disambiguation. In: Proceedings of COLING-ACL 2006, pp. 89–96 (2006)

5. Chan, Y.S., Ng, H.T.: Domain adaptation with active learning for word sense disambiguation. In: Proceedings of ACL 2007, pp. 49–56 (2007)
6. Chang, C.C., Lin, C.J.: LIBSVM: a library for support vector machines (2001), Software. http://www.csie.ntu.edu.tw/~cjlin/libsvm
7. Chen, T., Xu, R., He, Y., Wang, X.: Improving distributed representation of word sense via wordnet gloss composition and context clustering. In: Proceedings of ACL-IJCNLP 2015, pp. 15–20 (2015)
8. Clinchant, S., Csurka, G., Chidlovskii, B.: A domain adaptation regularization for denoising autoencoders. In: Proceedings of ACL 2016, pp. 26–31 (2016)
9. Daumé III, H.: Frustratingly easy domain adaptation. In: Proceedings of ACL 2007, pp. 256–263 (2007)
10. Daumé III, H., Kumar, A., Saha, A.: Frustratingly easy semi-supervised domain adaptation. In: Proceedings of the 2010 Workshop on Domain Adaptation for Natural Language Processing, ACL 2010, pp. 23–59 (2010)
11. Escudero, G., rquez, L.M., Rigau, G.: An empirical study of the domain dependence of supervised word sense disambiguation systems. In: Proceedings of EMNLP/VLC 2000, pp. 172–180 (2000)
12. Ganin, Y., Lempitsky, V.: Unsupervised domain adaptation by backpropagation. In: Proceedings of the 32nd ICML, pp. 1180–1189 (2015)
13. Hashida, K., Isahara, H., Tokunaga, T., Hashimoto, M., Ogino, S., Kashino, W.: The RWC text databases. In: Proceedings of the First International Conference on Language Resource and Evaluation, pp. 457–461 (1998)
14. Izquierd, R., Suárez, A., Rigau, G.: Word vs. class-based word sense disambiguation. J. Artif. Intell. Res. **54**, 83–122 (2015)
15. Jiang, J., Zhai, C.: Instance weighting for domain adaptation in NLP. In: Proceedings of ACL 2007, pp. 264–271 (2007)
16. Komiya, K., Okumura, M.: Automatic determination of a domain adaptation method for word sense disambiguation using decision tree learning. In: Proceedings of IJCNLP 2011, pp. 1107–1115 (2011)
17. Komiya, K., Okumura, M.: Automatic domain adaptation for word sense disambiguation based on comparison of multiple classifiers. In: PACLIC 2012, pp. 77–85 (2012)
18. Kouno, K., Shinnou, H., Sasaki, M., Komiya, K.: Unsupervised domain adaptation for word sense disambiguation using stacked denoising autoencoder. In: Proceedings of PACLIC-29, pp. 224–231 (2015)
19. Kunii, S., Shinnou, H.: Combined use of topic models on unsupervised domain adaptation for word sense disambiguation. In: Proceedings of PACLIC-27, pp. 224–231 (2013)
20. Maekawa, K.: Balanced corpus of contemporary written japanese. In: Proceedings of the 6th Workshop on Asian Language Resources (ALR), pp. 101–102 (2008)
21. Mikolov, T., Chen, K., Corrado, G., Dean, J.: Efficient estimation of word representations in vector space. IN: Proceedings of ICLR Workshop 2013, pp. 1–12 (2013)
22. Mikolov, T., Sutskever, I., Chen, K., Corrado, G., Dean, J.: Distributed representations of words and phrases and their compositionality. Proceedings of NIPS 2013, pp. 1–9 (2013)
23. Mikolov, T., tau Yih, W., Zweig, G.: Linguistic regularities in continuous space word representations. In: Proceedings of NAACL 2013, pp. 746–751 (2013)
24. National Institute for Japanese Language: Linguistics: Word List by Semantic Principles. Shuuei Shuppan (1964) (in Japanese)

25. Nishio, M., Iwabuchi, E., Mizutani, S.: Iwanami Kokugo Jiten Dai Go Han. Iwanami Publisher (1994) (in Japanese)
26. Okumura, M., Shirai, K., Komiya, K., Yokono, H.: Semeval-2010 task: Japanese WSD. In: Proceedings of the SemEval-2010, ACL 2010, pp. 69–74 (2010)
27. Postma, M., Izquierdo, R., Agirre, E., Rigau, G., Vossen, P.: Addressing the MFS bias in WSD systems. In: Proceedings of the 10th Language Resources and Evaluation Conference, LREC 2016, pp. 1695–1700 (2016)
28. Rothe, S., Schutze, H.: Autoextend: Extending word embeddings to embeddings for synsets and lexemes. In: Proceedings of ACL 2015, pp. 1793–1803 (2015)
29. Shinnou, H., Onodera, Y., Sasaki, M., Komiya, K.: Active learning to remove source instances for domain adaptation for word sense disambiguation. In: Proceedings of PACLING-2015, pp. 224–231 (2015)
30. Shinnou, H., Sasaki, M., Komiya, K.: Learning under covariate shift for domain adaptation for word sense disambiguation. In: Proceedings of PACLIC-29, pp. 215–223 (2015)
31. Sugawara, H., Takamura, H., Sasano, R., Okumura, M.: Context representation with word embeddings for WSD. In: Proceedings of PACLING 2015 (2015)
32. Sun, B., Feng, J., Saenko, K.: Return of frustratingly easy domain adaptation. In: Proceedings of AAAI-16, pp. 2058–2065 (2016)
33. Taghipour, K., Ng, H.T.: Semi-supervised word sense disambiguation using word embeddings in general and specific domains. In: Proceedings of NAACL-HLT 2015, pp. 314–323 (2015)
34. Tang, D., Wei, F., Yang, N., Zhou, M., Liu, T., Qin, B.: Learning sentiment-specific word embedding for twitter sentiment classification. In: Proceedings of ACL 2014, pp. 1555–1565 (2014)
35. Vu, T., Parker, D.S.: K-embeddings: Learning conceptual embeddings for words using context. In: Proceedings of NAACL-HLT 2016, pp. 1262–1267 (2016)

Reference and Coreference Resolution

"Show Me the Cup": Reference with Continuous Representations

Marco Baroni[1]([✉]), Gemma Boleda[1], and Sebastian Padó[2]

[1] Center for Mind/Brain Sciences, University of Trento, Trento, Italy
{marco.baroni,gemma.boleda}@unitn.it
[2] Institut für Maschinelle Sprachverarbeitung, Universität Stuttgart, Stuttgart, Germany
sebastian.pado@ims.uni-stuttgart.de

Abstract. One of the most basic functions of language is to *refer* to objects in a shared scene. Modeling reference with continuous representations is challenging because it requires *individuation*, i.e., tracking and distinguishing an arbitrary number of referents. We introduce a neural network model that, given a definite description and a set of objects represented by natural images, points to the intended object if the expression has a unique referent, or indicates a failure, if it does not. The model, directly trained on reference acts, is competitive with a pipeline manually engineered to perform the same task, both when referents are purely visual, and when they are characterized by a combination of visual and linguistic properties.

1 Introduction

Humans use language to talk about the world, and one of its most basic functions is to *refer* to objects [1]. This makes reference one of the fundamental devices to *ground* linguistic symbols in extralinguistic reality [2].[1] For successful reference, the speaker must choose an expression allowing the hearer to pick the right referent. For instance, assume that Adam and Barbara are in the context of Fig. 1, and consider the dialogues in (1).

(1) *Adam*: Can you please give me...
    a. ...the mug?
        Barbara: Sure.
    b. ...the pencil?
        Barbara (searching): Ahem, I can't see any pencil here...
    c. ...the book?
        Barbara: Sorry, which one?

In dialogue (1-a), reference is successful. It fails in (1-b) and (1-c), but for different reasons: in (1-b), the word "pencil" does not apply to any object in the

---

[1] We ignore the thorny philosophical issues of reference, such as its relationship to reality. For an overview and references (no pun intended), see [3].

© Springer Nature Switzerland AG 2018
A. Gelbukh (Ed.): CICLing 2017, LNCS 10761, pp. 209–224, 2018.
https://doi.org/10.1007/978-3-319-77113-7_17

Fig. 1. Example scene.

scene; in (1-c), the use of singular "the" implies that Adam refers to a unique object, while the scene contains three matching objects. These examples show how reference involves both *characterization* mechanisms that capture object properties, mainly through the use of content words (e.g. "mug" vs. "pencil"), and *individuation* mechanisms, prominently encoded in function words and morphology (e.g., "the" vs. "some", singular vs. plural), which allow us to track and distinguish referents.

Existing computational approaches to meaning account for one of these aspects at the expense of the other: Data-driven approaches, including distributional semantic and neural network models, typically model the conceptual level [4], accounting well for characterization, but not for individuation. The converse holds for logics-based approaches [5].

In this paper, we propose a neural network model aimed at both aspects of reference, and that can be trained directly on reference acts. Just like in the typical reference scenario, the model works across modalities, looking for the referent of a verbal expression in the visual world, or in a setting in which entities are characterized by joint visual and linguistic information. The model, *Point-or-Protest (PoP)*, behaves like Barbara: It identifies (*points* to) the image that corresponds to a given linguistic expression, or *protests* in case of reference failure. While the model is generic and could be extended to other reference types, our starting point in this paper is reference to (concrete) entities using single-entity denoting noun phrases (as in (1)). This case clearly illustrates the joint workings of characterization (reference requires recognizing the right sort of entity in the scene) and individuation (reference succeeds only if there is *exactly one* entity of the right kind: in (1), Barbara cannot simply recognize the presence of some "pencil mass", but she must check that there is only one pencil to unambiguously refer to). We show, in two experiments, that PoP is competitive with a state-of-the-art pipeline requiring specific heuristics.

2   Models

Point-or-Protest Point-or-Protest (PoP) is a feed-forward neural network learning from examples how to react to successful and failed reference acts.[2] Given

---

[2] For neural network design and training see, e.g., [6].

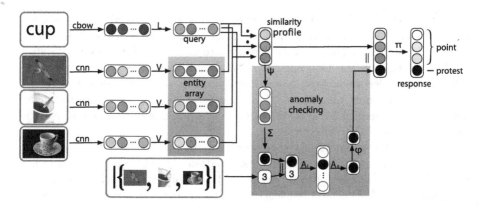

Fig. 2. The point-or-protest (PoP) model. Network inputs are marked with salmon-pink frames. Uppercase Latin letters represent linear transformations and lowercase Greek letters nonlinearities. ‖ stands for vector concatenation, the period represents dot products, Σ stands for summation across the values of a vector. Vectors containing ellipses may have different dimensionality than the one depicted; among such vectors, those with the same color belong to the same space (have the same number of dimensions). The intensity of a cell's fill is informally meant to express the size of the value it contains. (Color figure online)

a variable-length sequence of objects depicted in images (possibly coupled with other information characterizing them, e.g., verbal attributes) and a natural language query, PoP must either *point* to the object denoted by the query, returning its index in the sequence, or *protest* if the query phrase is not an appropriate referring expression. The PoP architecture builds an "entity array" whose entries are vectors storing information about the objects in the scene, and uses similarity-based reasoning about the vectors in the array and the query to decide its response. We currently focus on singular definite article semantics, as in (1), with failure if there is no possible referent (missing-referent anomaly) or if there is more than one (multiple-referent anomaly). We discussed above the linguistic appeal of this case. From a machine-learning perspective, one-entity individuation requires a non-linear separation of the anomalous reference acts (0 or more than 1) from the felicitous ones.

We use the diagram in Fig. 2 to introduce PoP. In this example, the input set contains a harrier and two cups, with the corresponding linguistic query being *cup*.[3] PoP should thus raise the anomaly flag.

The linguistic query is first mapped to a dense space by using pre-compiled cbow embeddings, whereas images are mapped to vector representations by passing them through a pre-trained convolutional neural network (cnn), and extracting the activation patterns in one of the top layers of the network (see Sect. 4

---

[3] We do not enter the determiner in the query, since it does not vary across data points: our setup is equivalent to always having "the" in the input. The network learns the intended semantics through training.

for further details). If the input consists of objects with linguistic attributes, we simply concatenate the corresponding cnn and cbow vectors to get their input representation, and analogously we concatenate cbow vectors to represent multiword linguistic phrases. Conceptually, using cbow embeddings means that the listener we model already possesses large amounts of unembodied knowledge about word meaning, as gathered from linguistic co-occurrence patterns independently of reference. This assumption is unrealistic, and we abandon it with the TRPoP model described below.

PoP maps the input object representations in the sequence to an array of entity vectors by applying a linear transformation. The corresponding mapping matrix $\mathbf{V}$ is shared across objects, as the position of objects in the input sequence is arbitrary, and PoP should not learn associations between objects and specific sequence slots (e.g., from the Fig. 2 example, it should not learn to associate cups with positions 2 and 3 in general). Each vector in the entity array corresponds to one input object. In parallel, PoP maps the linguistic expression to a "query" vector through a separate linear transformation $\mathbf{L}$. The query vector lives in the same space as the entity vectors to enable pairwise similarity computations. We can thus interpret the matrices $\mathbf{V}$ and $\mathbf{L}$ as mapping input vectors into a shared multimodal space, in which it is possible to probe visual (or mixed) entities with linguistic queries. Next, the network takes the dot product of the query with each entry in the entity array. The resulting vector (containing as many dimensions as dot products, and thus objects) encodes the similarity profile of the query with the entity vectors: the larger the value in dimension n, the more likely it is that the n-th object in the input sequence is a good referent for the query.

PoP also needs to assess whether the reference act was felicitous. The cumulative "similarity mass" across entity vectors should provide the network with good evidence to reason about anomaly. For the specific aim of modeling singular reference, the network should discover that, when cumulative similarity is too low or too high, the reference is not appropriate for the current sequence: in the first case, because no object matches the query; in the second, because there is more than one object that matches the query. More precisely, along the "anomaly pathway" shown in grey in Fig. 2, we first pass the similarity vector through a nonlinearity ψ to sharpen the contrasts, particularly zeroing out low similarities. For example, a relu transformation might set all low similarities to 0, making it easier to detect anomalies: in Fig. 2, the whitening of the harrier similarity cell is meant to suggest this process. We then sum across all values in the resulting vector, obtaining a cumulative similarity score. We concatenate it with the *cardinality* of the input sequence and feed them, via a linear transformation $\mathbf{A}_i$, to a vector of "anomaly sensor" cells. Cardinality enables the model to take the number of inputs into account when assessing the cumulative similarity score: the same score that looks suspiciously high for two objects is bound to be low for ten objects. More specifically, through cardinality the model can compare the average similarity to arbitrary thresholds, and subsets of anomaly sensor cells can learn different thresholds to pick up anomalies (the presence of multiple anomaly sensor cells allows the model to pick up "non-linear" patterns,

Fig. 3. Example sequences from the Object-Only dataset: Multi-referent anomaly (top) and successful reference (bottom, correct image marked with blue dashed frame). (Color figure online)

such as the one for single-entity reference we are addressing here). Their output is linearly combined via matrix $\mathbf{A}_o$ into a single value. The latter is passed through nonlinearity ϕ, that is bounding the anomaly score to approximate a discrete yes/no response.

We finally concatenate the similarity profile with the cell containing the anomaly score, and pass the resulting vector through a softmax nonlinearity (π). The model output for an input sequence of n objects will thus contain a probability distribution over $n+1$ indices. We take the index with the maximum value for this distribution as PoP's response: if it is one of the first n indices, then PoP "pointed" at the corresponding object, whereas if PoP assigned maximum probability to the $n+1$th cell, that means that it "protested". In the figure diagram, PoP has correctly raised the anomaly flag. We train PoP by backpropagating the error of the log-likelihood cost function when comparing its output (either the index of the correct object, or the anomaly flag) with the ground truth for the training reference acts.

Pipeline. As a strong competitor, we implemented a method that performs our task by manual pipelining of a set of separately trained/tuned components. The Pipeline first induces a set of multimodal embeddings by optimizing similarities between matched pairs of queries and objects, compared to random confounders. It uses a max-margin cost function forcing query representations to be (much) more similar to the objects they denote than to irrelevant ones. This has been shown to produce excellent multimodal embeddings [7–9]. Once these embeddings have been separately trained, the model computes similarities between the query and each of the objects in each referential act in our test sets, picking the object with largest similarity as candidate object to point at. Then, two separately-tuned heuristics are used to catch anomalous acts: Missing reference is predicted if no query-object similarity is above an (optimized) threshold. Multiple reference is guessed if the difference between the two largest similarities is below another optimized threshold.

Convolutional Neural Network. Since PoP uses input image embeddings based on a pre-trained convolutional neural network (CNN), we also test a model matching the categorical labels produced by the same CNN for the input images against the query. For the example of Fig. 2, it would pass each of the images through the full CNN, obtaining 3 labels. We take a lax approach to label matching, in which the model scores a hit even when, e.g., the gold label is a substring of the model-predicted one. Anomaly detection is straightforward (although again implemented ad hoc): CNN deems a reference act anomalous if no produced label matches the query, or if more than one does. Thus, the CNN would be successful if it predicted a synonym of *cup* for both image 2 and 3.

Tabula Rasa PoP. Through the `cbow` vectors, PoP can rely on pre-acquired text-induced word similarity knowledge. The assumption that word meanings are first learned separately, purely from language statistics, and then fine-tuned in the referential setup, is unrealistic. Ideally, we would want a model that learns word representations in parallel from reference acts *and* language statistics. For the time being, we consider instead the other extreme, where word representations are entirely induced from the reference acts during training. The "Tabula Rasa" PoP model (TRPoP) is identical to the one in Figure 2, except that input query representations (and attributes in the Object+Attribute setup explained below) are one-hot vectors. This model will thus induce distributed representations from scratch when estimating the weights of matrices $\mathbf{L}$ and $\mathbf{V}$. Such representations will then depend entirely on the role of words as queries or attributes in the referential acts we model.

3   Data

We test our model in two experiments, for each of which we have automatically created a large-scale dataset. Both datasets contain 40,000 sequences for training, 5,000 for validation and 10,000 for testing, each with 15% missing-referent and 15% multi-referent anomalies. The sequences are of varying length, from 2 to 5 candidate referents. Appendices A and B contain the algorithms used to generate the dataset. Appendix C reports detailed statistics.

Object-Only Experiment. Our first experiment represents a base case of reference, namely matching noun phrases consisting of single nouns with visually represented entities. Figure 3 shows two examples. The objects and images are sampled uniformly at random from a set of 2,000 objects and 50 ImageNet[4] images per object, itself sampled from a larger dataset used in [10]. As the examples show, we use natural objects and images, which makes the task very challenging (even humans might wonder which image in the second row depicts a *darling*). We generate data with an algorithm sampling sets of sequences with uniform distributions over sequence lengths (2 to 5) and indices of the queried object within a sequence.

Object+Attribute Experiment. Our second experiment, illustrated in Fig. 4, goes one step further in testing the model's individuation capabilities. In the scene

---

[4] http://imagenet.stanford.edu/.

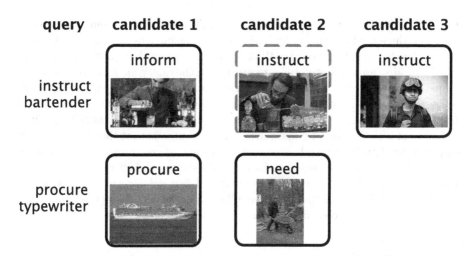

Fig. 4. Short Object+Attribute example sequences: successful reference (top) and missing-referent anomaly (bottom).

from Sect. 1, imagine that Adam points to the book on top and says "I recommend this book". This linguistically conveyed information will be associated to Barbara's representation of the entity, together with its visual features. Crucially, it can be used to identify the first book if later on Adam asks her "Can you bring the book I recommended?". We test this situation in a simplified form. Each referent is associated with both an image and a linguistically-expressed *attribute*, more specifically a verb (the only word class from which we could sample a sufficient number of attributes with the characteristics outlined below). The query and the sequence items are all pairs like *spend:bill*, where we interpret the attribute analogously to an object relative clause, that is, a *bill* that is being *spent* (we ignore tense for simplicity).

We restrict the attributes under consideration for each object to the 500 highest-associated syntactic neighbors of the object according to the DM resource [11], such that the attributes be compatible with the objects (to exclude nonsensical combinations such as *repair:dog*). Of these, we retain only verbs taking the target item as direct object, in line with the "relative clause" interpretation sketched above. Moreover, we focus on (relatively) abstract verbs, for two reasons. First, a concrete verb is more likely than an abstract one to have strong visual correlates that do not match what is actually depicted in an image (cf. *groom:dog* vs. *like:dog*). Second, successful reference routinely mixes concrete and abstract cues (e.g., a noun referring to a concrete object combined with a modifier recording an event associated to it: *the book I lent you*), and we are interested in simulating this scenario. We thus filter verbs through the concreteness norms of [12], retaining only those with a concreteness score of at most 2.5 (on a 1–5 scale).

The object-attribute structure of the stimuli in this experiment also enables us to introduce challenging confounders into the sequences – namely, pairs that share either the attribute or the object with the query. For each sequence, we start by picking an attribute-object query. Given the query, we generate two more compatible attributes for the query object, and alternative objects compatible to these attributes as well as the initial attribute. Starting from all attribute-object combinations, we randomly drop as many as necessary to obtain the final sequences of 2 to 5 items. A consequence of this design is that the objects within sequences tend to be somewhat related since they share compatible attributes, and vice versa. The first sequence in Fig. 4 illustrates the effect: For the query object *bartender*, we generate the confounder object *soldier*, connected through the attributes *instruct* and *inform*. The full sequence also includes the confounder object *emperor*, not shown in the figure.

4   Experiments

Method. PoP and Pipeline's input word representations are 400-dimensional cbow embeddings from [13], trained on about 2.8 billion tokens of unannotated text. These models, as well as TRPoP, use 4096-dimensional vectors as input visual representations, which are produced by passing images through the pre-trained VGG 19-layer CNN of [14] (trained on the ILSVRC-2012 data), and extracting the corresponding activations on the topmost fully connected layer.[5] The same pre-trained network was used to generate the labels of our CNN competitor model. The parameters of PoP/TRPoP and of the Pipeline max-margin embeddings are estimated by online stochastic gradient descent on the training portions of the two datasets. For Pipeline, we extract all possible pairs of

Table 1. Results. Figure of merit is percentage accuracy. See text for details.

	Exp 1: Object Only				Exp 2: Object+Attribute			
	Total	Pointing	MissRef	MultRef	Total	Pointing	MissRef	MultRef
PoP	66	71	57	51	69	77	57	46
TRPoP	65	70	58	44	62	70	38	48
Pipeline	67	75	51	45	65	74	37	55
CNN	35	9	100	94	-	-	-	-
Random	17	17	17	17	17	17	17	17
Majority	30	0	100	100	30	0	100	100
Probability	22	18	30	30	22	18	30	30
AttrRandom	-	-	-	-	47	64	16	0
ImgShuffle	-	-	-	-	50	58	31	32

[5] We use the MatConvNet toolkit, http://www.vlfeat.org/matconvnet/.

positive and negative query-object tuples from each reference act in the relevant training data. Details on model hyperparameter tuning are in appendix D. We consider three baselines for both experiments. **Random** assigns all labels randomly. **Majority** assigns the most frequent output label, namely anomaly, accounting for 30% of the sequences (the non-anomalous labels are distributed among predicted indices). **Probability** randomly assigns labels based on their relative frequency in the training data.

Experiment 1: Object-Only Results are reported on the left-hand side of Table 1. Besides overall accuracy (*Total*), we show accuracy itemized by successful reference acts (*Pointing*), missing-referent (*MissRef*) and multiple-referent anomalies (*MultRef*).

(TR)PoP and Pipeline are clearly above the baselines (Majority and Probability reach deceptively high anomaly-detection scores by over-raising the anomaly flag, at the cost of pointing performance). PoP's absolute performance is close to that of the manually-crafted Pipeline. By jointly learning to point and handling anomalies in reference acts, PoP loses some performance in pointing, but in exchange it does better on anomaly detection. As could be expected, MissRef is more difficult than MultRef for all three models. Interestingly, TRPoP, which does not rely on pre-trained word embeddings, performs comparably to PoP (but it requires more than twice as many epochs to converge, see supplementary materials). This suggests that useful representations of word meaning can be learned solely from examples of successful and failed reference acts.

CNN performance is barely above baseline, and, like Majority and Probability, it trivially reaches high performance on anomaly cases because it raises the anomaly flag whenever it fails to produce the name of the target object (and it rarely produces the right label). For instance, in the first example in Fig. 3, CNN can get a hit as long as it doesn't produce "cup". As for its extremely low pointing performance, note that CNN, unlike PoP, cannot make reasonable pointing guesses for objects it did not see during training. The large performance asymmetry between these two models sharing the same visual processing network shows that PoP generalizes well beyond the knowledge it inherited from this pre-trained network. Importantly, PoP reasons about similarity in multimodal space, rather than assigning hard labels. For example, the CNN, when presented with an image associated to the out-of-training query *academician*, tags it as *academic gown* – not unreasonably but incorrectly. PoP points to the correct slot because its multimodal *academician* query vector is most similar to the correct entity vector than to the other candidates, with no need to perform explicit label matching. Intriguingly, even when considering the subset of test data that CNN is trained on, we still observe an asymmetry: CNN reaches 58% accuracy, while PoP's performance is at 67%. This suggests that reference-based training has fine-tuned better representations also for the objects the CNN was explicitly trained for.

Experiment 2: Object+Attribute Results are shown on the right side of Table 1. CNN is not tested here, as it does not handle attributes. PoP's results are slightly higher than in the previous experiment, while those of TRPoP and Pipeline are slightly lower, such that now PoP is clearly above them. The three models are exploiting both visual and verbal information, as shown by their comparison to two additional baselines, shown at the bottom of the table. AttrRandom randomly picks one of the objects that shares the attribute with the query, if any, and raises the anomaly flag otherwise. This baseline has, by construction, 0% accuracy on MultRef anomalies, and it performs at random in MissRef detection. However, even in the pointing case, its performance is still well below that of the models. ImgShuffle is a variant of PoP trained after shuffling image vectors, so that each image ID is (consistently) associated with the CNN representation of another image (mostly depicting objects that do not match the image label). The only reliable signal that this baseline can then exploit is attribute information. Again, its performance is clearly below that of the models.

As for anomaly handling, while PoP still finds MissRef easier than MultRef, this time TRPoP and Pipeline actually perform worse on MissRef. Comparing the MissRef cases in which the models failed to raise the anomaly flag, we observe that Pipeline and TRPoP wrongly pointed to an entity sharing the query attribute much more often than PoP (943 and 883 vs. 629). They are thus overrelying on matching attributes, assigning too high a similarity to pairs that are simply sharing the verbal attribute. This also explains their higher performance on MultRef: Attribute sharing makes both repeated referents very similar to the query, triggering the relevant heuristic. Thus, PoP seems better at integrating verbal and visual cues than them. Compared to TRPoP, PoP has an important prior in the semantics encoded in its pre-trained word embeddings, which helps it discover systematic relations between words and objects while keeping the attribute information apart from that of the head noun. Compared to Pipeline, by jointly learning to point and to spot anomalies, it might be able to attain a better balance between visual and verbal information.

To conclude, our model PoP and its variant TRPoP can learn to refer directly from examples. While PoP is not clearly superior to Pipeline, it has a fundamental advantage: It learns to refer in one integrated architecture. Pipeline (as well as CNN) learns to *characterize* objects (e.g., to recognize cups as referents for "cup"), but uses an ad hoc strategy, needing a manually coded heuristic, to simulate *individuating* capabilities (distinguishing cases where there are several or no cups). As soon as the referential setup gets more complex, as in the Object+Attribute experiment in which visual and verbal information need to be combined, the heuristics break down.

5   Related Work

Modeling. The PoP model "reasons" about the similarity between a query and a set of candidates in vector space, akin to soft attention mechanisms in recent neural network architectures [15,16]. While attention is standardly used to retrieve auxiliary information when producing an output, we directly expose the similarity vector as (part of the) output, in order to obtain a model that learns to point robustly across input sequence orders and lengths. The idea of exposing an attention mechanism functioning as a pointer over the input has recently been employed by [17] in the context of sequence-to-sequence RNNs. PoP's entity array emulates traditional memory locations within a fully differentiable architecture. This is akin to the memory vectors of the recently proposed Memory Networks framework [18,19]. However, the Memory Networks array has fixed size, whereas our entity array adapts to input object cardinality.

Multimodal Reference Resolution. Our task is a special case of reference resolution. Various studies in this area have proposed multimodal approaches jointly handling vision and language, e.g., [20–23]. These papers focus on aspects of the resolution process we are not currently addressing, such as full compositionality or gesture, but they work with very limited perceptual input, such as simple shapes and colours. Probably the most relevant study in this area is the one by [24]. They consider visual scenes with more objects than our sequences, but more limited in nature (tables with 36 puzzle pieces). They handle spatial relations and flexible compositionality. However, they must train a separate classifier for each word in their set, which means that their method can't process unseen words, and would probably perform badly for words that are not observed frequently enough during training. Moreover, they do not present an integrated architecture for the whole resolution process, as we do, but separate components that are manually combined. Crucially, they assume referring expressions are always felicitous. We are not aware of prior work that, like our Object+Attribute setup, considers referents disambiguated by a mixture of perceptual and verbally-expressed abstract properties.

Referring Expression Generation and Other related Work. The task of referring expression generation [25] has recently received new impulses from the study of multimodal language/vision scenarios. The task is converse to ours: given a scene, generate the optimal linguistic expression to pick out a given referent. The focus is generally on considerably more complex (but artificial or heavily controlled) scenes than our sequences, and correspondingly on linguistically more complex referring expressions. Some recent efforts collect and analyze large corpora of referring expressions for multimodal tasks [26,27]. A method to generate unambiguous referring expressions for objects in natural images has been recently proposed by [28]. Our task is more distantly related to visual question answering [29–31], in the sense that we model one specific type of question that could easily be asked about an image. Even more generally, our approach fits into the multimodal distributional semantics paradigm. See [32] for a discussion of how the problem of reference is addressed in that line of work. There is of course a large body of work on modeling reference with symbolic/logical methods [33], that provides the framework for our problem, but is not directly relevant to our empirical aims. Our task can finally also be seen as a special case of the much broader problem of content-based image retrieval [34].

Acknowledgments. We are grateful to Elia Bruni for the CNN baseline idea, and to Angeliki Lazaridou for providing us with the visual vectors used in the paper. This project has received funding from the European Union's Horizon 2020 research and innovation programme under the Marie Sklodowska-Curie grant agreement No 655577 (LOVe) and ERC grant agreement No 715154 (AMORE); ERC 2011 Starting Independent Research Grant n. 283554 (COMPOSES); DFG (SFB 732, Project D10); and Spanish MINECO (grant FFI2013-41301-P). This paper reflects the authors' view only, and the EU is not responsible for any use that may be made of the information it contains.

A   Data Creation for the Object-Only Dataset (Experiment 1)

The process to generate a object sequence is shown in Algorithm 1. We start with an empty sequence and sample the length of the sequence uniformly at random from the permitted sequence lengths (l. 2). We fill the sequence with objects and images sampled uniformly at random (l. 4/5). We assume, without loss of generality, that the the object that we will query for, q, is the first one (l. 6). Then we sample whether the current sequence should be an anomaly (l. 7). If it should be a missing-anomaly (i.e., no matches for the query), we overwrite the target object and image with a new random draw from the pool (l. 9/10). If we decide to turn it into a multiple-anomaly (i.e., with multiple matches for the query), we randomly select another position in the sequence and overwrite it with the query object and a new image (l. 12/13). Finally, we shuffle the sequence so that the query is assigned a random position (l. 14).

Algorithm 1. Creation of Object-Only dataset

Input: Sequence length interval $[i \geq 2, j]$; Set of objects $O = \{o_1, \ldots, o_n\}$ and sets of associated images $I(o)$ for each object o; probability of missing-anomalies P_0; probability of multiple-anomalies P_m.

Output: $\langle$ object query q, object-image sequence $S \rangle$

1: $S \leftarrow []$
2: $l \sim U(i, j)$
3: **for** $k = 1$ to l **do**
4:     $o \sim O, i \sim I(o)$
5:     $S[k] = \langle o, i \rangle$
6: $q \leftarrow S[1]$
7: $r_0 \sim \text{Bern}(p_0), r_m \sim \text{Bern}(p_0 + p_m)$
8: **if** r_0 **then**
9:     $o' \sim O, i' \sim I(o)$ so that $o' \neq q$
10:     $S[0] \leftarrow \langle o, i \rangle$
11: **else if** r_m **then**
12:     $i \sim U(1, l)$
13:     $S[i] \leftarrow \langle o, i' \rangle$ where $S[1] = \langle o, i \rangle$, $i' \sim I(o)$
14: shuffle(S)

B   Data Creation for the Object+Attribute Dataset (Experiment 2)

Figure 5 shows the intuition for sampling the Object+Attribute dataset. Arrows indicate compatibility constraints in sampling. We start from the query pair (object 1 – attribute 1). Then we sample two more attributes that are both compatible with object 1. Finally, we sample two more objects that are compatible both with the original attribute 1 and one of the two attributes.

Algorithm 2 defines the sampling procedure formally. We sample the first triple randomly (1. 2). Then we sample two two compatible attributes for this object (1. 3), and one more object for each attribute (1. 4). This yields a set of six confounders (1. 5–10). After sampling the length of the final sequence l (1. 11), we build the sequence from the first triple and $l - 1$ confounders (1. 12–13), with the first triple as query (1. 14). The treatment of the anomalies is exactly as before.

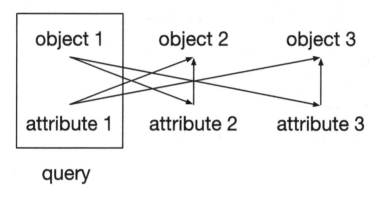

Fig. 5. Sampling intuition for Object+Attribute

Table 2. Statistics on Object-Only and Object+Attribute datasets. O: object, A: attribute, I: image.

	Train set avg. frequency				Test set avg. frequency				Unseen in test set (%)			
	O	O+I	O+A	O+A+I	O	O+I	O+A	O+A+I	O	O+I	O+A	O+A+I
Object-Only	90.0	2.0	–	–	22.5	1.2	–	–	0.0	23.1	–	–
Object+Attribute	90.9	2.2	8.2	1.1	23.1	1.3	2.7	1.0	0.0	20.2	0.9	82.9

C   Statistics on the Datasets

Table 2 shows statistics on the dataset. The first line covers the Object-Only dataset. Objects occur on average 90 times in the train portion of Object-Only, specific images only twice; the numbers for the test set are commensurately lower. While all objects in the test set are seen during training, 23% of the images are

Algorithm 2. Creation of Object+Attribute dataset

Input: Sequence length interval $[i \geq 2, j]$; Set of objects $O = \{o_1, \ldots, o_n\}$, sets of associated images $I(o)$ and associated abstract attributes $A(o)$ for each object o; probability of missing-anomalies P_0; probability of multiple-anomalies P_m.

Output: $\langle$object-attribute query q, object-image-attribute sequence $S\rangle$

1: $S \leftarrow [], S_c \leftarrow []$
2: $o_1 \sim O, a_1 \sim A(o_1), i_1 \sim I(o_1)$
3: $a_2, a_3 \sim A(o_1)$ so that $a_1 \neq a_2 \neq a_3$
4: $o_2 \sim A^{-1}(m_2), o_3 \sim A^{-1}(m_3)$
5: $S_c[1] \leftarrow \langle a_2, o_1, i \sim I(o_1)\rangle$
6: $S_c[2] \leftarrow \langle a_1, o_2, i \sim I(o_2)\rangle$
7: $S_c[3] \leftarrow \langle a_2, o_2, i \sim I(o_2)\rangle$
8: $S_c[4] \leftarrow \langle a_3, o_1, i \sim I(o_1)\rangle$
9: $S_c[5] \leftarrow \langle a_1, o_3, i \sim I(o_3)\rangle$
10: $S_c[6] \leftarrow \langle a_3, o_3, i \sim I(o_3)\rangle$
11: $l \sim U(i, j)$
12: $S[1] \leftarrow \langle o_1, a_2, i_1\rangle$
13: $S[2..l] \leftarrow$ sample candidates from S_c w.o. replacement
14: $q \leftarrow S[1]$
15: $r_0 \sim \text{Bern}(p_0), r_m \sim \text{Bern}(p_0 + p_m)$
16: **if** r_0 **then**
17:     $o' \sim O, a' \sim A(o), i' \sim I(o)$ so that $\langle o', a'\rangle \neq q$
18:     $S[0] \leftarrow \langle a', o', i\rangle$
19: **else if** r_m **then**
20:     $i \sim U(1, l)$
21:     $S[i] \leftarrow \langle a, o, i'\rangle$ where $S[1] = \langle a, o, i\rangle, i' \sim I(o)$
22: shuffle(S)

---

not. Due to the creation by random sampling, a minimal number of sequences is repeated (5 sequences occur twice in the training set, 1 four times) and shared between training and validation set (1 sequence). All other sequences occur just once.

The second line covers the Object+Attribute dataset. The average frequencies for objects and object images mirror those in Object-Only quite closely. The new columns on object-attribute (O+A) and object-attribute-image (O+A+I) combinations show that object-attribute combinations occur relatively infrequently (each object is paired with many attributes) but that the combination is considerably restricted (almost no combinations are new in the test set). The full entity representations (object-attribute-image triples), however, are very infrequent (average frequency just above 1), and more than 80% of these are unseen in the test set. A single sequence occurs twice in the test set, all others once; one sequence is shared between train and test.

D   Hyperparameter Tuning

We tuned the following hyperparameters on the Object-Only validation set and re-used them for Object+Attribute without further tuning (except for the Pipeline heuristics' thresholds). Chosen values are given in parentheses.

- **PoP**: multimodal embedding size (300), anomaly sensor size (100), nonlinearities ψ (`relu`) and ϕ (`sigmoid`), learning rate (0.09), epoch count (14).
- **TRPoP**: same settings, except epoch count (36).
- **Pipeline**: multimodal embedding size (300), margin size (0.5), learning rate (0.09), maximum similarity threshold (0.1 for Object-Only, 0.4 for Object+Attribute), top-two similarity difference threshold (0.05 and 0.07).

Momentum was set to 0.09, learning rate decay to 1E-4 for all models, based on informal preliminary experimentation.

References

1. Russell, B.: On denoting. Mind **14**, 479–493 (1905)
2. Harnad, S.: The symbol grounding problem. Physica D **42**, 335–346 (1990)
3. Reimer, M., Michaelson, E.: Reference. In: Zalta, E.N. (ed.) The Stanford Encyclopedia of Philosophy. Winter 2014 edn. (2014)
4. Turney, P.D., Pantel, P.: From frequency to meaning: vector space models of semantics. J. Artif. Intell. Res. **37**, 141–188 (2010)
5. Bos, J., Clark, S., Steedman, M., Curran, J.R., Hockenmaier, J.: Wide-coverage semantic representations from a CCG parser. In: Proceedings of the COLING, Geneva, Switzerland, pp. 1240–1246 (2004)
6. Nielsen, M.: Neural Networks and Deep Learning. Determination Press, New York (2015). http://neuralnetworksanddeeplearning.com/
7. Frome, A., et al.: DeViSE: A deep visual-semantic embedding model. In: Proceedings of NIPS, Lake Tahoe, NV, pp. 2121–2129 (2013)
8. Lazaridou, A., Dinu, G., Baroni, M.: Hubness and pollution: delving into cross-space mapping for zero-shot learning. In: Proceedings of ACL, Beijing, China, pp. 270–280 (2015)
9. Weston, J., Bengio, S., Usunier, N.: WSABIE: scaling up to large vocabulary image annotation. In: Proceedings of IJCAI, Barcelona, Spain, pp. 2764–2770 (2011)
10. Lazaridou, A., Pham, N., Baroni, M.: Combining language and vision with a multimodal skip-gram model. In: Proceedings of NAACL, Denver, CO, pp. 153–163 (2015)
11. Baroni, M., Lenci, A.: Distributional memory: a general framework for corpus-based semantics. Comput. Linguist. **36**, 673–721 (2010)
12. Brysbaert, M., Warriner, A.B., Kuperman, V.: Concreteness ratings for 40 thousand generally known English word lemmas. Behav. Res. Methods **46**, 904–911 (2014)
13. Baroni, M., Dinu, G., Kruszewski, G.: Don't count, predict! a systematic comparison of context-counting vs. context-predicting semantic vectors. In: Proceedings of ACL, Baltimore, MD, pp. 238–247 (2014)

14. Simonyan, K., Zisserman, A.: Very deep convolutional networks for large-scale image recognition. In: Proceedings of ICLR Conference Track, San Diego, CA (2015). http://www.iclr.cc/doku.php?id=iclr2015:main
15. Bahdanau, D., Cho, K., Bengio, Y.: Neural machine translation by jointly learning to align and translate. In: Proceedings of ICLR Conference Track, San Diego, CA (2015). http://www.iclr.cc/doku.php?id=iclr2015:main
16. Xu, K., et al.: Show, attend and tell: Neural image caption generation with visual attention. In: Proceedings of ICML, Lille, France, pp. 2048–2057 (2015)
17. Vinyals, O., Fortunato, M., Jaitly, N.: Pointer networks. In: Proceedings of NIPS, Montreal, Canada, pp. 2692–2700 (2015)
18. Sukhbaatar, S., Szlam, A., Weston, J., Fergus, R.: End-to-end memory networks (2015). http://arxiv.org/abs/1503.08895
19. Weston, J., Chopra, S., Bordes, A.: Memory networks. In: Proceedings of ICLR Conference Track, San Diego, CA (2015). http://www.iclr.cc/doku.php?id=iclr2015:main
20. Gorniak, P., Roy, D.: Grounded semantic composition for visual scenes. J. Artif. Intell. Res. **21**, 429–470 (2004)
21. Larsson, S.: Formal semantics for perceptual classification. J. Logic Comput. **25**, 335–369 (2015)
22. Matuszek, C., Bo, L., Zettlemoyer, L., Fox, D.: Learning from unscripted deictic gesture and language for human-robot interactions. In: Proceedings of AAAI, Quebec City, Canada, pp. 2556–2563 (2014)
23. Steels, L., Belpaeme, T.: Coordinating perceptually grounded categories through language: a case study for colour. Behav. Brain Sci. **28**, 469–529 (2005)
24. Kennington, C., Schlangen, D.: Simple learning and compositional application of perceptually grounded word meanings for incremental reference resolution. In: Proceedings of ACL, Beijing, China, pp. 292–301 (2015)
25. Krahmer, E., van Deemter, K.: Computational generation of referring expressions: a survey. Comput. Linguist. **38** (2012)
26. Kazemzadeh, S., Ordonez, V., Matten, M., Berg, T.: ReferItGame: referring to objects in photographs of natural scenes. In: Proceedings of EMNLP, Doha, Qatar, pp. 787–798 (2014)
27. Tily, H., Piantadosi, S.: Refer efficiently: use less informative expressions for more predictable meanings. In: Proceedings of the CogSci Workshop on the Production of Referring Expressions, Amsterdam, The Netherlands (2009)
28. Mao, J., Huang, J., Toshev, A., Camburu, O., Yuille, A., Murphy, K.: Generation and comprehension of unambiguous object descriptions. In: Proceedings of CVPR, Las Vegas, NV (2016) (in Press)
29. Geman, D., Geman, S., Hallonquist, N., Younes, L.: Visual turing test for computer vision systems. Proc. Nat. Acad. Sci. **112**, 3618–3623 (2015)
30. Malinowski, M., Fritz, M.: A multi-world approach to question answering about real-world scenes based on uncertain input. In: Proceedings of NIPS, Montreal, Canada, pp. 1682–1690 (2014)
31. Ren, M., Kiros, R., Zemel, R.: Exploring models and data for image question answering. In: Proceedings of NIPS, Montreal, Canada (2015). https://papers.nips.cc/book/advances-in-neural-information-processing-systems-28-2015
32. Baroni, M.: Grounding distributional semantics in the visual world. Lang. Linguist. Compass **10**, 3–13 (2016)
33. Abbott, B.: Reference. Oxford University Press, Oxford (2010)
34. Datta, R., Joshi, D., Li, J., Wang, J.: Image retrieval: ideas, influences, and trends of the new age. ACM Comput. Surv. **40**, 1–60 (2008)

Improved Best-First Clustering for Coreference Resolution in Indian Classical Music Forums

Joe Cheri Ross[✉] and Pushpak Bhattacharyya

Department of Computer Science and Engineering, Indian Institute of Technology
Bombay, Mumbai, India
{joe,pb}@cse.iitb.ac.in

Abstract. Clustering step in the mention-pair paradigm for coreference resolution, forms the chain of coreferent mentions from the mention pairs classified as coreferent. Clustering methods including best-first clustering considers each antecedent candidate individually, while selecting the antecedent for an anaphoric mention. Here we introduce an easy-to-implement modification to best-first clustering to improve coreference resolution on Indian classical music forums. This method considers the relation between the candidate antecedents along with the relation between the anaphoric mention and the candidate antecedent. We observe a modest but statistically significant improvement over the best-first clustering for this dataset.

Keywords: Coreference resolution · Information extraction
Indian classical music

1 Introduction

Coreference resolution is the task of finding mentions in a discourse referring to the same entity and grouping them into a set [1]. The motivation behind improving coreference resolution on Indian classical music forums is to be improve relation extraction from these forums, thus contributing to meta information in knowledge base for Indian classical music. Many of the forums and blogs on Indian classical music are rich source of information. Rasikas.org [2] forum considered for this study, has discussions in English on different topics in Carnatic music (sub-genre of Indian classical music). Considering the relevance of extractable information from this forum to the knowledge base for Indian classical music, coreference resolution is vital in improving extraction of relations.

The coreference resolution approach described in this paper is based on mention-pair model [3,4], where the classification of mention pairs is followed by clustering to form chain of coreferent mentions. The classification approach is hybrid with a rule-based sieve and machine learning based classifier. Pair wise classification decisions are utilized for partitioning coreferent mentions in clustering [5]. There are a few existing approaches for clustering. To find the antecedent

© Springer Nature Switzerland AG 2018
A. Gelbukh (Ed.): CICLing 2017, LNCS 10761, pp. 225–232, 2018.
https://doi.org/10.1007/978-3-319-77113-7_18

of an anaphoric mention, best-first clustering considers all the mention pairs classified as coreferent with the anaphoric mention. The best mention pair is picked to find the right antecedent, based on the classification confidence associated with the mention pair [4, 6]. The closest-first approach selects the closest preceding coreferent mention in the discourse as the antecedent [7]. Aggressive-merge approach selects all coreferent mentions to the anaphoric mention and make it part of the same coreferent chain [3]. Our method introduces an improvement over best-first clustering.

In the mention-pair model, mention pairs are formed between an anaphoric mention (m_{ana}) and candidate antecedent mentions which precede the anaphoric mention in the discourse. Mention pair classification classifies these mention pairs as coreferent or not. From the coreferent mention pairs involving the anaphoric mention, best-first clustering selects the antecedent (m_{ant}) from the mention pair having the highest classification confidence score associated with it. The probability estimate of mention-pair classification serves for the confidence score.

$$m_{ant} = \underset{m_c \in candidate\ antecedents}{argmax} P((m_c, m_{ana})) \qquad (1)$$

where $P((m_c, m_{ana}))$ denotes the classification probability estimate associated with the mention pair (m_c, m_{ana}). The modification to best-first clustering proposed in this paper, modifies the confidence score associated with a mention pair (m_c, m_{ana}), based on the cues obtained from other candidate antecedents in support to this coreferent decision. Other candidate antecedents which support the coreferent relation of this mention pair are called *support* mentions.

2   Improved Best-First Clustering

This method is motivated by the fact that when an anaphoric mention is found coreferent with multiple candidate antecedents, the candidate mentions which are coreferent to each other are more likely to be the antecedent, compared to another mention which has no coreferent relation with other candidates. Consider this sample forum post with mentions in bold.

Snehapriya is the topic of this thread. Has this forum discussed rAga snEhapriya. There is one composition in this raga AFAIK. kamalabhava sannuta by citraveeNa ravikiraN. Is this raga known by another name vaiSh-Navi?

Figure 1 shows the anaphoric mention *this raga* in this text (last sentence) and the candidate antecedents classified as coreferent with it during mention pair classification step (dotted line→coreference relation, bold line→strong coreference relation). The strong coreference relation between the candidates *Snehapriya* and *raga snehapriya* makes them better candidates over others. Here for the candidate *Snehapriya*, mention *raga snehapriya* is a support mention, making it a highly probable antecedent to *this raga*. While clustering, a candidate antecedent having a coreferent relation with other candidate antecedents of an anaphoric mention makes it a better candidate. This is the basement of the proposed modification to best-first clustering.

While best-first clustering depends solely on probability estimate associated with mention pair classification to determine confidence score, we propose to look for a method which finds the support for a candidate antecedent from other candidate antecedents and utilize this for computing confidence score along with probability estimate. Candidate antecedent having support from other candidate antecedents has better chances of getting accepted as the antecedent of the anaphoric mention (like *Snehapriya* in the example). The mention pair involving the candidate antecedent and support mention (another candidate antecedent) is termed as *support mention pair*. A mention is considered for support only if the classification confidence between the mention and the candidate antecedent is greater than the defined threshold (*conf_thresh*). For *raga snehapriya* to be a support to *Snehapriya* while resolving the antecedent for *this raga*, the classification confidence of the pair (*Snehapriya, raga snehapriya*) has to be greater than *conf_thresh*.

As mentioned our mention pair classification follows a hybrid approach combining a rule-based approach with a machine learning based approach. The rule-based sieve classifies mention pairs which can be easily classified with a set of defined rules like coreference due to lexical similarity. Rest of the mention-pairs depends on machine learning based classification. Rule-based classifications are done with a higher confidence and a high confidence value (1) is attached to these classifications as probability estimate value. Such mention pairs play a crucial role in this approach, as support decision is dependent on the classification confidence between the candidate antecedent and the support mention. In the example, the mention pair (*Snehapriya, raga snehapriya*) is classified by the rule-based sieve with a probability estimate value 1, making it a strong support mention pair for this case.

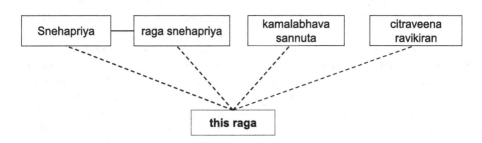

Fig. 1. An example scenario of antecedent selection taken from a forum post

This clustering method identifies all such support mentions for a candidate antecedent and computes the **support score** (refer Algorithm 1). The new confidence score (non-probabilistic value) associated with a mention pair, combines the classification confidence (probability estimate) and the support score. This is computed as the linear combination of classification probability estimate and the support score associated with this mention-pair (refer Eq. 2). This confidence score replaces the probability estimate in Eq. 1 to find the best antecedent for an anaphoric mention.

Algorithm 1. Compute coreferent support score

Require: mention pair for which support score has to be computed($(\mathrm{m}_{ant}, \mathrm{m}_{ana})$), coreferent mention pairs from the document(all_mpairs), confident mention pair threshold($conf\_thresh$)
Ensure: Support score($supp$)
1: $supp \leftarrow 0$
2: $confident\_mpairs \leftarrow$ mention pairs in all_mpairs classified coreferent with prob. est. $> conf\_thresh$
3: **for all** (m_i, m_j) in all_mpairs **do**
4:     **if** $(m_j == \mathrm{m}_{ana})$ **AND** $(((m_i, m_{ant}) \in confident\_mpairs$ **OR** $(m_{ant}, m_i) \in confident\_mpairs$) **then**
5:         $supp \leftarrow supp + P((m_i, m_{ant}))$

2.1   Algorithm

Algorithm 1 describes the method to compute the support score for a candidate antecedent given an anaphoric mention (m_{ana}). The support score $(supp)$ is computed for all candidate antecedents of this anaphoric mention. The method takes the mention pair involving a candidate antecedent (ex. (*Snehapriya, this raga*)) and all the coreferent mention pairs in the document as input. Mention pairs with a probability estimate greater than pre-defined threshold are considered for identifying the support (step 2). Step 4 defines the condition to be satisfied for a coreferent mention pair to be considered as a support mention pair for the candidate antecedent (ex. *Snehapriya*). The condition says that, the second mention of the pair must be m_{ana}. The latter part of the condition (after first AND) makes sure that m_i is coreferent with m_{ant} with classification probability estimate greater than the defined threshold $(conf\_thresh)$, by checking if this pair belongs to $confident\_mpairs$. Support score $(supp)$ is the sum of the classification probability estimate associated with all such support mention pairs $(P((m_i, m_{ant}))$ or $P((m_{ant}, m_i)))$. In the example, taking the candidate antecedent as *Snehapriya*, the former part of the condition assures the identified support mention is coreferent with *this raga*. *raga snehapriya* is one candidate that satisfies this. All the other 3 mentions shown in Fig. 1 also satisfy this. Latter part checks whether *raga snehapriya* has a coreferent relation $(> conf\_thresh)$ with the candidate antecedent *Snehapriya*. This is satisfied for this instance; hence *raga snehapriya* is a support mention to candidate antecedent *Snehapriya* for the anaphoric mention *this raga*.

The confidence score is now computed using

$$confidence\ score = \lambda P_e + (1 - \lambda)supp, \lambda \in (0, 1) \qquad (2)$$

where P_e is the probability estimate associated with the mention pair classification and *supp* is the support score associated with the mention-pair. λ decides the weightage of P_e in the confidence score.

2.2 Dynamic λ

The confidence score computation is modified to have different λ values depending on the mention pair instance. This is based on the assumption, λ is directly proportional to the classification confidence associated with the mention pair. The method in Eq. 3 takes the probability estimate value associated with the mention pair classification as its classification confidence.

$$\lambda = kP_e, k \in (0, 1) \tag{3}$$

where k is a constant. An alternate method is devised to decide classification confidence. Here classification confidence is computed using n different classifiers on the test data. Training data is partitioned to train these n classifiers. Testing is done on the actual test data and the variance of the classification result on a test mention pair instance is considered as its confidence of classification. Intuitively, higher variance should adversely affect classification confidence, hence λ is computed as

$$\lambda = \frac{1}{1 + clsf\_var} \tag{4}$$

where $clsf\_var$ is the variance of classification results from n classifiers. To maintain λ between 0 and 1, 1 is added to $clsf\_var$ in the denominator.

3   Dataset: Rasikas.org

The coreference annotated dataset contains forum posts from Rasikas.org. This is a prominent discussion forum for Carnatic music, which is the classical music of south India. The main topics of discussion in the forum includes raga [8], *tala* (rhythm), *vidwans & vidushis* (musicians), *vaggeyakaras* (composers), *kutcheri* (concert) reviews & recordings, album reviews, etc. Table 1 shows the details of this dataset. This forum is a rich source of information and listeners' opinions in the mentioned topics.

Table 1. Details of annotated posts.

Forum	#Posts	#Sent.	#Mentions
Raga & Alapana	300	2093	3630
Vidwans & Vidushis	587	3045	10884
Vaggeyakaras	325	2339	4421

Each forum post is a short discourse text comprising 4–5 sentences on an average. The content comprises mixture of written and spoken discourse reflecting the orality of online communication styles. This is attributed also with a few grammatical errors, less structuring and spelling discrepancies especially with the named entities.

Table 2. Results with different classifiers (P, R, F)→ (Precision, R:Recall, F:F-measure), CoNLL:CoNLL Score. CoNLL score of significant improvements are in bold.

Experiments		MUC			B^3			CEAFe			CoNLL
		P	R	F	P	R	F	P	R	F	
Neural Net	BF	55.45	62.35	58.38	54.84	65.36	59.44	50.62	60.75	54.88	57.56
	supp-BF	55.67	62.81	58.70	54.92	65.91	59.70	50.75	60.76	54.96	**57.79**
	supp-BF-1	55.78	62.71	58.72	55.00	65.86	59.74	50.74	60.90	55.02	57.83
	supp-BF-2	55.54	62.71	58.57	54.89	65.71	59.61	50.71	60.71	54.93	57.70
SVM (RBF)	BF	48.42	64.96	55.28	49.66	66.02	56.56	54.83	57.09	55.45	55.76
	supp-BF	48.93	65.57	55.84	49.77	67.01	57.00	55.01	57.29	55.64	**56.16**
	supp-BF-1	48.92	65.56	55.83	49.76	67.00	57.00	55.01	57.29	55.64	56.16
	supp-BF-2	48.77	65.35	55.65	49.73	66.73	56.88	54.99	57.25	55.61	56.05

4 Experiments and Results

Our system follows the mention-pair model with a machine learning approach. Conventional features and the features which are found to be more important for this domain are employed [9]. We employ k-fold (5 folds) cross-validation to make the maximum utilization of available annotated dataset. The consistency of the methods is validated across 2 different classifiers, *viz.*, Multi-layered Feed-Forward Neural Network (Neural Net) and Support Vector Machine (SVM). Effectively, validation of the system is done with predicted mentions. Results are reported with MUC [10], B^3 [11] and CEAFe [12] metrics. The average of F-measures from all these metrics is taken as CoNLL Score.

Table 2 compares the accuracy between the modifications to best-first clustering method on predicted mentions. 'BF' shows the result with best-first clustering with no modification, '*supp-BF*' with the proposed modification, '*supp-BF-1*' and '*supp-BF-2*' with the dynamic λ variations of our method. The results are reported with the best performing values for the parameters; supp-BF: $\lambda = 0.5$ $conf\_thresh = 0.9$ supp-BF-1: k: 0.5 $conf\_thresh$: 0.8 supp-BF-2: n classifiers $= 9$ $conf\_thresh$: 0.8. Parameter tuning is done taking neural network as the mention-pair classifier with the development set.

With the two classifiers, experiment supp-BF produces a noticeable improvement in accuracy compared to best-first clustering. Figure 2 shows the reduction in recall errors for nominal and pronoun anaphora types in supp-BF compared to BF. As mention-pairs involving proper noun (NAM) anaphoric mentions are handled by the rule-based sieve with higher classification confidence, there is no improvement with supp-BF on this anaphora type. The improvement in accuracy of supp-BF-1 over supp-BF is very small. supp-BF-2 produces no improvement in accuracy compared to supp-BF and supp-BF-1, but better compared to the baseline best-first.

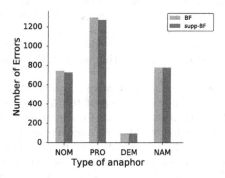

Fig. 2. Anaphora type wise comparison of errors between BF and supp-BF (Produced by Cort error analyzer [13])

The significance of the accuracy improvement is tested with a paired-t test on CoNLL scores [14]. For this, the test set is divided into 20 sub-samples and CoNLL score is computed for each sub-sample. There is a significant improvement in CoNLL score for all the variants of our method over the baseline ($p < 0.05$) with SVM and neural network. Evaluation is also done with gold mentions of the same dataset. Here also, there is a significant improvement in accuracy with supp-BF.

5   Conclusion and Future Work

This paper discussed an approach that refines best-first clustering, utilizing the candidate antecedent's relation with the other candidate mentions. In a way, this approach utilizes cues from the context in discourse, rather than just depending on the candidate mentions for coreference decision. This proposed method gives better accuracy on the rasikas.org dataset which is statistically significant, whereas the variations give improvement over baseline but not significant over the basic variant.

In this method, the mentions considered for finding a support for a candidate antecedent confines to other candidate antecedents. For future, we plan to explore how other mentions and words in the context can be utilized better for improved clustering.

References

1. Cai, J., Strube, M.: End-to-end coreference resolution via hypergraph partitioning. In: Proceedings of the 23rd International Conference on Computational Linguistics, pp. 143–151. Association for Computational Linguistics (2010)
2. Rasikas (2005). http://rasikas.org
3. McCarthy, J.F., Lehnert, W.G.: Using decision trees for coreference resolution. In: Proceedings of the International Joint Conference on Artificial Intelligence (1995)
4. Aone, C., Bennett, S.W.: Evaluating automated and manual acquisition of anaphora resolution strategies. In: Proceedings of the 33rd Annual Meeting on Association for Computational Linguistics, pp. 122–129. Association for Computational Linguistics (1995)
5. Ng, V.: Supervised noun phrase coreference research: the first fifteen years. In: Proceedings of the 48th Annual Meeting of the Association for Computational Linguistics, pp. 1396–1411. Association for Computational Linguistics (2010)
6. Ng, V., Cardie, C.: Improving machine learning approaches to coreference resolution. In: Proceedings of the 40th Annual Meeting on Association for Computational Linguistics, pp. 104–111. Association for Computational Linguistics (2002)
7. Soon, W.M., Ng, H.T., Lim, D.C.Y.: A machine learning approach to coreference resolution of noun phrases. Comput. Linguist. **27**(4), 521–544 (2001)
8. Bhagyalekshmy, S.: Ragas in Carnatic music. South Asia Books (1990)
9. Ross, J.C., Bhattacharyya, P.: Coreference resolution to support IE from indian classical music forums. Recent Adv. Nat. Lang. Process., 91 (2015)
10. Vilain, M., Burger, J., Aberdeen, J., Connolly, D., Hirschman, L.: A model-theoretic coreference scoring scheme. In: Proceedings of the 6th Conference on Message Understanding, pp. 45–52. Association for Computational Linguistics (1995)
11. Bagga, A., Baldwin, B.: Algorithms for scoring coreference chains. In: The First International Conference on Language Resources and Evaluation Workshop on Linguistics Coreference, vol. 1, pp. 563–566. Citeseer (1998)
12. Luo, X.: On coreference resolution performance metrics. In: Proceedings of the Conference on Human Language Technology and Empirical Methods in Natural Language Processing, pp. 25–32. Association for Computational Linguistics (2005)
13. Martschat, S., Göckel, T., Strube, M.: Analyzing and visualizing coreference resolution errors. In: Proceedings of the 2015 Conference of the North American Chapter of the Association for Computational Linguistics: Demonstrations, pp. 6–10 (2015)
14. Cai, J., Strube, M.: Evaluation metrics for end-to-end coreference resolution systems. In: Proceedings of the 11th Annual Meeting of the Special Interest Group on Discourse and Dialogue, pp. 28–36. Association for Computational Linguistics (2010)

A Robust Coreference Chain Builder for Tamil

R. Vijay Sundar Ram and Sobha Lalitha Devi[(⊠)]

AU-KBC Research Centre, MIT Campus of Anna University,
Chennai 600044, India
sobha@au-kbc.org

Abstract. We present on extraction of Co-reference chains from a document. Co-reference chains show cohesiveness of the document. The cohesiveness in the document is marked by cohesive markers such as Reference, Substitution, Ellipsis, Conjunction, and Lexical cohesion. In this work we will take up Pronominal, Reflexives, R-expressions and form Co-reference chains for each of the above markers. The Co-reference chains are very essential in building sophisticated natural language processing applications such as information extraction, profile generator, entity specific text summarization etc. It is also needed in machine translation and information retrieval task. Though pronominal resolution is dealt in few Indian languages such as Tamil, Hindi, Bengali, Malayalam, extraction of co-reference chains in Indian languages is not attempted. We extract co-reference chains from Tamil language text. We have evaluated the system with real time data and results are encouraging.

Keywords: Co-reference chains · Tamil · Pronominal resolution
Noun-noun anaphora

1 Introduction

Natural language text is cohesive whole. Cohesion is defined as a semantic relationship between an element in the text and some other element that is crucial to the interpretation. In this work we extract co-reference chains from a Tamil text. The co-reference chains show cohesiveness of the document. Cohesion in text is brought by various phenomena namely reference, substitution, ellipsis, conjunction and lexical cohesion [5]. Reference is use of words, which do not having meaning and its meaning can be inferred by referring to other constituent in the text such as pronouns, definite descriptions. Bloor and Bloor [2] defines substitution as follows, 'A speaker or a writer wishes to avoid the repetition of a lexical item and is able to draw on one of the grammatical resources of the language to replace the item'. Examples for substitution are one anaphors and verb phrase (VP) anaphors. Ellipsis is very similar to substitution; however in ellipsis the substitution is by zero (null). Conjunction markers specify relation between clauses and sentences. These conjunctions are classified as additive, adversative, causal and temporal. Lexical cohesion refers to cohesiveness obtained by selection of vocabulary. It is achieved by reiteration of words and collocation. Reiteration includes repetition of lexical items or their synonyms or its related words. In the present work, we mention reiteration of words as noun-noun anaphora. On analysing

© Springer Nature Switzerland AG 2018
A. Gelbukh (Ed.): CICLing 2017, LNCS 10761, pp. 233–246, 2018.
https://doi.org/10.1007/978-3-319-77113-7_19

the various Tamil texts, we found noun-noun anaphors, pronominal anaphors, zero pronouns, reflexives and definite descriptions are the commonly occurring phenomena. In this work, we take up all the commonly occurring phenomena namely, pronominal, zero pronouns, reflexives, definite descriptions and noun-noun anaphors and form the co-reference chains. Co-reference chains are very essential in building sophisticated natural language processing applications such as information extraction, profile generator, entity specific text summarization etc. Though pronominal resolution is dealt in few of Indian languages such as Tamil, Hindi, Bengali, Malayalam etc., there is oddly any published work in extraction of co-reference chains in Indian languages. We present a robust methodology to extract co-reference chains from real-time Tamil text such as News dailies and web logs. Our evaluation of the system shows encouraging results.

The early works in anaphora resolution were focussed on pronominal resolution. One of the earliest promising works was by Hobbs, which relies on the semantic information [6]. The early works in pronominal resolution can be classified into knowledge rich and knowledge poor approach [11]. Salience measure weights based approach by Lappin and Leass [8] and Kennedy and Boguraev [7], indicators based scoring by Mitkov [10] are few of the notable works. With the use of machine learning techniques, researches started to attempt noun-noun anaphora along with pronominal resolution to come up with co-reference chains. Few of the notable works are Soon et al. [26] using decision tree, Ng and Cardia [12] has worked using clustering approach, Van de Bosh using TiMBL etc. A detailed survey on coreference resolution task is presented by Ng [13].

In Indian languages most of the works in anaphora resolution were on pronominal resolution. One of the earliest published works in Indian languages is Vasisth, a multilingual anaphora resolution platform. The authors have exploited the morphological richness of Indian languages. It was initially developed for Malayalam and tested for Hindi [21]. There are published works on anaphora resolution in few Indian languages namely, Tamil, Hindi, Bengali, Malayalam and Punjabi. The shared task on anaphora resolution in ICON 2011 accelerated the work in Indian languages [24]. Table 1 presents an overview of anaphora resolution in Indian languages.

Table 1. An overview of anaphora resolution in indian languages

S. no	Languages	Methodology used
1	Tamil	Salience weight based approach [23], CRFs [1] TreeCRFs [19]
2	Bengali	GuiTAR tool for Bengali [20], BART for Bengali [21]
3	Hindi	Hobb's algorithm [4], Hybrid approach [3], Centering theory [15], Salience weight based approach [22]
4	Malayalam	Salience weight based approach [22]

2 Tamil Characteristics

Tamil belongs to the South Dravidian family of languages. It is a verb final language and allows scrambling. It has post-positions, the genitive precedes the head noun in the genitive phrase and the complementizer follows the embedded clause. Adjective, participial adjectives and free relatives precede the head noun. It is a nominative-accusative language like the other Dravidian languages. The subject of a Tamil sentence is mostly nominative, although there are constructions with certain verbs that require dative subjects. Tamil has Person, Number and Gender (PNG) agreement. Tamil is a relatively free word order language. Clausal constructions are introduced by non-finite verbs. Subject drop occurs in Tamil. Tamil also has copula drop, accusative drop, and genitive drop. Copula Drop is phenomenon where the copula verb in the sentences can be dropped. In certain constructions, accusative case marker with the noun phrase can be dropped. Similarly genitive case markers are dropped din certain constructions. Consider the example, "ramanvittu" (Raman house), here the genitive case marker 'utaiya' is dropped.

We have described co-reference chain in Tamil text with example 1.

Ex 1.
a, pirathamar   narenthara moodi   maaNavarkaLin kalvi paRRi
    Prime_minister(N) Narandra_Modi(N) student(N)+gen education(N) adv
    pecinaar.
    speak(V)+past+3sh
    (Prime Minister Narandra Modi spoke about education.)
b, avarkaL   aRivu   athikarikka   kalvi   uthava   veNtum
    They(PN) knowledge(N) improve(V)+inf education help(V)+nf want(V)_AUX
    ena   moodi   kuuRinaar.
    that Modi(N) say(V)+past+3sh
    (Modi said that education should improve their knowledge.)

In the above example 1, there are two sentences and it has two co-reference chains namely,

1, "pirathamar", "narenthara moodi", "moodi" ('Prime Minister', 'Narandhra Modi', 'Modi')
2, "maaNavarkaLin", "avarkaL" ('Students', 'they')

In the 1st co-reference chain, "pirathamar" has Definite Description relation with "narenthara moodi" and "narenthara moodi" and "moodi" has noun-noun anaphoric relation, where portion of the noun phrase 'moodi' is reiterated. These two pairs together form the first co-reference chain. In the 2nd co-reference chain, the pronoun 'avarkaL' (they/their) is refers to 'maaNavarkaLin' (students) and it forms the second co-reference chain.

Further section of the paper is presented as follows. In second section, we have presented details about Tamil language and examples on anaphoric expression in Tamil text. Third section describes the detail implementation of the various anaphora resolution engines. In section four we discussed on the corpus annotated for this task.

We present the experiment, result and analysis in fifth section. The paper concludes with the conclusion section.

3 Implementation of the Coreference Resolution Engine

Unlike many of the published works where one technique is used to resolve various types of anaphoric expressions, we have come-up with specific approach for each type of anaphoric expression. Resolution of pronominals, and reflexives relies on the syntactic features of the possible candidate antecedents. Noun phrase anaphora and Definite-Description relies on semantic features. We have built the system in five different stages.

Following are the processes done in each stage.

Stage1: Processing with Sentence Splitter and Tokenizer
Stage 2: Processing with Syntactic Processing Modules and Named Entity Recognizer
Stage 3: Classification of Anaphoric and Non-anaphoric Pronouns and Identification of Zero Pronouns
Stage 4: Resolution of various Anaphoric Expressions
Stage 5: Co-reference Chain Building

In the following section we have explained in detail each module in each stage.

3.1 Stage1: Sentence Splitter and Tokenizer

Sentence splitter splits the raw text into sentences based on cues such as period (.) or a question mark (?), exclamation mark (!) etc. These sentences are further split into tokens with white space as the delimiter. Once the tokens are split, they are placed one token per line. Each line has a token number and token which are tab-separated.

3.2 Stage 2: Syntactic Processing and Named Entity Recognition Modules

The tokenised and sentence splitted text is processed with the following syntactic processing modules namely, morphological analyser, part-of-speech tagger, chunker, pruner, clause boundary identifier. Morphological analysis of a word is the process of segmenting the word into component morphemes and assigning the correct morphosyntactic information. We have used a finite state automaton and paradigm rule based morphological analyser [16]. Part of Speech (POS) tagger is context sensitive and gives appropriate part-of-speech tag to each word based on its contextual words. We have used POS tagging engine develop using standardized BIS tagset [25]. Chunking is the task of grouping grammatically related words into chunks such as noun phrase, verb phrase, adjectival phrase etc. We have used a Chunker developed using CRFs technique [17]. Pruner is required to resolve the multiple analysis produced by the morphological analyser using POS tags. A clause is defined as a words sequence which contains a subject and a predicate. This subject can be explicit or implied. Clause boundary identifies and markers the boundaries of these clauses in the sentence automatically. We have a clause boundary identification engine built using machine learning technique [18]. The text enriched with syntactic information is fed to Named Entity recognition (NER) engine, where the Named Entities (NE) in the text are identified and classified. We have used NE engine based on CRFs techniques with linguistic features [9].

3.3   Stage3: Pre-processing to Enrich Anaphora Resolution

After processing the text with syntactic and NER, we try to identify anaphoric entities by filtering out the Non-anaphoric pronouns and identifying the zero pronouns. Pronouns can be anaphoric or non-anaphoric. Pronouns also occur as subject slot filler and they do not refer to the entities mentioned in the text. These are non-anaphoric pronouns. Subject drop(PRO drop) is one of the phenomenon in Tamil. This introduces zero pronouns. We need to identify these dropped pronouns and bring them back to the sentence.

In the following sections we describe the Anaphoric-Non Anaphoric Pronoun identification and identification of zero pronouns.

3.3.1   Anaphoric-Non Anaphoric Pronouns

All pronouns are not anaphoric. Consider the following example 2, where the pronouns occur as non-anaphoric.

Ex 2: athu   oru kuLir   kaalam.
   It   one winter season
(It was a winter season.)

Here 'athu' is a non-anaphoric pronoun, similar to a pleonastic 'it'. In the following paragraphs we have described about the anaphoric, non-anaphoric pronoun identification engine.

3.3.1.1 Anaphoric and Non-anaphoric Pronoun Identification Engine

Anaphoric and Non-anaphoric Pronoun Identification engine is built using Conditional Random Fields, a machine learning (ML) approach. Here we consider the following features from a window of five words with pronoun as the centre. We have used syntactic processed text. In the training phase these features are extracted from each pronoun which is annotated with Anaphoric and Non-anaphoric information and is fed to the machine learning engine to generate the language models.

In the testing phase with pronoun as centre, the features are collected from a window of five words and fed to the ML engine. Those pronouns which are identified as anaphoric are considered for further processing with the pronominal resolution engine.

Features Used in Anaphoric and Non-Anaphoric Pronouns

The pre-processed output data is given as input to the anaphoric and non-anaphoric classifier. The sentence boundaries in the document are identified and each sentence is given as input. The pronouns can be identified as anaphoric or non-anaphoric based on the sentence structure, finite verb of the sentence, presence of weather related nouns. For the machine learning engine to classify it properly, we need to present the distinct features. The features are explained as follows.

a. *Positional Feature:* The position of the pronoun places an important clue. If the pronoun occurs in the starting or middle of the sentence.
b. *Case marker of the pronoun:* The case marker affixed with the pronoun. The case marker suggests the role of the noun in that sentence.
c. *Finite Verb:* The type of finite verb is a clue. The verb can be an existential verb.
d. *Type of Nouns:* The root forms of the noun are presented as features along with their case markers. This feature will be combined with the verb feature and presented as a combinational feature.
e. *Number of verbs:* The number of verbs in the sentence gives the information as the number of clauses in the sentences.

Using the above mentioned features and annotated data with anaphoric and non-anaphoric information, CRFs engine is trained. In the testing phase, using the model generated and features extracted from the pre-processed input file, the non-anaphoric pronouns are identified.

3.3.2 Identification of Zero Pronouns

When a set of sentences have same focus, the subject (the focus) is introduced in the initial sentence and it may be dropped in the subsequent sentences. This introduces a zero pronoun. Consider example 3, which has two sentences describing about the same entity.

Ex 3: a. thalaivar   kuRiththa neraththil   vizaviRku   vanthaar.
      The leader(N) exact   time(N)+loc function(N)+dat come(V)+past+3sh
      (The leader came to the function at exact time.)
     b.PRO   thaane   kaarai   ootti   vanthaar.
      He(Pn) himself(reflexive) car(N)+acc drive(V) come(V)+past+3sh
      (He himself drove the car and came.)

In the above example, the second sentence has a reflexive and the reflexive will always refer to the Subject noun of the sentence. Here in this sentence the Subject is dropped.

Zero pronouns are common in complimentizer clause sentences and it is explained with example 4.

Ex4:a, sithaavin   thanthaiyai   kaNteen.
    Sita(N)+gen father(N)+acc   see(V)+past+1s
    (I met Sita's father.)
  b1. 'sithaa naalai   varuvaaL   enru   PRO
    Sita(N) tomorrow(N) come(V)+past+3sf that(complimentizer) he
    kuRinaar.'
    say(V)+past+3sh
    (He said tomorrow Sita will come.)
  b2. 'sithaa   naalai   varuvaaL   enRaar.'
    Sita(N) tomorrow(N)   come(V)+past+3sf that(complimetizer) +(he said)
    (He said tomorrow Sita will come.)

Example 4 has three sentences and 4.b.1 and 4.b.2 are the same sentences written in different styles. In example 4.b.1, the main clause is only the verb phrase 'kuRinaar' and the subject is dropped, giving rise to a zero pronoun. In example 4.b.2, the complimentizer is frozen with the verb phrase 'enraar'. Here 'enraar' has occurred in the place of 'enru avar kuRinaar'. Here the pronoun is not explicit and it occurs as a zero pronoun.

We have attempted to identify zero pronouns in the sentences with reflexives and in complimentizer clause sentences. We have used a rule based engine to identify zero pronouns in these two sentence structures. The algorithms are described below.

Algorithm to identify zero pronouns in sentences with reflexives is as follows.

Step 1: Check for reflexive pronoun in the sentence. If YES, go to step 2.
Step 2: Check for nominative noun phrase in the sentence, if NO, go to step 3.
Step 3: Extract PNG information of the finite verb.
Step 4: Introduce a nominative pronoun based the PNG information.

Algorithm to identify zero pronouns in complimentizer clause sentence is as follows.

Step 1: Check for sentence with complimentizer clause.
Step 2: If the sentence do not have complimentizer clause embedded in the main clause, go to step 3.
Step 3: If the main clause has enraar/enraaL/enrathu, replace it with 'enru avar kuRinaar'/'enru avaL kuRinaaL'/'enru athu kuRiyathu' Else Check for nominative noun phrase in the main clause. If does not exists then go to step 4.
Step 4: Extract PNG information of the finite verb in the main clause.
Step 5: Introduce a nominative pronoun based the PNG information in the main clause.

3.4   Stage 4: Resolution of Various Anaphoric Expressions

In this stage we feed the text enriched with syntactic and Named Entity information to various anaphora resolution engines. We have described the various resolution engines in the following sub sections.

3.4.1 Pronominal Resolution Engine

Pronominal resolution Engine does the task of identifying the antecedents of the anaphoric pronouns. We have come up with different engines to resolve 1st, 2nd, 3rd person singular pronouns, 1st, 2nd, 3rd person plural pronouns, and reflexives.

3.4.1.1 Singular Pronoun Resolution Engine

Singular Pronoun resolution engine is developed using Conditional Random Fields (CRFs), a machine learning technique [27]. In both training and testing phase Noun phrases with match in PNG of the pronoun are considered. The features are extracted from these NPs. In the training phase the positive and negative pairs are marked and fed to the ML engine for generating a language model. In the testing phase these NPs with its features are presented to the ML engine along with the language model to identify the antecedent of a pronoun.

Features Selection

The features required for machine learning are identified from shallow parsed input sentences. The features for all possible candidate antecedent and pronoun pairs are obtained the input sentences processed with in-depth morphological analyser, POS tagger, and chunker, clause boundary identifier and semantic parsing module namely Named Entity recognizer. The features identified can be classified as positional features, syntactic features and constraint features.

1. **Positional Features:** The occurrence of the candidate antecedent is noted in the same sentence where the pronoun occurs or in the prior sentences or in prior four sentences from the current sentence.
2. **Syntactic Features:** Syntactic Role: The syntactic role of the candidate noun phrases in the sentence is a key feature. The syntactic role of the noun phrases such as subject, object, indirect object, are obtained from the case suffix affixed with the noun phrase. We use morphological marking for the above.
   (i) Linguistic Characteristics: POS tag and chunk information of Candidate NP, suffixes affixed with the noun.
   (ii) Verb Suffixes: The suffixes which show the gender which gets attached to the verb.
   (iii) Nature of NP: Whether the candidate NP (probable antecedent) is Possessive or Existential.
3. **Constraint Features:** The constraint features are obtained from clause boundary and named entities recognized. The position of the candidate NP with respect to clause boundary such as whether the candidate NP occurs in current clause or immediate clause or non-immediate clause. The Named Entity tags associated with the candidate NPs help the learning algorithm to learn the constraints that the particular type of NEs can be its possible antecedents.

3.4.1.2 Plural Pronoun Resolution Engine

Plural pronoun resolution engine is developed using a rule-based approach. The antecedent for a plural pronoun can be a plural Noun phrase, and co-ordinated NPs.

We have developed the plural pronoun resolution engine using salience factor weights measurement, as we required to weigh each of the Noun phrase matching in gender with the plural pronoun. The features for the salience factors are obtained from the syntactic parsing output. We have mentioned the salience factors and its weights were as per Sobha (2007). Following is the algorithm used in resolving plural pronouns.

Step 1: If a plural pronoun occurs then Step 2.
Step 2: Collect all Noun phrases in the current sentence and previous four sentences which match with the gender of the plural pronoun.
Step 3: Each Noun phrase (NP) in the collection of possible antecedent set is scored with salience factor weights.
Step 4: The NPs re-sorted in descending order with their weights.
Step 5: If the highest scored NP is a plural NP, then it is selected as the Antecedent. Else step 6.
Step 6: If the highest scored NP is singular, check if this NP is part of co-ordinated NP or split antecedent, then choose the co-ordinated NP or the split antecedent as the antecedent.

Check for Co-ordinated NP: Co-ordinated NPs are those NPs which have the same scores as the highest score NP.

3.4.2   Definite Description Relation Identification Engine

Definite Description (DD) is a denoting phrase of an entity. Consider the example, Indian Prime Minister Narendra Modi. Here the phrase "Indian Prime Minister" describes about an Entity 'Narendra Modi'. In Philosophy of language, Definite Descriptions were dealt in-depth by Russell (1919). In text, Definite Descriptions can be used to refer the Entity, so DD relation needs to be identified. Here we used a CRFs technique to identify the DD relations.

We have used the POS, NE features of the two NPs (possible definite description NP and Entity NP) and two preceding and following words.

3.4.3   Noun-Noun Anaphora Resolution

Noun-Noun Anaphora resolution is the task of identifying the referent of the noun which has occurred earlier in the document. Proper name referring to proper name, partial name referring to proper name, acronyms referring to proper names, definite description referring to proper name is included in Noun-Noun anaphora resolution. The engine to resolve the noun anaphora is built using Conditional Random Fields technique. Features used in Noun-Noun Anaphora Resolution are discussed below.

We consider the noun anaphor as NP_i and the possible antecedent as NP_j. Unlike pronominal resolution, Noun-Noun anaphora resolution requires such as commonality between NP_i and NP_j. We consider the word, head of the noun phrase, Named entity tag and Definite Description tag, gender, sentence position of the NPs and the distance between the sentences with NP_i and NP_j as features. The features used in the CRFs techniques are presented below. The features are divided into two types.

Individual Features

1. Single Word: is NP_i is a single word; is NP_j is a single word
2. Multiple Words: Number of Words in NP_i; Number of Words in NPj
3. POS Tags: POS tags of both NP_i and NP_j.
4. Case Marker: Case marker of both NP_i and NP_j.
5. Presence of Demonstrative Pronoun: check for presence of Demonstrative pronoun in NP_i and NP_j.

Comparison Features

1. Full String Match: Check is both the noun phrase NP_i and NP_j are same.
2. Partial String Match: Calculate the percentage of commonality between NP_i and NP_j.
3. First Word Match: Check is the first word of both the NP_i and NP_j are same.
4. Last Word Match: Check is the last word of both the NP_i and NP_j are same.
5. Last Word Match with first Word is a demonstrator: If the last word is same and is there a demonstrative pronoun as the first word.
6. Acronym of Other: Check is NP_i is an acronym of NP_j and vice-versa.

3.5 Stage 5: Co-reference Chain Building

If both the antecedent and anaphor are used as referring expressions and has the same referent in the real world, then they are termed as co-referential. Co-reference chain is formed by connecting entities referring to same entity. To identify the coreference chain, the anaphoric entities discussed above have to be identified. Hence this system has dependency with all the anaphora resolving modules. Using these anaphor-antecedent pairs, we try to build the co-reference chains by combining the pairs having common NPs. This is performed using heuristic rules.

4   Corpus Description

We have collected 210 News articles from Tamil News dailies online versions. After scrapping the text from the web pages, we feed the text into sentence splitter, followed by a tokeniser. The sentence splitted and tokenised text is pre-processed with syntactic processing tools namely morphanalyser, POS tagger, chunker, pruner clause boundary identifier. The text enriched with shallow parsed information is fed to Named entity recogniser and the Named entities are identified. Detailed explanation on syntactic processing modules and Named Entity Recogniser is presented in the next chapter. The News articles are from Sports, Disaster and General News.

The anaphoric expressions are annotated along with its antecedents using graphical tool, PAlinkA, a highly customisable tool for Discourse Annotation [14]. We have used two tags MARKABLEs and COREF. Details on the distribution of the anaphoric expressions are presented in the following Table 2.

A detailed statistics of the pronouns is given Table 3. Here we have presented the statistics of singular and plural pronouns and the number of anaphoric and

Table 2. Distribution of anaphoric expressions in the corpus

S. no	Type	Number of occurrence
1	Noun-noun anaphora	2,387
2	Anaphoric pronominal	870
3	Definite-description	378
4	Zero pronoun	87
5	Reflexives	4
6	One-anaphora	3
7	Distributives	1

Table 3. Statistics of anaphora, non-anaphoric pronouns

S. no	Pronoun type	Total no of occurrence	Anaphoric	Non-ananphoric
1	Singular	1018	575	443
2	Plural	482	308	212
3	Question	26	0	26
4	Event pronoun	78	0	78
	Total	1604	870	734

non-anaphoric pronouns in both the types of pronouns. We have also presented the statistics of Wh-pronouns such as 'ethu' (which), 'etho' (where) etc. Event pronouns are those anaphoric pronouns which refer to events such as 'aththutan' (following that), 'athanpati' (based on that) etc. We have not handle event anaphora resolution in this work.

Table 3 has the statistics of anaphoric and non-anaphoric pronouns in singular pronouns. The masculine and feminine pronouns such as 'avan' (he), 'avaL' (she), 'avar' (he), 'than' (his) occur mostly as anaphoric. While 3rd person, neuter pronouns such as 'athu' (it), 'athan' (genitive form of it) occurs more as non-anaphoric. We have also considered pronouns referring to events as non-anaphoric.

5   Experiments and Result

Evaluation of anaphora resolution engines and coreference chain builder was performed using 40 documents from Tamil News dailies. The corpus is first pre-processed with sentence splitter, tokeniser followed by shallow parsing modules and then fed to Named Entity Recogniser. The text enriched with shallow parsing and Named Entity information is fed to various anaphora resolution engines. The anaphors and their antecedents identified using these engines were given to the coreference chain builder. The performance measures of each module are presented in Table 4.

On evaluating the output of Anaphoric-Non Anaphoric identification module, there were more errors in identification of 3rd person neuter pronoun 'atu'. These pronouns are the maximum number of non-anaphoric pronouns. In Zero pronoun identification,

Table 4. Performance of individual modules

S. no	Task	Precision (%)	Recall (%)	F-Measure (%)
1	Anaphoric-non anaphoric	91.45	91.45	91.45
2	Zero-pronouns	89.34	79.34	84.05
2	Singular pronoun resolution	79.04	62.87	70.03
3	Plural pronoun	81.41	64.70	72.09
4	Definite-description	92.98	70.00	79.87
5	Noun-noun anaphora resolution	86.14	66.67	75.16

zero pronouns occurring in complimentizer sentence construction are identified properly. One of the major challenges zero pronoun identification is genitive drop phenomenon in Tamil. When genitive marker is dropped, the noun phrase is nominative and it brings an ambiguity to identifying the subject noun, which is nominative noun phrase. In singular pronoun resolution, the resolution 'atu' has less accuracy. In plural pronoun, identifying the type of 'avarkal' is challenging as it can also occur as honorific anaphor referring to singular noun phrase. Definite Description engines relies more on Named Entity identification and the errors in NER affect this module. In Noun-Noun anaphora, the recall is less as due to the errors in identification of proper noun and NER.

We have evaluated the Coreference chains with standard metrics namely MUC, B^3, CEAFm, CEAFe and BLANC. The results are presented in Table 5.

Table 5. Performance measures of coreference engine

S. no	Metric	Precision (%)	Recall (%)	F-Measure (%)
1	MUC	51.21	35.5	41.94
2	B-CUB	74.8	52.71	61.84
3	CEAFm	46.31	46.31	46.31
4	CEAFe	30.2	44.73	36.06
5	BLANC	64.35	56.74	57.80
	Average	**53.37**	**47.19**	**48.79**

6 Conclusion

We have presented a robust methodology to extract co-reference chains in Tamil documents. Co-reference chains show cohesiveness of the document. Co-reference chains are required essentially in Natural language understanding systems. We have used different approaches for various anaphoric expressions. We have identified anaphoric and non-anaphoric pronoun and zero pronouns to improve the performance of the pronominal resolution engines. The system is evaluated with real time data such as online News dailies and the results are comparable with start-of-art systems in other languages.

References

1. Akilandeswari, A., Devi, S.L.: Conditional random fields based pronominal resolution in Tamil. Int. J. Comput. Sci. Eng. **5**(6), 601–610 (2013)
2. Bloor, T., Bloor, M.: The Functional Analysis of English: A Hallidayan Approach. Arnold, London (1995)
3. Dakwale, P., Mujadia, V., Sharma, D.M.: A hybrid approach for anaphora resolution in Hindi. In: Proceedings of International Joint Conference on Natural Language Processing, Nagoya, Japan, pp. 977–981 (2013)
4. Dutta, K., Prakash, N., Kaushik, S.: Resolving pronominal anaphora in Hindi using Hobbs "algorithm". Web J. Formal Comput. Cogn. Linguist. **1**(10), 5607 (2008)
5. Halliday, M.A.K., Hasan, R.: Cohesion in English. Longman, London (1976)
6. Hobbs, J.: Resolving pronoun references. Lingua **44**, 339–352 (1978)
7. Kennedy, C., Boguraev, B.: Anaphora for everyone: pronominal anaphora resolution without a parser. In: 16th International Conference on Computational Linguistics COLING 1996, Copenhagen, Denmark, pp. 113–118 (1996)
8. Lappin, S., Leass, H.J.: An algorithm for pronominal anaphora resolution. Comput. Linguist. **20**(4), 535–561 (1994)
9. Malarkodi, C.S., Devi, S.L.: Automatic identification of named entities in Tamil using CRF. In: Proceedings of International Seminar on Current Trends in Dravidian Linguistics, 27–29 May 2013 (2013)
10. Mitkov, R.: Robust pronoun resolution with limited knowledge. In: 17th International Conference on Computational Linguistics (COLING 1998/ACL 1998), Montreal, Canada, pp. 869–875 (1998)
11. Mitkov, R.: Anaphora resolution: the state of the art. Working paper (Based on the COLING 1998/ACL 1998 tutorial on anaphora resolution) (1999)
12. Ng, V., Cardie, C.: Improving machine learning approaches to coreference resolution. In: 40th Annual Meeting of the Association for Computational Linguistics, pp. 104–111 (2002)
13. Ng, V.: Supervised noun phrase coreference research: the first fifteen years. In: ACL 2010, pp. 1396–1411, July 2010
14. Orasan, C.: PALinkA: a highly customisable tool for discourse annotation. In: Proceedings of the 4th SIGdial Workshop on Discourse and Dialogue, ACL 2003, pp. 39–43 (2003)
15. Prasad, R., Strube, M.: Discourse salience and pronoun resolution in Hindi. In: Penn Working Papers in Linguistics, vol. 6.3, pp. 189–208 (2000)
16. Ram, R.V.S., Menaka, S., Devi, S.L.: Tamil morphological analyser. In: Parakh, M. (ed.) Morphological Analysers and Generators, pp. 1–18. LDC-IL, Mysore (2010)
17. Ram, R.V.S., Devi, S.L.: Noun phrase chunker using finite state automata for an agglutinative language. In: Proceedings of the Tamil Internet - 2010 at Coimbatore, India, 23–27 June 2010, pp. 218–224 (2010)
18. Ram, R.V.S., Bakiyavathi, T., Sindhujagopalan, R., Amudha, K., Devi, S.L.: Tamil clause boundary identification: annotation and evaluation. In: Proceedings of LREC 2012, Istanbul (2012)
19. Ram, R.V.S., Devi, S.L.: Pronominal resolution in Tamil using tree CRFs. In: Proceedings of 6th Language and Technology Conference, Human Language Technologies as a challenge for Computer Science and Linguistics - 2013, Poznan, Poland (2013)
20. Senapati, A., Garain, U.: GuiTAR-based pronominal anaphora resolution in Bengal. In: Proceedings of 51st Annual Meeting of the Association for Computational Linguistics, Sofia, Bulgaria, pp. 126–130 (2013)

21. Sikdar, U.K., Ekbal, A., Saha, S., Uryupina, O., Poesio, M.: Adapting a state-of-the-art anaphora resolution system for resource-poor language. In: Proceedings of International Joint Conference on Natural Language Processing, Nagoya, Japan, pp. 815–821 (2013)
22. Sobha, L., Patnaik, B.N.: Vasisth: an anaphora resolution system for indian languages. In: Proceedings of International Conference on Artificial and Computational Intelligence for Decision, Control and Automation in Engineering and Industrial Applications, Monastir, Tunisia (2000)
23. Sobha, L.: Resolution of pronominals in Tamil. In: Computing Theory and Application, pp. 475–479. The IEEE Computer Society Press, Los Alamitos (2007)
24. Sobha, L., Bandyopadhyay, S., Ram, V.S.R., Akilandeswari, A.: NLP tool contest @ICON2011 on anaphora resolution in indian languages. In: Proceedings of ICON 2011 (2011)
25. Devi, S.L., Pattabhi, R.K., Rao, T.: Hybrid approach for POS tagging for relatively free word order languages. In: The Proceedings of Knowledge Sharing Event on Part-of-Speech Tagging, 25–26th March 2010, LDC-IL, CIIL, Mysore (2010)
26. Soon, W., Ng, H., Lim, D.: A machine learning approach to coreference resolution of noun phrases. Comput. Linguist. 27(4), 521–544 (2001)
27. Kudo, T.: CRF++, an open source toolkit for CRF (2005). http://crfpp.sourceforge.net

Named Entity Recognition

Structured Named Entity Recognition by Cascading CRFs

Yoann Dupont[1,2(✉)], Marco Dinarelli[1], Isabelle Tellier[1], and Christian Lautier[2]

[1] Laboratoire Lattice, UMR 8094 CNRS,
1 rue Maurice Arnoux, 92120 Montrouge, France
`yoa.dupont@gmail.com`
[2] Expert System France, 207 rue de Bercy, 75012 Paris, France

Abstract. NER is an important task in NLP, often used as a basis for further treatments. A new challenge has emerged in the last few years: *structured* named entity recognition, where not only named entities must be identified but also their *hierarchical components*. In this article, we describe a cascading CRFs approach to address this challenge. It reaches the state of the art while remaining very simple on a structured NER challenge. We then offer an error analysis of our system based on a detailed, yet simple, error classification.

Keywords: Machine learning · Structured named entity recognition
CRF · Quaero

1 Introduction

In this paper, we present a linear CRF cascade approach for structured named entity recognition (SNER) on Quaero v1 and v2 corpora, used in the ETAPE evaluation campaigns [10]. Named Entity Recognition (NER) is a fundamental NLP task, its structured variant being increasingly popular. We can overall distinguish two main approaches used to address this task, the first one being cascading multiple annotations with either the same or different methods. In this respect, we can cite [19], which cascaded rules in order to gradually build the structure. We can also cite [5], where a CRF and a PCFG were used, the former giving the leaves while the latter built the rest of the tree. And finally [22], the winner of ETAPE, used one CRF per entity type, for a total of 68 CRFs, and then aligned their annotations. The second approach to annotate tree-structured named entities is to directly retrieve the structure, as was done by [20], who used partial annotation rules for predicting beginnings and ends of entities and then built the tree in one pass. Finally, we can cite [8], who used a tree-CRF to learn nested biomedical entities on the GENIA corpus [14].

Cascading linear CRFs have also been applied for syntactic parsing, as did [25]. At each step, they retrieved chunks and then only kept their respective heads for the next iteration until only one chunk covering the whole sentence

© Springer Nature Switzerland AG 2018
A. Gelbukh (Ed.): CICLing 2017, LNCS 10761, pp. 249–263, 2018.
https://doi.org/10.1007/978-3-319-77113-7_20

was found (with the class "sentence"). The tree was then reconstructed by simply unfolding chunks at each step. In this paper, we design a new, more general and effective cascade of CRFs adapted to the ETAPE evaluation campaign (Sects. 2 and 3), evaluate its efficiency and analyse its errors (Sect. 4) and finally conclude (Sect. 5).

2   Structured Named Entity Recognition

2.1   Named Entity Recognition

NER is a very important NLP task, often used as the starting point of many others, such as relation extraction [2], entity linking and coreference resolution [4,7,12].

Since their definition in the MUC-6 [11], named entities have been integrated into more and more refined classifications, covering more elements of different nature and/or refining the grain of already defined typologies [6,24]. The need for structuration in named entities appeared early. The first available corpora where this need was taken into account came out with an imbrication structure where the same entity set was used along different annotation layers, applied to longer and longer sequences. It is for instance the case for the SemEval'2007 [18] task 9 corpora. To our knowledge, one of the first corpus providing real structured named entities is Quaero [23], which we will use for our experiments.

2.2   Quaero Corpus

The Quaero corpus is made of French transcribed oral broadcast news. Two annotation variants (v1 and v2) have been applied to the same data. Their main characteristics are given in the Table 1, from which we can see that there are 60% more annotations in Quaero v2 compared to Quaero v1 (v1 annotations thus probably keep silent on many entities). The specificity of the Quaero typology is that it integrates two kinds of annotations: *types* (that we will call *entities* for sake of clarity) and *components*. *Entities* follow the common named entity definition: they can be a location, a person, an organisation, an amount, etc. The different Quaero *entities* are shown in Fig. 1. *Components*, as their name suggests, are parts of an entity. For example, a person has a first and/or last name, an absolute date may have a year and/or a day and/or a month. This means that a *component* cannot be at the top level of the hierarchy. There are 27 of them, 10 of which are transversal, meaning that they can be components of different entity types.

The main Quaero difficulties lie in its wide coverage named entity definition, Quaero considering a lot of common nouns as named entities, its tree structure named entity and the fact that it is oral transcription.

Some differences between the typologies of Quaero v1 and v2 are shown in Fig. 2. Amongst the most notable differences between the two versions, there is the disappearance of organisation sub-types, namely *org.ent* (companies),

Table 1. Statistics on the Quaero train and test sets

		Training	Test
Documents		188	18
Tokens		1,291,225	108,010
Components	v1	146,405	8,902
	v2	255,498	13,612
Entities	v1	113,885	5,523
	v2	161,984	8,399

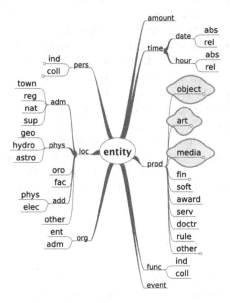

Fig. 1. Quaero v1 *Entities*

org.adm (organisations) and *org.other* (other organisations), replaced by *org.ind* (individual organisation) and *org.coll* (a collection of organisations). Many *kind* components were refined: functions, for example, are components on their own in v2. Some changes go along those previously cited: in v1, *function* and *person* were two different entities, in v2 they are one. This echoes the change of some *kind* components to *function*, a function being a component of a person in v2, whose spans were enlarged accordingly.

Quaero offers a very large number of annotations of very different natures, many entities being noun phrases without a proper name. It is for example the case for amounts, like in *deux incendies*[1] or *des historiens*[2], but not in sport

---

[1] French for *two wildfires*.

[2] French for *some historians*.

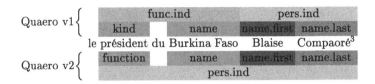

Fig. 2. Some differences between Quaero v1 et v2 (The President of Burkina Faso Blaise Compaore)

results or administrative language (e.g. *affirmation 22*[3] in Quaero guidelines). The generic nature of some entities make them sometimes hard to grasp.

While most Quaero entities have a depth of 2, there is no limit in Quaero's definitions of how deep an entity can be: we found that the deepest Quaero entity was of depth 9 (we cannot show it here for space issues). We also checked for overlapping entities having the same type, which is an argument for using cascading annotations. We found about 300 examples in the training set, a little more than 1 per thousand annotations. The system used by the winner of the challenge [22], who used binarized CRFs (one per type), is unable to model this kind of structures. Given that most components are entity-specific and given the two phenomena previously mentioned, cascaded annotation approaches are an effective way to deal with Quaero specificities, while allowing to recognize such embedded structures. In this work we obtain such an effective modeling using a cascade of linear-chain CRFs.

3   Linear-Chain CRF Cascade

Linear-chain CRFs [15] are discriminative probabilistic graphical models modeling sequential dependencies. One of the most effective implementations of linear-chain CRFs is Wapiti [16], which was used for our experiments.

The principle of a linear-chain CRF (or of any other linear-chain model) cascade for structured annotation is very simple, yet has proven to be effective for syntactic parsing [21, 25]. A basic overview is that one or multiple chunking models are used repeatedly until no more additional information is found. Taking syntactic parsing as an example, it means that, at each step, one chunk if found (NP, VP, etc.), until there is only one chunk called "S" (for sentence) left, that spans over the entire sequence. Our contribution is an adaptation of this CRF parsing technique for structured named entities to better fit the particularities of the task at hand. The main problem a classic parsing algorithm has to deal with when applied to named entities is the overabundance of "out" labels (words that are not part of an entity). Using previous algorithms as they are, a lot of passes would consist in parsing "out" labels, which would be suboptimal. We adapted the algorithm as follows: since we do not want to fully parse the sentence, we will stop as soon as no new entities are found at a given step. For the Quaero

---

[3] French for *assertion 22*, Quaero annotation guidelines being written in French.

corpus, the simplest instance of cascade would consist in training two CRFs, one for *components* and the other one for *types*, used alternatively to annotate entities layer by layer. An example of a layered annotation is illustrated in Fig. 2. This approach can be generalized to any number of CRFs. This approach was not proposed by any ETAPE constestant. The closest systems would be: 1. one constestant used fixed-depth CRF for *entities* and retrieved *components* using a rule-based approach and 2. the system of ETAPE's winner, who used binarized CRFs (one per type). Neither system use any kind of recursion, making ours more general and closer to Quaero's entity structure.

Our method is "cumulative", being in this respect somewhat comparable to the ones described in [21,25]. At each step, when an entity of length two or more is found, it is merged into a single token. Previous methods, typically used for syntactic parsing, substitute the sequence of tokens by the head of the chunk, so they only keep the most relevant token. For named entity chunking, the concept of head word does not seem so natural and useful information could easily be lost. For an overview of heuristics used for cumulating tokens, see Sect. 4.1.

Quaero entities may be very deep, we then need some recursion in our annotation scheme. The simplest way to achieve this is to have one model that would annotate components and one that would annotate entities. However, entities may be components of other entities. To model full annotations, we use two main passes: the first one being a "no context" annotation, where a first annotation has to be made with no additional information. The second one is a "context aware" annotation, where a context can be seen by the current CRF. Quaero entities also have the property that a component will always have a type as an ancestor in a parse tree. To model this property, we divide each pass into two annotations, each one being done with a specific CRF. This gives us a total of four CRFs that will be launched, following the Algorithm 1 (for sake of simplicity, we left out the entity aggregation to one token and rebuilding of base text). The first two CRFs (leaf) are called once to give a starting context to the other two (upper), which will be successively called until there is no more additional annotation. For specific features used in our models, see Sect. 4.1. We have observed in our experiments that using this approach we were able to manage annotations up to a depth of 6. Our approach is thus able to model recursion, improving the more naive fixed-depth CRF used during the ETAPE evaluation campaign.

To illustrate with the example of Fig. 2, CRF1 and CRF2 would annotate the first two levels as illustrated. CRF3 and CRF4 would not find any *component* or *entity* above, and would then stop.

4 Results

In this section, we present the results reached with our method. We will first compare the results we obtained on Quaero v1 with those of the contestants of the ETAPE evaluation campaign, as a first evaluation of our method. We will then analyse the errors it made on Quaero v2, for which no other result has been published yet.

Algorithm 1. The base algorithm for CRF cascade

---

function CRFCASCADE(*Corpus, leafC, leafE, upperC, upperE*)

                               ▷ *C are models for components. *E are models for entities.

    annotations ← ∅;

    currentAnnotations ← ∅;

    annotations ← annotations ∪ annotate(Corpus, leafC);

    annotations ← annotations ∪ annotate(Corpus, leafE) ;

    newAnnotations ← (annotations ≠ ∅);

    while newAnnotations **do**

        annotations ← annotations ∪ currentAnnotations;

        currentAnnotations ← ∅;

        currentAnnotations ← currentAnnotations ∪ annotate(Corpus, upperC);

        currentAnnotations ← currentAnnotations ∪ annotate(Corpus, upperE);

        newAnnotations ← (currentAnnotations ∩ annotations ≠ ∅);

    end while;

    return annotations;

end function;

4.1 Features and Performances

Every feature detailed here is applied on a window of two words before to two words after. We considered different sets of features to evaluate the importance that some have compared to others.

For our baseline, we only used word-specific features: not a single lexicon is used, no tagging or lemmatisation is performed. The features used are the shapes of the words, their prefixes and suffixes up to a length 5 and a variety of boolean features such as "does the word start with an uppercase?" or "is the word a number?". This baseline will then be enriched with other (more or less specific) features, to measure their impact.

We first added the outputs of basic syntactic analyses, namely lemmatisation, PoS and chunking. This model is called "+syntax". As can be seen, adding this information leads to an important quality loss. This is probably due to the fact that they do not provide any new information (lemmas) or are not precise enough (PoS, chunking).

It is commonly known that the verb is the most important syntactic unit of a sentence. Verbs could be used to disambiguate between various entities and help improving recall on unknown entities, as the same verbs could be used for entities of the same type. We added, for each word, the previous and next verb found in the sentence. French uses auxiliaries in some tenses, which precede the verb: in this case, we took the first non-auxiliary verb. This provides the "+verb" model.

We then used a full set of features, containing all the previous features described above. We also added "word classes": these classes are obtained by substituting uppercase letters by "A", lowercase letters by "a", numbers by "0" and everything else by "x". The "brief" alternative version of this feature consists in applying the same substitutions, but on contiguous sequences of characters of the same class. For example, the first name "Jean-Pierre" would become "AaaaxAaaaaa" as a word classe and "AaxAa" as brief word class. This allows to represent words in a condensed fashion that is far more general than lemmas. We also have some basic chunk-based patterns (sequences of prepositional phrases following some keyword) which simulate "rules-based" entity recognizers. We used some gazetteers extracted either from Wikipedia or from internal tools, mainly first names, last names, locations and companies. Quaero being an oral corpus, we also removed discourse markers using the list defined by [3], but only the non ambiguous ones such as "euh"[4] or "enfin bref"[5]. We did not remove, for example, "ben"[6] as it could also be a part of an Arab name. We removed repeated words with the exception of "nous nous"[7] or consecutive numbers. We considered those markers as part of an entity if they were in the middle of it, but not otherwise.

When doing accumulation, a lot of interesting information may be lost. To limit this loss, we defined some heuristic rules based on which information the feature is supposed to extract. Examples of such rules are given in Table 2

We also tried a top-down approach: finding entities first, then components. While it is relatively easy to retrieve entities when their components have been identified, components themselves may be difficult to identify: some components, such as *kind, name, extractor, range-mark, object*, tend to be ambiguous as they can either cover entities of very different natures, or be very contextual and appear in conjunction with others (an *extractor* is never isolated, for example). Their identification could be eased by first retrieving the entities that cover them, giving more useful context to the CRF.

Table 2. Examples of heuristic rules of accumulation

Feature	Example
Word	12 January → 12_January
Character classes	00 Aaaaaaa → 00_Aaaaaaa
First is upper?	12 January → false
Has number?	12 January → true
Is number?	12 January → false

The metric used to measure performances in ETAPE evaluation campaign is a modified version of the Slot Error Rate (SER) [17], which is the ratio between errors made by the system and the number of slots in the reference (N). The errors in the original SER are divided into three categories: substitutions (S), deletions (D) and insertions (I). Deletions measure the silence of a system (slots

---

[4] French for "err".

[5] French for "anyway".

[6] which can stand for "well" in French oral discourses.

[7] which can be a correct sequence in French.

in the reference which cannot be aligned to suggestions of the system), while insertions measure its noise (slots in the system's suggestions which cannot be aligned to a reference slot), substitutions are the rest of precision errors. ETAPE used a weighted SER: pure type errors (S_t) and pure boundary errors (S_b) were counted as half an error, while type and boundary errors (S_{t+b}) were counted as a full error, which gives the Eq. 1. It is the measure we used.

$$SER = \frac{D + I + S_{t+b} + 0.5 * (S_t + S_b)}{N} \tag{1}$$

The results reached with our cascade of CRFs with different sets of features are compared with those of the top 5 contestants of the ETAPE campaign in the Table 3 (SER being an error rate, the lower the best). Had we participated in ETAPE campaign, our model would have reached second position with our baseline CRF cascade, which does not include any kind of morphosyntactic analysis, dictionary or any other external resource. Top competitors in the ETAPE campaign used some external tools. [5] used WMatch [1,9], ETAPE's winner [22] used dictionaries along mined trigger words (words that have high mutual information with output classes) and a number discretiser. Our approach is competitive, as our baseline would have ranked second without using any such resource. We also have a significant quality improvement using our cascade compared to using only a naive *two levels* CRF cascade. We did not manage to improve our baseline on Quaero v1, going from slightly worse to significantly worse, the worst being when the full set of features was used. That last experiment had roughly twice the noise of the baseline. As seen in Sect. 2.2, this noise may actually be corrections of the silence due to incomplete annotations.

Table 3. On the left, results of ETAPE contestants. On the right, our results.

Contestant	method	SER
3	CRF	**33.8**
8	CRF+PCFG	36.4
10	rules	42.9
5	CRF	43.6
4	rules	55.6

Our results	SER
baseline	**35.5**
+syntax	37.0
+verbs	37.4
full set	43.3
two levels	37.0
top-down	37.1

As can be seen in Table 4, we obtain better results on Quaero v2 than on Quaero v1, due to improved typology and a more thorough human annotation. We also see that adding neighboring verbs has a detrimental effect on the quality of the annotation, no matter the experiment. Dictionaries, surprisingly, also had a detrimental effect on our results, but far smaller on Quaero v2 than on Quaero v1, which shows that a lot of the noise induced by dictionaries in Quaero v1 were actually entities missed by the annotators (SER penalizes more systems that are noisy). Looking at macro F1-scores in Table 4, we can see that the full

Table 4. Our best results on Quaero v2.

Experience	SER	micro F1	macro F1
Our baseline	**33.2**	**73.1**	**54.2**
+verb	33.7	72.9	51.4
Full set	34.8	72.3	53.2

set of features yields better results, and that the lower micro F1-score is due to the imbalance in data set, as told in Sect. 2.2. Table 5 shows some examples of difference in terms of F1-scores between the baseline and the other experiments, displaying why using the full set leads to worse results: the quality on *amount*, which is disproportionately represented, decreased significantly.

Table 5. Some entity-specific F1-score differences compared to the baseline

	+verbs	Full set
Town	−0.6	+7.1
org.ind	−1.4	+1
Amount	+0.8	−2.2

Despite these results, it is obvious that our system can still be improved. Since SER as a unique measure is not very informative, we make a more detailed analysis in the next section, trying to find some hints on where we can get improvements. Since there are some papers on Quaero v1, but none to our knowledge on Quaero v2, we will focus our error analysis on the latter.

4.2 Error Analysis

SER, as well as micro F1-score, is a measure that tends to favor most frequent entities as they carry more weight on the global metric than less frequent ones. Displaying scores by entity may allow to know on which ones the system performs better, but does not give an accurate view on where best gains can be made. To make up for this, we suggest a quantification of the shortfalls of our system in Table 6. These shortfalls are the number of F1-score points the system would gain if it were 100% accurate on a specific entity.

We see in the tables that shortfalls gather on leaves or on major types (*amount, org, pers*). If we perfectly annotated the entities of these tables, we would obtain about 90 in absolute F1-score. We can see that gains are hard to make by focusing on a single entity: if we wanted to gain a single F1-score point that way, this would equate to gaining 10 points on *name* components, 20 on *org.ind* or 24 or *kind*. However, just as errors propagate, corrections would also propagate: *name* being a component of multiple entities, corrections on that component would also spread on the entities above it and would improve scores on multiple entities (for example *org*, *loc* and *pers* where most ambiguities happen at the component level).

Table 6. Entities with the highest shortfalls on global F1-scores

Entity	F1-score	F1 gain if perfect
Name	81.48	2.28
org.ind	65.12	2.25
Amount	75.48	2.17
Kind	51.20	1.97
Qualifier	49.51	1.73
Object	76.03	1.7
pers.coll	59.91	1.68
pers.ind	78.05	1.4

To ease error analysis, we capitalized on the Quaero refinements on SER, giving us 5 kinds of errors: type errors, boundary errors, type+boundary errors, noise and silence.

Table 7. Raw percentage of the various errors

Error kind	Proportion (%)
Type	8.0
Boundary	11.7
Type+boundary	6.2
Noise	21.6
Silence	52.5

As illustrated on the table of Table 7, the main problem of our system is its silence, amounting to more than 50% of the system's errors, 19% of reference annotations not having a suggestion made by the CRF.

Now, we detail the most common errors made by our system. First, examples of such errors are given in the Table 8 for *components*. Errors on entities being mainly propagated, we will focus on component errors.

As illustrated on the chart of Fig. 3, most type errors involve either *func*, *kind* or *name*. Going from Quaero v1 to Quaero v2, some *kind* components have been replaced by *func* (cf. Fig. 2): they are closely related but there are also some possible human errors which could explain in part the confusion between the two. Some errors come from *name* morphing into *kind* in presence of other

Table 8. Error overview on components

Error	Description
Boundary	-unfrequent variation of frequent entity
	-adjective or prepositional phrase
Type	*kind* vs *func* (+human errors?)
	name/kind : some *name* become *kind* with other components (in gold)
Noise	*val*: wrong PoS on "des", "de" et "d"[a] (+human errors?)
	object: common nouns and sports results
	name: known components → country, val (numbers), relative time
Silence	*-val* : not numbered amounts
	-qualifier : missed if qualified component is missed
	-name : forgotten on relative times
	-kind : polysemic common nouns

[a] "de" and "des" in French may be partitive, possessive (not annotated) or complement (annotated).

components (ex: a country's government). CRFs seem to have trouble modeling this "isolation" phenomenon. Still on the government example, there is a volume disparity between gold annotations and what the CRF yields: while it is mostly annotated *name*, this annotation only amounts to 20% of the CRF's output. Maybe some post-processing rules could help correct this kind of errors.

Silence errors are mainly made on Quaero components, amounting to nearly 60% of all silence errors. They are mainly made on *val, object, kind* and *qualifier*. *Object* being an *amount* component, it is accompanied with a *val*, most likely those errors are linked to each other, even though we did not manage to quantify the phenomenon. Errors on *qualifier* are nearly always contextual, as it never appears alone. Most silent qualifiers are so because the component they qualify was not identified either. This allows to think that the CRF managed to "understand" this structural constraint, meaning that those silences will most likely be corrected if we manage to catch the component they qualify.

Boundary errors on components are usually of length 1 or 2 and seem equally distributed between additions and deletions, they mostly are adjectives or prepositional phrases. When it comes to entities, boundary errors tend to be larger on overall, this is due to the propagation of two kinds of errors: first the boundary errors on some components will cause an entity to have a boundary error also. Second, a silence error on a component can lead to a boundary error in an entity, for example when a first/last name is not identified at a component level, but the person is still identified (Fig. 4).

Most noise errors are on components such as *val, object, kind, qualifier* and *name*. Nearly 80% of those noise errors are on components whose form was observed on the training corpus. While some are most likely human errors, such as countries and proper names, some others are more contextual and may indicate an overfitting of the CRF, that just took those components "at face value".

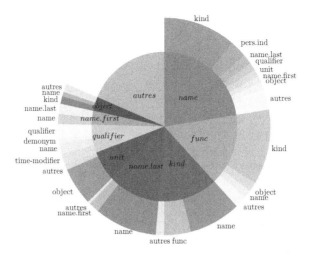

Fig. 3. Type errors on components

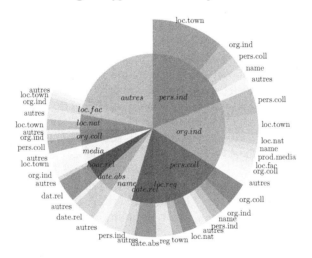

Fig. 4. Type errors on entities

As previously showed, most errors on entities seem to originate from errors made on previous levels. To check this assertion, we tried a run using the reference components instead of using a CRF to annotate leaf level components in Algorithm 1: the SER dropped to 6.3%, a result coherent with the one stated by [5], who made the same test (Table 4). We plan to isolate non-propagated type-specific errors to analyse them specifically in further research. This last test provides a strong proof that we should focus more on components of the first level, especially for common nouns that tend to be more ambiguous than proper nouns. We also need to model some "horizontal" structuration better:

some components work "symbiotically" with others, such as *val* and *object*. Type errors also showed the need for a better disambiguation between the different *name* components.

5   Conclusion

In this article, we have described a general method for structured named entity recognition using a cascade of linear-chain CRFs. We have given a generic procedure that we adapted to best fit the architecture of Quaero named entities. We showed why this specific architecture was justified; it gave promising results, while remaining simple. While we did not manage to improve the current state-of-the-art on Quaero v1, we nonetheless showed that our approach has competitive performances.

We tried to characterize the most common errors and quantified the different shortfalls of our system, which gave us some insights on how to improve it and even found potential human errors. This process sadly lacks in automation. We could compute an estimation of the propagated errors on types by checking if a component below it has the same error type. We could also check the merits of our approach by comparing it to a single CRF that would learn only the top-level entities: by comparing the two, we could see errors made by one and not the other, or by both of them. We could also compare the SER score with a recent metric named "Entity Tree Error Rate" (ETER) [13], a metric based on the SER but aims to better take into account structuration.

We plan to continue our research, especially by integrating more efficient models, by focusing on annotating common nouns and how to model context, which we think are the two most important tasks if we want to improve results on Quaero. We also plan to use the hierarchy of the Quaero entity types to our advantage: we could first learn a coarse grain CRF (ex: *pers* instead of *pers.ind* and *pers.coll*) which would be followed by a fine grain CRF that would assign the various subclasses. This could improve the disambiguation between the different subtypes of *name*.

References

1. Bernard, G., Rosset, S., Galibert, O., Adda, G., Bilinski, E.: The LIMSI participation in the QAst 2009 track: experimenting on answer scoring. In: Peters, C., et al. (eds.) CLEF 2009. LNCS, vol. 6241, pp. 289–296. Springer, Heidelberg (2010). https://doi.org/10.1007/978-3-642-15754-7_33
2. Bunescu, R.C., Mooney, R.J.: A shortest path dependency kernel for relation extraction. In: Proceedings of the conference on Human Language Technology and Empirical Methods in Natural Language Processing, pp. 724–731. Association for Computational Linguistics (2005)
3. Chanet, C.: Fréquence des marqueurs discursifs en français parlé: quelques problèmes de méthodologie. Recherches sur le français parlé **18**, 83–106 (2003)
4. Denis, P., Baldridge, J.: Global joint models for coreference resolution and named entity classification. Procesamiento del Lenguaje Natural **42**(1), 87–96 (2009)

5. Dinarelli, M., Rosset, S.: Models cascade for tree-structured named entity detection. In: Proceedings of 5th International Joint Conference on Natural Language Processing, Chiang Mai, Thailand, pp. 1269–1278. Asian Federation of Natural Language Processing (2011)

6. Doddington, G.R., Mitchell, A., Przybocki, M.A., Ramshaw, L.A., Strassel, S., Weischedel, R.M.: The automatic content extraction (ace) program-tasks, data, and evaluation. In: LREC, vol. 2, p. 1 (2004)

7. Durrett, G., Klein, D.: A joint model for entity analysis: coreference, typing, and linking. Trans. Assoc. Comput. Linguist. **2**, 477–490 (2014)

8. Finkel, J.R., Manning, C.D.: Nested named entity recognition. In: Proceedings of the 2009 Conference on Empirical Methods in Natural Language Processing: Volume 1-Volume 1, pp. 141–150. Association for Computational Linguistics (2009)

9. Galibert, O.: Approches et méthodologies pour la réponse automatique à des questions adaptées à un cadre interactif en domaine ouvert. Université Paris Sud-Paris XI, Diss (2009)

10. Gravier, G., Adda, G., Paulson, N., Carré, M., Giraudel, A., Galibert, O.: The etape corpus for the evaluation of speech-based tv content processing in the french language. In: LREC-Eighth International Conference on Language Resources and Evaluation (2012)

11. Grishman, R., Sundheim, B.: Message understanding conference-6: a brief history. In: COLING, vol. 96, pp. 466–471 (1996)

12. Hajishirzi, H., Zilles, L., Weld, D.S., Zettlemoyer, L.S.: Joint coreference resolution and named-entity linking with multi-pass sieves. In: EMNLP, pp. 289–299 (2013)

13. Ben Jannet, M., Adda-Decker, M., Galibert, O., Kahn, J., Rosset, S.: Eter: a new metric for the evaluation of hierarchical named entity recognition. In: Ninth International Conference on Language Resources and Evaluation, pp. 3987–3994 (2014)

14. Kim, J.-D., Ohta, T., Tateisi, Y., Tsujii, J.: Genia corpus-a semantically annotated corpus for bio-textmining. Bioinformatics **19**(suppl 1), i180–i182 (2003)

15. Lafferty, J., McCallum, A., Pereira, F.: Conditional random fields: probabilistic models for segmenting and labeling sequence data. In: Proceedings of ICML 2001, pp. 282–289 (2001)

16. Lavergne, T., Cappé, O., Yvon, F.: Practical very large scale CRFs. In: Proceedings of ACL 2010, pp. 504–513. Association for Computational Linguistics, July 2010

17. Makhoul, J., Kubala, F., Schwartz, R., Weischedel, R., et al.: Performance measures for information extraction. In: Proceedings of DARPA Broadcast News Workshop, pp. 249–252 (1999)

18. Màrquez, L., Villarejo, L., Martí, M.A., Taulé, M.: Semeval-2007 task 09: multilevel semantic annotation of catalan and spanish. In: Proceedings of the Fourth International Workshop on Semantic Evaluations (SemEval-2007), Prague, Czech Republic, pp. 42–47. Association for Computational Linguistics (2007)

19. Maurel, D., Friburger, N., Antoine, J.-Y., Eshkol, I., Nouvel, D.: Cascades de transducteurs autour de la reconnaissance des entités nommées. Traitement automatique des langues **52**(1), 69–96 (2011)

20. Nouvel, D., Antoine, J.-Y., Freeburger, N., Soulet, A.: Fouille de règles d'annotation partielles pour la reconnaissance des entités nommées. In: Proceedings of TALN (2013)

21. Ratnaparkhi, A.: A linear observed time statistical parser based on maximum entropy models. arXiv preprint arXiv: cmp-lg/9706014 (1997)

22. Raymond, C.: Robust tree-structured named entities recognition from speech. In: 2013 IEEE International Conference on Acoustics, Speech and Signal Processing, pp. 8475–8479. IEEE (2013)

23. Rosset, S., Grouin, C., Zweigenbaum, P.: Entités nommées structurées: guide d'annotation Quaero. LIMSI-Centre national de la recherche scientifique (2011)

24. Sekine, S., Nobata, C.: Definition, dictionaries and tagger for extended named entity hierarchy. In: LREC (2004)

25. Tsuruoka, Y., Ananiadou, S.: Fast full parsing by linearchain conditional random fields. In: Proceedings of EACL (2009)

Arabic Named Entity Recognition: A Bidirectional GRU-CRF Approach

Mourad Gridach[1(✉)] and Hatem Haddad[2]

[1] High Institute of Technology, Ibn Zohr University, Agadir, Morocco
m.gridach@uiz.ac.ma
[2] Department of Computer and Decision Engineering, Université Libre de Bruxelles,
City of Brussels, Belgium
Hatem.Haddad@ulb.ac.be

Abstract. The previous Named Entity Recognition (NER) models for Modern Standard Arabic (MSA) rely heavily on the use of features and gazetteers, which is time consuming. In this paper, we introduce a novel neural network architecture based on bidirectional Gated Recurrent Unit (GRU) combined with Conditional Random Fields (CRF). Our neural network uses minimal features: pretrained word representations learned from unannotated corpora and also character-level embeddings of words. This novel architecture allowed us to eliminate the need for most of handcrafted engineering features. We evaluate our system on a publicly available dataset where we were able to achieve comparable results to previous best-performing systems.

1 Introduction

[1] stated that named entity recognition is the NLP task that consists of identifying, labeling and tagging atomic elements in any text with a set of named entities such as Organization, Person and Location. Much research has been done by the NLP community to build NER systems because many NLP applications could use their output results and also improve their performances by integrating NER systems. NLP applications like entity coreference resolution [2], syntactic parsing [3], Question Answering (QA) [4], machine translation [5] and text clustering [6] are becoming more robust, in part because they integrated output information of NER systems.

Developing an Arabic NER system is a challenging task given the complex nature of this language compared to other languages like English or French. The first challenge is that Arabic has complex and rich morphology where words are highly inflectional and derivational entities. The second challenge is the absence of capitalization where it cannot be used as a feature to recognize named entities unlike other languages. Another challenge came from the problem of agglutination: building Arabic words consists of combining prefixes, stem and suffixes. As a result, Arabic morphology is highly complex and raises the out-of-vocabulary problem. Finally, unlike other languages (English, French, German, etc.), most of the Arabic linguistic resources are not available for the research community.

© Springer Nature Switzerland AG 2018
A. Gelbukh (Ed.): CICLing 2017, LNCS 10761, pp. 264–275, 2018.
https://doi.org/10.1007/978-3-319-77113-7_21

Most traditional high performance Arabic NER models are linear statistical models, including Conditional Random Fields [7], which rely heavily on hand-crafted engineering features and large gazetteers. For example, in their study, [8] used a combination of rules in addition to a set of features such as orthographic, morphological, POS tag, word length, and dot (i.e. if a word has an adjacent dot) features. We believe that using heavy set of hand-crafted engineering features is costly to develop and make recognizing Arabic named entities difficult to adapt to new tasks or new domains.

In the past few years, non-linear neural networks combined with word representations have been widely applied to NLP problems with great success. [9] used a feed-forward neutral network to build a system for English NER by using context window approach. More recently, recurrent neural networks (RNN) [10] were widely used in handling sequences of variable length using a recurrent hidden unit whose activation at each time step is dependent on the previous one. Various RNN models, like Long-Short Term Memory (LSTM) [11,12] and Gated Recurrent Unit (GRU) [13], have shown great success in modeling sequential data like speech recognition [14] and POS tagging [15].

In this paper, we propose a novel neural network architecture for Arabic named entity recognition. Our model takes advantages from the recent success of deep neural networks in various NLP applications. We use gated recurrent unit (GRU) as the main building block for our model combined with conditional random fields on the top of the network. In addition, we initialize our word vectors using pretrained word representations, which showed great success in sequence labeling tasks [9]. We add character-level representations of words to our model to handle the out-of-vocabulary issues and allow our system to model rare morphological variants of Arabic words. The contributions of this work are:

- Proposing novel neural network architecture for Arabic named entity recognition.
- Investigating the use of character-level representations for morphologically rich languages such as Arabic and also show that using pretrained word representations improve the system performance.
- Giving empirical evaluations of this system on publicly available dataset for modern standard Arabic NER.
- We are the first to investigate gated recurrent unit as the main building block to recognize named entities for Arabic. We leave the investigation of GRU for other languages as future work.
- Our system based on GRU and CRF is competitive on ANERcorp dataset.

2   Neural Network Architecture

In this section, we describe the different components (layers) of our neural network architecture. We present the GRU as the main component. Then, we introduce the other components: bidirectional GRU without and with CRF layer.

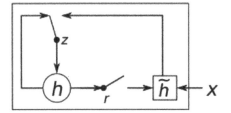

Fig. 1. Gated Recurrent Unit.

2.1   Gated Recurrent Unit (GRU)

Recurrent neural networks (RNNs) are an extension of the main architecture of feed-forward neural networks. They are powerful models that capture time dynamics via cycles in the graph. The main advantage over feed-forward neural networks is their ability to handle sequences of variable length. They use a recurrent hidden unit whose activation at each time step is dependent on that of the previous one. The main drawback of recurrent neural networks is the gradient vanishing/exploding problems, which make them difficult to train and scale to large machine learning problems [16,17].

More recently, [13] proposed Gated Recurrent Unit (GRU), which is a family of RNNs to cope with these gradient vanishing problems. GRU is a powerful and simpler alternative of Long Short-Term Memory (LSTM) networks [11]. Similarly to the LSTM unit, GRU was designed to adaptively update or reset its memory content by using a reset gate r_t^j and an update gate z_t^j which are reminiscent of the forget and input gates of the LSTM. The main difference between LSTM and GRU is in the memory content. To compute the state h_t^j of the j-th GRU at timestep t, we use the following equation [13]:

$$h_t^j = (1 - z_t^j)h_{t-1}^j + z_t^j \tilde{h}_t^j \tag{1}$$

where $\tilde{h}_t^j$ and h_{t-1}^j respectively correspond to the new candidate memory content and the previous memory content. z_t^j represents the update gate where its main goal is to control how much of the previous memory content is to be forgotten and how much of the new memory content is to be added. To compute the update gate, we use the previous hidden states h_{t-1} and the current input x_t in the following equation:

$$o_t = \sigma(W_z x_t + U_z h_{t-1}) \tag{2}$$

The new memory content $\tilde{h}_t^j$ is computed as follow:

$$\tilde{h}_t = \tanh(W x_t + r_t \odot U h_{t-1}) \tag{3}$$

where $\odot$ is the element-wise product. The final equation concerns the reset gate r_t^j. This gate allows a GRU unit to ignore the previous hidden states whenever it is deemed necessary considering the previous hidden states and the current input:

$$r_t = \sigma(W_r x_t + U_r h_{t-1}) \tag{4}$$

A graphical illustration of GRU unit is depicted in Fig. 1. GRU has simplified architecture with fewer parameters than an LSTM. It has been shown that using GRU reduces the generation complexity while conserving the quality [18,19]. More recently, the idea of gating was used by [20] with convolutional networks for English language modeling. They achieved a new state of the art performance on WikiText-103 as well as a new best single-GPU result on the Google Billion Word benchmark.

2.2 Bidirectional GRU Networks

One drawback of GRU networks is that they are only able to make use of previous context. For NER task, it is useful to have access to both past and future features at each timestep that will be a powerful modeling idea. We use bidirectional GRUs to cope with this problem. A bidirectional GRU can process data in both directions where the output layer receives results from the two separate hidden layers.

For a given Arabic sentence $Z = \{z_1, z_2,..., z_n\}$ in the dataset containing n words, the bidirectional GRU computes two representations: the left context of the sentence at every word t denoted by $\overleftarrow{h}_t$ and the right context of the sentence denoted by $\overrightarrow{h}_t$ by using a second GRU unit reading the same sentence in the opposite direction. It should be noted that Arabic language is different from the European languages where it starts from right to left which results in a slightly different use of a bidirectional GRU networks in order to deal with Arabic sentences. In summary, every word is represented by concatenating its left and right context representations $h_t = [\overleftarrow{h}_t; \overrightarrow{h}_t]$.

2.3 Bidirectional GRU with CRF

In this context, we consider Arabic NER as a sequence labeling task where, for a given sentence, it is useful to take in consideration the correlations between adjacent labels and decode them jointly by choosing the best sequence of them. For example, in standard BIO format, I-ORG cannot follow I-PER [21]. In order to capture these correlations between labels, we model sequences using a conditional random field (CRF) [22], instead of decoding each label independently. To model a CRF layer, we used a state transition matrix, which will be used in order to predict the current tag given the past and future tags. We denote this transition matrix by $X_{i,j}$ representing the transition score from the i-th tag to the j-th tag. This matrix will represent the parameters of this layer. It should be noted that we used dynamic programming to compute this matrix.

Our final model combines the bidirectional GRU with the CRF layer. Given a sentence $Z = \{z_1, z_2,..., z_n\}$, we denote $Y(Z_{1,T})_i^t$ to be the matrix of scores output by the bidirectional GRU network for the sentence $Z_{1,T}$ and the i-th tag at the t-th word. The sum of the transition scores of the CRF layer and the scores from the bidirectional GRU network constitutes the final score for a sentence $Z_{1,T}$ along with a sequence of tags $i_{1,T}$. The following equation summarize this final score F_S:

$$F_S(Z_{1,T}, i_{1,T}) = \sum_{t=1}^{T} (X_{i_{t-1}, i_t} + Y(Z_{1,T})_{i_t, t}) \tag{5}$$

Standard softmax function is used to get probabilities over all possible tag sequences:

$$p(y|Z_{1,T}) = \frac{exp(F_S(Z_{1,T}, i_{1,T}))}{\sum_{w \in I_Z} exp(F_S(Z_{1,T}, w))} \tag{6}$$

where I_Z represents all possible tag sequences for a given sentence $Z_{1,T}$. During training, we maximize the log-probability $log(p(y|Z_{1,T}))$ of the correct tag sequence:

$$log(p(y|Z_{1,T}))) = log(\frac{exp(F_S(Z_{1,T}, i_{1,T}))}{\sum_{w \in I_Z} exp(F_S(Z_{1,T}, w))}) \tag{7}$$

$$= F_S(Z_{1,T}, i_{1,T}) - log(\sum_{w \in I_Z} exp(F_S(Z_{1,T}, w))) \tag{8}$$

2.4   Main Features

In this section, we present the main features used to build our Arabic NER system. We begin by introducing the word embeddings which played an important role to increase the performance system. Then, we describe the character-based embeddings.

Word Embeddings. Continuous word embeddings trained on unannotated corpora have been evaluated in various NLP tasks for their ability to capture syntactic and semantic word similarity [23,24]. For this reason, they were used for different NLP tasks like parsing, semantic role labeling, part-of-speech tagging, named entity recognition, dependency parsing, chunking, and sentiment classification.

[25] showed that using word embeddings in parsing English text improved the system performance. [26] argued that adding word embeddings as features for English part-of-speech (POS) tagging task helped the model to increase its performance. For chunking, which is another syntactic sequence labeling task, [27] showed that adding word embeddings allows the chunker for English to increase its F1-score. [28] used simpler version of word embeddings features for English dependency parsing where they employed flat (non-hierarchical) cluster IDs and binary strings obtained via sign quantization ($1[x > 0]$) of the vectors.

In our model, we use pretrained word embeddings to initialize our word vectors. It should be noted that Arabic pretrained word embeddings developed by [29] are publicly available for research purpose. In our system, we used Arabic pretrained word embeddings developed using word2vec model [30].

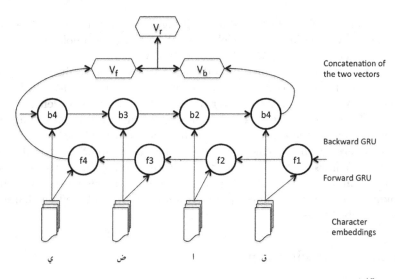

Fig. 2. The character-based embeddings for the Arabic word "قاضي".

Character-Level Embeddings. The second features we investigated in this paper are character-level embeddings. There are many reasons behind adding these embeddings to our system: the first reason lies in their success to increase the performance of many systems in various NLP applications. The second reason is their importance for complex languages and also languages with rich morphology (Czech, Arabic, etc.) [31].

Recently, many NLP applications used character-level representations in their systems. For language modeling, [32] used character-level representations to encode words, which allowed their system to capture rich semantic and orthographic features. As a result, they were able to outperform the baseline models. In parsing, [33] used character-level representations to overcome the main problem in parsing which is the out-of-vocabulary problem and without any additional resources. Using character-level representations allowed the parser to increase the performance when out-of-vocabulary rates are high. Finally, they argued that it is very useful to incorporate character-level representations for morphologically rich languages.

In neural machine translation (NMT), [31] used character-level representations features for English to Czech translation task. They demonstrated that their character models learned many useful features: well-formed words for Czech, highly-inflected language with a very complex vocabulary and was able to construct correct representations for English source words.

Arabic language is a highly inflectional and derivational language with complex and rich morphology. Therefore, any Arabic NER system will suffer from the out-of-vocabulary problem on the word level. Thus, using character-level embeddings for representing Arabic words to build a NER system will be very useful to increase the system performance and deal the out-of-vocabulary problem.

In addition, character-level embeddings will highly generalized to rarely seen or unseen words by being able to model different and rare morphological variants of a word, which is the main drawback of word-level.

Figure 2 shows the main idea to add character-level embeddings to our system. Character-level embeddings of Arabic words are computed using bidirectional GRUs. We read words character by character from right to left (from the character "ق" to the character "ي") to compute the first vector embedding (Vf). Then, we computed the second vector embeddings (Vb), where we start from the last character (from the character "ي" to the character "ق"), using the same process. The final vector Vr represents the concatenation of the previous two vectors. We concatenate this final vector with word embeddings vector taken from the lookup-table to get the final representation of any word based on character- and word- embeddings.

3   Training Mechanism

In this section, we provide details about training our neural network. Figure 3 illustrates the main architecture of our model. The input for our model will be sentences from the dataset. For each word in the sentence, it is represented by the concatenation of two vectors: the corresponding pretrained word embeddings downloaded from [29] and the vector obtained from the character-based embeddings. The vectors of words in each sentence are fed to the bidirectional GRU to get the final vector, which we feed to the CRF layer to get predictions for each word in a given sentence. In the rest of this section, we present the dataset used for training and testing our model and we provide training details.

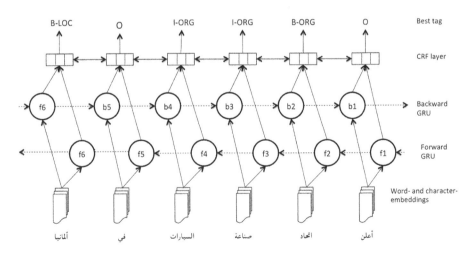

Fig. 3. The final architecture of our model.

3.1 ANERcorp Dataset

ANERcorp is the dataset used in this paper for training and testing our system. It was developed by [7] for Arabic NER task. In this dataset, sentences are represented in the standard IOB format (Inside, Outside, Beginning) where they labelled every token as B-label if the token is the beginning of a named entity, I-label if it is inside a named entity but not the first token within the named entity, or O otherwise. They used three types of named entities for tagging: Person, Location and Organization, which conformed to Linguistics Data Consortium ACE tagging guidelines. This dataset consists of 316 articles where the authors decided to choose articles from different newspapers containing many topics so as to build a generalized dataset.

ANERcorp contains more than 150 000 tokens and 32 114 types annotated for this task where the proportion of the tokens that are named entities reaches 11% of the dataset.

3.2 Training Details

We initialized our character-embeddings with uniform samples from the interval $[-\sqrt{\frac{2}{dim}}, +\sqrt{\frac{2}{dim}}]$, where we set the dimension $dim = 25$. As we discussed in previous section, we initialized our word vectors with Arabic pretrained word embeddings derived using word2vec model, developed by [29] where we set the dimension of our word embeddings to be 100.

The algorithm used to train our neural networks is backpropagation through time (BBTT) [34]. We used mini-batch stochastic gradient descent (SGD) to perform parameter optimization with a fixed learning rate of 0.01 and a gradient clipping of 5.0. It should be noted that we explored more advanced optimization algorithms such as Adadelta [35], Adam [36] and RMSProp [37]. Experimental results showed that none of these algorithms was able to perform better than SGD.

4 Experimental Results and Discussions

In this section, we present the experimental results obtained by using different architectures of our model applied on ANERcorp dataset. We run two parts of the experiments: we run the first part to select the best architecture among some models and the second part of the experiments concerns the comparison between our best selected model and the best previous systems [38–40].

Firstly, we investigated the use and the combination of different architectures: bidirectional GRU networks, the CRF layer, character-based (char-emb) models and word embeddings (word-emb). Table 1 shows the results obtained by combining different architectures. Related to the experimental results presented in Table 1, we can observe that the combination of bidirectional GRU, CRF, pretrained word embeddings (word-emb) and character-based embeddings (char-emb) gives us the best F1-score where we reached 89.74 points. From these

Table 1. Experimental results of our different models on ANERcorp dataset.

Models	Precision	Recall	F1-score
B-GRU + CRF	82.15	80.27	81.18
B-GRU + CRF + char-emb	85.42	84.50	84.95
B-GRU + CRF + word-emb	90.11	86.24	88.13
B-GRU + CRF + word-emb + char-emb	91.00	88.52	**89.74**

experiments, we can see that pretrained word embeddings gave us the remarkable improvement in overall performance of +6.95 points in F1 score compared to the model using just randomly initialized embeddings (B-GRU + CRF). In addition, adding character embeddings to the model "B-GRU + CRF" improved the system by +3.77 points in F1 score.

In the second part of the experiments, we compare our system with the previous Arabic NER systems. We selected three systems: the first system developed by [38], the second system developed by [40] and the last system developed by [39].

[40] used a feed-forward neural network with pretrained word embeddings to build an Arabic NER system where they reached 88.64 F1 score. [38] used an hybrid approach to develop their system called NERA. It yields an F1-score of 88.77%. The state-of-the-art results were produced by [39] where their system called J48, reaches 90.66 in F1 score. This system used a rule-based NER component producing NE labels with six categories of engineering features: rule-based features, morphological features, POS features, gazetteer features, contextual features and word-level features. Table 2 shows the comparison between our model and these three systems. We were able to achieve comparable results to previous best-performing system (J48) and outperformed the two other systems (Gridach, 2016; NERA).

Table 2. Comparison between our system and three Arabic NER systems on ANERcorp dataset.

Systems	Precision	Recall	F1
Gridach, 2016	**95.76**	82.52	88.64
NERA	90.58	87.05	88.77
J48	–	–	**90.66**
Our system	91.00	**88.52**	89.74

As far as we know, we are the first to explore the impact of character-based embeddings, pretrained word embeddings and contextual features (CRF) to develop an Arabic NER system. Using these features allow our model to learn interesting morphological and orthographic features instead of hand-engineering them. In addition, we are the first to use bidirectional GRU combined with CRF to build a named entity recognition system. Studying the impact of bidirectional GRU in developing NER systems for other languages should be an interesting direction of future work.

5 Conclusion

In this paper, we present a novel neural network architecture based on bidirectional GRU, CRF, word embeddings and character-based embeddings where the overall model does not rely on lots of hand-engineering features. Word embeddings learned from unlabeled corpora give us a remarkable improvement in the overall system performance. Because Arabic is a high inflectional and derivational language with complex and rich morphology, using character-level models allow our model to reduce out-of-vocabulary issues and being able to model different and rare morphological variants of Arabic words. Therefore, our system uses minimal features and does not rely on any dictionary or large gazetteers. We achieved comparable results to previous best-performing systems on ANERcorp dataset. One of the potential direction for future work will be the integration of our system to do Arabic sentiment analysis task.

References

1. Nadeau, D., Sekine, S.: A survey of named entity recognition and classification. Lingvisticae Invest. **30**, 3–26 (2007)
2. Ng, V.: Supervised noun phrase coreference research: the first fifteen years. In: Proceedings of the 48th Annual Meeting of the Association for Computational Linguistics, pp. 1396–1411. Association for Computational Linguistics (2010)
3. Chen, D., Manning, C.D.: A fast and accurate dependency parser using neural networks. In: EMNLP, pp. 740–750 (2014)
4. Ferrández, S., Toral, A., Ferrández, Ó., Ferrández, A., Muñoz, R.: Applying Wikipedia's multilingual knowledge to cross–lingual question answering. In: Kedad, Z., Lammari, N., Métais, E., Meziane, F., Rezgui, Y. (eds.) NLDB 2007. LNCS, vol. 4592, pp. 352–363. Springer, Heidelberg (2007). https://doi.org/10.1007/978-3-540-73351-5_31
5. Nikoulina, V., Sandor, A., Dymetman, M.: Hybrid adaptation of named entity recognition for statistical machine translation. In: Proceedings of the Second Workshop on Applying Machine Learning Techniques to Optimise the Division of Labour in Hybrid MT (ML4HMT-2012), Mumbai, Inde (2012)
6. Cao, T.H., Tang, T.M., Chau, C.K.: Text clustering with named entities: a model, experimentation and realization. In: Holmes, D.E., Jain, L.C. (eds.) Data Mining: Foundations and Intelligent Paradigms, pp. 267–287. Springer, Heidelberg (2012). https://doi.org/10.1007/978-3-642-23166-7_10
7. Benajiba, Y., Rosso, P.: Arabic named entity recognition using conditional random fields. In: Proceedings of Workshop on HLT & NLP Within the Arabic World, LREC, vol. 8, pp. 143–153. Citeseer (2008)
8. Shaalan, K., Raza, H.: NERA: named entity recognition for arabic. J. Am. Soc. Inf. Sci. Technol. **60**, 1652–1663 (2009)

9. Collobert, R., Weston, J., Bottou, L., Karlen, M., Kavukcuoglu, K., Kuksa, P.: Natural language processing (almost) from scratch. J. Mach. Learn. Res. **12**, 2493–2537 (2011)

10. Goller, C., Kuchler, A.: Learning task-dependent distributed representations by backpropagation through structure. In: 1996 IEEE International Conference on Neural Networks, vol. 1, pp. 347–352. IEEE (1996)

11. Hochreiter, S., Schmidhuber, J.: Long short-term memory. Neural Comput. **9**, 1735–1780 (1997)

12. Gers, F.A., Schmidhuber, J., Cummins, F.: Learning to forget: continual prediction with LSTM. Neural Comput. **12**, 2451–2471 (2000)

13. Cho, K., Van Merriënboer, B., Bahdanau, D., Bengio, Y.: On the properties of neural machine translation: encoder-decoder approaches. arXiv preprint arXiv:1409.1259 (2014)

14. Graves, A.: Generating sequences with recurrent neural networks. arXiv preprint arXiv:1308.0850 (2013)

15. Huang, Z., Xu, W., Yu, K.: Bidirectional LSTM-CRF models for sequence tagging. arXiv preprint arXiv:1508.01991 (2015)

16. Bengio, Y., Simard, P., Frasconi, P.: Learning long-term dependencies with gradient descent is difficult. IEEE Trans. Neural Netw. **5**, 157–166 (1994)

17. Pascanu, R., Mikolov, T., Bengio, Y.: On the difficulty of training recurrent neural networks. ICML **28**(3), 1310–1318 (2013)

18. Chung, J., Gulcehre, C., Cho, K., Bengio, Y.: Empirical evaluation of gated recurrent neural networks on sequence modeling. arXiv preprint arXiv:1412.3555 (2014)

19. Karpathy, A., Johnson, J., Fei-Fei, L.: Visualizing and understanding recurrent networks. arXiv preprint arXiv:1506.02078 (2015)

20. Dauphin, Y.N., Fan, A., Auli, M., Grangier, D.: Language modeling with gated convolutional networks. arXiv preprint arXiv:1612.08083 (2016)

21. Sang, E.F., Veenstra, J.: Representing text chunks. In: Proceedings of the Ninth Conference on European Chapter of the Association for Computational Linguistics, pp. 173–179. Association for Computational Linguistics (1999)

22. Lafferty, J., McCallum, A., Pereira, F.: Conditional random fields: probabilistic models for segmenting and labeling sequence data. In: Proceedings of the Eighteenth International Conference on Machine Learning, ICML, vol. 1, pp. 282–289 (2001)

23. Huang, E.H., Socher, R., Manning, C.D., Ng, A.Y.: Improving word representations via global context and multiple word prototypes. In: Proceedings of the 50th Annual Meeting of the Association for Computational Linguistics: Long Papers, vol. 1, pp. 873–882. Association for Computational Linguistics (2012)

24. Mikolov, T., Sutskever, I., Chen, K., Corrado, G.S., Dean, J.: Distributed representations of words and phrases and their compositionality. In: Advances in Neural Information Processing Systems, pp. 3111–3119 (2013)

25. Bansal, M., Gimpel, K., Livescu, K.: Tailoring continuous word representations for dependency parsing. In: ACL, vol. 2, pp. 809–815 (2014)

26. Huang, F., Ahuja, A., Downey, D., Yang, Y., Guo, Y., Yates, A.: Learning representations for weakly supervised natural language processing tasks. Comput. Linguist. **40**, 85–120 (2014)

27. Turian, J., Ratinov, L., Bengio, Y.: Word representations: a simple and general method for semi-supervised learning. In: Proceedings of the 48th Annual Meeting of the Association for Computational Linguistics, pp. 384–394. Association for Computational Linguistics (2010)

28. Hisamoto, S., Duh, K., Matsumoto, Y.: An empirical investigation of word representations for parsing the web. In: Proceedings of ANLP, pp. 188–193 (2013)
29. Zahran, M.A., Magooda, A., Mahgoub, A.Y., Raafat, H., Rashwan, M., Atyia, A.: Word representations in vector space and their applications for arabic. In: Gelbukh, A. (ed.) CICLing 2015. LNCS, vol. 9041, pp. 430–443. Springer, Cham (2015). https://doi.org/10.1007/978-3-319-18111-0_32
30. Mikolov, T., Kombrink, S., Burget, L., Černocký, J., Khudanpur, S.: Extensions of recurrent neural network language model. In: 2011 IEEE International Conference on Acoustics, Speech and Signal Processing (ICASSP), pp. 5528–5531. IEEE (2011)
31. Luong, M.T., Manning, C.D.: Achieving open vocabulary neural machine translation with hybrid word-character models. arXiv preprint arXiv:1604.00788 (2016)
32. Kim, Y., Jernite, Y., Sontag, D., Rush, A.M.: Character-aware neural language models. arXiv preprint arXiv:1508.06615 (2015)
33. Ballesteros, M., Dyer, C., Smith, N.A.: Improved transition-based parsing by modeling characters instead of words with LSTMs. arXiv preprint arXiv:1508.00657 (2015)
34. Boden, M.: A guide to recurrent neural networks and backpropagation. The Dallas project, SICS technical report (2002)
35. Zeiler, M.D.: Adadelta: an adaptive learning rate method. arXiv preprint arXiv:1212.5701 (2012)
36. Kingma, D., Ba, J.: Adam: a method for stochastic optimization. arXiv preprint arXiv:1412.6980 (2014)
37. Hinton, G., Tieleman, T.: Lecture 6.5-rmsprop: divide the gradient by a running average of its recent magnitude. COURSERA Neural Netw. Mach. Learn. **4**, 26–31 (2012)
38. Abdallah, S., Shaalan, K., Shoaib, M.: Integrating rule-based system with classification for arabic named entity recognition. In: Gelbukh, A. (ed.) CICLing 2012. LNCS, vol. 7181, pp. 311–322. Springer, Heidelberg (2012). https://doi.org/10.1007/978-3-642-28604-9_26
39. Shaalan, K., Oudah, M.: A hybrid approach to arabic named entity recognition. J. Inf. Sci. **40**, 67–87 (2014)
40. Gridach, M.: Deep learning approach for arabic named entity recognition. In: Gelbukh, A. (ed.) CICLing 2016. LNCS, vol. 9623, pp. 439–451. Springer, Cham (2018). https://doi.org/10.1007/978-3-319-75477-2_31

Named Entity Recognition for Amharic Using Stack-Based Deep Learning

Utpal Kumar Sikdar and Björn Gambäck[⊠]

Department of Computer Science, Norwegian University of Science and Technology,
7491 Trondheim, Norway
{sikdar.utpal,gamback}@ntnu.no

Abstract. In order to improve the performance of a deep-learning neural network, the paper outlines a stack-based approach incorporating various information sources. A named entity recognition system for Amharic was implemented using a recurrent neural network, a bi-directional long short term memory model. Word vectors based on semantic information were built using an unsupervised learning algorithm, word2vec, while a Conditional Random Fields (CRF) classifier was trained on language independent features to predict each token's named entity class. The predictions, features and word vectors were fed to the deep neural network to assign labels to the words. This stack-based approach reached an 74.26% F-score, outperforming various other deep-learning set-ups, as well as a baseline CRF classifier, and an ensemble method incorporating the same information sources.

1 Introduction

Named entity recognition plays an important role in helping to make semantic search more efficient. For example, if the search term is 'apple', it may refer to a common noun or proper noun (the company). If the user intention of using the search term is only to capture the proper noun reading, named entity information helps by guiding the search for the term as a proper noun. This paper will focus on named entity recognition for Amharic, the main language for country-wide communication in Ethiopia. The task of named entity recognition (NER) is to identify proper names and classify them into some predefined categories. Most current NER research utilises machine learning approaches [13, 19, 20], heavily based on the availability of resources on the web, a route which obviously cannot always be taken for resource-poor languages, such as Amharic.

Recently, deep neural networks have been shown to effectively solve several language processing tasks such as part-of-speech tagging, sentiment analysis, and NER. A feed-forward neural network to identify named entities was designed by Collobert et al. [6] using a fixed number of context words, while Chiu and Nichols [5] further improved performance using character and word embeddings. Huang et al. [11] proposed a more complex method based on a recurrent neural network, a bi-directional long short term memory (LSTM) model. Their model included

© Springer Nature Switzerland AG 2018
A. Gelbukh (Ed.): CICLing 2017, LNCS 10761, pp. 276–287, 2018.
https://doi.org/10.1007/978-3-319-77113-7_22

a layer based on a Conditional Random Field (CRF) classifier having a state transition matrix as parameters to use past and future tags to predict the current tag. A bi-directional LSTM deep learner is also used in the present paper, stacked together with a supervised CRF classifier and a feature extractor. A large set of different, language independent features was developed for identifying and classifying the names, and the classifiers were trained on an annotated Amharic named entity (NE) corpus using these features. A one-hot vector for each token was generated using the NE class predicted by the CRF classifier, and concatenated with the feature vectors and the word embeddings from word2vec. The combined result was used as input to a bi-directional LSTM neural network which was trained to predict the tag for each token.

The rest of the paper is organized as follows: Sect. 2 introduces the Amharic language and the annotated Amharic named entity corpus that has been utilized, while some previous work on named entity recognition for Amharic is discussed in relation to our basic strategy in Sect. 3. Then Sect. 4 describes the stacked-based deep learning-based named entity recognition system for Amharic, including the name identification and classification methodology, and the different features used. Experimental results are given in Sect. 5, comparing the stack-based deep learning approach to some other potential set-ups. Finally, Sect. 6 summarizes the discussion and gives some suggestions for possible future extensions.

2   Amharic and Amharic Corpora

Amharic is the working language of the Ethiopian government and spoken by about 40 million people as a first or second language,[1] making it the second most spoken Semitic language in the world (after Arabic), the second largest language in Ethiopia (after Oromo), and possibly one of the five largest languages on the African continent. Like many other Semitic languages, Amharic has a rich verbal morphology based on tri-consonantal roots, with vowel variants describing modifications to, or supplementary detail and variants of the root form.

Unlike Arabic or Hebrew, Amharic is written from left to right, using a unique script (shared with the closely related Tigrinya language) which lacks capitalization and in total has 275 characters (mainly consonant-vowel pairs). The script originates from the Ge'ez alphabet (the liturgical language of the Ethiopian Orthodox Church) and can be traced back to at least the 4th century A.D. The writing system uses multitudes of ways to denote compound words and there is no agreed upon spelling standard for compounds. As a result of this—and of the size of the country leading to vast dialectal dispersion—lexical variation and homophony is very common.

In spite of the relatively large number of speakers, Amharic is still a language for which very few computational linguistic resources have been developed, and

---

[1] The CIA World Factbook estimates Ethiopia's population to currently be 102.4 million, with 27% having Amharic as first language (https://www.cia.gov/library/publications/the-world-factbook/geos/et.html), while Hudson [12] claimed Amharic to be understood by about 40% of the Ethiopians—at least at that time.

very little has been done in terms of making useful higher level Internet or computer-based applications available to those who only speak Amharic. The largest corpora for Amharic being a 3.5 million word untagged corpus [9] and a part-of-speech tagged corpus of 200,000 words retrieved from the webpages of the WALTA news agency [7,8]. The Polyglot project [2] has created word embeddings for the more than 100 languages that have at least 10,000 Wikipedia entries, which mainly include European and Asian languages (and some artificial languages), but in addition to Arabic also a few Sub-Saharan African languages such as Yoruba, Swahili, Africans and Amharic.[2] In addition, they have annotated named entities for 40 languages using a 4-class scheme: persons, organisations, locations and others (for non-named entity tokens) [1]. However, Arabic is the only language spoken in Africa which has been annotated so far.[3]

Table 1. Training and test data statistics

Fold	Training data			Test data				
	Sentences	Tokens	Named entities	Sentences	Tokens	Named entities		
						total	match	noMatch
1	3,784	99,095	5,056	453	10,581	424	236	188
2	3,801	98,293	4,894	436	11,383	586	249	337
3	3,859	99,379	4,828	378	10,297	652	313	339
4	3,743	96,282	4,878	494	13,394	602	291	311
5	3,895	100,197	5,018	342	9,479	462	200	262
6	3,862	100,963	5,027	375	8,713	453	209	244
7	3,902	100,877	5,012	335	8,799	468	218	250
8	3,730	96,620	4,788	507	13,056	692	275	417
9	3,832	98,300	4,951	405	11,376	529	215	314
10	3,725	97,078	4,868	512	12,598	612	307	305
Total				4,237	109,676	5,480		

In contrast, the datasets annotated within the SAY project at New Mexico State University's Computing Research Laboratory include Amharic and use a richer 6-class annotation scheme, with the categories person, location, organization, time, title and other (not named entity). The SAY Amharic annotations are available in 322 XML files from the Lexical Data Repository of the Ge'ez Frontier Foundation.[4] In the present work, these files were split into ten parts for ten-fold cross-validation. Table 1 shows the statistics of the training data for each fold (i.e., the sum of the other nine folds) in terms of number of sentences, tokens, and named entities, as well as the same information for the test data (i.e., each fold in itself), but in addition to the total number of named entities

---

[2] http://bit.ly/embeddings.
[3] https://bit.ly/polyglot-ner.
[4] https://github.com/geezorg/data/tree/master/amharic/tagged/nmsu-say.

for each fold also the number of NEs in the test set (fold) that also could be found in the training set ('match') and those that could not ('noMatch'). Hence the 'noMatch' column shows the number of NEs unique to that specific fold.

3 Named Entity Recognition for Amharic

Most named entity recognition efforts have focused on a few European and Asian languages, while African languages have been given little attention; however, three Master's Thesis projects at Addis Ababa University experimented with randomly extracted sentence subsets of the part-of-speech annotated WALTA corpus [7,8] using the version of the corpus transcribed into the Roman alphabet. Two of the Master students themselves annotated the WALTA corpus subsets with 4-class named entity tags for persons, organisations, locations and others (non-NE; roughly 90% of their datasets).

Both Mehamed [14] and Alemu [3] used about 90% of their data for training and 10% for testing (i.e., without performing any cross-validation), building Conditional Random Fields (CRFs) classifiers trained on different word and context features (word prefixes and suffixes, and the NE and part-of-speech tags of the word), with Mehamed [14] achieving recall, precision and F_1-measure values of 75.0%, 74.2%, and 74.6%, respectively, on a 10,405 word subset of the WALTA corpus, of which 961 words (incl. only 96 NEs) were used for testing. Alemu [3] experimented with context windows of up to two words before and after the current token on another 13,538 word WALTA corpus subset, of which 1,242 words were used for testing, improving recall to 84.9% and precision to 76.8% for an 80.7% F-score.

Belay [4] used a combination of decision trees, support vector machines and hand-crafted rules on part of the dataset annotated by Alemu [3], but artificially expanded it to get a better balance between the classes: since 9,899 of the 12,196 tokens he used were tagged as non-NEs, he inflated the NE classes so that the training dataset contained 31,347 instances. Although using 40% of the training data for testing, Belay [4] noted that his hybrid learning approach was outperformed by a straight-forward baseline method using only part-of-speech information and a binary flag for nominals. The small test sets make the evaluations in these three Master Theses fairly unreliable, as their conclusions also indicate: Mehamed [14] found that part-of-speech tagging improved the results, while using word prefixes did not. Alemu [3] on the other hand claimed that word prefixes contributed positively, while part-of-speech tagging did not.

In contrast, we have experimented with the substantially larger SAY dataset (Table 1) and several deep learning approaches to identify and classify Amharic named entities into six (rather than four) predefined classes: Person, Location, Organization, Time, Title, and Other (non-named entity tokens). The experimental set-ups utilize a recurrent neural network, a bi-directional long short term memory (LSTM) model with various features. Word vectors based on semantic information are built for all tokens using an unsupervised learning algorithm, word2vec. In a basic system set-up, the word vectors are merged with a set of specifically developed language independent features and together fed to the neural network model to predict the classes of the words. In order to improve performance, a stacked system approach utilizes a supervised CRF classifier trained on language independent features to first predict each token's named entity class, and the CRF predictions are then combined with the selected feature set and the word2vec word vectors before being fed to the deep neural network (LSTM model) which assigns labels to the words.

4   A Stacked Amharic Named Entity Recogniser

A stacked deep neural network approach was used to recognize named entities from Amharic text. However, in a first step two baseline systems were created: a plain Conditional Random Field (CRF) classifier and a non-stacked bidirectional long short term memory (LSTM) network.

4.1   CRF-Based Model

In the first baseline system, Amharic named entities were extracted using a supervised CRF machine learner, built with the C++ based CRF++ package,[5] a simple, customizable, and open source implementation of CRF for segmenting or labelling sequential data. The CRF classifier was trained on the following set of features:

Local context plays an important role in identifying names. As in the work by Alemu [3], two words before and after the focus word were used as local context (so there are four context features).
Part-of-speech tags extracted for each token using HornMorpho [10], which is one of the few resources that already are available for Ethiopian languages and provides some morphological processing for Amharic, Tigrinya and Afaan Oromo.
Word suffix and prefix obtained by stripping a fixed number (up to 4) of characters from the beginning and end of the current word.
Word frequency: Less frequent words were found to often belong to named entities. If the pre-calculated frequency from the training/test data of the current word is less than a certain threshold, this binary feature is set. The thresholds were empirically set to 10 and 4 for training and test data, respectively (the frequency count threshold for the test data has to be lower than that for the training data, since there are a lot fewer instances in the test data than in the training data).
Digit check: in particular to identify the 'time' class, it is helpful to mark if a token contains any digit(s). Hence this binary flag is set for tokens containing at least one digit.

---

[5] http://crfpp.sourceforge.net.

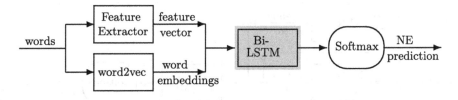

Fig. 1. Baseline LSTM model

4.2 Bi-directional Long Short Term Memory Model

In the second baseline system, a bi-directional LSTM (Long Short Term Memory) model was used. The deep learning method is divided into two parts: word embedding and bi-directional LSTM. The word embeddings were generated through word2vec, which takes inputs from large corpora and generates a word vector for each word. There are two types of embeddings: continuous-bags-of-words (CBOW) and skip-gram models [15,16]. In the CBOW architecture, the model predicts the current word from a window of surrounding context words. In the skip-gram model, it predicts the context words using the current word. The word2vec model can be trained using a softmax function [18] or negative sampling [18]. Since word2vec is a unsupervised approach where annotations are not needed, the entire Amharic corpus (without annotations) was taken as training data for a word2vec model using skip-grams and negative sampling.

The bi-directional LSTM model classifies the words following the NE prediction pipeline shown in Fig. 1. The network consists of an embedding/input layer with two hidden layers. In the output layer, the softmax [18] function assigns the words to six categories/labels. In the input layer, the word embeddings developed using word2vec are combined with the same non-context features as used by the CRF classifier (suffix and prefix, POS, frequency and digit-check). For the suffix and prefix features, 5-dimensional word vectors are generated for each length of suffix/prefix character(s) using word2vec. The suffix and prefix lengths are set for up to four characters, so that 40 (5 × 8) word vectors are generated for the suffix and prefix features. In addition, one-hot vectors are generated for each of the other features: a length 2 one-hot vector for frequency, a length 2 one-hot vector for the digit-check feature, and a length 5 one-hot vector for POS (encoding the five classes nouns, verbs, infinitives, copulas, and others).

4.3 Stack-Based LSTM Model

In order to optimise the system performance, a stack-based deep learner was built, combining the output of a supervised Conditional Random Field model with word embedding vectors and a set of language-independent features fed to the LSTM model to classify the words. The pipeline of the stack-based model is shown in Fig. 2, with the different parts further described below.

A word embedding of size 300 was created from word2vec using a skip-gram model. Different feature vectors of size 49 were extracted for each word in the

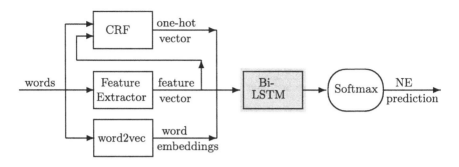

Fig. 2. Pipeline of stack-based model

training and test datasets, using the features described in Sect. 4.2. These feature vectors are concatenated with the word2vec output. In addition, the same features were used by the CRF classifier to predict the tag for each word, with length 6 (each bit represents one class) one-hot word vectors being generated for the CRF outputs, and concatenated with the word vectors generated by the word2vec model and the feature vectors. These three information sources (vector size 355) were then fed to a bi-directional LSTM neural network with two hidden layers in order to classify the tokens into one of the six different classes. The LSTM model was trained for 100 epochs with a batch size of 256 and with the maximum sentence length set to 70.

For comparison, another system was trained using the same setup, but without feeding feature vectors to the LSTM model. Hence that network only uses word embeddings generated by word2vec (size 300) concatenated with the CRF output vector (6 one-hot word vector), so a total input word dimension of 306.

5   Results

To establish a baseline, a supervised model was built using a CRF classifier based on the features mentioned in Sect. 4.1. After 10-fold cross-validation, the CRF classifier achieved the average precision, recall and F_1-scores of 85.02%, 61.67% and 71.44%, respectively. For each fold, the recall, precision and F-measure values are given on the left side of Table 2.

The second baseline model was the LSTM described in Sect. 4.2, utilizing feature vectors based on the language independent features (except the context feature) that were added to the word vectors built from skip-gram word2vec model. The average recall, precision and F-scores after 10-fold cross-validation are shown on the right side of Table 2: the model using both word vectors and features achieves an average precision of 77.2% and recall of 63.4%, for a 69.7% F_1-score. It thus slightly improves the recall compared to the CRF-based classifier, but at the price of a clearly lower precision.

The performance of the stacked-based LSTM model incorporating two information sources (word vectors and CRF outputs), but not the feature vectors

Table 2. Baseline systems: CRF classifier and LSTM with word2vec plus features

Fold	CRF classifier			word2vec + features		
	Precision	Recall	F_1-score	Precision	Recall	F_1-score
1	0.8726	0.6887	0.7698	0.7930	0.6667	0.7244
2	0.8070	0.6197	0.7010	0.7016	0.6490	0.6743
3	0.8499	0.6044	0.7064	0.8093	0.6290	0.7079
4	0.8608	0.6399	0.7341	0.7766	0.6516	0.7087
5	0.8492	0.5580	0.6738	0.7792	0.5974	0.6763
6	0.8335	0.6039	0.7004	0.7614	0.6073	0.6757
7	0.8611	0.6072	0.7122	0.7859	0.6156	0.6904
8	0.8733	0.5796	0.6968	0.7578	0.6360	0.6916
9	0.8144	0.6133	0.6997	0.7464	0.6467	0.6930
10	0.8804	0.6530	0.7498	0.8055	0.6633	0.7275
Avg.	0.8502	0.6167	0.7144	0.7717	0.6363	0.6970

reached the average precision, recall and F-measure values of 85.91%, 65.33% and 74.10%, respectively, as shown on the left side of Table 3 ('No-feat LSTM'), hence surpassing both baseline systems on all accounts.

The stacked-based model incorporating all three information sources (word vectors, features and CRF outputs) out-performed all previous models, with the 10-fold cross-validation results shown in the middle of Table 3, so reaching average precision, recall and F-measure values of 85.97%, 65.51% and 74.26%, respectively, using the outputs of the previous CRF learning classifier along with the feature vectors and the word vectors from the word2vec model. However, the improvements compared to the 'No-feat' LSTM model are small, indicating that the mileage stemming from the language-independent feature set is not very significant.

As an alternative to combining the CRF classifier and the LSTM in a stack-based model, a set of experiments were run where they were instead emsembled in a voting-based scheme. As can be seen in the right-most part of Table 3, a combination of CRF with an LSTM trained on the word2vec word embeddings and the language independent feature set performed better than the stack-based LSTM model in terms of recall, but worse than all the other models in terms of precision, for an average F-score which was slightly lower than both versions of the stacked LSTM (i.e., both with and without the features in the input set).

In order to investigate the cause of the low recall values of the stacked LSTM model, another set of experiments was run, exploring the effect of increasing the size of the training dataset. Here Fig. 3 represents the variation of F-measure values with respect to different sizes of training datasets. The three graphs in the figure compare the performance of the stack-based LSTM model to that of the plain CRF classifier and the LSTM model trained on word2vec word embeddings and the feature vectors, in each case showing how the performance increases as

Table 3. A stack-based LSTM without feature vector input (but using word2vec and CRF outputs), a stack-based LSTM using all three information sources, and a voting-based ensemble of a CRF classifier and an LSTM model using word2vec and features.

Fold	No-feat LSTM			Stack-based LSTM			Voting-based Ensemble		
	Precision	Recall	F_1-score	Precision	Recall	F_1-score	Precision	Recall	F_1-score
1	0.8716	0.7242	0.7911	0.8746	0.7267	0.7938	0.7511	0.8026	0.7760
2	0.8035	0.6661	0.7284	0.8048	0.6669	0.7293	0.7517	0.7052	0.7277
3	0.8760	0.6405	0.7400	0.8760	0.6405	0.7400	0.7787	0.7273	0.7521
4	0.8632	0.6726	0.7561	0.8643	0.6734	0.7567	0.7703	0.7356	0.7525
5	0.8750	0.5814	0.6986	0.8756	0.5847	0.7011	0.7557	0.6687	0.7096
6	0.8401	0.6338	0.7225	0.8431	0.6361	0.7251	0.7657	0.7123	0.7381
7	0.8807	0.6615	0.7555	0.8807	0.6615	0.7555	0.7295	0.7416	0.7355
8	0.8940	0.6169	0.7301	0.8917	0.6199	0.7313	0.7296	0.7024	0.7158
9	0.8195	0.6704	0.7375	0.8208	0.6722	0.7391	0.7325	0.7390	0.7357
10	0.8617	0.6656	0.7511	0.8658	0.6687	0.7546	0.7592	0.7551	0.7571
Avg.	0.8591	0.6533	0.7410	**0.8597**	0.6551	**0.7426**	0.7524	**0.7289**	0.7400

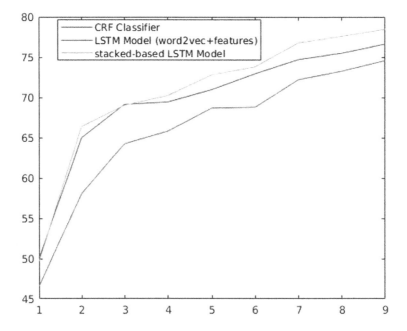

Fig. 3. F-measure vs. size of training data (each step adds on average 424 sentences)

more data (another fold) is added to the training dataset. As can be seen, all three systems improve rapidly as the first three folds are added, but (as could be expected) performance keeps on improving as more data is added, indicating that all systems would benefit from having access to even more training data.

6   Conclusion

Here we have experimented with a system for named entity recognition for Amharic, an under-resourced language. A set of language independent features was developed to extract Amharic named entities using a supervised CRF classifier and bi-directional LSTM model. Better performance was achieved after creating (unsupervised) word embeddings based on the output of the supervised model and the feature vectors together with word2vec word vectors, and then feeding the result to the neural network for training and classification. When the outputs of the CRF model were concatenated with the word embeddings, the recurrent network outperformed the other models. This may be since many of the tags/classes are identified already by the CRF model. However, the stack-based LSTM model utilizing the CRF output as an information source clearly improved on the CRF classifier itself, in particular in terms of recall. A voting-based ensemble solution using the CRF classifier and an LSTM built only on word embeddings and feature vectors showed further improvements to recall, but at the price of substantially lower precision.

Tools and resources that can help reduce language barriers and thereby provide people all over the world with improved access to information and services will have beneficial effects for most sectors of society and in the long-term contribute to the development of technology that will enable massive social and economic transformations. The present system takes a small but important step in the direction of developing such tools, but the error levels are still high and many names were not identified by the system or classified into the wrong NE categories. Notably though, one error source is that many names in the training data are annotated as non-entities, but in test data the names are annotated as named entities. However, the main cause of the low recall is most likely insufficient number of training instances, which are further reduced both by Amharic being an agglutinative language and by it lacking spelling standard for many names. This is in line with the results reported by Poostchi et al. [17] who carried out a similar NE task on Persian, another under-resourced language, using an almost equal-sized corpus (250K tokens, of which about 10% were named entities). They compared an SVM-HMM based approach to CRF and a recurrent neural network, with the SVM-based classifier performing best, potentially since their dataset also was too small for the neural network to be trained efficiently.

In the future it would be reasonable to also develop some language dependent features to improve the performance. A set of models can also be generated by using several different classifiers and ensemble these models with the help of an evolutionary algorithm. It might also be possible to utilize the word embeddings generated for Amharic in the Polyglot or HaBiT projects. Furthermore, the word2vec model used here was built on skip-grams that predict the context words using the current word. An alternative would be to use the continuous-bags-of-words (CBOW) model, which basically does the opposite and predicts the current word from a window of surrounding context words.

Acknowledgements. Thanks to Lars Bungum, Biswanath Barik, the anonymous reviewers, and our project partners at Masaryk University (Brno, Czech Republic), Addis Ababa University (Ethiopia), and University of Oslo (Norway). This work was carried out within the HaBiT project ("Harvesting big text data for under-resourced languages": http://www.habit-project.eu) funded by the Research Council of Norway (NFR) and the Czech Ministry of Education, Youth and Sports (MŠMT) through the CZ09 Czech-Norwegian Research Programme and the EEA/Norway Financial Mechanism under Project Contract 7F14047.

References

1. Al-Rfou, R., Kulkarni, V., Perozzi, B., Skiena, S.: POLYGLOT-NER: massive multilingual named entity recognition. In: Proceedings of the 2015 SIAM International Conference on Data Mining (SDM 2015). Society for Industrial and Applied Mathematics, Vancouver, June 2015
2. Al-Rfou, R., Perozzi, B., Skiena, S.: Polyglot: Distributed word representations for multilingual NLP. In: Proceedings of the 17th Conference on Computational Natural Language Learning (CONLL 2013), pp. 586–594. ACL, Sofia, August 2013
3. Alemu, B.: A named entity recognition for Amharic. Master's thesis, School of Information Science, Addis Ababa University, Addis Ababa, Ethiopia, June 2013
4. Belay, M.T.: Amharic named entity recognition using a hybrid approach. Master's thesis, School of Information Science, Addis Ababa University, Addis Ababa, Ethiopia, August 2014
5. Chiu, J.P.C., Nichols, E.: Named entity recognition with bidirectional LSTM-CNNs. Trans. ACL **4**, 357–370 (2016)
6. Collobert, R., Weston, J., Bottou, L., Karlen, M., Kavukcuoglu, K., Kuksa, P.: Natural language processing (almost) from scratch. J. Mach. Learn. Res. **12**, 2493–2537 (2011)
7. Demeke, G.A., Getachew, M.: Manual annotation of Amharic news items with part-of-speech tags and its challenges. ELRC Working Papers, vol. 2(1), pp. 1–17, March 2006
8. Gambäck, B.: Tagging and verifying an Amharic news corpus. In: Proceedings of the 8th International Conference on Language Resources and Evaluation, pp. 79–84. ELRA, Istanbul, May 2012. Workshop on Language Technology for Normalisation of Less-Resourced Languages
9. Gambäck, B., Asker, L.: Experiences with developing language processing tools and corpora for Amharic. In: Cunningham, P., Cunningham, M. (eds.) Proceedings of IST-Africa 2010, the 5th Conference on Regional Impact of Information Society Technologies in Africa. IIMC, Durban, May 2010
10. Gasser, M.: HornMorpho: a system for morphological processing of Amharic, Oromo, and Tigrinya. In: Proceedings of Conference on Human Language Technology for Development, Alexandria, Egypt, pp. 94–99, May 2011
11. Huang, Z., Xu, W., Yu, K.: Bidirectional LSTM-CRF models for sequence tagging. CoRR abs/1508.01991 (2015)
12. Hudson, G.: Linguistic analysis of the 1994 Ethiopian census. Northeast Afr. Stud. **6**(3), 89–107 (1999)
13. Lafferty, J.D., McCallum, A., Pereira, F.C.N.: Conditional random fields: probabilistic models for segmenting and labeling sequence data. In: Proceedings of the 18th International Conference on Machine Learning, ICML 2001, pp. 282–289. Morgan Kaufmann Publishers Inc., San Francisco (2001)

14. Mehamed, M.A.: Named entity recognition for Amharic language. Master's thesis, Department of Computer Science, Addis Ababa University, Addis Ababa, Ethiopia, November 2010
15. Mikolov, T., Chen, K., Corrado, G., Dean, J.: Efficient estimation of word representations in vector space. CoRR abs/1301.3781 (2013)
16. Mikolov, T., Sutskever, I., Chen, K., Corrado, G., Dean, J.: Distributed representations of words and phrases and their compositionality. In: Advances in Neural Information Processing Systems 26 (NIPS 2013), pp. 3111–3119. Curran Associates, Red Hook (2013)
17. Poostchi, H., Borzeshi, E.Z., Abdous, M., Piccardi, M.: PersoNER: persian named-entity recognition. In: Proceedings of the 26th International Conference on Computational Linguistics, pp. 3381–3389. ACL, Osaka, December 2016
18. Rong, X.: word2vec parameter learning explained. CoRR abs/1411.2738 (2014)
19. Takeuchi, K., Collier, N.: Use of support vector machines in extended named entity recognition. In: Proceedings of the 6th Conference on Natural Language Learning. COLING-2002, vol. 20, pp. 1–7. Association for Computational Linguistics, Stroudsburg (2002)
20. Zhou, G., Su, J.: Named entity recognition using an HMM-based chunk tagger. In: Proceedings of the 40th Annual Meeting of the Association for Computational Linguistics, ACL 2002, pp. 473–480. Association for Computational Linguistics, Stroudsburg (2002)

Semantics and Text Similarity

Idioms: Humans or Machines, It's All About Context

Manali Pradhan, Jing Peng, Anna Feldman[(✉)], and Bianca Wright

Department of Linguistics, Department of Computer Science, Montclair State
University, Montclair, NJ 07043, USA
feldmana@montclair.edu

Abstract. Expressions can be ambiguous between idiomatic and literal interpretation depending on the context they occur in ("sales hit
the roof" vs "hit the roof of the car"). Previous studies suggest that
idiomaticity is not a binary property, but rather a continuum or the so-
called "scalar phenomenon" ranging from completely literal to highly
idiomatic. This paper reports the results of an experiment in which
human annotators rank idiomatic expressions in context on a scale from
1 (literal) to 4 (highly idiomatic). Our experiment supports the hypothesis that idioms fall on a continuum and that one might differentiate
between highly idiomatic, mildly idiomatic and weakly idiomatic expressions. In addition, we measure the relative idiomaticity of 11 idiomatic
types and compute the correlation between the relative idiomaticity of
an expression and the performance of various automatic models for idiom
detection. We show that our model, based on the distributional semantics
ideas, not only outperforms the previous models, but also positively correlates with the human judgements, which suggests that we are moving
in the right direction toward automatic idiom detection.

1   Introduction

Philip Johnson-Laird once said: "If natural language had been designed by a
logician, idioms would not exist" [1]. According to [2], there are as many fixed
expressions as there are words in American English, roughly 80,000. This means
that people have at least 160,000 items memorized and available for use. What
sets idioms from most other fixed expressions is the absence of any observable
relation between their linguistic meaning and their idiomatic interpretation [1].
Researchers have not come up with a single agreed-upon definition of idioms
that covers all members of this class [3–8]. The common property ascribed to the
idiom is its relative non-compositionality. Additional properties include lexical
and syntactic flexibility, i.e., *kick the bucket* is not the same thing as *kick the
pail* and *the bucket was kicked* does not preserve the idiomatic meaning.

According to [9], the study of the identification and comprehension of ambiguous idiomatic expressions, like *sales **hit the roof*** vs. ***hit the roof*** *of the car*,
shares many of the issues that are involved in the study of lexical ambiguity.
One of the most important components involved in the comprehension of idioms

© Springer Nature Switzerland AG 2018
A. Gelbukh (Ed.): CICLing 2017, LNCS 10761, pp. 291–304, 2018.
https://doi.org/10.1007/978-3-319-77113-7_23

is context [9]. In particular, when ambiguous idioms are involved, local context seems to contribute to the selection of the particular sense of an idiom.

In this paper, we describe an experiment in which we use Amazon.com's Mechanical Turk (MTurk) to gather human subject rankings on 165 idiomatic expressions, from sixty raters. The purpose of the experiment was to determine whether subjects could rank idioms on a scale and whether the human rankings correlate with the performance of our automatic idiom classifier.

2   MTurk Experiment

2.1   Data

In both of the automatic classification and human judgement experiments, we use the VNC-Tokens dataset developed by [10], a resource of almost 3,000 English verb-noun combination (VNC) usages annotated as to whether they are literal or idiomatic. We selected expressions that were both idiomatic and ambiguous between idiomatic and literal interpretations. [10] report that in their analysis of 60 VNCs, approximately half of these expressions frequently appear in their literal sense in the British National Corpus (BNC) [11]. The original VNC-Tokens list was created by two annotators, both native English-speakers. According to [10], the annotators were presented with the single sentence containing the VNC usage. Sentences in the surrounding context were not included. If the annotator was unable to determine the class of a token based on the sentence in which it occurs, he or she could choose the unknown label.

An important observation that [10] make which is subsequently supported by [12] is that the idiomaticity of an expression is not binary. Expressions may be more or less idiomatic, falling on a continuum ranging from completely literal expressions to semantically opaque. While we do not agree that expressions which are completely literal can be still called idioms (perhaps, the authors meant "collocational continuum"), we do think that idiomaticity is a scalar property, and this observation is used in the experiments described below. [10] also notice that at the adjudication step, when the annotators were supposed to discuss the tokens on which the judges originally disagreed to achieve a consensus annotation, among the issues that arose were the expressions that fall in the middle of the literal-idiomatic continuum. For example, [10] mention that the idiomatic expression *have a word* is related to the literal meaning, as in *At the moment they only had the word of Nicola's husband for what had happened.*

[10] divide their data into three sets: development, test, and skewed. Skewed contains expressions for which one of the literal or idiomatic meanings is infrequent, while the expressions in the development and test sets are more balanced across the senses. [10] notice that while the observed agreement for all the sets is quite high (in the 80s), the kappa scores are low on the test and the skewed sets. [10] mention that the judges consistently disagreed on the label for *have words*, *hold fire*, and *make hit*. Eliminating these three expressions improves the unweighted Kappa score significantly. We address this issue in this paper as well.

2.2 Procedure

In our experiment, we wanted to see whether human annotators are capable of ranking the idioms on a scale and later correlate their judgement with the performance of our algorithm. Using Turktools [13], we randomized and formatted the target material into an html template compatible with Mechanical Turk. The 165 target items were split into three separate Mechanical Turk Human Intelligence Tasks (HITs), each of which contained five target idiomatic expressions presented in context, from all eleven idiom types. Each target item was to be assigned a ranking ranging from 1 to 4, or "not idiomatic" to "highly idiomatic". The purpose of rankings 2 and 3, was to allow for the possibility of an idiomatic expression to be perceived as neither strictly literal nor idiomatic. Prior to the beginning of the experiment, participants were presented with four example questions to aid in understanding the four possible rankings. In order to ensure the turkers were paying attention, each example question had instructions to select a specific ranking. Participant responses were primarily rejected if they consisted of numerous missing entries or an abnormally large number of low rankings. To increase the likelihood of participants being the native speakers of English, we required that all turkers had a high school diploma obtained in the US[1]

Here is an HIT excerpt:

Instructions
You will be presented with 55 text excerpts which contain various focus phrases (highlighted in bold). Your task is to rate how idiomatic each phrase is in its respective text excerpt. The contexts in which the phrases appear, will determine the degree of idiomaticity. There is no "correct" response, simply follow your native speaker intuitions. Below are some typical properties of idiomatic and literal phrases:

Idiomatic phrases tend to be:
- Abstract/complex
- Vague
- Commonly used by native speakers in casual speech and difficult for English learners

Literal phrases tend to be:
- Straightforward in meaning
- Basic

Here is a scale you can follow. Not idiomatic (1)[2]: The meaning sounds fairly straightforward. Little idiomatic (2): The meaning seems like it could be taken

---

[1] Naturally, this requirement does not guarantee a native speaker, but we had not a better option to control for it.

[2] We should clarify here that even though all items that we used were already marked as idiomatic in the [10]'s data, we decided to keep the option of ranking them as literal, just in case of a mistake or a different interpretation. Remember that [10]'s dataset is annotated by only two annotators.

literally, but not completely. It almost seems literal (or not idiomatic) but there's a hint of figurativeness. Somewhat idiomatic (3): The meaning seems to be figurative, but not completely. It almost seems idiomatic, but there is a hint of literalness. Idiomatic (4): It is figurative and cannot be taken literally. So, subjects were supposed to rank idiomatic expressions in context using the scale above. The context is a paragraph from BNC in which an idiomatic expression occurs. It's exactly the same context our automatic classifier uses to tell apart idioms from literal expressions. Here's an example:

> *We decided to go out to dinner the other day, but I was a little worried because I wasn't sure if she was still mad at me or not. So whatever, we still went and we got into the same argument we had last week. She ended up **making a huge scene** right there, in the middle of the restaurant!*

2.3 Results

The results show that a ranking of "4", or "highly idiomatic", was the most frequent among all sixty raters, while the average ranking was 3.2. Although all of the paragraphs presented to participants consisted of strictly idiomatic expressions, lower rankings were assigned consistently by all raters across the 165 target items. The ratings for each idiom type show that some expressions received lower rankings than others. One expression (*have word*) in particular, received a very high assignment of low rankings in comparison to the others, resulting in an average ranking of 2.29. This result is consistent with what was reported in [10]. Apart from this expression, the other ten were assigned a ranking of "4" most frequently. The agreement among the raters was low in terms of both the unweighted B-statistic [14] (0.31) and Cohen's Kappa [15](0.06). The weighted measures were 0.70 for the B-statistic and 0.11 for the Cohen's Kappa.

Table 1 summarizes the experiment. Table 2 reports the average ranking per idiom type.

Table 1. Human ranking experiment

Number of subjects	20 per experiment (60 total)
Each experiment	50 tokens, across 11 idiom types
Total number of tokens tested	165
Average ranking	3.2
Most frequently used ranking	4
Agreement	
B-stat unweighted	0.31
B-stat weighted	0.70
Cohen's K unweighted	0.06
Cohen's K weighted	0.11

As has been mentioned in Sect. 2.1, Cook et al. [10] report high observed agreement, but low kappa values on the data. They eliminate three expressions that the annotators consistently disagree on to improve the unweighted kappa score. Shankar and Baugdiwala [16] address the paradox earlier noticed by [17], namely, (1) low kappa values despite high observed agreement under highly symmetrically imbalanced marginals, and (2) higher kappa values for asymmetrical imbalanced marginal distributions. [16] examine the behavior of alpha, kappa and B-statistic [14] under different scenarios of marginal distributions, balanced or not, symmetrical or not. They show that while all statistics are affected by lack of symmetry and imbalances in the marginal totals, the B-statistic comes closest to resolving the paradoxes identified by [17]. Therefore, based on the B-statistic scores, we assume that the results of our human ranking experiment are reliable.

Table 2. Average human rankings of 11 idiom types

hold fire	3.28	hold horse	3.37	lose head	3.35
blow whistle	3.16	have word	**2.29**	make scene	3.02
give sack	3.33	take hear	3.30	blow top	**3.44**
hit wall	3.19	hit roof	3.34		

2.4 Related Work

A similar experiment was conducted by [18]. They use a dataset with human judgements of compositionality [19] and ask the subjects to judge the compositionality of verb-noun combinations. The focus of their experiment is the detection of the more non-compositional verb-noun combinations, but they do not pay attention to the ambiguity of the expressions. Their list is largely idiomatic, whereas our experiment only deals with ambiguous expressions which can only be disambiguated in context. We are also aware of [20]'s dataset of 1048 noun-noun compounds annotated as non-compositional, compositional, conventionalized and not-conventionalized. The reason why we chose to work with verb-noun constructions is that we wanted to compare our algorithms with the state-of-the-art.

3   Automatic Approach

Our approach is based on two hypotheses: (1) words in a given text segment that are representatives of the local context are likely to associate strongly with a literal expression in the segment, in terms of projection of word vectors onto the vector representing the literal expression; (2) the context word distribution for a literal expression in word vector space will be different from the distribution for an idiomatic one (similarly to [21, 22]).

3.1   Projection Based on Local Context Representation

To address the first hypothesis, we propose to exploit recent advances in vector space representation to capture the difference between local contexts [23,24].

A word can be represented by a vector of fixed dimensionality q that best predicts its surrounding words in a sentence or a document [23,24]. Given such a vector representation, our first proposal is the following. Let v and n be the vectors corresponding to the verb and noun in a target verb-noun construction, as in *blow whistle*, where $v \in \Re^q$ represents *blow* and $n \in \Re^q$ represents *whistle*. Let $\sigma_{vn} = v + n \in \Re^q$. Thus, σ_{vn} is the word vector that represents the composition of verb v and noun n, and in our example, the composition of *blow* and *whistle*. As indicated in [24], word vectors obtained from deep learning neural net models exhibit linguistic regularities, such as additive compositionality. Therefore, σ_{vn} is justified to predict surrounding words of the composition of, say, *blow* and *whistle* in a literal context. Our hypothesis is that on average, the projection of v onto $\sigma_{blowwhistle}$, (i.e., $v \cdot \sigma_{blowwhistle}$, assuming that $\sigma_{blowwhistle}$ has unit length), where vs are context words in a literal usage, should be greater than $v \cdot \sigma_{blowwhistle}$, where vs are context words in an idiomatic usage.

For a given vocabulary of m words, represented by matrix

$$V = [v_1, v_2, \cdots, v_m] \in \Re^{q \times m},$$

we calculate the projection of each word v_i in the vocabulary onto σ_{vn}

$$P = V^t \sigma_{vn} \tag{1}$$

where $P \in \Re^m$, and t represents transpose. Here we assume that σ_{vn} is normalized to have unit length. Thus, $P_i = v_i^t \sigma_{vn}$ indicates how strongly word vector v_i is associated with σ_{vn}. This projection forms the basis for our proposed technique.

Let $D = \{d_1, d_2, \cdots, d_l\}$ be a set of l text segments (local contexts), each containing a target VNC (i.e., σ_{vn}). Instead of generating a term by document matrix, where each term is *tf-idf* (product of term frequency and inverse document frequency), we compute a term by document matrix $M_D \in \Re^{m \times l}$, where each term in the matrix is

$$p \cdot idf. \tag{2}$$

That is, the product of the projection of a word onto a target VNC and inverse document frequency. That is, the term frequency (tf) of a word is replaced by the projection of the word onto σ_{vn} (1). Note that if segment d_j does not contain word v_i, $M_D(i, j) = 0$, which is similar to *tf-idf* estimation. The motivation is that topical words are more likely to be well predicted by a literal VNC than by an idiomatic one. The assumption is that a word vector is learned in such a way that it best predicts its surrounding words in a sentence or a document [23,24]. As a result, the words associated with a literal target will have larger projection onto a target σ_{vn}. On the other hand, the projections of words associated with an idiomatic target VNC onto σ_{vn} should have a smaller value.

We also propose a variant of $p \cdot idf$ representation. In this representation, each term is a product of p and typical $tf\text{-}idf$. That is,

$$p \cdot tf \cdot idf. \tag{3}$$

3.2 Local Context Distributions

Our second hypothesis states that words in a local context of a literal expression will have a different distribution from those in the context of an idiomatic one. We propose to capture local context distributions in terms of scatter matrices in a space spanned by word vectors [23,24].

Let $d = (w_1, w_2 \cdots, w_k) \in \Re^{q \times k}$ be a segment (document) of k words, where $w_i \in \Re^q$ are represented by a vectors [23,24]. Assuming w_is have been centered, we compute the scatter matrix

$$\Sigma = d^t d, \tag{4}$$

where Σ represents the local context distribution for a given target VNC.

Given two distributions represented by two scatter matrices Σ_1 and Σ_2, a number of measures can be used to compute the distance between Σ_1 and Σ_2, such as Choernoff and Bhattacharyya distances [25]. Both measures require the knowledge of matrix determinant. We propose to measure the difference between Σ_1 and Σ_2 using matrix norms. We have experimented with the Frobenius norm and the spectral norm. The Frobenius norm evaluates the difference between Σ_1 and Σ_2 when they act on a standard basis. The spectral norm, on the other hand, evaluates the difference when they act on the direction of maximal variance over the whole space.

4   Experiments

4.1   Methods

We carried out an empirical study evaluating the performance of the proposed techniques. The following methods are evaluated:

1. $p \cdot idf$: compute term by document matrix from training data with proposed $p \cdot idf$ weighting (2).
2. $p \cdot tf \cdot idf$: compute term by document matrix from training data with proposed p*tf-idf weighting (3).
3. $CoVAR_{Fro}$: proposed technique (4) described in Sect. 3.2, the distance between two matrices is computed using Frobenius norm.
4. $CoVAR_{Sp}$: proposed technique similar to $CoVAR_{Fro}$. However, the distance between two matrices is determined using the spectral norm.
5. $Context+$ $(CTX+)$: supervised version of the CONTEXT technique described in [26].
6. GMM: Gaussian Mixture Model as described in [27].

For methods **3** and **4**, we compute the literal and idiomatic scatter matrices from training data (4). For a test example, compute a scatter matrix according to (4), and calculate the distance between the test scatter matrix and training scatter matrices using the Frobenius norm for method **3**, and the spectral norm for method **4**. Method **5** corresponds to a supervised version of CONTEXT described in [26]. CONTEXT is unsupervised because it does not rely on the "gold-standard". Rather it uses knowledge about automatically acquired canonical forms (C-forms). Thus, the gold-standard is "noisy" in CONTEXT. Here we provide manually annotated training data. Therefore, CONTEXT+ is a supervised version of CONTEXT. For Method **6**, [27]'s work uses Normalized Google Distance to model semantic relatedness in computing features [28,29]. We use inner product between word vectors. The main reason is that Google's custom search engine API is no longer free.

4.2 Data Preprocessing

We use BNC and a list of VNCs [30] (described above) and labeled as L (Literal), I (Idioms), or Q (Unknown). For our experiments we only use VNCs that are annotated as I or L. We only experimented with idioms that can have both literal and idiomatic interpretations. Each document contains three paragraphs: a paragraph with a target VNC, the preceding paragraph and following one. Our data is summarized in Table 3.

Table 3. Datasets: Is = idioms; Ls = literals

Expression	Train	Test
BlowWhistle	20 Is, 20 Ls	7 Is, 31 Ls
LoseHead	15 Is, 15 Ls	6 Is, 4 Ls
MakeScene	15 Is, 15 Ls	15 Is, 5 Ls
TakeHeart	15 Is, 15 Ls	46 Is, 5 Ls
BlowTop	20 Is, 20 Ls	8 Is, 13 Ls
GiveSack	20 Is, 20 Ls	26 Is, 36 Ls
HaveWord	30 Is, 30 Ls	37 Is, 40 Ls
HitRoof	50 Is, 50 Ls	42 Is, 68 Ls
HitWall	90 Is, 90 Ls	87 Is, 154 Ls
HoldFire	20 Is, 20 Ls	98 Is, 6 Ls
HoldHorse	80 Is, 80 Ls	162 Is, 79 Ls

Since BNC did not contain enough examples, we extracted additional ones from COCA, COHA and GloWbE (http://corpus.byu.edu/). Two human annotators labeled this new dataset for idioms and literals. The inter-annotator agreement was relatively low (Cohen's kappa = .58); therefore, we merged the results keeping only those entries on which the two annotators agreed. For our experiments reported here, we obtained word vectors using the word2vec tool [23,24] and the text8 corpus. The text8 corpus has more than 17 million words, which can be obtained from `mattmahoney.net/dc/text8.zip`. The resulting vocabulary has 71,290 words, each of which is represented by a $q = 200$ dimension vector. Thus, this 200 dimensional vector space provides a basis for our experiments.

4.3 Datasets

Table 3 describes the datasets we used to evaluate the performance of the proposed technique. All these verb-noun constructions are ambiguous between literal and idiomatic interpretations.

5   Results

Table 4 shows the average precision, recall and accuracy of the competing methods on 11 datasets over 20 runs. (The average best performance is in bold face. We calculate accuracy by adding true positives and true negatives and normalizing the sum by the number of examples. The results show that the $CoVAR$ model outperforms the rest of the models overall.

Interestingly, the Frobenius norm outperforms the spectral norm. One possible explanation is that the spectral norm evaluates the difference when two matrices act on the maximal variance direction, while the Frobenius norm evaluates on a standard basis. That is, Frobenius measures the difference along all basis vectors. On the other hand, the spectral norm evaluates changes in a particular direction. When the difference is a result of all basis directions, the Frobenius norm potentially provides a better measurement. The projection methods ($p \cdot idf$ and $p \cdot tf \cdot idf$) outperform $tf \cdot idf$ overall but not as pronounced as $CoVAR$.

Finally, we have noticed that even the best model ($CoVAR_{Fro}$) does not perform as well on certain idiomatic expressions. We hypothesize that the model works the best on highly idiomatic expressions.

6   Is There a Correlation Between the Human Judgements and the Automatic Approach?

We measure the correlation between the human judgements and the competing algorithms in terms of Pearson's correlation coefficient. Figure 1 shows the plots of the correlation matrices between the average human judgements per idiom type shown in Table 2 and the judgements by the algorithms. The resulting correlation matrices show that the performance of the proposed algorithm $CoVar_{Fro}$ is highly correlated with the human judgements, followed by $CoVar_{Sp}$. This once again demonstrates that $CoVar_{Fro}$ is capable of exploiting context information.

Table 4. Average accuracy of competing methods on 11 datasets: BlWh (BlowWhistle), LoHe (LoseHead), MaSe (MakeScene), TaHe (TakeHeart), BlTo (BlowTop), GiSa (GiveSack), HaWo (HaveWord), HiRo (HitRoof), HiWa (HitWall), HoFi (HoldFire), and HoHo (HoldHorse).

	BlWh	LoHe	MaSe	TaHe	BlTo	GiSa	HaWo	HiRo	HiWa	HoFi	HoHo	Ave
Precision												
$p \cdot idf$	0.29	0.49	0.82	0.9	0.59	0.55	0.52	0.54	0.55	0.97	0.86	0.64
$p \cdot tf \cdot idf$	0.23	0.31	0.4	0.78	0.54	0.54	0.53	0.41	0.39	0.95	0.84	0.54
$CoVAR_{Fro}$	0.65	0.6	0.84	0.95	0.81	0.63	0.58	0.61	0.59	0.97	0.86	**0.74**
$CoVAR_{sp}$	0.44	0.62	0.8	0.94	0.71	0.66	0.56	0.54	0.5	0.96	0.77	0.68
$CTX+$	0.17	0.55	0.78	0.92	0.66	0.67	0.53	0.55	0.92	0.97	0.93	0.70
GMM	0.18	0.46	0.67	0.79	0.41	0.45	0.42	0.4	0.41	0.94	0.73	0.53
Recall												
$p \cdot idf$	0.82	0.27	0.48	0.43	0.58	0.47	0.53	0.84	0.92	0.83	0.81	0.63
$p \cdot tf \cdot idf$	0.99	0.3	0.11	0.11	0.53	0.64	0.53	0.98	0.97	0.89	0.97	0.64
$CoVAR_{Fro}$	0.71	0.78	0.83	0.61	0.87	0.88	0.49	0.88	0.94	0.86	0.97	**0.80**
$CoVAR_{sp}$	0.77	0.81	0.82	0.55	0.79	0.75	0.53	0.85	0.95	0.87	0.85	0.78
$CTX+$	0.56	0.52	0.37	0.66	0.7	0.83	0.85	0.82	0.57	0.64	0.89	0.67
GMM	0.55	0.48	0.54	0.36	0.49	0.47	0.41	0.55	0.73	0.72	0.57	0.53
Accuracy												
$p \cdot idf$	0.6	0.48	0.53	0.44	0.68	0.62	0.54	0.66	0.7	0.81	0.78	0.62
$p \cdot tf \cdot idf$	0.37	0.49	0.33	0.18	0.65	0.55	0.53	0.45	0.43	0.85	0.86	0.52
$CoVAR_{Fro}$	0.87	0.58	0.75	0.62	0.86	0.72	0.58	0.74	0.74	0.84	0.87	**0.74**
$CoVAR_{sp}$	0.77	0.61	0.72	0.56	0.79	0.73	0.58	0.66	0.64	0.84	0.73	0.69
$CTX+$	0.4	0.46	0.45	0.64	0.75	0.76	0.57	0.67	0.71	0.64	0.88	0.63
GMM	0.46	0.5	0.52	0.39	0.49	0.53	0.49	0.51	0.53	0.7	0.57	0.52

Interestingly, the supervised version of the CONTEXT technique described in [26] negatively correlates with the human rankings, suggesting that this model does not use contextual information in the most optimal way.

6.1   Related Work

Previous approaches to idiom detection can be classified into two groups: (1) type-based extraction, i.e., detecting idioms at the type level, e.g., [6,26,31,32]; (2) token-based detection, i.e., detecting idioms in context. Type-based extraction is based on the idea that idiomatic expressions exhibit certain linguistic properties such as non-compositionality that can distinguish them from literal expressions [6,26]. While many idioms do have these properties, many idioms fall on the continuum from being compositional to being partly unanalyzable to completely non-compositional [33]. [22,26,27,34–38], among others, notice that type-based approaches do not work on expressions that can be interpreted idiomatically or literally depending on the context and thus, an approach that

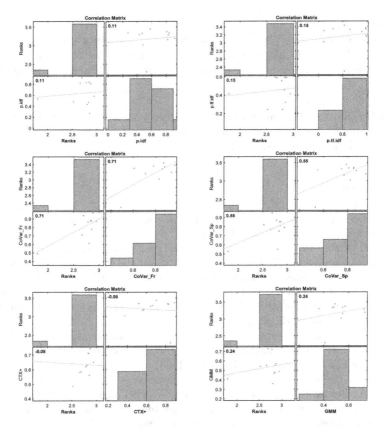

Fig. 1. Pairwise Pearson's correlation matrix between the human judgements and the competing algorithms. Top row: $p \cdot idf$ and $p \cdot tf \cdot idf$. Middle row: $CoVar_{Fro}$ and $CoVar_{Sp}$. Bottom row: $CTX+$ and GMM.

considers tokens in context is more appropriate for idiom recognition. To address these problems, [39] investigate the bag of words *topic* representation and incorporate an additional hypothesis–contexts in which idioms occur are more affective. Still, they treat idioms as semantic outliers. [40–45] explore a range of distributional vector-space models for semantic composition.

7   Conclusions

In this paper we reported the results of an experiment in which human annotators ranked idiomatic expressions in context on a scale from 1 (literal) to 4 (highly idiomatic). Our experiment supports the hypothesis that idioms fall on a continuum and that one might differentiate between highly idiomatic, mildly idiomatic and weakly idiomatic expressions. In addition, we measured the relative idiomaticity of 11 idiomatic types and computed the correlation between the relative idiomaticity of an expression and the performance of various automatic models for idiom detection. We have shown that our model, based on the distributional semantics ideas, positively correlates with the human judgements. This suggests that we are moving in the right direction toward automatic idiom detection.

Acknowledgments. This material is based upon work supported by the Air Force Office of Scientific Research under award number FA9550-16-1-0261 and a National Science Foundation grant IIS-1319846.

References

1. Glucksberg, S.: From Metaphors to Idioms. Oxford Psychology Series, vol. 36. Oxford University Press, Oxford (2001)
2. Jackendoff, R.: The boundaries of the lexicon. In: Everaert, M., van der Linden, E., Schenk, A., Schreuder, R. (eds.) Idioms: Structural and Psychological Perspectives, pp. 133–165. Lawrence Erlbaum Associates, Hillsdale (1995)
3. Glucksberg, S.: Idiom meanings and allusional content. In: Cacciari, C., Tabossi, P. (eds.) Idioms: Processing, Structure, and Interpretation, pp. 3–26. Lawrence Erlbaum Associates, Hillsdale (1993)
4. Cacciari, C.: The place of idioms in a literal and metaphorical world. In: Cacciari, C., Tabossi, P. (eds.) Idioms: Processing, Structure, and Interpretation, pp. 27–53. Lawrence Erlbaum Associates, Hillsdale (1993)
5. Nunberg, G., Sag, I.A., Wasow, T.: Idioms. Language **70**, 491–538 (1994)
6. Sag, I.A., Baldwin, T., Bond, F., Copestake, A., Flickinger, D.: Multiword expressions: a pain in the neck for NLP. In: Proceedings of the 3rd International Conference on Intelligence Text Processing and Computational Linguistics (CICLing 2002), Mexico City, Mexico, pp. 1–15 (2002)
7. Villavicencio, A., Copestake, A., Waldron, B., Lambeau, F.: Lexical encoding of MWEs. In: Proceedings of the Second ACL Workshop on Multiword Expressions: Integrating Processing, Barcelona, Spain, pp. 80–87 (2004)
8. Fellbaum, C., Geyken, A., Herold, A., Koerner, F., Neumann, G.: Corpus-based studies of german idioms and light verbs. Int. J. Lexicogr. **19**, 349–360 (2006)
9. Colombo, L.: The comprehension of ambiguous idioms in context. In: Cacciari, C., Tabossi, P. (eds.) Idioms: Processing, Structure, and Interpretation, pp. 163–200. Lawrence Erlbaum Associates, Hillsdale (1993)
10. Cook, P., Fazly, A., Stevenson, S.: The VNC-tokens dataset. In: Proceedings of the LREC Workshop Towards a Shared Task for Multiword Expressions (MWE 2008) (2008)
11. Burnard, L.: The British National Corpus Users Reference Guide. Oxford University Computing Services (2000)
12. Wulff, S.: Rethinking idiomaticity: a usage-based approach. A&C Black (2010)
13. Erlewine, M.Y., Kotek, H.: A streamlined approach to online linguistic surveys. Nat. Lang. Linguist. Theory **34**, 481–495 (2016)
14. Bangdiwala, S.: A graphical test for observer agreement. In: 45th International Statistical Institute Meeting, pp. 307–308 (1985)
15. Cohen, J.: A coefficient of agreement for nominal scales. Educ. Psychol. Meas. **20**, 37–46 (1960)

16. Shankar, V., Bangdiwala, S.I.: Observer agreement paradoxes in 2×2 tables: comparison of agreement measures. BMC Med. Res. Methodol. **14**, 100 (2014)
17. Feinstein, A.R., Cicchetti, D.V.: High agreement but low kappa: I. The problems of two paradoxes. J. Clin. Epidemiol. **43**, 543–549 (1990)
18. McCarthy, D., Keller, B., Carroll, J.: Detecting a continuum of compositionality in phrasal verbs. In: Proceedings of the ACL-SIGLEX Workshop in Multiword Expressions: Analysis, Acquisition and Treatment, Sapporo, Japan, pp. 73–80 (2003)
19. Venkatapathy, S., Joshi, A.K.: Measuring the relative compositionality of verb-noun (VN) collocations by integrating features. In: Proceedings of the Conference on Human Language Technology and Empirical Methods in Natural Language Processing, pp. 899–906. Association for Computational Linguistics (2005)
20. Farahmand, M., Smith, A., Nivre, J.: A multiword expression data set: annotating non-compositionality and conventionalization for English noun compounds. In: Proceedings of NAACL-HLT, pp. 29–33 (2015)
21. Firth, J.R.: A synopsis of linguistic theory, 1930–1955 (1957)
22. Katz, G., Giesbrecht, E.: Automatic identification of non-compositional multiword expressions using latent semantic analysis. In: Proceedings of the ACL/COLING 2006 Workshop on Multiword Expressions: Identifying and Exploiting Underlying Properties, pp. 12–19 (2006)
23. Mikolov, T., Chen, K., Corrado, G., Dean, J.: Efficient estimation of word representations in vector space. In: Proceedings of Workshop at ICLR (2013)
24. Mikolov, T., Sutskever, I., Chen, K., Corrado, G., Dean, J.: Distributed representations of words and phrases and their compositionality. In: Proceedings of NIPS (2013)
25. Fukunaga, K.: Introduction to Statistical Pattern Recognition. Academic Press, New York (1990)
26. Fazly, A., Cook, P., Stevenson, S.: Unsupervised type and token identification of idiomatic expressions. Comput. Linguist. **35**, 61–103 (2009)
27. Li, L., Sporleder, C.: Using Gaussian mixture models to detect figurative language in context. In: Proceedings of NAACL/HLT 2010 (2010)
28. Cilibrasi, R., Vitányi, P.M.B.: The Google similarity distance. IEEE Trans. Knowl. Data Eng. **19**, 370–383 (2007)
29. Cilibrasi, R., Vitányi, P.M.B.: Normalized web distance and word similarity. CoRR abs/0905.4039 (2009)
30. Cook, P., Fazly, A., Stevenson, S.: The VNC-tokens dataset. In: Proceedings of the LREC Workshop: Towards a Shared Task for Multiword Expressions (MWE 2008), Marrakech, Morocco (2008)
31. Widdows, D., Dorow, B.: Automatic extraction of idioms using graph analysis and asymmetric lexicosyntactic patterns. In: Proceedings of the ACL-SIGLEX Workshop on Deep Lexical Acquisition, DeepLA 2005, Stroudsburg, PA, USA, pp. 48–56. Association for Computational Linguistics (2005)
32. Hearst, M.A.: Automatic acquisition of hyponyms from large text corpora. In: Proceedings of the 14th Conference on Computational Linguistics - Volume 2, COLING 1992, Stroudsburg, PA, USA, pp. 539–545. Association for Computational Linguistics (1992)
33. Cook, P., Fazly, A., Stevenson, S.: Pulling their weight: exploiting syntactic forms for the automatic identification of idiomatic expressions in context. In: Proceedings of the ACL 2007 Workshop on A Broader Perspective on Multiword Expressions, pp. 41–48 (2007)

34. Birke, J., Sarkar, A.: A clustering approach to the nearly unsupervised recognition of nonliteral language. In: Proceedings of the 11th Conference of the European Chapter of the Association for Computational Linguistics (EACL 2006), Trento, Italy, pp. 329–226 (2006)
35. Sporleder, C., Li, L.: Unsupervised recognition of literal and non-literal use of idiomatic expressions. In: EACL 2009: Proceedings of the 12th Conference of the European Chapter of the Association for Computational Linguistics, Morristown, NJ, USA, pp. 754–762. Association for Computational Linguistics (2009)
36. Bu, F., Zhu, X., Li, M.: Measuring the non-compositionality of multiword expressions. In: Proceedings of the 23rd International Conference on Computational Linguistics, pp. 116–124. Association for Computational Linguistics (2010)
37. Boukobza, R., Rappoport, A.: Multi-word expression identification using sentence surface features. In: Proceedings of the 2009 Conference on Empirical Methods in Natural Language Processing: Volume 2, vol. 2, pp. 468–477. Association for Computational Linguistics (2009)
38. Reddy, S., McCarthy, D., Manandhar, S.: An empirical study on compositionality in compound nouns. In: IJCNLP, pp. 210–218 (2011)
39. Peng, J., Feldman, A., Vylomova, E.: Classifying idiomatic and literal expressions using topic models and intensity of emotions. In: Proceedings of the 2014 Conference on Empirical Methods in Natural Language Processing (EMNLP), Doha, Qatar, pp. 2019–2027. Association for Computational Linguistics (2014)
40. Yazdani, M., Farahmand, M., Henderson, J.: Learning semantic composition to detect non-compositionality of multiword expressions. In: EMNLP, pp. 1733–1742 (2015)
41. Salehi, B., Cook, P., Baldwin, T.: A word embedding approach to predicting the compositionality of multiword expressions. In: HLT-NAACL, pp. 977–983 (2015)
42. Peng, J., Feldman, A., Jazmati, H.: Classifying idiomatic and literal expressions using vector space representations. In: RANLP, pp. 507–511 (2015)
43. Salton, G.D., Ross, R.J., Kelleher, J.D.: Idiom token classification using sentential distributed semantics. In: Proceedings of the 54th Annual Meeting on Association for Computational Linguistics, pp. 194–204 (2016)
44. Peng, J., Feldman, A.: Experiments in idiom recognition. In: COLING, pp. 2752–2762 (2016)
45. Cordeiro, S., Ramisch, C., Idiart, M., Villavicencio, A.: Predicting the compositionality of nominal compounds: giving word embeddings a hard time. In: Proceedings of the 54th Annual Meeting of the Association for Computational Linguistics (Volume 1: Long Papers), pp. 1986–1997. Association for Computational Linguistics (2016)

Dialogue Act Taxonomy Interoperability Using a Meta-model

Soufian Salim[(✉)], Nicolas Hernandez, and Emmanuel Morin

Université de Nantes, LS2N UMR 6004,
2 rue de la houssiniére, 44322 Nantes Cedex 03, France
{soufiansalim,nicolashernandez,emmanuelmorin}@univ-nantes.fr

Abstract. Dialogue act taxonomies, such as those of DAMSL, DiAML or the HCRC dialogue structure, can be incorporated into a larger meta-model by breaking down their labels into primitive functional features. Doing so enables the re-exploitation of annotated data for automatic dialogue act recognition tasks across taxonomies, *i.e.* it gives us the means to make a classifier learn from data annotated according to taxonomies different from the target taxonomy. We propose a meta-model covering several well-known taxonomies of dialogue acts, and we demonstrate its usefulness for the task of cross-taxonomy dialogue act recognition.

1 Introduction

Speech act theory [1] attempts to describe utterances in terms of communicative function (*e.g.* question, answer, thanks). Dialogue act theory extends it by incorporating notions of context and common ground, *i.e.* information that needs to be synchronized between participants for the conversation to move forward [2]. Dialogue acts are a fundamental part of the field of dialogue analysis, and the availability of annotations in terms of dialogue acts is essential to the machine learning aspects of many applications, such as automated conversational agents, e-learning tools or customer management systems. However, depending on the applicative or research goals sought, relevant annotations can be hard to come by. This work attempts to alleviate the costs of building systems based on dialogue act statistical learning and recognition. Supervised methods for classification are the norm for dialogue act recognition tasks, and since the annotation of new data is a costly and complicated endeavour, making annotated data reusable as much as possible would be a boon for many researchers.

Several corpora annotated in terms of dialogue acts are available to researchers, such as Switchboard, MapTask, MRDA, *etc.* [3–5]. Most of these corpora are annotated using taxonomies of varying levels of similarity. For example, the Switchboard corpus is annotated using a variation of the DAMSL scheme [6], MapTask and MRDA use their own taxonomies, and BC3 uses the MRDA tagset [5]. Intuitively, it makes sense that different researchers would use different taxonomies since not all information captured by such or such annotation

© Springer Nature Switzerland AG 2018
A. Gelbukh (Ed.): CICLing 2017, LNCS 10761, pp. 305–316, 2018.
https://doi.org/10.1007/978-3-319-77113-7_24

scheme is relevant to each of every possible task, domain, and modality. In a similar way, general-purpose taxonomies may ignore information that can be crucial to a given task, or specific to a particular domain. This is also why many researchers develop their own taxonomies, or alternatively use a variant or simplification of an existing taxonomy. These taxonomies are then applied to some data used in a few experiments, and often the data isn't even published.

This is all very wasteful, and at the source of an important issue. Annotating data in terms of dialogue acts is expensive and time-consuming, yet most of the resulting annotations aren't used as much as they could be because everyone uses a different taxonomy, or is interested in different domains. There is a need in the community for the availability of diverse corpora sharing the same annotations, as demonstrated by the significant efforts that were recently put in the development of the Tilburg DialogBank [7]. This project aims at publishing annotations for several common corpora using the ISO standard 24617-2 for DiAML [8]. While it is a very useful and commendable venture, it is important to remember that DiAML is not the answer to every task and every problem; there is too much potential information to annotate in dialogues to hope for a comprehensive and complete domain-independent annotation scheme. Even though DiAML is a standard, no standard will ever be sufficient to cover all possible situations of dialogue, and no standard can be useful to all dialogue analysis tasks. Even though ISO 24617-2 does provide guidelines for extending the standard, mainly by extending or reducing sets of annotations, the end result of applying them would always be the creation of a new albeit similar taxonomy.

Thus, rather than attempting to solve the problem of the inter-usability of corpora by proposing a better or more exhaustive standard, which is beyond our capabilities, we propose a different approach: the adoption of a meta-model for the abstraction of dialogue act taxonomies. The meta-model is built by breaking down dialogue act labels into primitive functional features, which are postulated to be aspects of dialogue acts captured by various labels across taxonomies. In this work, we demonstrate that it is possible to use a meta-model of taxonomies for annotation conversion, but also that such a model can be used to train a dialogue act classifier on a corpus annotated with a taxonomy different from the one it is designed to output annotations for.

This article is organized as follows. In Sect. 2 we discuss standardization efforts and the separation of dialogue act primitive features. We detail our meta-model in Sect. 3, before presenting our experimental framework in Sect. 4. In Sect. 5 we report the results of two sets of experiments. The first one evaluates methods for converting annotations from one taxonomy to another using the meta-model. The second demonstrates that it is possible to train a classifier to output annotations for a taxonomy different than the one used for the data it was trained on. We also experiment with complex taxonomies and show that at least some information can be identified without any annotation by training a DiAML classifier on DAMSL data and evaluating it on the Switchboard corpus. We conclude this article in Sect. 6.

2   Related Work

As we mentioned previously, one approach to the lack of interoperability of dialogue schemes is the development of new standards and their assorted resources. From this perspective, the DialogBank [7] is the most recent effort to bring reliable and generic annotated data to the community. It is essentially a language resource containing dialogues from various sources re-segmented and re-annotated according to the ISO 24617-2 standard. Dialogues come from various corpora, such as HCRC MapTask, DIAMOND and Switchboard.

The authors' efforts are based on their conviction that DiAML is more complete semantically, application-independent and domain-blind. However, we believe that the standardization approach would benefit from efficient tools to improve the interoperability of existing annotations that do not conform to the DiAML recommendations. Firstly, because while it is true that DiAML is more complete semantically and less dependant on application and domain than the other existing annotation schemes, as demonstrated by Chowdhury *et al.* [9], it is not universal. For example, someone working with conversations extracted from internet forums will miss important features of online discourse by using DiAML, such as document-linking or channel-switching, all the while being burdened by a significant number of dimensions and communicative functions that are near absent from his or her data, such as functions of the time or turn dimensions. Secondly, we believe that dialogues are so complex and so rich that we cannot realistically expect a single annotation scheme to capture all of the information that may be relevant to any dialogue analysis system. There will always be missing information that would have been useful for something, and the pursuit of exhaustivity in annotation can sometimes lead to the development of cumbersome and impractical tools. Such ambitions may lead to the phenomenon known as *feature creep*, which is the continuous addition of extra features that are only useful for specific use-cases and go beyond the initial purpose of the tool, which can result in over-complication rather than simple and efficient design.

Perhaps it is preferable to build different taxonomies for different purposes, and focus the efforts put in the standard on making it more interoperable. Petukhova *et al.* [10] provide a good example of such efforts by providing a method to query the HCRC MapTask and the AMI corpora through DiAML. They notably report that the multi-dimensionality of the scheme makes it more accurate: *i.e.*, coding dialogue acts with multiple functions is a good way to make the taxonomy more interoperable. Indeed, the fact that utterances can generally have multiple functions is well known. Traum [11] notes that there are two ways to capture this multiplicity in a taxonomy: either annotate each function separately, which requires that each utterance can bear several labels, or group these functions into coherent sets and code utterances with complex labels.

The first option is the one preferred by DiAML, as it has the advantage of reducing the size of the tagset considered for each tagging decision, and better capture the multi-functionality of utterances. The idea behind this is that it is better to use several mutually exclusive tagsets than one tagset in which labels may often share functional features. For example, let us consider the following dialogue

(Speaker 1) Now take a left
(Speaker 1) And then uuh
(Speaker 2) Turn right?
(Speaker 1) Yeah

With DiAML, it would be possible to annotate the second utterance with both the INSTRUCT and the STALLING labels. However, in the HCRC coding scheme, the INSTRUCT tag is separate from the UNCODABLE tag, and therefore the utterance can only be coded with one or the other. The issue here is that it can be difficult to decide how to code an utterance that shares some features with several labels. In effect, what multi-dimensional taxonomies do is separate function features to resolve such problems. But this separation is only meant to ease the annotation of utterances within a single taxonomy: in order to make a coding scheme more compatible with others, we believe that even function features within labels of the same dimension can be identified.

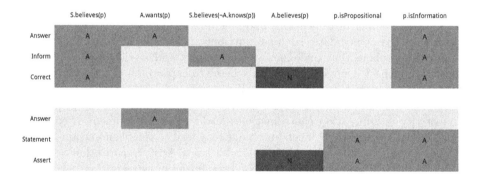

Fig. 1. Example of a meta-model for six labels from DiAML (top) and DAMSL (bottom). Medium dark (green), "A", is always present in utterances (the definition includes the feature), dark (red), "N", is never present in utterances (the definition includes the negation of the feature), light (blue) is sometimes present in utterances (the definition does not include the feature). Feature designations use several shorthands: S stands for "Speaker", A for "Addressee", p for "(the uttered) proposition" and ¬ for "not". Therefore, S.believes(p) could be rewritten as "the speaker believes the uttered proposition to be true", and represents a single feature.

3   The Meta-Model

The purpose and manner in which dialogue acts (DA) should be defined has been discussed at length in the literature. Traum [11] raises many questions about the different aspects that should be considered when defining DA, such as "should taxonomies used for tagging dialogue corpora given formal semantics?"

or "should the same taxonomy be used for different kind of agents?". The purpose of this work is not to promote or depreciate one approach over another, but to suggest a way to join them together.

We postulate that most taxonomies of dialogue acts can be generalized using primitive features as defining attributes of their labels. For example, an ANSWER in DiAML can't have an action-discussion aspect[1], but an ANSWER in DAMSL can. In both cases, the label can only be applied to an utterance elicited by the addressee. We could thus identify a few features of these labels to define the ANSWER label of these two taxonomies. The fact that the answer must be wanted by the addressee would be a common feature, and the fact that the answer cannot have an action-discussion aspect would be a differentiating feature.

We define a meta-model as the set of all features that can be used to define all the labels of a given set of taxonomies. A few benefits of such a tool are illustrated in Fig. 1. The figure displays the manifestation of primitive features in utterances according to their label. A few acts are described, for the DiAML and the DAMSL schemes. Going back to our previous example, we see that the ANSWER labels are easy to compare when defined as sets of features, and doing so requires no human discernment: in the columns "S.believes(p)" and "p.isInformation", the cells are green for DiAML but blue for DAMSL. This means that an answer must be genuine in DiAML, but answers that are lies are accepted in DAMSL. Moreover, in DiAML an answer must be informational - *i.e.* it cannot be an action-discussion utterance, nor a declarative act - which is not the case in DiAML. For example, answering to a request for action can be an ANSWER in the sense of DAMSL but not for DiAML. A computer could compare them, which would be impossible if presented with written definitions. We can observe in Fig. 1 that when two labels share colour-codes everywhere, they are essentially the same label, when in places a square that is blue in one is red or green in the other, the second label is a specialization of the first one, and when there are opposing green and red squares, they are mutually exclusive.

For the purposes of this work, we built a meta-model including labels from the SWBD-DAMSL annotation scheme, the DiAML standard for the annotation of dialogue acts, and the HCRC dialogue structure coding system.

3.1 Feature Formatting

We chose to format the features using a few basic components that can be linked together: *participants* ((S)peaker, (A)ddressee) use *verbs* (*e.g.* provides(), wants(), believes()) on *objects* (*e.g.* (p)roposition, (f)eedback, (a)ction), and these objects have *properties* (*e.g.* isPositive). The following example lists the features of the AUTO NEGATIVE FEEDBACK label in DiAML, meant for utterances providing negative feedback, such as "I don't get it" for example:

S.provides(f) ∧ f.isAuto ∧ ¬ f.isPositive

---

[1] *i.e.* It can't discuss the planning of an action, such as the utterance "ok I'll reboot my computer then".

Features are separated by conjunction symbols. The first feature means "the speaker provides feedback", the second "the feedback concerns the speaker's understanding of an utterance", and the third "the feedback is negative".

This way of formatting features offers multiple advantages. Notably, it helps to avoid redundancy in features, and it allows for the use of logical operators (*e.g.* not ¬, or ∨, and ∧). Moreover, using such a format makes it possible to break down features into learnable bits that can be used to train a classifier (for example, the presence of the object (a)ction in the feature). We also chose to make it similar to logical predicates so that it can be parsed and evaluated: although representing dialogue within a logical framework is an idea that has been explored in the literature before [12], we did not come across any work attempting to utilize the individual representation of dialogue act classes for data analysis and recognition. This aspect of our research however - parsing utterances to match logical definitions - is out of the scope of this paper. At the moment, each feature is treated as a boolean by the algorithms and the naming convention does not impact the experiments, *i.e.* "S.provides(f) ∧ f.isAuto ∧ ¬ f.isPositive" is equivalent to "$feature_a = true, feature_b = true, feature_c = true$".

However, the main advantage of this formulation is that it allows us to use concepts such as "belief" or "feedback" across multiple features, and implement theoretically grounded notions in the meta-model's building blocks. These elements reflect the conceptual foundation of the taxonomies comprised within the meta-model. In the meta-model used in this work, the primitive features used hint at the fact that the researchers behind DAMSL, DiAML and HCRC subscribed to a certain vision of dialogue structure. Indeed, the features are predominantly built around the notions of belief, desire and intention [13,14], as well as the linguistic notion of *grounding* [2]. However it is important to note that the meta-model itself is not linguistically motivated, and could incorporate elements from any theory. For example, should a meta-model integrate Verbal Response Modes [15], its features would necessarily capture notions such as the frame of reference or the source of experience. In effect, primitive features can describe characteristics of knowledge, intention and belief of the speaker and the addressee, as well as characteristics of action and acknowledgement.

3.2   Feature Extraction

We based our features on the exact written definitions of their labels, as published in the literature by their authors. For example, the AUTO-NEGATIVE FEEDBACK label used in our earlier example, the written definition as found in the ISO 24617-2 guidelines is the following:

"Communicative function of a dialogue act performed by the sender, S, in order to inform the addressee, A that S's processing of the previous utterance(s) encountered a problem."

Theoretically, any number of features can be extracted from such a definition. Perhaps a feature signifying that the utterance bears an information, another one to signal that it is not information related to the task, another to mark the utterance as potentially non-verbal *etc.* Our formalization of the label is "S.provides(f) ∧ f.isAuto ∧ ¬ f.isPositive". To reach that result from the definition, we used a simple principle: new features should only be introduced as a mean to distinguish the label from its parent or siblings[2].

All three of these features are therefore used to distinguish AUTO-NEGATIVE FEEDBACK from other labels. "S.provides(f)" means that the utterance informs the processing of a previous utterance's execution, and in doing so distinguishes feedback functions from general-purpose functions[3]. "f.isAuto" means that the feedback pertains to the speaker's own processing, and is used to distinguish the label from ALLO-NEGATIVE FEEDBACK, which pertains to the addressee's processing of an utterance. "¬ f.isPositive" means that the feedback signals a problem; this feature is used to distinguish it from AUTO-POSITIVE FEEDBACK. No more than these three features are required to efficiently distinguish each of the feedback labels. This method aims at building a meta-model suited to label comparison, not at capturing all the information contained in an annotation.

4   Experimental Framework

The experiments detailed in this paper deal with the conversion and recognition of dialogue acts across taxonomies. First we present the corpora we perform the experiments on, and then our implementation of the meta-model.

4.1   Corpora and Taxonomies

Two corpora seem most relevant for our task: the Switchboard corpus [3][4] and the MapTask corpus [4][5].

Switchboard [3][6] is a very large corpus (over 200 000 annotated utterances) annotated with the SWBD-DAMSL coding scheme [16]. DAMSL is the first annotation scheme to implement a multidimensional approach (*i.e.* which allows multiple labels to be applied to a single utterance) and is a *de facto* standard in dialogue analysis. SWBD-DAMSL is a DAMSL variant meant to reduce the multidimensionality of the latter [6]. A portion of the Switchboard corpus, about 750 utterances, has also been annotated with the ISO standard 24617-2 for DiAML [7]. The standard is inspired by DAMSL, but expands on it and attempts to annotate dialogue with a more theoretically-grounded approach.

The MapTask corpus [4][7] is also a relatively large corpus (over 2 700 annotations) annotated using the HCRC dialogue structure coding system [17], which

---

[2] If the taxonomy is "flat", *i.e.* not hierarchical, all labels are treated as siblings.
[3] While not specified in the guidelines, INFORM and in some cases ANSWER could arguably be considered a parent of all feedback labels.
[4] https://catalog.ldc.upenn.edu/ldc97s62.
[5] http://groups.inf.ed.ac.uk/maptask/.
[6] https://catalog.ldc.upenn.edu/ldc97s62.
[7] http://groups.inf.ed.ac.uk/maptask/.

comprises twelve labels. A portion of this corpus, a little over 200 utterances, has also been annotated using the DiAML scheme, which makes it an ideal candidate for our first task, converting annotations from one taxonomy to another.

4.2   Experimental Meta-Model

We built a meta-model for the labels in the taxonomies of SWBD-DAMSL, DiAML and the HCRC coding system in the manner described in Sect. 3.2. It contains 108 different features built from 2 participant types, 19 verbs, 6 object types and 32 object properties.

5   Experiments

First, we experiment with annotation conversion within the same corpus to demonstrate the ability of the meta-model to act as an effective bridge between taxonomies. Then, we present our results with cross-taxonomy classifiers, that are trained on data annotated with a different taxonomy than the one they output annotations for.

5.1   Annotation Conversion

In the context of the construction of the Tilburg DialogBank, significant efforts were put towards the re-annotation of corpora with DiAML annotations, such as the Switchboard corpus [18]. Such endeavours were met with some difficulties [19]. Some automation was employed, in the form of manually defined mappings between labels that had a many-to-one or one-to-one relation. Our experiment explores a new automated method for label conversion.

For this experiment we do not apply any supervised algorithm for dialogue act classification. We merely attempt to use the meta-model to convert annotations from one taxonomy to another on the same data. Since some data from the Switchboard corpus is annotated with both SWBD-DAMSL and DiAML tags, we use it in this experiment. We also use the utterances from the MapTask corpus that are annotated with both the HCRC dialogue structure coding system and the ISO 24617-2 annotation scheme.

Annotations of the source taxonomy are first converted to primitive features (the set of all features of all labels for the utterance), then reassembled into new annotations for the target taxonomy (including the NONE label). We first attempted to perform the second step by computing the cosine similarity between the features of the original label and the features of labels in the target taxonomy. The system would choose the label with the feature set most similar to that of the original label. We then repeated the experiment using a NaiveBayes algorithm. The system would classify sets of features into target labels. This system was evaluated through cross-validation, over ten folds. Results for both methods are reported in Table 1.

We compare our results to a simple baseline, called the direct conversion approach. It consists of using a NaiveBayes classifier trained on the combinations of tags from the source and target taxonomy. The baseline classifier does not make use of the meta-model at all.

Results were evaluated on a sample of 746 DA for the Switchboard (SWBD) corpus and 675 DA for the MapTask corpus. They are reported in Table 1.

Table 1. Label conversion scores.

Corpus	Source → Target	Accuracy
Baseline: direct conversion approach		
MapTask	DiAML → HCRC	0.60
MapTask	HCRC → DiAML	0.70
SWBD	DiAML → SWBD-DAMSL	0.60
SWBD	SWBD-DAMSL → DiAML	0.78
Labels recovered with similarity algorithm		
MapTask	DiAML → HCRC	0.60
MapTask	HCRC → DiAML	0.76
SWBD	DiAML → SWBD-DAMSL	**0.65**
SWBD	SWBD-DAMSL → DiAML	0.87
Labels recovered with NaiveBayes algorithm		
MapTask	DiAML → HCRC	**0.71**
MapTask	HCRC → DiAML	**0.82**
SWBD	DiAML → SWBD-DAMSL	0.64
SWBD	SWBD-DAMSL → DiAML	**0.93**

We see that both methods outperform the direct conversion baseline. We also observe that a simple classifier trained on very little data can have stronger performances for the task of converting annotations than using a similarity metric. The exception being the DiAML to SWBD-DAMSL conversion, for which results are almost identical. This confirms that the meta-model has value for the task of automated annotation conversion.

5.2 Cross-Taxonomy Classification

Three sets of results are reported for this experiment. The first one is our baseline: it comprises results for a straightforward DA recognition task: over ten folds of a corpus, a model is trained on nine tenth of the data and evaluated on the rest. This method requires data annotated with the target taxonomy to function. The next two sets of results are those of systems that attempt to reach similar levels of accuracy, but this time using data from annotations in a different taxonomy from the output annotations.

The first of those systems, system A, works as follows: (1) a model is trained on correct labels from the source corpus annotated according to the source taxonomy, (2) labels from the source taxonomy are projected on data from the target corpus, (3) projected labels are converted into labels from the target taxonomy according to the method described in Subsect. 5.1.

The second system, system B, attempts to learn primitive features instead of labels: (1) a model is trained on correct *primitive features* from the source corpus annotated according to the source taxonomy, (2) the target corpus is automatically annotated in terms of primitive features, (3) labels from the target taxonomy are recognized from primitive features according to the method described in Subsect. 5.1.

5.3   Method

For classification, we use an SVM for our algorithm and tokens, lemmas and parts-of-speech tags as features. Each feature type is used to build a bag-of-n-grams model. The SVM classifier was implemented using the *liblinear* library for text classification and analysis (Fan *et al.* 2008). We use a bigram model without stopword removal. We use a heuristic based on WordNet [20] for lemmatization and the Stanford toolkit [21] for part-of-speech tagging.

Since one of our taxonomies is multidimensional, allowing each instance to be tagged separately (and optionally) in several different dimensions (*i.e.* categories), a system that would attempt to pick one tag out of a tagset comprising all labels for the taxonomy would not be appropriate. Rather than using a multi-class SVM on the entire set of labels, which would not be entirely appropriate either since in DiAML only one label per dimension can be applied to an utterance, we chose to split them into dimensional tagsets. We then added the NONE label to each tagset to capture utterances that should not receive any label. Therefore, for DiAML the provided results are averaged over the results obtained over each dimension. If some results seem high for DiAML, it's because a few dimensions - such as ALLO FEEDBACK for example - will mostly be annotated with the NONE label. This is not an issue for our evaluation since the systems used as baselines also benefit from it.

5.4   Results

Results are provided in Table 2. We observe that system B has much weaker performances than system A. Its accuracy is 22 and 13 points behind the direct dialogue act classifier, for DIAML and HCRC respectively. System A, by contrast, is only outperformed by 9 and 8 points. This suggests that many features are hard to learn, comparatively to DA classes.

We can see that while the system B performs poorly, the system A is fairly efficient, less than ten points behind the results of a direct dialogue act recognition classifier. Accuracy loss can be attributed to two factors: (1) error rates of label conversion, and (2) increased error rates from the classifier due to structural and linguistic differences between the corpora used in this experiment.

Table 2. Macro and micro accuracies of a baseline classifier (label-to-label) and an indirect cross-taxonomy dialogue act classifier (label-to-features-to-label).

Source	Target	Accuracy
Baseline: direct dialogue act recognition		
SWBD (DiAML)	SWBD (DiAML)	0.83
MapTask (HCRC)	MapTask (HCRC)	0.59
A: DA recognition, decomposition then recomposition		
SWBD (DAMSL)	SWBD (DiAML)	0.74 (-0.09)
SWBD (DAMSL)	MapTask (HCRC)	0.51 (-0.08)
B: DA decomposition, recognition then recomposition		
SWBD (DAMSL)	SWBD (DiAML)	0.61 (-0.22)
SWBD (DAMSL)	MapTask (HCRC)	0.46 (-0.13)

6   Conclusion

In this paper, we presented a meta-model for the abstraction of dialogue act taxonomies. We believe the meta-model to have many useful applications for dialogue analysis and taxonomical research. The main contribution of this work is to provide a method to build supervised dialogue act recognition systems that do not require data annotated with the target taxonomy, but merely data annotated with a taxonomy which captures relevant information. We showed that a classifier trained on SWBD-DAMSL annotations could output DiAML or HCRC annotations at an accuracy not much lower than a regular classifier.

In future work, we will start a more data-driven approach to primitive feature identification by experimenting with clustering methods on annotated data. We believe an automated method will remove author bias in feature selection and allow for greater reproducibility. In order to further establish the relevance of the system, we also plan to replicate methods used in state-of-the-art dialogue act recognition systems to better understand how well a classifier can perform without a large corpus of data annotated in the appropriate taxonomy.

References

1. Austin, J.L.: How to Do Things With Words. Oxford University Press, Cambridge (1975)
2. Traum, D.R., Hinkelman, E.A.: Conversation acts in task-oriented spoken dialogue. Comput. Intell. **8**, 575–599 (1992)
3. Godfrey, J.J., Holliman, E.C., McDaniel, J.: SWITCHBOARD: telephone speech corpus for research and development. In: Proceedings of the IEEE International Conference on Acoustics, Speech, and Signal Processing (ICASSP-1992), vol. 1, San Francisco, CA, USA, pp. 517–520. IEEE (1992)
4. Anderson, A.H., Bader, M., Bard, E.G., Boyle, E., Doherty, G., Garrod, S., Isard, S., Kowtko, J., McAllister, J., Miller, J.: The HCRC map task corpus. Lang. Speech **34**, 351–366 (1991)

5. Shriberg, E., Dhillon, R., Bhagat, S., Ang, J., Carvey, H.: The ICSI meeting recorder dialog act (MRDA) corpus. Technical report, DTIC Document (2004)
6. Core, M., Allen, J.: Coding dialogs with the DAMSL annotation scheme. In: Working Notes of AAAI Fall Symposium on Communicative Action in Humans and Machines, Boston, MA, USA, pp. 28–35 (1997)
7. Bunt, H., Petukhova, V., Malchanau, A., Fang, A., Wijnhoven, K.: The Dialog-Bank. In: Proceedings of the 2016 International Conference on Language Resources and Evaluation (LREC 2016), Portorož, Slovenia, pp. 3151–3158 (2016)
8. Bunt, H., Alexandersson, J., Choe, J.W., Fang, A.C., Hasida, K., Petukhova, V., Popescu-Belis, A., Traum, D.R.: ISO 24617-2: a semantically-based standard for dialogue annotation. In: Proceedings of the 2012 International Conference on Language Resources and Evaluation (LREC 2012), Istanbul, Turkey, pp. 430–437 (2012)
9. Chowdhury, S.A., Stepanov, E.A., Riccardi, G.: Transfer of corpus-specific dialogue act annotation to ISO standard: is it worth it? In: Proceedings of the 2016 International Conference on Language Resources and Evaluation (LREC 2016), Portorož, Slovenia, pp. 132–135 (2016)
10. Petukhova, V., Malchanau, A., Bunt, H.: Interoperability of dialogue corpora through ISO 24617-2-based querying. In: Proceedings of the 2014 International Conference on Language Resources and Evaluation (LREC 2014), Reykjavik, Iceland, pp. 4407–4414 (2014)
11. Traum, D.R.: 20 questions on dialogue act taxonomies. J. Semant. **17**, 7–30 (2000)
12. Sadek, M.D.: Dialogue acts are rational plans. In: Proceedings of the ESCA/ETRW Workshop on The Structure of Multimodal Dialogue (VENACO II), Maratea, Italy, pp. 19–48 (1991)
13. Grosz, B.J., Sidner, C.L.: Attention, intentions, and the structure of discourse. Comput. Linguist. **12**, 175–204 (1986)
14. Georgeff, M., Pell, B., Pollack, M., Tambe, M., Wooldridge, M.: The belief-desire-intention model of agency. In: Müller, J.P., Rao, A.S., Singh, M.P. (eds.) ATAL 1998. LNCS, vol. 1555, pp. 1–10. Springer, Heidelberg (1999). https://doi.org/10.1007/3-540-49057-4_1
15. Stiles, W.B.: Verbal response modes and dimensions of interpersonal roles: a method of discourse analysis. J. Person. Soc. Psychol. **36**, 693 (1978)
16. Jurafsky, D., Shriberg, E., Biasca, D.: Switchboard SWBD-DAMSL shallow-discourse-function annotation coders manual. Technical report (1997)
17. Carletta, J., Isard, A., Kowtko, J., Doherty-Sneddon, G.: HCRC dialogue structure coding manual. Technical report (1996)
18. Fang, A.C., Cao, J., Bunt, H., Liu, X.: The annotation of the Switchboard corpus with the new ISO standard for dialogue act analysis. In: Proceedings of the 8th Joint ISO-ACL SIGSEM Workshop on Interoperable Semantic Annotation, Pisa, Italy, p. 13 (2012)
19. Bunt, H., Fang, A.C., Liu, X., Cao, J., Petukhova, V.: Issues in the addition of ISO standard annotations to the Switchboard corpus. In: Proceedings of the 9th Joint ISO ACL SIGSEM Workshop on Interoperable Semantic Annotation (ISA-9), Potsdam, Germany, pp. 59–70 (2013)
20. Miller, G.A.: WordNet: a lexical database for English. Commun. ACM **38**, 39–41 (1995)
21. Manning, C.D., Surdeanu, M., Bauer, J., Finkel, J.R., Bethard, S., McClosky, D.: The Stanford CoreNLP natural language processing toolkit. In: Proceedings of 52nd Annual Meeting of the Association for Computational Linguistics: System Demonstrations, Baltimore, MD, USA, pp. 55–60 (2014)

Textual Entailment Using Machine Translation Evaluation Metrics

Tanik Saikh[1]($\boxtimes$), Sudip Kumar Naskar[2], Asif Ekbal[1],
and Sivaji Bandyopadhyay[2]

[1] Indian Institute of Technology, Patna, India
{tanik.srf17,asif}@iitp.ac.in
[2] Jadavpur University, Kolkata, India
{sudip.naskar,sbandyopadhyay}@cse.jdvu.ac.in

Abstract. In this paper we propose a novel approach to determine Textual Entailment (TE) relation between a pair of text expressions. Different machine translation (MT) along with summary evaluation metrics and polarity feature have been used as features for different machine learning classifiers to take the entailment decision in this study. We consider three machine translation evaluation metrics, namely BLEU, METEOR and TER and a summary evaluation metric namely ROUGE as similarity metrics for this task. We also used the negation polarity feature in combination with the similarity measure features. We performed experiments on the datasets released in the shared tasks on textual entailment organized in RTE-1, RTE-2, RTE-3, RTE-4 and RTE-5. The best classification accuracies obtained by our system on the RTE-1, RTE-2, RTE-3, RTE-4 and RTE-5 datasets are 54%, 55%, 60%, 52% and 51% respectively.

Keywords: Textual entailment
Machine translation evaluation metrics
Summary evaluation metrics · Machine learning

1 Introduction

The term textual entailment (TE) can be defined in many ways. In natural language processing a text (T) entails a hypothesis (H) if H is true for every circumstances in which T is true. It can also be defined as a directional relationship between a pair of sentences, T and H and T entails H if the meaning of H can be derived from the meaning of T. In probabilistic terminology, T probabilistically entails H if P(H is true | T) > P (H is true), i.e., T increases the likelihood of H being true. The task of determining the TE relation between a pair of text expressions can be considered as a classification task. In this task for a given T-H pair, we need to determine whether T implies H or not. We need to learn a classifier from annotated corpora to accomplish this task. In any machine learning framework, we need a suitable feature space. To determine the TE relation between a pair of text expressions possible features can be the

© Springer Nature Switzerland AG 2018
A. Gelbukh (Ed.): CICLing 2017, LNCS 10761, pp. 317–328, 2018.
https://doi.org/10.1007/978-3-319-77113-7_25

ones which capture similarity between pair of text expressions. Here we consider lexical similarity measures between two text fragments as features.

The proposed approach is based on supervised machine learning algorithms such as Support Vector Machine (SVM) [1,2], multilayer perceptron model [3,4] and RBF model [5]. We make use of three machine translation (MT) evaluation metrics namely BLEU, METEOR and TER, and a summary evaluation metric - ROUGE. BLEU, METOR and TER are popular MT evaluation metrics which are used to evaluate the quality of machine translated text. The MT evaluation metrics are applied on the machine translated text and human generated reference translation(s) to find out how close the machine-translated text is to human translation(s).

On the other hand, ROUGE measures similarity between machine generated summary and gold standard summary by comparing n-gram lexical matches between these two summaries. As far our knowledge goes, use of combination of such features into supervised machine learning frameworks is new and first of its kind except the work of [6] that made use of conventional similarity metrics like, Cosine, Dice, Jaccard, etc. and two MT evaluation metrics namely BLEU, METEOR and made a comparative study. The same set of features were employed in a machine learning framework by [7] in the shared task on *"Detecting Paraphrases in Indian Languages (DPIL)"* organized in the *"Forum for Information Retrieval Evaluation (FIRE)"- 2016* for the Indian languages, namely Hindi, Punjabi, Tamil and Telugu.

2   Related Work

Literature shows that there are significant numbers of works available in textual entailment on several datasets including RTE1, RTE2, RTE3, RTE4 and RTE5 datasets. In RTE-1 the best result was reported by [8] using the word overlap method. The output of BLEU was taken as confidence score and it was used to give TRUE or FALSE value to each entailment pair. They performed an optimization procedure for the development set that chose the best threshold according to the percentage of success of correctly recognized entailments and got a particular value, if the BLEU's output is higher than that of threshold value then the entailment relation is TRUE for that T-H pair, otherwise the entailment relation is FALSE. They obtained an accuracy of 70% on this dataset. In RTE-2, the best result of 75% was obtained by [9], using lexical relation and syntactic matching. The work of [10] produced the best accuracy of 80% on RTE-3 using discourse commitments, lexical alignment and knowledge extraction methods. The task defined in [11] used cosine similarity along with causal non-symmetric measure and obtained 63.5% accuracy using Naïve Bayes classifier on the RTE-3 dataset. The work reported in [12] used seven features, namely lexical semantic similarity, named entities, dependent content word pairs, average distance, negation, task, length and produced an accuracy of 62.7% on the RTE-3 dataset. The authors of [13] computed the similarity between two sentences in terms of the degree of overlapping between the semantic contents of the two sentences and obtained 61.5% accuracy on the RTE-3 dataset. The study of [14] use support vector machine (SVM) technique to take entailment decision between each T - H pair and they achieved an accuracy of 61.75% on the RTE-3 dataset. In all the above cases conventional lexical and semantic features have been employed.

On the other hand, the motive of MT evaluation metrics is to determine how close the translation hypothesis is to a reference translation, the closer they are the better the translation hypothesis and hence the MT system. Over the years researchers of MT have discovered several MT evaluation metrics like word error rate (WER) [15], position-independent WER (PER) [16], BLEU [17], NIST [18], Meteor [19], Translation error/edit rate (TER) [20], General Text Matcher (GTM) [21], etc. Among these MT evaluation metrics BLEU is perhaps the most widely used MT evaluation metric among the MT researchers. The task described in [8] demonstrates a comparative evaluation between this BLEU based algorithm and a Latent Semantic Analysis (LSA) based system for recognizing textual entailments. The study described in [22] applies METEOR to T - H pairs assuming that they are two different translations of the same source sentence. However, there has not been any notable work which made use of MT evaluation metrics such as BLEU, METEOR and TER, and summary evaluation metric like ROUGE, and negation polarity feature in machine learning framework. In our proposed approach, we exploit MT and summary evaluation metrics along with negation polarity as features. The experiments reveal a new direction of research where we establish that MT and summary evaluation metrics together with negation polarity feature can be used to recognize textual entailment for the binary class classification problem (in RTE-1, RTE-2 and RTE-3) and into the *Entailed, Unknown* and *Contradiction* classes for the ternary class classification problem (in RTE-4 and RTE-5). It can be observed from the RTE datasets that the length of T is usually larger than that of H. Length of T and H does not matter in the context of taking TE decision. However, MT evaluation metrics penalizes the similarity score when the two pieces of text (hypothesis and reference) are of different length; as a result it effects the TE decision. Therefore, we also make use of the ROUGE metric which does not suffer from the penalty score in case of length mismatch. Negative polarity feature is also taken into consideration since it plays a crucial role in TE in general and particularly in identifying the 'contradiction' class in 3-class TE. We believe that use of these MT and summary evaluation metrics together with the negative polarity feature captures the essence of TE properly.

3   Feature Analysis

Features play a pivotal role in any machine learning approach. Appropriate feature combination is very important for achieving good performance. The features which are used in our experiments are described below.

3.1   Machine Translation Evaluation Metrics

MT evaluation metrics essentially measure how close a translation hypothesis (i.e., translation output) is to a human translation (i.e., reference translation). The closer the translation hypothesis is to the reference translation, the better the translation system. Over the years, MT researchers proposed several MT evaluation metrics like word error rate (WER), position-independent word error rate (PER), BLEU, NIST, METEOR, Translation error/edit rate (TER), GTM, etc. Among these MT evaluation metrics, BLEU is perhaps the most widely used in the MT community. In the present work we consider three MT evaluation metrics, namely BLEU [17], METEOR [23] and TER [20] as similarity measures in deciding about entailment by a classifier.

BLEU: BLEU (BiLingual Evaluation Understudy) [17] is a metric used to judge the quality of a MT output from one natural language to another and most probably the most demanding metric in MT community. The core idea behind this metric is the closer a MT output to a human translation the better the MT output. This metric produce the score by comparing the n-gram of the MT output with the n-gram of the reference translation and normalized by number of n-gram in the MT output. It does not take recall into account, it only calculate precision separately for each n-gram ordering and combine them following geometric means.

METEOR: METEOR (Metric for Evaluation of Translation with Explicit Ordering) [19,23] is another metric which is also used to judge the quality of MT output. This metric match MT output against one or multiple references. It basically creates a word alignment (mapping between words) between two given strings, such that every word in each string maps to at most one word in the other string. This alignment is incrementally produced by a sequence of word mapping modules in which exact, stem (Porter stemming algorithm), and synonymy (Wordnet Synonymy) matching are taken in account. It then calculate the Unigram precision $P = m/t$ and recall $R = m/r$, where m: Number of mapped unigram found between two string and t: Number of unigram in translation and m: Number of unigram in reference. It then computes a parameterized harmonic mean of P and R by the Eq. 1

$$F_{mean} = P.R/\alpha * P + (1 - \alpha) * R \qquad (1)$$

It's gives a penalty score based on the number of cross links. Finally, the Meteor score for the alignment between the two strings is calculated as Eq. 2

$$Score = (1 - Pen) * F_{mean} \qquad (2)$$

TER: TER (Translation Error Rate) [20] is an automatic metric for machine translation evaluation. It is based on edit distance. It produce the error rate by measuring the number of edits required to transform a MT output sentence into human translated reference sentence. So the complement of this error rate are taken in account for similarity score. It can measure by Eq. 3

$$TER = \#of\,edits\,required/Total\#of\,words\,in\,reference\,sentence \qquad (3)$$

3.2 Summary Evaluation Metric

The MT evaluation metrics have drawbacks of suffering from penalty score that occur due to different text lengths. We use summary evaluation metric (ROUGE) which does not consider text length. Summary evaluation metric generally measures how close the machine generated summary is to human generated summary (or reference summary). ROUGE is the available standard metric which is used to evaluate the summary. In this study this metric is used as a feature.

ROUGE: Recall-Oriented Understudy for Gisting Evaluation (ROUGE) [24] is an automatic metric for machine generated summary evaluation which is based on the notion of unigram matching between the candidate summary and human summary. It is an automatic metric which determines the quality of summary by comparing with ideal summary generated by human. This metric counts the number of overlapping units such as n-grams, word sequences, and word pairs between computer generated summary and ideal summary which is created by human, normalized by the number of n-grams in references. It may be termed as recall version of BLEU.

3.3 Polarity Feature

The presence of negative polarity context like *"no/not"* can make a sentence meaning different. Consider the pair of sentences below.

> *Example*1   *Example 2*
> *T1 : Oil price Surged.*   *T2 : I live in India.*
> *H1 : Oil price did not grow. H2 : I don't live in India.*

In Example 1, two sentences are textually entailed, but when we consider them for n-gram matching they might provide different results. In case of Example 2, the sentences are not textually entailed, but if we consider them for lexical matching they might yield high similarity score, and consequently the system will consider them as textually entailed. We introduce this feature in this study to address this kind of obstacles. In this study we search for a negation word *("no/not")* in a particular T and H pair, if it is found in T but not found in H and vice versa a score of '1' has been assigned for that T-H pair, otherwise if it is absent in both the pairs or present in both the pairs, a score of '0' is assigned.

4 Experimental Setup and Results

In this section we describe the preprocessing step, datasets, experimental setup, results and analysis.

4.1   Preprocessing Module

The system extracts a pair of T and H from the development and test set also from five RTEs datasets. The datasets contains T-H pairs as given below.
<*pair id="13" value="TRUE" task="IR"*>
    <*t*>*iTunes software has seen strong sales in Europe.* </*t*>
    <*h*>*Strong sales for iTunes in Europe.*</*h*>
</*pair*>
From this XML data, we extract T and its corresponding H part and remove the stop words and white spaces (if any) from both text and hypothesis.

4.2   Dataset

We carried out our experiments on the datasets released in the shared tasks on textual entailment organized in RTE-1, RTE-2, RTE-3, RTE-4 and RTE-5. Table 1 shows the statistics of the five datasets. The table yields the number of T-H pairs (THP), average text length (ATL) and average hypothesis length (AHL) for the development and the test set belonging to each dataset. ATL and AHL provide average sentence length in words. The model predicts textual entailment relation between a pair of text expressions and we want to estimate how accurately our predictive model performs in practice. In prediction problem, a model is generally given a dataset of known data on which training is performed, and a dataset of unknown data against which a model is tested. The Table 1 shows that the length of H is very less compared to the length of T.

Table 1. The Statistics of the datasets

Dataset	Development set			Test set		
	THP	ATL	AHL	THP	ATL	AHL
RTE-1	567	23	9	800	25	10
RTE-2	800	26	9	800	27	8
RTE-3	800	34	8	800	29	7
RTE-4	0	0	0	1000	39	7
RTE-5	600	97	7	600	96	7

4.3   Experiments

We calculate several similarity scores with the help of above mentioned features for each T-H pair contained in the development and test sets of five datasets namely RTE-1, RTE-2, RTE-3, RTE-4 and RTE-5 which were extracted with the help of preprocessing module using the aforementioned MT, summary evaluation metrics and polarity feature. These scores are used as feature values to build a classification model. We build three models based on SVM, Multilayer perceptron model and Radial basis function (RBF) network. We use the implementation as available in Weka toolkit[1]. The classifiers are trained with the features as discussed earlier and summarized in Table 2. The classifier assigns a prediction class to each T-H pair in test dataset of unknown class. Based on the comparisons between the gold standard output of that particular dataset and the predicted output of the classifier, confusion matrix is generated, which yields the system accuracy.

Table 2. Different sets of features and the corresponding models.

Feature set	Model
BLEU, METEOR and TER	Model1
BLEU, METEOR, TER and ROUGE	Model2
BLEU, METEOR, TER, ROUGE and Polarity	Model3

4.4   Results and Discussions

The results on the five datasets are shown below one by one. We plot the results of three different models on the RTE-1 dataset using different machine learning approaches. In RTE-1 third model gives the best performance in Multilayer perceptron, i.e., 54% which is depicted in Fig. 1. However, the best result reported in the literature for this dataset is 70% by the system of [8]. Results obtained on RTE-2 are depicted in Fig. 2. We obtain the highest score of 55.5% using SVM.

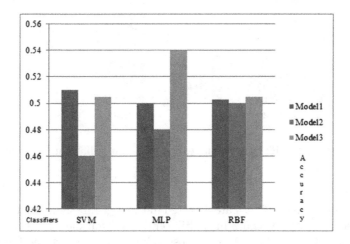

Fig. 1. Results on the RTE-1 on different models using different machine learning approaches

---

[1] http://www.cs.waikato.ac.nz/ml/weka/.

However, the best accuracy in this dataset reported is 75% [9]. Results obtained on RTE-3 dataset are presented in Fig. 3. Here we achieve the highest score of 60.37% in Model 3 by SVM. The best result obtained [10] in this dataset is 80%.

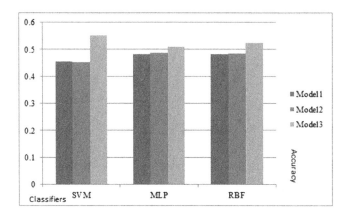

Fig. 2. Results on the RTE-2 on different models using different machine learning approaches

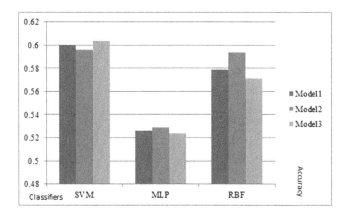

Fig. 3. Results on the RTE-3 on different models using different machine learning approaches

Results obtained on RTE-4 are depicted in the Fig. 4. It shows that we obtain the highest accuracy of 52.5% in multilayer perceptron model. As there is no development set we perform 10-fold cross validation to report the final result. Literature shows that highest accuracy for this task is around 68.5% by [25] that makes use of semantic knowledge bases like WordNet, Verb Ocean, Wikipedia, Acronyms database etc.

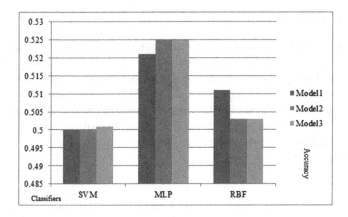

Fig. 4. Results on the RTE-4 on different models using different machine learning approaches

In RTE-5, we obtain the best results with multilayer perceptron model that shows an accuracy of 51.87% in Fig. 5. RTE-5 is ternary class classification problem with the following classes: *ENTAILMENT, UNKNOWN, CONTRA-DICTION*. It is to be noted that the best accuracy is 68.33% as reported in [26]. Overall it is observed that, in our datasets with these feature combination Multilayer perceptron performs well compared to SVM and RBF. Multilayer perceptron is a kind of classifier which classifies instances based on back propagation, perform training on multiple layers.

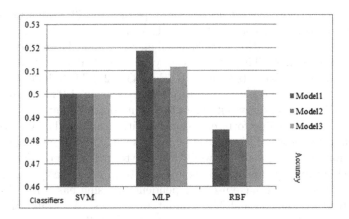

Fig. 5. Results on the RTE-5 on different models using different machine learning approaches

4.5 Error Analysis

In our work we make an attempt to establish relationships between MT evaluation metric, summary evaluation metric and polarity features in detecting entailment relation between a pair of text snippets. The results obtained in all the datasets used need to be improved. All the MT evaluations which have been used in this study provide similarity score by considering n-gram lexical matching. Lexical matching has some drawbacks, it does not capture the negation words. Following example describes the problem.

T: John Loves Merry
H: Merry Loves John
Here,
Unigram matching score = 3/3 = 1 and Bigram matching score = 0/2 = 0

According to unigram match, the above sentences are textually entailed, however in realty they should not.

Consider the example below,

textitT: I live in India.
textitH: I do not live in India.

For TE problem length should not be considered. In all our experimental settings, H is much shorter than T. The MT metrics that we have used considers text length. They generally assign a penalty score, which substantially decreases the matching score between text and hypothesis. This, in turn, affects the overall performance.

5 Conclusions and Future Works

In this experiment we have set up relationships among MT evaluation metrics, summary evaluation metric and polarity feature (no/not) for textual entailment. We develop three classification models based on SVM, multilayer perceptron and RBF network classifiers. These models have been evaluated for textual entailment on five different datasets, namely RTE-1, RTE-2, RTE-3, RTE-4 and RTE-5. Unfortunately due to the unavailability of the datasets of RTE-6, and RTE-7, we could not be able to carry out the experiments on these datasets.

In future we plan to perform experiments with other similarity metrics. We have a plan to use other MT evaluation metrics such as GTM, NIST, and CDER as features for our experiment. We would also like to consider dependency parsing based matching to overcome the drawback of n-gram lexical matching. It is also to be noted that we will make use of dependency parsing based MT evaluation metric instead of conventional MT evaluation metrics. We would also like to explore deep learning based techniques for textual entailment.

Acknowledgements. The research work has received funding from the project "Development of Cross Lingual Information Access (CLIA)" funded by The Department of Electronics and Information Technology (DeitY), Ministry of Communication and Information Technology, Government of India.

References

1. Vapnik, V.N.: The Nature of Statistical Learning Theory. Springer, New York (1995). https://doi.org/10.1007/978-1-4757-3264-1
2. Chang, C.C., Lin, C.J.: LIBSVM: a library for support vector machines. ACM Trans. Intell. Syst. Technol. 2(3), 27:1–27:27 (2011)
3. Becerra, R., Joya, G., García Bermúdez, R.V., Velázquez, L., Rodríguez, R., Pino, C.: Saccadic points classification using multilayer perceptron and random forest classifiers in EOG recordings of patients with Ataxia SCA2. In: Rojas, I., Joya, G., Cabestany, J. (eds.) IWANN 2013. LNCS, vol. 7903, pp. 115–123. Springer, Heidelberg (2013). https://doi.org/10.1007/978-3-642-38682-4_14
4. Costa, W., Fonseca, L., Körting, T.: Classifying grasslands and cultivated pastures in the brazilian cerrado using support vector machines, multilayer perceptrons and autoencoders. In: Perner, P. (ed.) MLDM 2015. LNCS (LNAI), vol. 9166, pp. 187–198. Springer, Cham (2015). https://doi.org/10.1007/978-3-319-21024-7_13
5. Rivera, A.J., Charte, F., Pérez-Godoy, M.D., de Jesús, M.J.: Multi-label testing for CO^2RBFN: a first approach to the problem transformation methodology for multi-label classification. In: Cabestany, J., Rojas, I., Joya, G. (eds.) IWANN 2011. LNCS, vol. 6691, pp. 41–48. Springer, Heidelberg (2011). https://doi.org/10.1007/978-3-642-21501-8_6
6. Saikh, T., Naskar, S.K., Giri, C., Bandyopadhyay, S.: Textual entailment using different similarity metrics. In: Gelbukh, A. (ed.) CICLing 2015. LNCS, vol. 9041, pp. 491–501. Springer, Cham (2015). https://doi.org/10.1007/978-3-319-18111-0_37
7. Saikh, T., Naskar, S.K., Bandyopadhyay, S.: JU_NLP@DPIL-FIRE 2016: paraphrase detection in indian languages - a machine learning approach. In: Working Notes of FIRE 2016 - Forum for Information Retrieval Evaluation, Kolkata, India, pp. 275–278 (2016)
8. Alfonseca, E.: Application of the BLEU Algorithm for Recognising Textual Entailments. In: Proceedings of the PASCAL Challenges Workshop on Recognising Textual Entailment, pp. 9–12 (2005)
9. Hickl, A., Bensley, J., Williams, J., Roberts, K., Rink, B., Shi, Y.: Recognizing textual entailment with lcc's groundhog system. In: Proceedings of the Second PASCAL Challenges Workshop (2005)
10. Hickl, A., Bensley, J.: A discourse commitment-based framework for recognizing textual entailment. In: Proceedings of the ACL-PASCAL Workshop on Textual Entailment and Paraphrasing, RTE 2007, pp. 171–176. Association for Computational Linguistics, Stroudsburg (2007)
11. Ríos Gaona, M.A., Gelbukh, A., Bandyopadhyay, S.: Recognizing textual entailment using a machine learning approach. In: Sidorov, G., Hernández Aguirre, A., Reyes García, C.A. (eds.) MICAI 2010. LNCS (LNAI), vol. 6438, pp. 177–185. Springer, Heidelberg (2010). https://doi.org/10.1007/978-3-642-16773-7_15
12. Li, B., Irwin, J., Garcia, E.V., Ram, A.: Machine learning based semantic inference: experiments and observations at RTE-3. In: Proceedings of the ACL-PASCAL Workshop on Textual Entailment and Paraphrasing, RTE 2007, pp. 159–164. Association for Computational Linguistics, Stroudsburg (2007)
13. Ferrés, D., Rodríguez, H.: Machine learning with semantic-based distances between sentences for textual entailment. In: Proceedings of the ACL-PASCAL Workshop on Textual Entailment and Paraphrasing, RTE 2007, pp. 60–65. Association for Computational Linguistics, Stroudsburg (2007)

14. Malakasiotis, P., Androutsopoulos, I.: Learning textual entailment using SVMs and string similarity measures. In: Proceedings of the ACL-PASCAL Workshop on Textual Entailment and Paraphrasing, RTE 2007, pp. 42–47. Association for Computational Linguistics, Stroudsburg (2007)

15. Mohri, M.: Finite-state transducers in language and speech processing. Comput. Linguist. **23**(2), 269–311 (1997)

16. Tillmann, C., Vogel, S., Ney, H., Zubiaga, A., Sawaf, H.: Accelerated Dp Based Search for Statistical Translation. In: European Conference on Speech Communication and Technology, pp. 2667–2670 (1997)

17. Papineni, K., Roukos, S., Ward, T., Zhu, W.J.: BLEU: a method for automatic evaluation of machine translation. In: Proceedings of the 40th Annual Meeting on Association for Computational Linguistics, ACL 2002, pp. 311–318 (2002)

18. Doddington, G.: Automatic evaluation of machine translation quality using n-gram co-occurrence statistics. In: Proceedings of the Second International Conference on Human Language Technology Research, HLT 2002, pp. 138–145. Morgan Kaufmann Publishers Inc., San Francisco (2002)

19. Lavie, A., Agarwal, A.: Meteor: an automatic metric for mt evaluation with high levels of correlation with human judgments. In: Proceedings of the Second Workshop on Statistical Machine Translation, StatMT 2007, pp. 228–231. Association for Computational Linguistics, Stroudsburg (2007)

20. Snover, M., Dorr, B., Schwartz, R., Micciulla, L., Makhoul, J.: A study of translation edit rate with targeted human annotation. In: Proceedings of Association for Machine Translation in the Americas, pp. 223–231 (2006)

21. Turian, J.P., Shen, L., Melamed, I.D.: Evaluation of machine translation and its evaluation. In: Proceedings of MT Summit, pp. 386–393 (2003)

22. Volokh, A., Neumann, G.: Using MT-based metrics for RTE. In: Proceedings of the Fourth Text Analysis Conference, TAC 2011, Gaithersburg, Maryland, USA, 14–15 November 2011 (2011)

23. Lavie, A., Agarwal, A.: METEOR: an automatic metric for MT evaluation with improved correlation with human judgments. In: Proceedings of the Second Workshop on Statistical Machine Translation, pp. 65–72 (2005)

24. Lin, C.Y.: ROUGE: a package for automatic evaluation of summaries. In: Proceedings of ACL Workshop on Text Summarization Branches Out (2004)

25. Iftene, A.: UAIC Participation at RTE4. In: TAC. NIST (2008)

26. Iftene, A., Moruz, M.A.: UAIC Participation at RTE5. In: TAC. NIST (2009)

Supervised Learning of Entity Disambiguation Models by Negative Sample Selection

Hani Daher, Romaric Besançon$^{(\boxtimes)}$, Olivier Ferret, Hervé Le Borgne,
Anne-Laure Daquo, and Youssef Tamaazousti

CEA, LIST, Vision and Content Engineering Laboratory, Gif-sur-Yvette, France
{hani.daher,romaric.besancon,olivier.ferret,herve.le-borgne,
anne-laure.daquo,youssef.tamaazousti}@cea.fr

Abstract. The objective of Entity Linking is to connect an entity mention in a text to a known entity in a knowledge base. The general approach for this task is to generate, for a given mention, a set of candidate entities from the base and determine, in a second step, the best one. This paper focuses on this last step and proposes a method based on learning a function that discriminates an entity from its most ambiguous ones. Our contribution lies in the strategy to learn efficiently such a model while keeping it compatible with large knowledge bases. We propose three strategies with different efficiency/performance trade-off, that are experimentally validated on six datasets of the TAC evaluation campaigns by using Freebase and DBpedia as reference knowledge bases.

1 Introduction

In the domain of Information Extraction, the Entity Disambiguation task (or Entity Linking) consists in connecting an entity extracted from a text to known entities in a knowledge base [11,17], which is useful for further extraction tasks (relation extraction or event detection, for instance) or to provide a unique normalization of the entities in an Information Retrieval context. This task is sometimes part of a more general framework that globally disambiguates all the concepts in a document with respect to a knowledge base, whether they are named entities or nominal expressions (e.g. Wikify [12] or Babelfy [13]).

An Entity Disambiguation system is usually based on three main steps [8]. First, it analyzes an input (query) to identify an "entity mention" that needs to be linked to the knowledge base. Second, for each mention, the system generates several candidate entities from the knowledge base and finally, it selects the best entity among the candidates. For such systems, one of the main challenge is to deal with the very large number of entities present in the knowledge base.

Our contribution focuses on the last step of this process. We propose to learn a discriminative model between each candidate and its most ambiguous entities: this model provides a score reflecting the association between the entity mention and each candidate entity, that we call the *Discriminative Disambiguation Score*.

© Springer Nature Switzerland AG 2018
A. Gelbukh (Ed.): CICLing 2017, LNCS 10761, pp. 329–341, 2018.
https://doi.org/10.1007/978-3-319-77113-7_26

For the sake of computational tractability, we adopt a linear model. The core of our contribution lies in the strategy for choosing the right set of negative training examples to be able to learn several millions of models in a tractable time. We evaluate this model using Freebase and DBpedia as reference knowledge bases and six test datasets from the TAC evaluation campaigns.

2 Related Work

One way of classifying the different approaches for Entity linking is the degree of supervision they required. Unsupervised methods generally rely on the definition of a similarity score between the entity mention and the entity in the knowledge base: selecting the correct entity simply corresponds to maximizing this score. Such similarity scores are usually based on the overlap of contexts [4] and can combine several measures: for instance, Han and Zhao [7] combine similarities between words and Wikipedia concepts. These methods are usually simple and easy to implement but they have a lower performance, when compared to the supervised methods as it was shown in past evaluation campaigns [3].

Supervised methods are generally based on binary classifiers [10,18] or ranking models [2,16], specifically dedicated to entity disambiguation. In both cases, the difficulty lies in building labeled data, which is time consuming, especially for large knowledge bases like Freebase or DBpedia. Among these approaches, some studies [6,20] use ambiguous entities to learn the models that allow to select the correct entity but, as far as we know, none of them proposes to build a discriminative model for each entity. Zhang et al. [20] focuses on the unambiguous mentions of entities in DBpedia to automatically create training examples for their disambiguation model. idea is to generate disambiguation examples by replacing in documents the mentions of an entity that are not ambiguous by alternate names of this entity that are ambiguous. The positive examples are built by associating the modified documents with the entity while the negative examples are produced by associating these documents with the entities referred by the alternate names. Zhang et al. [19] uses an iterative learning algorithm to select the most informative entities that are close to the separating hyperplane. Fan et al. [6] uses a one-vs-all strategy to disambiguate entities. Instead of training one classifier per entity in Freebase, they propose a strategy to merge all of them into one generic classifier, which consists in using, for an entity, its unique Freebase identifier as an extra descriptor. All the positive examples of a given entity have thus the same Freebase identifier. The negative examples are randomly chosen from entities with a similar name.

Finally, some studies use a semi-supervised approach, such as [21], that address the problem of data acquisition and labeling, using a two sets of labeled and unlabeled data. In an iterative approach, a model is learned using positive examples extracted from Wikipedia pages that contain an unambiguous entity (thus providing reference data) and negative examples taken randomly from other entities. The learned model is used to annotate the unlabeled documents, which will then be used in the learning phase of the next iteration.

3 Discriminative Disambiguation Score

In the context of supervised systems, we present in this section the new score we propose for Entity Linking, named Discriminative Disambiguation Score (DDS). The idea of DDS is to reflect the likelihood of an entity mention to be disambiguated by a given candidate entity. It represents the posterior probability $P(candidate_i|mention)$ that a candidate is appropriate to disambiguate a given entity mention, computed from a classifier score [15].

The novelty of our proposal is to learn a classifier for each candidate entity (as long as the required data are available). Such an approach has sometimes been dismissed and considered impossible because of intractable computational issues [6]. Indeed, one difficulty of our approach lies in the capacity of learning such classifiers for several millions of entities while still keeping a relevant discriminative power for each of them. First, for the sake of computational tractability, both at learning and testing time, we restrain our approach to linear classifiers (in practice, we used logistic regression models, but we obtained comparable results with SVMs, that are not reported in this paper).

For each candidate, we must select both positive and negative examples, extract features then learn the classifier. In all cases, the vector representation of examples is based on a *tf-idf* model relying on the same vector space, built from the complete collection of the Wikipedia pages associated with the entities in the knowledge base. Regarding the positive examples, the textual context of each entity in the knowledge base is considered by using the following information:

- Abstract of the Wikipedia page associated with the entity;
- Paragraphs explicitly containing the entity in the Wikipedia page of the entity;
- Paragraphs, from other Wikipedia pages, that contain a wikilink pointing to the entity.

As suggested by [6] a direct one-versus-all strategy would be computationally intractable. To solve this problem, we propose three approaches, illustrated in Fig. 1, to select a subset of negative examples containing representative *tf-idf* vectors.

- **DDS-Rand**: In the Random approach, the negative examples are randomly selected from the positive examples of all the other entities in the knowledge base. There is no constraint on whether the negative examples should only be selected from ambiguous entities.
- **DDS-Ambig**: In the Ambiguous approach, the negative examples are randomly selected from the positive examples of the ambiguous entities. The ambiguous entities are generated by using the same approach as for the candidate entity generation (see Sect. 4.1): for each known form of the entity in the knowledge base (normalized form or variation), ambiguous entities are the entities that share a common form or have a close form (inclusion or string edit distance ≤ 2). Since the set of ambiguous entities can be large, the negative examples are actually selected from a random subset.

– **DDS-Ambig-NN**: In the Ambiguous Nearest Neighbor approach, the entity
for which we want to compute a discriminative model is represented by the
centroid of the *tf-idf* vectors that constitute its positive examples. We then
select as negative examples the *tf-idf* vectors that have the closest Cosine
similarities to this representation among all the examples from the ambiguous
entities. These negative examples are considered as the most informative data
instances and the most relevant for discrimination because they are the most
ambiguous with the entity.

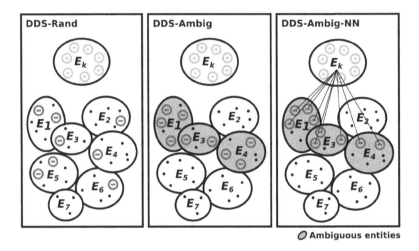

Fig. 1. Methods for selecting negative examples

4   Overall Entity Linking System

To test the DDS, it is integrated into a standard Entity Linking architecture [8]
composed of two main steps: for a given entity mention and its textual context,
a first module generates possible candidate entities for the linking and a second
one selects the best one.

4.1   Generation of Candidate Entities

The generation of the candidate entities relies on both the analysis of the entity
mention and its textual context. In this study, we focus mainly on the disambiguation of entities, not their recognition. Therefore, the entity mentions
to disambiguate are given as input to the system. A complementary analysis
of these entity mentions in the text is carried out, in order to associate a type
(Person, Location, Organization) with the entity mentions[1] and define their context in terms of surrounding entities (we consider only the explicit named entity

---

[1] We used the tool MITIE for this step (https://github.com/mit-nlp/MITIE).

mentions and we ignore the nominal and pronominal mentions). Two forms of entity mention expansion are performed, which can be considered as simple co-reference approaches: (i) if an entity mention is an acronym, we search the text for entity mentions of the same type whose initials correspond to the acronym (ii) we search the text for other entity mentions whose expression includes the target entity mention as a substring. These other forms are added as variations for the entity mention.

After the intrinsic analysis of the entity mention, candidate entities are generated by comparing the entity mention with the entities of the knowledge base, using the following four strategies [5]:

- Equality between the forms of the entity mention and an entity in the knowledge base;
- Equality between the form of the entity mention and a variation (alias or translation) of an entity in the knowledge base;
- Inclusion of the form of the entity mention in one of the forms of the variations of an entity in the knowledge base;
- Similarity between the form of the entity mention and a variation of an entity in the knowledge base. We use the Levenshtein distance, which is well suited to overcome the spelling errors and name variations. In the experiments, we considered an entity in the knowledge base as a candidate entity if its form or any of its variations have a distance with the form of the entity mention ≤ 2. For better efficiency, we exploited a BK-tree structure [1] for this selection.

The candidate entities are also filtered in order to keep only entities that have one of the expected named entity types (e.g. Person, Location, Organization).

4.2   Selection of the Best Candidate Entity

The goal of this step is to find the correct candidate entity in the set of generated candidate entities. To this purpose, a classifier is trained to recognize the best entity among the entity candidates, using training data composed of disambiguated entity mentions. More precisely, each candidate entity is associated with a set of features:

- Four binary features indicating which strategy was used for the generation of this candidate entity;
- Two general scores comparing the context of the entity mention with the context of the candidate entities in the knowledge base. The first score focuses on their lexical context. It compares, with the Cosine similarity, a vector representation of the textual context of the entity mention (we considered the whole document as the context of the entity mention in these experiment) and the vector representation of the Wikipedia page of the candidate entity. The second score focuses on a context based on surrounding entities and compares, once again with the Cosine similarity, the vector representation of the textual context of the entity mention[2] and a vector representation of

---

[2] For the entity mention, we took the whole lexical context as we did not have an entity recognizer for all the entity types of the knowledge base.

the entities in relation with the candidate entity in the knowledge base. The vector space supporting these representations is built from all the Wikipedia pages of the entities in the knowledge base;
– One of the DDS scores computed as presented in Sect. 3.

A binary classifier is then trained to associate such a set of features with a decision whether the candidate entity is the correct one for the entity mention. Using the training data, we generate the candidate entities from the entity mentions: the positive examples for the training are then formed by the (entity mention, candidate) pairs that correspond to the expected link in the reference. The negative examples are pairs with wrong candidates generated for the entity mentions. Since the classes are imbalanced (in our experiments, we have between 1 and 460 generated candidates), we undersample the negative examples to limit their number to 10 times the number of positive examples. Each decision of the classifier is weighted by the probability estimate of the classifier and the candidate entity with the highest probability is selected as the final disambiguated entity. In the standard entity disambiguation task, the system must also be able to determine when an entity mention does not link to any entity in the knowledge base (referred as NIL entities). In our approach, this occurs if no candidate is generated or if all candidates are rejected by the classifier.

5   Experiments and Analysis

5.1   Datasets

To validate our approach, we use the 2009–2013 and 2015 datasets of the TAC-KBP evaluation campaign. For TAC 2015, we consider the monolingual English Diagnostic Task, where the entity mentions in the query texts are already given as input, in order to use the same evaluation framework as for the other datasets. We report in Table 1 the main features of these datasets. For the 2009–2013 campaigns, the reference knowledge base is extracted from Wikipedia infoboxes (similarly to DBpedia) [14]. It contains 818,741 entities, which are all associated with a Wikipedia page. In the 2015 campaign, the knowledge base was built from Freebase [9]. The whole Freebase snapshot contains $43M$ entities but a filter is applied to remove some entity types that are not relevant to the campaign (such as music, book, medicine and film), which reduces it to $8M$ entities. Among them, only 3,712,852 (46%) have an associated content in Wikipedia and are thus subject to provide positive examples to learn the DDS. In the 2015 campaign, the purpose was to link all the entities of a restricted set of documents. On the contrary, the former campaigns aimed at linking a restricted number of entity per document; hence, the number of entities and documents is approximately the same for the 2009–2013 campaigns.

Table 1. Description of the datasets used in the evaluation process

	Dbpedia			Freebase	
	Nb. docs.	Nb. entities		Nb. docs.	Nb. entities
TAC 2009	3,688	3,904	TAC 2015 train	168	12,175
TAC 2010	2,231	2,250	TAC 2015 test	167	13,587
TAC 2011	2,231	2,250			
TAC 2012	2,016	2,226			
TAC 2013	1,820	2,190			

5.2   Results on Candidate Generation and DDS

Generated Candidate Entities. We present in Table 2 some statistics on the queries and the generated candidate entities. In particular, the candidates recall, defined by the percentage of non-NIL queries for which the expected candidate is in the candidate list, is quite good for the 2009–2013 datasets, using simple candidate generation strategies that generate a reasonable number of candidates per query (150 in average). For the 2015 dataset, the KB contains 10 times more entities and the number of generated candidates is much larger. In addition, the candidate recall is lower (77%): an analysis of the missing entities showed that the variations contained in the KB should be enriched for a better coverage (e.g., links between nationalities and countries are missing, such as French → France).

Table 2. Candidate statistics for the DBpedia and Freebase datasets (TAC 2009-2013 and 2015)

Dbpedia						
Dataset	Nb queries	NIL queries	Nb cand	NIL cand	Avg. cand	cand. recall
2009	3,904	2,229	208,060	949	70.41	84.0%
2010	2,250	1,230	232,672	601	141.10	89.4%
2011	2,250	1,126	329,508	388	176.96	87.9%
2012	2,226	1,049	420,179	117	199.23	92.4%
2013	2,190	1,007	394,217	395	219.62	83.5%
Freebase						
2015 train	12,175	3,215	5,844,592	1,282	458.08	76.0%
2015 test	13,587	3,379	6,141,369	1,255	480.32	77.6%

Discriminative Disambiguation Score. The extraction of the textual context (Sect. 3) from the Wikipedia pages is performed for the 818,741 entities in DBpedia and the 3,712,852 entities in Freebase. A vector space model of 169,647 dimensions, built from the whole Wikipedia dump, is used to convert these set of paragraphs into *tf-idf* vectors. When applied on DBpedia, a total number of 32,939,218 examples are generated. On average, an entity has around 40.18 examples and between 1 and 119,178 examples per entity. When applied on Freebase, a total number of 97,157,120 examples are generated. On average, an entity has around 26.16 examples and the number of examples per entity is between 1 and 119,197. The candidate entities for TAC 2009-2013 (Table 1, Nb. Candidates) represent 41,313 unique entities in DBpedia. For Freebase, the candidate entities that are associated with a Wikipedia page represent 124,456 unique entities, cumulated on training and test datasets. For each candidate entity in DBpedia and Freebase, we train a classifier based on the approaches described in Sect. 3: in our experiments, we used a L2-regularized logistic regression classifier, from the LIBLINEAR library[3], whose complexity is $O(n)$, where n is the number of features.

We report in Table 3 the minimum, maximum and average time in seconds needed to train a classifier for the different approaches. *DDS-Rand* is the simplest approach in complexity. This is why it needs less time to select the negative examples and train the classifier. *DDS-ambig-NN* takes more time because of the negative examples selection module where we have to compute the distance between the centroid of the entity and each *tf-idf* vector of its ambiguous entities.

To test the relevance of the DDS scores, we first selected a subset of entities from the DBPedia KB that have at least 100 positive examples and evaluated the performance of the trained classifiers for these entities using a 5-fold cross validation on this subset. The accuracy results of the classifiers trained on these 26,819 entities are reported in Table 4, showing a good performance in differentiating a particular entity from the others. With *DDS-ambig-NN*, we discriminate against the closest negative examples, which makes the disambiguation task harder and the results a bit lower.

Table 3. Minimum, Maximum and Average time (in seconds) needed for each approach to select the negative examples and train the classifier on DBpedia and Freebase

	Dbpedia			Freebase		
	Min.	Max.	Avg.	Min.	Max.	Avg.
DDS-Rand	0.003	109.59	1.55	0.002	49.29	1.13
DDS-Ambig	0.01	398.47	11.65	0.006	270.45	6.49
DDS-Ambig-NN	0.027	2551.62	146.88	0.014	2102.31	85.49

Table 4. Cross-validation results of the classifiers trained on 26,819 datasets having each at least 100 positive examples

	Precision	Recall	F-score
DDS-Rand	0.987 ± 0.015	$0,969 \pm 0.042$	0.977 ± 0.028
DDS-Ambig	0.963 ± 0.050	$0,919 \pm 0.112$	0.937 ± 0.086
DDS-Ambig-NN	0.954 ± 0.058	$0,798 \pm 0.188$	0.857 ± 0.151

---

[3] https://www.csie.ntu.edu.tw/~cjlin/liblinear/.

5.3   Entity Linking Results

In this section, we present the results of the DDS scores in a full entity linking
system. We compare the results obtained by the Baseline system (as presented
in Sect. 4.2, without the DDS) with the results obtained when adding the DDS
feature (this DDS feature is the score given by the classifier of the considered
candidate entity for the classification of the entity mention). In order to verify
the interest of the discriminative models compared to the simple addition of more
textual contexts for the entity, we also considered a score computed as the Cosine
similarity between the centroid of the positive examples of a candidate entity
and the textual context of the entity mention. We name this score DDS-baseline.

Evaluation Measures and Datasets. As no training data was provided for
the TAC 2009-2013 datasets, we used for each year the data from the other
years as training data. For the evaluation measures, we adopted the standard
precision/recall/F-score measures on three criteria: the correct recognition of
the reference entity when it exists (*link*), the correct recognition of an entity
without a reference entity (*nil*) and the combined results (*all*). These measures
correspond to the measures named *strong_link_match, strong_nil_match and the
strong_all_match* in the TAC 2015 evaluation campaign [9]. We did not take into
account the type of the named entities in this evaluation.

Results and Analysis. Table 5 reports for each approach the F-score for
respectively the *strong_nil_match, strong_link_match and the strong_all_match*
evaluation measures. These results show that including the DDS score to the
set of features used for the entity disambiguation clearly improves results. The
best results are obtained with the *DDS-Ambig-NN* for the TAC 2009-13 datasets,
whereas *DDS-Ambig* gives the best results for TAC 2015. Interestingly, we note
that even if *DDS-Ambig-NN* does not perform as well as the other approaches
in the pure classification task, its usage in full system is beneficial. This result
suggests that the *DDS-Ambig-NN* model learns more discriminant information
that complements better the information provided by the other features.

In order to further assess the influence of the DDS, we tested a disambiguation
approach using only this score: we trained a classifier for the final disambiguation
using the DDS as single feature (which allows learning automatically a threshold
on the DDS). Table 6 reports the results obtained with this method, for each
variant of DDS. With the exception of the 2011 and 2015 evaluation campaigns,
the *DDS-Ambig-NN* score used as single feature produces the best results and
is often not very far from the results obtained by the Baseline approach. For the
2012 dataset, the DDS alone produces a *strong_all_match* F-score of 63.8%, which
is even better than the one obtained when combined with the other features.

Finally, we present a comparison of the results we obtained with the other
teams that participated in the TAC evaluation campaigns in Table 7. We only
show the results for 2009, 2010 and 2015 because the official measure for these
years – the micro-average KB accuracy – corresponds to the measure we use
in this paper. For 2011–2013, the official measure is the $B^3 + F1$ score, which

Table 5. F-score results obtained with the addition of DDS scores. We report the strong_nil_match (top, best in italic), strong_link_match (middle, best in italic), strong_all_match evaluation (bottom, best is bold) criteria on TAC datasets

		2009	2010	2011	2012	2013	2015
Baseline	*nil*	0.851	0.863	0.808	0.649	0.801	0.668
	link	0.707	0.743	0.645	0.441	0.717	0.588
	all	0.795	0.813	0.735	0.533	0.761	0.601
DDS-Baseline	*nil*	0.851	0.859	0.807	0.649	0.8	0.667
	link	0.709	0.736	0.639	0.446	0.705	0.603
	all	0.796	0.808	0.734	0.535	0.754	0.611
DDS-Rand	*nil*	0.856	0.858	0.817	*0.651*	0.801	0.679
	link	0.72	0.751	0.646	0.436	0.704	*0.659*
	all	0.803	0.813	0.741	0.531	0.753	0.654
DDS-Ambig	*nil*	0.858	0.867	0.812	0.643	0.799	*0.694*
	link	0.73	0.762	0.647	0.454	0.722	0.654
	all	0.808	0.824	0.741	0.537	0.763	**0.656**
DDS-Ambig-NN	*nil*	*0.874*	*0.884*	*0.821*	0.649	*0.82*	0.687
	link	*0.754*	*0.796*	*0.663*	*0.468*	*0.756*	0.644
	all	**0.828**	**0.848**	**0.752**	**0.547**	**0.789**	0.646

Table 6. F-score results obtained using *only* the DDS as a disambiguation feature. We report the strong_nil_match (top, best in italic), strong_link_match (middle, best in italic), strong_all_match evaluation (bottom, best is bold) criteria on TAC datasets

		2009	2010	2011	2012	2013	2015
DDS-Baseline	*nil*	0.749	0.718	0.655	0.639	0.634	0,405
	link	0.289	0.222	0.278	0.22	0.182	0,002
	all	0.609	0.56	0.493	0.489	0.47	0,245
DDS-Rand	*nil*	0.828	0.838	*0.771*	0.818	0.757	0.611
	link	0.622	0.687	*0.546*	0.156	0.585	*0.508*
	all	0.749	0.772	**0.67**	0.537	0.672	**0.541**
DDS-Ambig	*nil*	0.815	0.833	0.662	0.565	0.748	*0.665*
	link	0.568	0.666	0.243	0.221	0.56	0.443
	all	0.73	0.768	0.481	0.395	0.667	0.517
DDS-Ambig-NN	*nil*	*0.84*	*0.849*	0.756	*0.754*	*0.772*	0.633
	link	*0.625*	*0.703*	0.429	*0.446*	*0.614*	0.433
	all	**0.771**	**0.797**	0.645	**0.638**	**0.708**	0.503

Table 7. Ranking of our approaches compared to the official results of the campaigns, based on the strong_all_match evaluation criterion.

	2009	2010	2015
Nb. of teams	18	21	10
Median	0.67	0.683	0.634
Min.	0.0085	0.345	0
Max.	0.822	0.864	0.875
Baseline	0.783 (8)	0.798 (9)	0.601 (5)
DDS-Baseline	0.791 (6)	0.81 (8)	0.611 (5)
DDS-Rand	**0.803 (2)**	**0.821 (3)**	0.654 (4)
DDS-Ambig	0.794 (5)	0.815 (6)	**0.656 (4)**
DDS-Ambig-NN	0.795 (3)	0.82 (4)	0.641 (4)

requires a clustering of the NIL entities that we do not perform here. Note as well that the official scores in 2015 takes into account the type of the entity while we only consider the accuracy of the disambiguation. For the 2009 and 2010 datasets, the *DDS-Rand* approach achieves results that are not far from the best participant: it would be ranked respectively 2nd out of 18 participants and 3rd out of 21 participants[4]. For the 2015 dataset, our system still needs some tuning on the candidate generation to obtain better candidate recall but the general trend of the results provides solid ground to indicate that the DDS is a good feature for entity disambiguation.

6   Conclusion

We proposed a new feature for the entity disambiguation task based on a supervised approach to learn discriminative models for each entity in a knowledge base. By combining this feature with a set of features commonly used in the literature, the scores, expressed as percentages, increased by more than 4 points. Our entity disambiguation method showed its stability on several Entity Linking evaluation datasets, using two different knowledge bases. We also addressed the problem of the selection of negative examples by proposing three approaches. The *DDS-Rand* and *DDS-Ambig* approaches provide improvements over our baseline system with a very low computation time and a linear complexity. *DDS-Ambig-NN* gives better results but with a higher computation time. Most importantly, we showed that individual binary classifiers can be trained for each entity of a large knowledge base for a disambiguation task. We plan to improve the performance of the DDS by using dense vector representations (word embeddings) to represent the positive and negative examples, this type of representations having proven their efficiency for various classification tasks.

---

[4] The comparison is not absolutely fair since we used the data from other years for training, which were not available to the participants.

Acknowledgments. This work was partly supported by the F1409071Q CuratedMedia project.

References

1. Burkhard, W.A., Keller, R.M.: Some approaches to best-match file searching. Commun. ACM **16**(4), 230–236 (1973)
2. Cao, Z., Tao, Q., Tie-Yan, L., Ming-Feng, T., Hang, L.: Learning to rank: from pairwise approach to listwise approach. In: 24th International Conference on Machine Learning (ICML 2007), pp. 129–136. Corvalis, Oregon (2007)
3. Cassidy, T., et al.: CUNY-UIUC-SRI TAC-KBP2011 entity linking system description. In: Text Analysis Conference (TAC 2011) (2011)
4. Cucerzan, S.: Large-scale named entity disambiguation based on Wikipedia data. In: 2007 Joint Conference on Empirical Methods in Natural Language Processing and Computational Natural Language Learning (EMNLP-CoNLL 2007), Prague, Czech Republic, pp. 708–716 (2007)
5. Dredze, M., McNamee, P., Rao, D., Gerber, A., Finin, T.: Entity disambiguation for knowledge base population. In: 23rd International Conference on Computational Linguistics (COLING 2010), Beijing, China, pp. 277–285 (2010)
6. Fan, M., Zhou, Q., Zheng, T.F.: Distant supervision for entity linking. In: 29th Pacific Asia Conference on Language, Information and Computation (PACLIC 29), Shanghai, China, pp. 79–86 (2015)
7. Han, X., Zhao, J.: NLPR_KBP in TAC 2009 KBP track: a two-stage method to entity linking. In: Text Analysis Conference (TAC 2009) (2009)
8. Ji, H., Nothman, J., Hachey, B.: Overview of TAC-KBP2014 entity discovery and linking tasks. In: Text Analysis Conference (TAC 2014) (2014)
9. Ji, H., Nothman, J., Hachey, B., Florian, R.: Overview of TAC-KBP2015 tri-lingual entity discovery and linking. In: Text Analysis Conference (TAC 2015) (2015)
10. Lehmann, J., Monahan, S., Nezda, L., Jung, A., Shi, Y.: LCC approaches to knowledge base population at TAC 2010. In: Text Analysis Conference (TAC 2010) (2010)
11. Ling, X., Singh, S., Weld, D.: Design challenges for entity linking. In: Transactions of the Association for Computational Linguistics (TACL), vol. 3, pp. 315–328 (2015)
12. Mihalcea, R., Csomai, A.: Wikify! linking documents to encyclopedic knowledge. In: Proceedings of the Sixteenth ACM Conference on Conference on Information and Knowledge Management, pp. 233–242. ACM, Lisbon(2007)
13. Moro, A., Raganato, A., Navigli, R.: Entity linking meets word sense disambiguation: a unified approach. Trans. Assoc. Comput. Linguist. (TACL) **2**, 231–244 (2014)
14. Namee, P.M., Simpson, H., Dang, H.T.: Overview of the TAC 2009 knowledge base population track. In: Text Analysis Conference (TAC 2009) (2009)
15. Platt, J.C.: Advances in kernel methods. Fast Training of Support Vector Machines Using Sequential Minimal Optimization, pp. 185–208. MIT Press, Cambridge (1999)
16. Shen, W., Jianyong, W., Ping, L., Min, W.: LINDEN: linking named entities with knowledge base via semantic knowledge. In: 21st International Conference on World Wide Web (WWW 2012), Lyon, France, pp. 449–458 (2012)
17. Shen, W., Wang, J., Han, J.: Entity linking with a knowledge base: issues, techniques, and solutions. IEEE Trans. Knowl. Data Eng. **27**(2), 443–460 (2015)

18. Varma, V., et al.: IIT Hyderabad at TAC 2009. In: Text Analysis Conference (TAC 2009) (2009)
19. Zhang, W., Chuan, S.Y., Jian, S., Lim, T.C.: Entity linking with effective acronym expansion, instance selection and topic modeling. In: Twenty-Second International Joint Conference on Artificial Intelligence (IJCAI-11), Barcelona, Catalonia, Spain, pp. 1909–1914 (2011)
20. Zhang, W., Jian, S., Lim, T.C., Ting, W.W.: Entity linking leveraging: automatically generated annotation. In: 23rd International Conference on Computational Linguistics (COLING 2010), Beijing, China, pp. 1290–1298 (2010)
21. Zheng, Z., Xiance, S., Fangtao, L., Y, C.E., Xiaoyan, Z.: Entity disambiguation with freebase. In: 2012 IEEE/WIC/ACM International Joint Conferences on Web Intelligence and Intelligent Agent Technology (WI-IAT 2012), pp. 82–89 (2012)

The Enrichment of Arabic WordNet Antonym Relations

Mohamed Ali Batita[(⊠)] and Mounir Zrigui[(⊠)]

Research Laboratory in Computer Science and Communication Technologies
and Electrical Engineering LaTICE, University of Monastir, Monastir, Tunisia
BatitaMohamedAli@gmail.com, Mounir.Zrigui@fsm.rnu.tn

Abstract. Arabic WordNet (AWN) is a lexical database, freely available, and useful resource to Natural Language Processing (NLP) research and applications (Information Retrieval, Machine Translation...). This project is built following the methods developed for Princeton WordNet (PWN) and EuroWordNet (EWN). However, this database needs more intention to improve NLP applications. Compared with others wordnets, AWN has a very poor content in both, quantity and quality levels. This paper concentrate on the quality plan, especially on the antonym relations. Therefore, the authors propose a pattern-based approach to extend these relations, using Arabic Corpus and a corpus analysis tool. This proposed method relies on two steps: patterns definition and automatic antonym pair extraction. The evaluation of this approach has given good results.

Keywords: Arabic WordNet · Semantic relation
Antonym enrichment · Arabic corpus · Patterns generation

1 Introduction

Natural Language Processing (NLP) is an important part of computational linguistics, computer science and artificial intelligence. There are plenty of researched tasks in NLP such as automatic summarization, machine translation, event recognition and relationship extraction that need lexical and semantic resources to proceed. The more the resource is inclusive the more the results are accurate. But the lack of resources, particularly in the Arabic language, has always been problematic.

Arabic is one of the most widely spoken Semitic languages today (300 million) followed by Amharic (22 million) and Tigrinya (7 million)[1]. The development of efficient Arabic NLP systems, therefore, is important. For instance, after its release, Arabic WordNet (AWN) [9,13] has quickly gained attention and became known in the Arabic NLP community as one of the exceptional and freely available lexical and semantic resources [1]. AWN is a lexical database for Modern

---

[1] Semitic languages, https://en.wikipedia.org/wiki/Semitic_languages#cite_note-Owens300-3.

© Springer Nature Switzerland AG 2018
A. Gelbukh (Ed.): CICLing 2017, LNCS 10761, pp. 342–353, 2018.
https://doi.org/10.1007/978-3-319-77113-7_27

Standard Arabic (MSA) [4] in which words that have a common meaning are grouped together in a so-called synsets. Those synsets contain general concepts and need to be extended to cover more specific domains.

AWN enrichment concern the number of synsets and the relations between them. Many researchers have attempted to improve AWN's content through different approaches by extending its coverage [1,24,25] or adding semantic relations [10]. The recent work that is related to the enrichment of the semantic relations, is the work of Boudabous et al. [10] in which they suggested a hybrid linguistic method based on morphological patterns to determine new relations between nominal synsets. They relied on a corpus contracted from Arabic articles in Wikipedia to define new morphological patterns. To get new pairs of words, they applied the NooJ grammar. This approach was validated by a number of linguistic experts.

Although it is fundamental in many NLP applications, it can be noticed that the semantic relation enrichment does not get much attention. Thereby, the authors studied the semantic relation existence in AWN and found out that it lacks the antonymous relation. Thus, this work is going to be the automatic enhancement of semantic relations.

On the other hand, semantic relation identification is a tough task that can improve the degree of accuracy in NLP applications, if it has been treated well. There are different kinds of semantic relations like synonym, homonym, hyponym etc. and to know the difference between all of them, one needs a coordinated approach. It is undeniable that manual approaches are time-consuming and intensives, some researchers argued that an automatic approach might be helpful [4,5,22,28].

This paper is organized as follows: Sect. 2 represents AWN description (statistics, advantages, weaknesses, domain of application...). Section 3 studies the related work on semantic relation extraction. The adopted approach the tools used has been detailed in Sect. 4. The authors presented the obtained results and evaluation in Sect. 5. Section 6 deals with the conclusion and perspective.

2   Arabic WordNet Description

AWN has been released in 2007 [13]. It has been developed based on the top-down method, which consists of two steps: PWN core translation and extension downwards to more specific concepts. Now, the AWN v2, contains 47% of the PWN core[2]. Note that the core is the most frequently used word sense in any language which is around 5,000 words [11].

The first version of AWN contains 9,698 synsets, corresponding to 21,813 words. These synsets are related by 6 different kinds of semantic relations (hyponymy, meronymy...), in a total of 143,715 links [12]. The words are distinguished by their part of speech (POS). The POS can be either a noun, verb, adverb, adjective or adjective satellite[3]. At this very moment, AWN version 2

---

[2] Open Multilingual WordNet, http://compling.hss.ntu.edu.sg/omw/.

[3] 'Steaming' is the adjective satellite of 'cold'.

contains 11,296 synsets corresponding to 23,841 words with 22 types of semantic relations and totaling 161,705 links[4]. Those words include Named Entities (NE) and other forms like roots and broken plurals. Comparing with PWN that has 117,659 synsets, AWN still needs improvements.

As previously noted, many works attempted to enhance the AWN content which are increasing over the time where researchers are trying to find suitable solutions to reach a higher level. One of the remarkable work is the work of Abouenour et al. [1]. They studied the AWN in two terms, coverage and usability. Concerning its coverage, they compared its content with other Arabic lexicon, Princeton WordNet (PWN) [14,15] and the Spanish WordNet [8]. Concerning its usability, They affirm that AWN has many weaknesses and that is why it is has been used in just few Arabic NLP applications. After this study, they improved its coverage by adding new verbs, nouns, Named Entities (NE) and the form of broken plurals.

Later on, They released their work and made it available online for the NLP community, structured under the Lexical Markup Framework[5] LMF [23]. This new version is not included in the browser (AWN v2), so the authors parsed it to extract its content. Below is Table 1 comparing the v2 of the AWN and the content of the LMF file.

Table 1. Description of AWN.

	V2	LMF
Noun	7,960	16,432
Verb	2,538	42,298
Adverb	110	771
Adjective	500	270
Adjective satellite	161	386
Total	**11,269**	**60,154**
Named Entity	16,819	17,097
Broken Plural	180	3,060

In conclusion, the LMF file contains much more words in different categories. Despite its enormity, it does not have many links between synsets as expected. The version 2 of AWN has 22 links between synsets, while the LMF file contains only 5; *similar* 412 links, *hyponym/hypernym* 19,806 links, *hasInstance/isInstance* 549 links and finally *antonym* with only 14. Thus, this work will be based upon that file. There is not much work to cite about the enhancement of the semantic relations in AWN. However, there have been many attempts to extract semantic relations from texts.

---

[4] Those statistics are extracted from the AWN browser.
[5] Lexical Markup Framework is the ISO standard for NLP and machine-readable dictionary (MRD) lexicons.

3   Related Work

The complexity of the Arabic language and the lack of resources (annotated corpora, analysis tools, lexical resource...) are two factors that affect the semantic relation extraction and made it a tough task. Harmain [16] claimed that the Arabic language is complex in term of morphology, syntax and semantics; this is due to the great amount of grammar rules. Besides, most articles found on the internet do not have diacritics, which make it more ambiguous, not to mention many other peculiarities. All these factors make every automatic task in the NLP more difficult, especially semantic relation extracting.

It is hard to know what is the relationship between a 'chair' and 'legs' in a given context, knowing that most of the living beings have legs and there are some kind of chairs that do not have legs (zaisu chair), which also include word sense disambiguation. Generally speaking, the extraction of semantic relations can be divided into three approaches: supervised, semi-supervised, and bootstrapping [4]. The supervised approach uses machine learning methods. The problem is treated as a binary classification and a classifier is going to be trained by a selection of negative and positive examples of specific semantic relations. Thus, more examples get you more accuracy and precision. On the other side, semi-supervised and bootstrapping approaches are pattern-based; therefore, they need a few set of patterns, specifically handcrafted ones.

Most commonly used methods for the identification of semantic relation are pattern-based methods [4]. A pattern is a schematic construction that any two semantically related words can fill in. For example the pattern لا ... ولا^6 $l\bar{a}$ ...$wl\bar{a}$[6] (Neither... nor...) can host the antonym pair (فوز، خسارة) *(fwz, ḥsārt)* (victory, defeat) or (ناعم، خشن) *(nāʿm, ḥšn)* (soft, rough). Hearst [17] is the first one who explored this type of method in which she manually used five crafted patterns to automatically extract the hyponym relation from texts.

Following the approach of Hearst, Lin et al. [20] also used these patterns to identify synonym among distributionally similar words. Their work depends on the hypothesis that if two words w and w' appear in the same patterns, then they are strongly synonymous. Their dependency triples, extracted from a corpus, are triples $||w, r, w'||$ consisting of two words and the grammatical relationship between them. Their goal was to create an automatically thesauri. They compared the work with two others resources, WordNet 1.5 and Roget Thesaurus [26].

In addition to being time consuming, handcrafted pattern does not let you predict all possible patterns. So, other solutions existed to detect them, either by bootstrapping or using machine learning algorithms to automatically detect them from a text. However, machine learning algorithms need many resources

---

[6] From now on, the examples written in Arabic language are followed by a transliteration using the transliteration system of LATEX and an English translation of it in a parenthesis.

and tools to proceed, but languages like Arabic cannot afford this. Thus, bootstrapping technique is a better choice.

Wang et al. [29] presented an automatic approach to extract verb synonyms and antonyms patterns. To maximize the precision, they used many ones like 'X or Y', 'either X or Y' and 'can either X or Y'. Those patterns are automatically generated by seeds extracted from WordNet and a corpus which are presented to the corpus to bootstrap patterns.

Similar methods are used to extract semantic relations for the Arabic language. Al-Rajebah et al. [2] used Arabic Wikipedia as a resource to generate triples. Like [10] they also used linguistic approach that comprises features and semantic relations *infoboxes* from which they extracted 760,000 triples to finally achieve a 65% precision by the system.

Amar et al. [6] used the LMF to create an Arabic ontology in the field of astronomy. They used a rule-based system that depends on lexico-syntactic patterns. LMF has a defined structure to identify entities, so they relied on that point and some rules to create the ontology concepts and the relation between them.

Last but not least, AlHedayani [5] did a great job on the antonym in the Modern Standard Arabic (MSA). She claimed that antonyms in Arabic are similar to those found in other languages. She used two on-line corpuses to extract them and a Sign-Based Construction Grammar to capture the syntagmatic and paradigmatic properties of antonyms.

To conclude, the work in the Arabic language is still not enough to be compared with other languages like English. The coverage is still narrow and cannot cover all the need of NLP application. This method is a pattern-based bootstrapping approach in which the authors used *arTenTen* [7], an Arabic corpus, the analysis tool *Sketch Engine* [19] and *lexical contrast information* [21] to distinguish antonym pairs from others. The work described below is going to be concentrated on the adjectives of the AWN, but it can be applied to other POS.

4   Proposed Approach

As cited previously, this study focuses on the antonym relation between adjectives in AWN where a pair of antonym is defined by a relationship between two words that are basically different, the contrast between these words can be binary or not. For instance, binary as in *cold* and *hot*, and not binary as in *chilly* and *warm*. To get rid of this ambiguity, the authors will only be working on binary contrast.

The method is classified as follows: *pattern-based* because it relies on patterns to extract pairs of antonym and *bootstrapping* because it automatically generates those patterns from a corpus using Sketch Engine [19]. To get the patterns, the authors use a set of adjective pairs from AWN, and chose the seeds according to their co-occurrence in the corpus. Then, they arrange the patterns according to their co-occurrence in the corpus and the number of antonym pair co-occurrence with it. Once they have the relevant patterns, the authors use them to extract new pairs of antonym. Sketch Engine provides an advanced way to search for new pairs by using the corpus query language (CQL). Finally, the authors obtain several new pairs which they need to distinguish from one another, in which they rely on the contrast hypothesis [21] (see Sect. 4.3).

4.1 Sketch Engine

Sketch Engine is a corpus analysis tool, which makes it possible for researchers to seek in a large text through advanced query system, due to its large collection of corpora in different languages, including Arabic where you can also add your own corpus to work with it. Sketch Engine offers different kind of features like word sketch, thesaurus, concordance and n-grams. It has been considered as a robust analysis tool for corpus [18], a reason why it has been choose to work with.

In addition to adding your own corpus, Sketch Engine offers several Arabic corpuses, multilingual or monolingual. One of the remarkable monolingual corpuses is *Arabic Web 2012* or shortly *arTenTen12* [3]. It has 7,475,624,779 words (8,322,097,229 tokens). It belongs to the generation of *Tenten* corpus. This collection of corpuses in many languages, requires that the corpus has at least 10 billion words and that's where the name *TenTen* comes from. The corpus is crawled from the internet, cleaned from anything related the internet (links, headers, footers...), and reduplicated. After that, the corpus is tokenized, lemmatised, and POS tagged[7]. A larger corpus is an appropriate choice to extract new antonyms, despite the type of the genera of the texts (MSA or Classical Arabic). On the other hand, Sketch Engine offers many statics like Dice and logDice [27].

4.2 Statistics in Sketch Engine

One of the major statics used in Sketch Engine is the frequencies of triples. This measure is based on the work of Lin [20] concerning the dependency triples $||w, r, w'||$. It is a score accorded to a word w with a grammatical relation R with another word w'. Lexicographic association score is another measure to calculate the frequency of collocation between words, present in Sketch Engine with many varieties, logDice [27] is one of them. This score is stable, independent from the language, and the size of the corpus. Below is the Equation of the LogDice, knowing that $||w_1, R, *||$ is the frequency of w_1 with R with any other word w_2 and $||*, *, w_2||$ is the frequency of w_2.

$$LogDice = 14 + \log_2 \frac{2 \cdot ||w_1, R, w_2||}{||w_1, R, *|| + ||*, *, w_2||} \tag{1}$$

---

[7] Sketch Engine; https://www.sketchengine.co.uk/artenten-corpus/.

4.3   Lexical Contrast Information

The lexical contrast information [21] measure allows us to know the degree of importance of the word features. A feature f of a word is the word that co-occurs with it in a context. This measure is based on the hypothesis of [21] that says if two words A and B are antonyms, then there is another antonym pair C and D such that A and C are fully connected as well as B and D. Considering the example of جميل *ǧmyl* (beautiful) and قبيح *qbyḥ* (ugly) (A and B), such that العدل *ālʿdl* (justice)(C) is related to جميل *ǧmyl* and قبيح *qbyḥ* (D) to الظلم *ālẓlm* (injustice). Below is the Eq. 2 of the LMA:

$$weight(w, f) = \frac{1}{\#(w,u)} \sum_{u \in W(f) \cap S(w)} sim(w, u) -$$
$$\frac{1}{\#(w',v)} \sum_{w' \in A(w)} \sum_{v \in W(f) \cap S(w')} sim(w', v) \qquad (2)$$

Equation 2 is the difference between the similarities sim of the synonyms ($sim(w, u)$, with $u \in S(w)$) and the antonyms ($sim(w', v)$, with $w' \in A(w)$ and $v \in S(w')$) of a target word w. Back to the previous example of جميل *ǧmyl* (beautiful) and قبيح *qbyḥ* (ugly),

The feature f العدل *ālʿdl* (the justice) only occurs with w جميل *ǧmyl* (beautiful) and its synonyms $S(w)$, so the weight $weight(w, f)$ will be positive. The feature الظلم *ālẓlm* (the injustice) only occurs with the antonym of w قبيح *qbyḥ* (ugly) w' and with its synonyms $S(w')$, therefore the weight will be negative. On the other hand, the feature وجه *wǧh* (face) can be with both جميل *ǧmyl* (beautiful) and قبيح *qbyḥ* (ugly) and with their synonyms too, so the weight should be close to zero.

4.4   Process

This approach is divided into tree major parts: Pattern Identification, Pairs Extraction, and Antonyms Distinction.

4.4.1   Pattern Identification

In the first step, the authors identify the most frequent patterns in the chosen corpora (see Sect. 4.1). In order to say that a pattern is a *good* one, it has to occur only with many seed pairs [4]. They choose the most frequent adjectives from AWN in arTenTen and affect to each one of them its appropriate antonym, which is done by a linguistic expert to make sure that each adjective has its own antonym. Table 2 shows a sample of the most frequent antonym seeds along with their frequencies in arTenTen their logDice.

The patterns are generated from those seeds by putting the expression 1:[word ='...'] 2:[]{1,4} 3:[word='...'] in the CQL search bar in sketch engine. The two *words (1: an 3:)* are a pair of antonyms where the order of

Table 2. Antonym seeds.

	Antonym seed	Translation	Frequency	LogDice
1	بعيد ، قريب *qryb, bʿyd*	Near, Far	54,914	10.847
2	كثير ، قليل *qlyl, kṯyr*	Few, Many	29,257	8.581
3	أصغر ، أكبر *ʾakbr, ʾaṣġr*	Bigger, Smaller	9,096	7.235
4	قديم ، جديد *ğdyd, qdym*	New, Old	9,071	7.350
5	ميت ، حيّ *ḥya, myt*	Alive, Dead	5,878	8.050
6	طويل ، قصير *qṣyr, ṭwyl*	Tall, Short	5,130	8.374
7	غربي ، شرقي *šrqy, ġrby*	Eastern, Western	3,913	9.277
8	فقير ، غني *ġny, fqyr*	Rich, Poor	1,613	9.070
9	مرتفع ، منخفض *mnḫfḍ, mrtfʿ*	Low, High	1,012	8.345
10	جبان ، شجاع *šğāʿ, ğbān*	Brave, Coward	312	7.186

expression matters, for example the frequency of (بعيد ، قريب *qryb, bʿyd*) (Near, Far) is 20,326 and (قريب ، بعيد *bʿyd, qryb*) (far, near) 2,229. Those results are obtained with a maximum of 4 words between the two antonyms. With 2 and 3 words, the authors got quite a few results and with 5 they got a lot, so they choose the average 4. This step helps to get only *good* patterns and to avoid patterns that have a high frequency and co-occur with a few antonyms like Translate («... و ... كلا» *'klā ...w ...'* ('both ... and ...').

4.4.2   Pairs Extraction

Once the patterns are prepared, the authors transfer them into an CQL expression, like for example, («... إلى ... من» *'mn ... ʾilā ...'* ('from ... to ...') will be 1:[word="من *mn*"] 2:[tag="(DT)?JJ.*"] 3:[word ="إلى *ālā*"] 4:[tag="(DT)?JJ.*"] & 2.word != 4.word. The *tag="J.*"* means that only adjectives are extracted and *2.word != 4.word* means that the two adjectives are different. Of course, *word* can be used instead of *tag* but this will give random results such as nouns, propositions, and numbers. The patterns give hundreds of pairs, so they have to be filtered.

4.4.3   Antonyms Distinction

As expected, many odd pairs are found such as, non-antonym pairs, noun-noun pairs..., so it has to be filtered and extract only antonym pairs. To do so, the authors used the *logDice* and the *weight (w,f)* measures. The logDice measure calculates word collocations that exist with the pattern in question. they defined a threshold to 6.0 to minimize the results, so each pair with a logDice above 6.0 is considered as an antonym pair; otherwise it needs to be eliminated. The threshold is defined after many experiences. Despite the stability of the logDice, it gives antonym and synonym pairs with a high score. For example, the logDice of من رائع إلى أروع *mn rāyʿ ʾilā ʾarwʿ* (from wonderful to excellent) equals to 9.574, which is expected because logDice is all about probability and measures only the appearance of a pair (x, y) in the same context. LogDice is used only as a filter to reduce the obtained results.

Here comes the role of the *lexical contrast information* to distinguish the rest of the results. It determines whether two words are antonyms or not by their features, which can be presented as a strong one for the synonym of a word and weak for its antonyms, like in the example cited above in Sect. 4.3. The feature العدل *ālʿdl* occurs only with the synonym of the word (adjective in this case) جميل *ǧmyl* but not with its antonym قبيح *qbyḥ* . After the application of logDice, the $weight(w, f)$ has been calculated for the rest of the results.

5   Results and Evaluation

First, the authors choose 50 adjective seeds from AWN to work with, and then they reduced them to 30 due to their frequency in the corpus. The total obtained patters is 72, some of which had a high frequency but co-occurs with only few seeds, so they had to be decrease to only 4 that occurred with almost all the seeds (see Table 3).

Table 3. Results of the chosen patterns.

	Patterns	Translation	Frequency	Number of seeds
1	«... و ...» '... *w* ...'	('... and ...')	10,832,270	30
2	«... أو ...» '... *ʾaw* ...'	('... or ...')	773,256	30
3	«بين ... و ...» '*byn* ... *w* ...'	('between ... and ...')	41,478	7
4	«من ... إلى ...» '*mn* ... *ʾilā* ...'	('from ... to ...')	20,826	11
5	«إما ... أو ...» '*ʾimā* ... *ʾaw* ...'	('either ... or ...')	4,145	22
6	«كلا ... و ...» '*klā* ... *w* ...'	('both ... and ...')	104	5

The patterns «... و ...» '... *w* ...' ('... or ...') and «... أو ...» '... *ʾaw* ...' ('... and ...') co-occurred with all the seeds with very high frequency. While, the patterns «بين ... و ...» '*byn* ... *w* ...' ('between ... and ...') and «من ... إلى ...» '*mn* ... *ʾilā* ...' ('from ... to ...') co-occurred less, with only 7 and 11 seeds respectively. In the meantime, the patterns, «إما ... أو ...» '*ʾimā* ... *ʾaw* ...' ('either ... or ...') co-occurred with 22 seeds. In the end, the pattern «كلا ... و ...» '*klā* ... *w* ...' ('both ... and ...') did not scores very well. That is why the patterns number 2, 3, 4, and 5 has been chosen to work with.

After running the chosen patterns, 9,712,449 pairs are obtained. The authors did not take them all, only the frequent pairs are chosen. Sketch Engine offers a way to choose only the frequent pairs in an ordered list. For example, the authors choose from the first pattern the top 50 pairs and test their semantic relations. To do so, they used the *lexical contrast information*, then calculated the weight of each pair. Some of these pairs had a negative value which means that it is not an antonym but rather a synonym. Table 4 shows some samples of the extracted pairs with the pattern «... أو ...» '... ʾaw ...' ('... or ...') with their weight and their semantic relations.

Table 4. The weight of some extracted pairs.

	Pairs	Translation	Weight (w,f)	LogDice	semantic relations
1	حزين، سعيد ḥzyn, sʿyd	Sad Happy	-0.41	9.254	antonym
2	سهل، صعب shl, ṣʿb	Easy Hard	-0.57	8.986	antonym
3	طاهر، نجس ṭāhr, nǧs	Pure Unclean	-0.39	7.294	antonym
4	صادق، كاذب ṣādq, kāḏb	Honest Liar	-0.54	8.254	antonym
5	نظيف، طاهر nẓyf, ṭāhr	clean Pure	0.64	7.214	synonym
6	قوي، شديد qwy, šdyd	Strong Intense	0.69	8.012	synonym
7	خصوصي، عمومي ḥṣwṣy, ʿmwmy	Private Public	-0.43	7.004	antonym
8	ساخن، حار sāḥn, ḥār	Hot Spicy	0.51	8.021	synonym
9	أبيض، أسمر ʾabyḍ, ʾasmr	White Black	-0.25	8.267	antonym
10	ضعيف، هزيل ḍʿyf, hzyl	Weak Skinny	0.60	7.014	synonym

The appearance of the synonym instead of an antonym relation is due to the similarity of the context. That's why after the application of the *lexical contrast information*, an expert has accepted to verify the results manually. At the end, the total new antonym pairs is 800, some of them already exist in AWN and others they has just been added.

Despite the great Sketch Engine tool, it has not been able to identify any semantic relations since it only provides a measure for similarity distribution. LogDice offered great results to determine whether a pair has a relation or not, by giving them a high score if they had a similar context, a relation that could not be determined. The *weight* in the lexical contrast, information can help determine the type of the relation.

6 Conclusion and Future Work

In this paper, the authors presented a pattern-based approach to extract Arabic antonym pairs from a corpus. The semantic relations between those pairs has been able to identify using some specific measure. This approach yielded promising results, but it extracted some noisy pairs along with antonym ones. The authors were able to filter them but it was not enough because even if they used the *lexical contrast information* to get more specific antonym pairs, Sketch Engine does not have all the sufficient semantic relationships between words.

The work on Arabic WordNet is still going forward, the authors will move towards improvement to be at the same level of other wordnets. They will examine the existing semantic relation, try to improve them, and add more if it is possible. They are also thinking about creating a browser to the LMF file just like version 2 of the Arabic WordNet.

References

1. Abouenour, L., Bouzoubaa, K., Rosso, P.: On the evaluation and improvement of Arabic wordnet coverage and usability. Lang. Resour. Eval. **47**(3), 891–917 (2013)
2. Al-Rajebah, N.I., Al-Khalifa, H.S.: Extracting ontologies from Arabic wikipedia: a linguistic approach. Arab. J. Sci. Eng. **39**(4), 2749–2771 (2014)
3. Al-Thubaity, A.O.: A 700 m + Arabic corpus: Kacst arabic corpus design and construction. Lang. Resour. Eval. **49**(3), 721–751 (2015)
4. ALdhubayi, L., Alyahya, M.: Automated Arabic antonym extraction using a corpus analysis tool. J. Theoret. Appl. Inf. Technol. **70**(3) (2014)
5. AlHedayani, R.: Antonymy in modern standard Arabic. Ph.D. thesis, University of Sussex (2016)
6. Ben Amar, F.B., Gargouri, B., Ben Hamadou, A.: Domain ontology enrichment based on the semantic component of LMF-standardized dictionaries. In: Wang, M. (ed.) KSEM 2013. LNCS (LNAI), vol. 8041, pp. 404–419. Springer, Heidelberg (2013). https://doi.org/10.1007/978-3-642-39787-5_33
7. Arts, T., Belinkov, Y., Habash, N., Kilgarriff, A., Suchomel, V.: arTenTen: Arabic corpus and word sketches. J. King Saud Univ. Comput. Inf. Sci. **26**(4), 357–371 (2014)
8. Atserias, J., Villarejo, L., Rigau, G.: Spanish Wordnet 1.6: porting the Spanish WordNet across Princeton versions. In: LREC (2004)
9. Black, W., et al.: Introducing the Arabic WordNet project. In: Proceedings of the Third International WordNet Conference, pp. 295–300 (2006)
10. Boudabous, M.M., Kammoun, N.C., Khedher, N., Belguith, L.H., Sadat, F.: Arabic WordNet semantic relations enrichment through morpho-lexical patterns. In: 2013 1st International Conference on Communications, Signal Processing, and their Applications (ICCSPA), pp. 1–6. IEEE (2013)
11. Boyd-Graber, J., Fellbaum, C., Osherson, D., Schapire, R.: Adding dense, weighted connections to WordNet. In: Proceedings of the Third International WordNet Conference, pp. 29–36. Citeseer (2006)
12. Cavalli-Sforza, V., Saddiki, H., Bouzoubaa, K., Abouenour, L., Maamouri, M., Goshey, E.: Bootstrapping a WordNet for an Arabic dialect from other WordNets and dictionary resources. In: 2013 ACS International Conference on Computer Systems and Applications (AICCSA), pp. 1–8. IEEE (2013)
13. Elkateb, S., et al.: Building a WordNet for Arabic. In: Proceedings of the Fifth International Conference on Language Resources and Evaluation (LREC 2006), pp. 22–28 (2006)

14. Fellbaum, C.: WordNet. Wiley Online Library (1998)
15. Fellbaum, C.: Wordnet. In: Poli, R., Healy, M., Kameas, A. (eds.) Theory and Applications of Ontology: Computer Applications, pp. 231–243. Springer, Dordrecht (2010). https://doi.org/10.1007/978-90-481-8847-5_10
16. Harmain, H.M., El Khatib, H., Lakas, A.: Arabic text mining. In: IADIS International Conference Applied Computing, pp. 23–27 (2004)
17. Hearst, M.A.: Automatic acquisition of hyponyms from large text corpora. In: Proceedings of the 14th Conference on Computational Linguistics, vol. 2, pp. 539–545. Association for Computational Linguistics (1992)
18. Kilgarriff, A., Kosem, I.: Corpus tools for lexicographers. na (2012)
19. Kilgarriff, A., Rychly, P., Smrz, P., Tugwell, D.: Itri-04-08 the sketch engine. Inf. Technol. **105**, 116 (2004)
20. Lin, D., Zhao, S., Qin, L., Zhou, M.: Identifying synonyms among distributionally similar words. IJCAI **3**, 1492–1493 (2003)
21. Mohammad, S.M., Dorr, B.J., Hirst, G., Turney, P.D.: Computing lexical contrast. Computational Linguistics **39**(3), 555–590 (2013)
22. Nguyen, K.A., Walde, S.S.i., Vu, N.T.: Integrating distributional lexical contrast into word embeddings for antonym-synonym distinction. arXiv preprint arXiv:1605.07766 (2016)
23. Regragui, Y., Abouenour, L., Krieche, F., Bouzoubaa, K., Rosso, P.: Arabic WordNet: new content and new applications
24. Rodríguez, H., et al.: Arabic WordNet: current state and future extensions. In: Proceedings of The Fourth Global WordNet Conference, Szeged, Hungary (2008)
25. Rodríguez, H., Farwell, D., Ferreres, J., Bertran, M., Alkhalifa, M., Martí, M.A.: Arabic WordNet: semi-automatic extensions using Bayesian inference. In: LREC (2008)
26. Roget, P.M.: Roget's Thesaurus of English Words and Phrases... TY Crowell Company (1911)
27. Rychlý, P.: A lexicographer-friendly association score. In: Proceedings of Recent Advances in Slavonic Natural Language Processing, RASLAN, pp. 6–9 (2008)
28. Schropp, G., Lefever, E., Hoste, V.: A combined pattern-based and distributional approach for automatic hypernym detection in Dutch. In: RANLP, pp. 593–600 (2013)
29. Wang, W., Thomas, C., Sheth, A., Chan, V.: Pattern-based synonym and antonym extraction. In: Proceedings of the 48th Annual Southeast Regional Conference, p. 64. ACM (2010)

Designing an Ontology for Physical Exercise Actions

Sandeep Kumar Dash[1]([✉]), Partha Pakray[2], Robert Porzel[3], Jan Smeddinck[3], Rainer Malaka[3], and Alexander Gelbukh[4]

[1] National Institute of Technology Mizoram, Aizawl, India
sandeep.cse@nitmz.ac.in
[2] National Institute of Technology Silchar, Silchar, India
partha@cse.nits.ac.in
[3] Digital Media Lab, University of Bremen, Bremen, Germany
{porzel,smeddinck,malaka}@tzi.de
[4] CIC, Instituto Politécnico Nacional, Mexico City, Mexico
gelbukh@gelbukh.com

Abstract. Instructions for physical exercises leave many details under-specified that are taken for granted and inferred by the intended reader. For certain applications, such as generating virtual action visualizations from such textual instructions, advanced text processing is needed, requiring interpretation of both implicit and explicit information. This work presents an ontology that can support the semantic analysis of such instructions in order to support the identification of matching action constructs. The proposed ontology lays down a hierarchical structure following the human body structure along with various type of movement restrictions. This facilitates flexible yet adequate representations.

Keywords: Ontology design · Text processing
Virtual action generation

1 Introduction

The modeling of real world entities by means of formal ontologies has been an active part of research in natural language processing (NLP). As semantic annotation identifies possible roles for predicates containing pertinent information, ontologies provide the conceptualization with respect to the domain that links the meaning of natural language expressions to an underlying logical calculus. In this work we propose a design of an ontology which will help in the annotation and explication of instructional texts of physical exercises. The proposed ontology will depict information on the human body structure and the various actions or poses that it can attain. The requirement for the ontology emerged from the larger picture of converting textual exercise instructions to virtual actions. The said system will be built along with the information base of the proposed ontology, which will help to generate permissible virtual executions of exercise actions.

© Springer Nature Switzerland AG 2018
A. Gelbukh (Ed.): CICLing 2017, LNCS 10761, pp. 354–362, 2018.
https://doi.org/10.1007/978-3-319-77113-7_28

Ontology design has been mostly seen as an integral part of information retrieval (IR), but the hierarchy provided through it can always help in other parts of NLP such as in semantic role labeling where the roles of various part of sentences are determined based on the considered predicate that may be a verb or noun. The ontology will address the task of semantic role labeling of exercise instructions by providing information about exercise actions, body parts movement in particular activities, the duration of maintaining postures etc. We propose the use of the ontology in multiple domains related to human body structure as multiple rehabilitation physiotherapies, virtual human representation for variety of applications and general exercise related activities as well. In the task of ontology design the concepts or classes has been proposed are seen with an inclination towards re-usability of the overall design.

The structure of the paper is as follows; Sect. 2 describes the related developments in the area, Sect. 3 describes the considerations for the ontology design, Sect. 4 describes the exercise ontology, Sect. 5 implies the proposed use of the ontology before we conclude in Sect. 6.

2   Related Work

Domain engineering particularly aims at modeling domain knowledge that is, ideally, employed in multiple applications [2]. This has been the main reason behind ontology development. Gutierrez et al. [3] have developed a semantics-based method for organizing the various types of data for the synthesis, animation and functionalities of virtual humans in terms of ontology. Their approach aims at effectively modeling virtual humans through shape analysis and segmentation combined with anthropometric knowledge and large sets of acquired data. Our proposed design will differ as it seeks to represent information related to human body movement for virtual action generation. There have been also few systems which annotate texts based on ontologically modeled entities that make accessing of information easier. Satoru et al. have developed a health advice system [4] using an ontology and inference rules. The system measures vital data of user through wearable sensor device which is sent to database along with user input of exercise and meal detail. The health advices are derived based on the data of the user, the ontology and the inference rules. The ontology contains concepts which suggests which exercise the user should or should not do with respect to his health condition, which resembles our design of suggesting the right manner in which the exercise should be performed. The 'TrhOnt' [5] ontology focuses on the rehabilitation of the glenohumeral joint, however its general nature makes it reproducible to model any other body structure deserving rehabilitation. The ontology was developed following the NeOn Methodology [6]. It integrates knowledge from ontological (e.g. FMA ontology [7]) and non-ontological resources (e.g. a database of movements, exercises and treatment protocols) as well as additional physiotherapy-related knowledge.

Apart from the semantic annotation of text, ontologies have also been helpful in other applications. For example, 'newsEvents' [8] is such a system that describes relevant entities and events in the business news context. They have used OpenCalais[1] for obtaining annotations which were further used as definition of concepts represented in the ontology. Based on those concepts patterns were identified, that described the events and entities of the ontology. Athanasiadis et al. [9] have adopted a modular ontology infrastructure for the representation of the knowledge, to be used in a generic analysis scheme to semantically interpret and annotate multimedia content.

There have been various approaches to ontology development through semi-automatic acquisition of resources from text or corpus. This involves six steps: (i) term extraction. (ii) disambiguation and synonyms (iii) finding concepts (iv) establishing concept hierarchies (v) finding relations between concepts (vi) finding rules in the ontology. The system by Kietz et al. [10] describes semi-automatic ontology acquisition from a corporate intranet of an insurance company. Razika et al. [11] automatically extract terms from domain specific texts collected from medical reports, which are sets of concepts or relations which have either taxonomic or non-taxonomic relationships among them to develop an extensive and detailed ontology in the field of gynaecology.

3   Ontology Development

Ontologie are developed based on the analysis of the particular domain of interest and established modeling principles [13]. The terms and assertions employed therein provide an schematic view of the particulars involved in the domain. The terms may be physical entities, processes, qualities or abstract properties. Further the relationships established between the terms focuses on the real world constraints among the various particulars of the ontology.

Ontology development has various approaches that may be either extending upon existing ontologies or building a new one altogether. However both follow the following sequence of processes for finding out;

- Classes in the domain of interest.
- The foundational properties they inherit.
- The relations they have to each other.

The domain and scope of an ontology is defined through some basic questions which the ontology should be able to answer. These are called competency questions. The development of ontology starts with finding ways to answer these questions

3.1   Competency Questions

Every Ontology is based upon the necessities which are expressed through the competency questions. As Ontology development is a continuous process which grows along its implementation with the overall system, we have started the design based on the following set of questions:

---

[1] http://www.opencalais.com.

- CQ1: Domain or Application Oriented
  - CQ1.1: Is it applicable for physiotherapy related movements of body-parts?
  - CQ1.2: Which types of actions are described through this?
  - CQ1.3: Do the actions involve any machineries?
- CQ2: Concepts used
  - CQ2.1: How many types of exercise activity exists?
  - CQ2.2: What are the movements the exercise activities involve?
  - CQ2.3: Which bodyparts are mostly involved in exercise activities?
  - CQ2.4: What are the joints considered for hand?
  - CQ2.5: What types of posture are considered?
  - CQ2.6: What types of intensity level are involved for a stretch activity?
- CQ3: Features involved
  - CQ3.1: What is the minimum duration to maintain a squat posture in stretching exercise?
  - CQ3.2: How much we can bend the hand in y-plane?
  - CQ3.3: For a leg stretch exercise what should be the starting and ending position of legs?
  - CQ3.4: What are the movements possible for hand?
  - CQ3.5: What is the maximum permissible angle for flexion movement of leg?

The answer to these are the basis on which the ontology has been designed. However, as the development takes shape it will be necessary to answer more implicit questions.

3.2 Defining the Class Hierarchy

The proposed ontology design follows top-down approach as it is easier to identify the expected semantic roles given in the textual instructions to be the main ontology classes and that can further easily be specialized to the variety of each available type of actions or terms of the exercise terminology. The building process started with identifying four main areas as (i) activity (ii) bodypart (iii) posture (iv) position

The 'activity' class groups the types of activities that falls under the exercise domain with its subclasses as 'aerobic', 'bone strengthening', 'muscle strengthening', and 'stretching'. The 'bodypart' class similarly has subclasses which are members of and related to various exercise related activities. Here the design do not include the bodyparts which are not mostly a direct part of any of the mentioned activities. The duration class has an instance 'time' which provides the information of necessary duration for maintaining a posture during an activity sequence. The 'intensity' class will specify the speed or intensity of doing the activity. It has three instances namely, 'high', 'light' & 'moderate' as some of the activities vary in terms of their intensity. This is also one of the vital necessity for producing the correct type of action representation. Similarly 'joints' class also holds useful information about bodyparts, which may be accessed for the movement related representation of each bodypart. The two further classes in the hierarchy 'rotationPlane' and 'rotationDegree' provide comparatively detailed information about almost all the listed activities, as different bodyparts have different planes of movement and limitations for rotation. The 'movement' class further provides the various movements that the bodyparts and joints generally attain during the exercise activities. The list is not at all exhaustive as different entities can be added in the hierarchy of the application in the process of application development. The proposed hierarchy is as shown in Fig. 1.

Fig. 1. The class hierarchy

3.3   Defining Conceptual Relations

The ontology will have several relations existing among the classes described above;

(i) hasPlaneOfMovement: This implies the possible plane of movement for those bodyparts which are involved in an activity so as to make necessary suggestions through the generated virtual action and also to apply a check on the permissible plane of movement for body parts (as hands & legs) in the generated action

(ii) takesPartIn: This implies the type of bodypart involved in a particular activity. This will stress on the particular movements of the involved bodypart in the action

(iii) canRotateAt: As per exercise guidelines of various physiotherapists this will establish min & max angle of rotation for individual bodyparts which have rotational capacity

(iv) hasJoints: This relation establishes whether a bodypart possesses any movable joints in consideration with the action generation

(v) consistsOf: This explains the postures involved in any type of activity

(vi) hasLevelOf: This explains the level of intensity required for any type of activity

(vii) maintainFor: Duration for maintaining the posture is mentioned through this

(viii) restsAt: This provides one of the necessary things for producing the action sequence such as starting & ending position of any bodypart during motion

(ix) hasMovements: This specifies zero or more bodyparts, may have more than one type of movement possible for them

(x) possibleFor, permissiblePlane, permissibleAngle: These relations carries the constraint for permissible movement construct for the bodyparts.

3.4 Considerations for Re-usability

As ontology design and development is a non-trivial process which demands time and effort, its re-usability should be given equal importance as its application orientation. In the proposed design most of the concepts related to human body structure have an implicit use in related domains. The classes 'Bodypart', 'Posture', 'Joints' & 'Position' can be used in such ontologies which contains such kinds of interactions with human body structure.

The Web Ontology Language (OWL) represents the concepts of ontologies and relations among them in a suitable format for interaction with any semantic web interface. The ontology can further be improved by the semantic web community which shares the OWL platform. The application queries can easily be mapped to the ontology for the required domain knowledge using the OWL representation.

4 Ontology Design

The exercise ontology focuses on permissible human body movements related information in different instances of exercise activity as explained through the ontology structure below. As it has been explained in the previous section, the conceptual relations are vital in terms of the action structure to be correctly derived. Each action structure is interpretable to be one particular action which

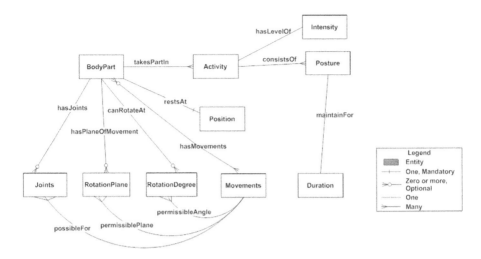

Fig. 2. Exercise ontology

can be used to deduce the virtual action of the text representation [12]. The classes or concepts in the proposed ontology are place holders for the semantic roles whereas their instances will represent the implicit detailed information about the action structure.

The ontology structure of Fig. 2 shows the various types of relationship among concepts. The one-to-many, zero-or-more etc. relationships explains about the possible instances of action structures that can be derived out of them. As seen from the structure the class 'Activity' has a one-to-one relationship with the class 'Intensity' implying that an activity has to be performed with only one type of intensity level. Similarly one-to-many relationship among 'BodyPart' and 'Activity' explains the possible involvement of a bodypart in multiple activities. These implications will be helpful in producing the representation of action structure with pre-specified constraints as per the domain requirement.

5   Ontology Integration with Application System

The proposed ontology carries the implicit terms of exercise instructions necessary for a proper virtual action representation system. The instances of the classes hold the important details that are rendered into an action sequence. The following figure shows an overview of the application system.

The system diagram in Fig. 3 shows the stepwise process of the proposed application for textual instruction to virtual action conversion. The 'Semantic Parsing' stage utilizes the exercise ontology for generating the action content for the input instructions, which undergoes a mapping process on to an intermediate representation. The intermediate representation will be mostly expressed using certain kind of mark-up language such as Behaviour Markup Langauge (BML) etc.

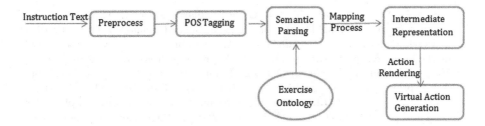

Fig. 3. Text-to-Action conversion system

6   Conclusion

In this work, a design of an ontology is proposed which incorporates the human body movement related information that will be helpful to analyze implicit constituents in exercise instructions. We have shown how this will be helpful in the application of converting textual instructions to virtual actions. The ontology carries various concepts and relations that will suffice to the need of any other related domain requirements. The design also gives importance to proper feedback about any exercise activity.

Our future work concerns with developing and evaluating the ontology with more detailed information along with the semantic structure derivation from the exercise instruction texts. The ontology module will be implemented for providing necessary integration of implicit information in the exercise instructions.

Acknowledgements. This research is supported by DST-DAAD Project Based Personnel Exchange Programme: an Indo-German Joint Research Collaboration (No. INT/FRG/DAAD/P-15/2016). The research reported in this paper has been partially supported by the German Research Foundation DFG, as part of Collaborative Research Center (Sonderforschungsbereich) 1320 "EASE - Everyday Activity Science and Engineering", University of Bremen (http://www.ease-crc.org/). The research was conducted in subproject P01 "Embodied semantics for the language of action and change".

References

1. Gruber, T.R.: A translation approach to portable ontology specifications. Knowl. Acquis. **5**(2), 199–220 (1993)
2. de Almeida Falbo, R., Guizzardi, G., Duarte, K.C.: An ontological approach to domain engineering. In: Proceedings of the 14th International Conference on Software Engineering and Knowledge Engineering (2002)
3. Gutiérrez, M., et al.: An ontology of virtual humans: incorporating semantics into human shapes. In: Proceedings of Workshop Towards Semantic Virtual Environments (SVE05), no. VRLAB-CONF-2007-004, pp. 56–57 (2005)
4. Izumi, S., Kuriyama, D., Itabashi, G., Togashi, A., Kato, Y., Takahashi, K.: An ontology-based advice system for health and exercise. System **3** (2006)

5. Berges, I., Antón, D., Bermúdez, J., Goñi, A., Illarramendi, A.: TrhOnt: building an ontology to assist rehabilitation processes. J. Biomed. Semant. **7**(1), 1–21 (2016)
6. Suárez-Figueroa, M.C., Gómez-Pérez, A., Fernández-López, M.: The NeOn methodology for ontology engineering. In: Suárez-Figueroa, M., Gómez-Pérez, A., Motta, E., Gangemi, A. (eds.) Ontology Engineering in a Networked World, pp. 9–34. Springer, Heidelberg (2012). https://doi.org/10.1007/978-3-642-24794-1_2
7. Rosse, C., Mejino Jr., J.L.V.: The foundational model of anatomy ontology. In: Burger, A., Davidson, D., Baldock, R. (eds.) Anatomy Ontologies for Bioinformatics, pp. 59–117. Springer, London (2008). https://doi.org/10.1007/978-1-84628-885-2_4
8. Lösch, U., Nikitina, N.: The newsEvents ontology: an ontology for describing business events. In: Proceedings of the 2009 International Conference on Ontology Patterns, vol. 516. CEUR-WS.Org (2009)
9. Athanasiadis, T., Tzouvaras, V., Petridis, K., Precioso, F., Avrithis, Y., Kompatsiaris, Y.: Using a multimedia ontology infrastructure for semantic annotation of multimedia content. In: Proceedings of 5th International Workshop on Knowledge Markup and Semantic Annotation (SemAnnot 2005), Galway, Ireland (2005)
10. Kietz, J.-U., Maedche, A., Vólz, R.: A method for semi-automatic ontology acquisition from a corporate intranet. In: EKAW 2000 Workshop "Ontologies and Text", Juan-Les-Pins, France (2000)
11. Houda, D.R.B., Nawel, K.: Domain Ontology Building Process Based on Text Mining from Medical Structured Corpus
12. Dash, S.K., Pakray, P., Gelbukh, A.: Representing action structure through semantic roles of physiotherapy instructions. In: The 20th International Conference on Asian Language Processing, IALP (2016)
13. Masolo, C., Borgo, S., Gangemi, A., Guarino, N., Oltramari, A., Schneider, L.: WonderWeb Deliverable D17: The WonderWeb Library of Foundational Ontologies (2003)

Visualizing Textbook Concepts: Beyond Word Co-occurrences

Chandramouli Shama Sastry[1]([✉]), Darshan Siddesh Jagaluru[1],
and Kavi Mahesh[2]

[1] Center for Knowledge Analytics and Ontological Engineering (KAnOE),
PES Institute of Technology, Bengaluru, India
chandramouli.sastry@gmail.com,
calltodarshan@gmail.com
[2] School of Computing and Decision Sciences,
Great Lakes International University, Sricity, India
drkavimahesh@gmail.com

Abstract. In this paper, we present a simple and elegant algorithm to extract and visualize various concept relationships present in sections of a textbook. This can be easily extended to develop visualizations of entire chapters or textbooks, thereby opening up opportunities for developing a range of visual applications for e-learning and education in general. Our algorithm creates visualizations by mining relationships between concepts present in a text by applying the idea of transitive closure rather than merely counting co-occurrences of terms. It does not require any thesaurus or ontology of concepts. We applied the algorithm to two textbooks - Theory of Computation and Machine Learning - to extract and visualize concept relationships from their sections. Our findings show that the algorithm is capable of capturing deep-set relationships between concepts which could not have been found by using a term co-occurrence approach.

Keywords: Concept extraction · Concept relations · Textbook visualization
Transitive closure · Term co-occurrences

1 Introduction

Identifying concept relationships and visualizing them are important for understanding the subject matter of a textbook. Textbooks are organized hierarchically into chapters and sections (and further into subsections) with each such unit being centered on one or more concepts. Depending on the style of the author, a section (or a subsection) can be considered to be the smallest unit of the textbook. Every section of the textbook is devoted to explaining certain aspects of one or more concepts. We observe that in a typical textbook no concept is independent but is explained in conjunction with other concepts. Any two concepts can be said to be related if some aspect of one concept can be understood or inferred using the other concept. We also observe that typically the author either reinforces some concept relationship already explained earlier in the text (or assumed to be known as a prerequisite) or introduces the reader to new concept

© Springer Nature Switzerland AG 2018
A. Gelbukh (Ed.): CICLing 2017, LNCS 10761, pp. 363–376, 2018.
https://doi.org/10.1007/978-3-319-77113-7_29

relationships. We can thus say that every section is characterized by the set of concept relationships that it expounds. The aim of our work is to extract such concept relationships given the text and present them explicitly and visually to help the student in reading the textbook and understanding the subject matter.

Previous attempts that aimed to statistically extract semantic relationships from texts followed the popular approach of counting term co-occurrences (or some variations of it). Although this may work quite well for nontechnical books, textbooks on technical subjects tend to be different in that mere co-occurrence of terms does not suffice to characterize the concept relationships present in the subject matter. Hence, we need an algorithm that can capture concept relationships going beyond co-occurrences.

We begin with the set of terms (concepts) mentioned in the glossary or the back-of-book index in the textbook. We apply our algorithm on every section of the textbook and extract the set of concept-relationships. We then visualize these relationships as a network. Our findings show that the relationships extracted from the sections effectively represent the concepts explained in the underlying text. For further analysis, we also create a concept timeline which visualizes how the concepts are developed as the reader goes through the chapters and sections of the textbook.

The following table describes the structure of two of the textbooks - Theory of Computation [11] and Machine Learning [12] - used for this study (Table 1):

Table 1. Description of data set

	Theory of computation	Machine learning
Total number of sections	121	95
Total number of terms/keywords	252	362
Average number of paragraphs per section	5	11
Average number of keywords per section	21	20

2 Related Work

First, we discuss previous work on extracting concept-relationships from texts. Automated knowledge mining from text has been a popular field. [7] attempts to do concept-level text analysis using ConceptNet Ontology [4] to determine relationships between the concepts extracted from the text. The ontology represents "commonsense knowledge" used in linking the concepts to create a concept map. However, the use of the ontology for determining concept relationships in technical texts may not be feasible for analyzing textbooks as a comprehensive ontology may not be available and the convention of terminology followed by each author in writing the text varies significantly across subjects. While a generic ontology fails to capture some relationships, building customized ontology is an unnecessary burden. Also, the ontology brings in inherent bias depending on how and for what purpose it was developed.

[8] tries to build links between textbook sections that present similar topics and concepts. This work was primarily done to help readers find potential sections of other textbooks which offer similar information but in an alternative and possibly better way. They use the Latent Dirichlet Allocation (LDA) [9] algorithm to extract the topics and

build topic models for every section of each textbook on a subject. LDA algorithm is a probabilistic model and provides a group of keywords which represents a topic. Based on this topic model, they build links between textbook sections and use the topical models to characterize the conceptual content of the section. [10] is similar to our work, in that, it too attempts to build a concept graph of a text. However, it uses a co-occurrence based approach to link concepts and as such, it may not be powerful enough to capture complex relationships between the concepts in a technical textbook.

It appears that contemporary approaches either use ontologies or co-occurrences to build a concept graph. We present a novel technique to build a concept graph of a text which is capable of capturing complex relationships between concepts beyond just co-occurrences without the need for a domain ontology.

Concept-level visualization of a text document has many applications. [2, 4, 6] show how we can use concept maps for various general-purpose applications.

3   The Algorithm

Our algorithm is an elegant application of the idea of transitive closure to extract concept-relationships. We construct a graph with sentences and concepts as nodes and apply transitive closure on this graph. The graph that we construct before transitive closure has only two types of links: sentence-to-sentence links and sentence-to-concept links. We claim that the concept-concept links that emerge from transitive closure denote valid concept-relationships put forth by the considered text.

Let the graph $G = (V, E) = (V_S \cup V_C, E_S \cup E_{S-C})$, where V_S is the set of sentences and V_C is the set of concepts. E_S denotes the set of edges quantifying the relationship between two sentences, generally a measure of similarity. E_{S-C} denotes the set of edges quantifying the relationship between a sentence and a concept, generally a measure of the relative importance of the concept in the sentence. We can represent this graph as a matrix:

$$A = \begin{bmatrix} Sent - Sent & Sent - Concept \\ Concept - Sent & Concept - Concept \end{bmatrix}$$

The Sent-Sent block represents the graph $G_S = (V_S, E_S)$, while the Sent-Concept block represents the graph $G_{SC} = (V_S \cup V_C, E_{S-C})$. The Concept-Sent block is just the transpose of the Sent-Concept block. The Concept-Concept block is initialized to zeroes in the beginning. The transitive closure can now be written as:

$$S = \sum_{2 \le i \le n-1} A^i, \ where \ n = |V_S| + |V_C| \tag{1}$$

We sum up to the $(n-1)^{th}$ power to obtain the transitive closure. It may also be observed that we do not consider any attenuation factor to attenuate weights of longer paths since we want to consider all possible ways in which any two concepts can get linked. The Concept-Concept block of the resulting matrix S denotes the conceptual relationships obtained through transitive closure.

Constructing the graph involves construction of the Sentence-Sentence graph G_S and the Sentence-Concept graph G_{SC}:

1. **Sentence-Sentence Graph**: This graph considers the strength of similarity between two sentences. For the purposes of this work, we considered cosine similarity between 2 sentences which can be computed as:

$$Similarity(S_1, S_2) = \frac{\overrightarrow{S_1}.\overrightarrow{S_2}}{\left|\overrightarrow{S_1}\right|\left|\overrightarrow{S_2}\right|} \tag{2}$$

where S_1 and S_2 are vectors of terms representing the sentences. If S_1 and S_2 represent the same sentence, then the similarity score will be 1.0. Note that this enables co-occurring concepts to get linked. More advanced methods of sentence similarity may be used (e.g., synonym overlap, verb/argument structure overlap, stem overlap, co-reference resolution, paraphrasing, and so on [1]).

2. **Sentence-Concept Graph**: This graph considers how important the concept is for a given sentence. We build this graph based on the idea put forth in [2]. TF-ISF is a measure that is closely related to the idea of TF-IDF (Term Frequency-Inverse Document Frequency). In TF-ISF (Term Frequency-Inverse Sentence Frequency), we measure the importance of a concept to a sentence instead of a document. The formula for computing the importance can be written as:

$$Importance(S, C) = tf_{SC} \times isf_C \tag{3}$$

where S is a sentence and C is a concept. tf_{SC} denotes the term frequency of concept C in sentence S. isf_C measures the inverse sentence frequency of concept C in the given text. It is defined by the measure $isf_C = \log\left(1 + \frac{|V_S|}{n_C}\right)$, where n_C is the number of sentences in which the concept C occurs. We then normalize this measure using $\sum_{C_i \in S} Importance(S, C_i)$. The equation can be rewritten incorporating the normalization as:

$$Importance(S, C) = \frac{tf_{SC} \times isf_C}{\sum_{C_i \in S} tf_{SC} \times isf_C} \tag{4}$$

It must be noted that we can consider certain other linguistic features as well for computing this as outlined in the next section.

We make a couple of observations:

1. Two concepts co-occurring in the same sentence need not get a strong edge since the strength of an edge depends on the relative importance of the two words with respect to the sentence.
2. Two concepts linked by a longer path need not have a weaker edge since it depends on the path followed.

4   Implementation

In this section, we outline the details of implementation our algorithm and the necessary steps in pre-processing:

1. ***Extract List of Concepts from the glossary/index:*** The first step of the implementation involves extracting the list of keywords (concepts) from the glossary/index of the textbook. This step also includes identifying abbreviations and mapping them to the corresponding expanded concept names. For example, SQL should be mapped to Structured Query Language. This information is usually available within the index at the back of the book. We can also use subject-specific keyword lists. For example, ACM Encyclopedia of Computer Science contains a list of all key terms in computer science.

2. ***Stem the words:*** This step involves converting each term of the concept to its stemmed form. We use NLTK's SnowBall stemmer for achieving this which does suffix-stemming without affecting the meaning of words like "antipattern" with a negative prefix. For example, "Structured Query Language" will be converted to "structur queri languag".

3. ***Extract sections from text:*** This step involves extracting all sections from the textbook. We identify section boundaries using the table of contents and certain heuristics involving font style and font size. For every section that we extract, we perform the following two steps.

4. ***Extract all sentences:*** This involves extracting all sentences of a section and creating a vector representation for each sentence. These vectors are used for computing sentence similarity and for computing the relative importance of every concept. We consider the following for constructing the vector: the concepts that are identified in Step 1 and certain linking words like "Section 1.7", "Fig. 4.5", "Example 4.7.3", "Eq. 9.1", and so on. We consider the linking words since they are used either when describing some object (figure, example or exercise problem) or when making forward references. They offer important cues for computing sentence similarity. These terms can be interpreted as the components of the sentence vector. We use term frequencies as magnitudes of the vector. Having obtained the sentence vectors, we create graph G_S using sentence similarity measure. The sentence vectors also capture the term frequencies which can be used for creating graph G_{SC}.

5. ***Execute the algorithm:*** Now that we have obtained the matrices G_S and G_{SC}, we can construct the graph G and apply transitive closure as outlined in the previous section.

6. ***Creating Visualizations:*** In order to picture what the section represents, we extract top 20% (or the 80th percentile) of the edges and draw out a network. In order to create better visualizations, we can increase the sizes of the nodes in proportion to their centrality. We used Eigen-vector centrality for this purpose. These visualizations were used for quantitative evaluation of the results.

5   Results and Discussion

We executed our algorithm on the textbooks [11, 12] and extracted concept-relationships from their various sections. For the purposes of qualitative evaluation, we created a concept-timeline that shows the various concept associations that a reader will develop as he reads along. We also found examples where there is a significant inferred relationship which could not have been identified with term co-occurrences alone. For an extensive qualitative analysis, we chose one of the textbooks- Theory of Computation [11].

Figures 1 and 2 show clippings from the concept timeline of the concept *automaton* in the textbook. The edges between various concepts indicate that they are related and their thickness quantifies the strength of the relation. The unexpanded nodes denote section numbers: for ex, in Fig. 1, 2.0 and 2.2 are section numbers and 2.1 is expanded to show the concept-relationships of *automaton* in that section.

Fig. 1. Timeline 1 of *automaton*

Figure 1 is derived from the section which introduces the idea of *computing* and a machine for computation known as *deterministic finite automation*. We can see that the thickest edge is between *automaton* and *computing* which perhaps means that *automaton* is presented from the perspective of *computing*. We also observe that the section relates *vending machine* and *automaton*. Certain basic aspects of the *automaton* like *input* and *output*, *start state* and *final state*, *acceptance* and *rejection*, *string*, *alphabet* and *symbol* are also found in this section.

Figure 2 shows the introductory section of *non-deterministic finite automaton*. The concepts of *non-deterministic finite automaton* and *backtracking* in *automata* are found here. An important observation to make is that some of the links in Figs. 1 and 2 are common. For example, links with *deterministic finite automaton, string, rejection, acceptance, start state, final state* and *symbol* are common to both figures.

As we had mentioned in the introduction, the author usually either reinforces existing relationships or introduces new relationships as the text book progresses. The ideas of *backtracking* in *automata,* *configuration* of an *automaton,* and *non-deterministic finite automaton* as a kind of *automata* are new to this section. If we observe closely, we can see that the author tries to refine the idea of *deterministic finite automaton* as an *automaton* in the context of

Fig. 2. Timeline 2 of automaton

non-deterministic finite automaton. We can see that the meanings of *start state, final state, rejection* and *acceptance* of an *automaton* are being relooked at in the context of *non-deterministic finite automaton.*

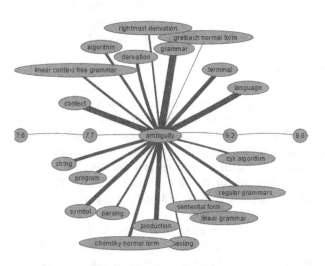

Fig. 3. Timeline of *ambiguity*

Figure 3 shows the concept-relationships of the term *ambiguity* in the chapter summary section. This section summarises the sections on *ambiguity* in *grammars* and gives an overview of various kinds of *grammars* like *regular grammars, context-free grammars, linear grammars,* and so on. However, one interesting link found by our algorithm is between *ambiguity* and *Chomsky Normal Form.* Close analysis of the textbook shows that *ambiguity* and *Chomsky Normal Form* do not co-occur in any sentence or section. After investigating, we found that *ambiguity* was talked about with respect to various terms like *grammar, parsing,* and *derivation.* The algorithm found a significant link between *Chomsky Normal Form* and *ambiguity* through transitive closure. The reader of the textbook, with the help of our visualizations, can conclude that this is indeed a valid relationship in the following sense: If a *grammar* is in *Chomsky Normal Form*, it helps in identifying example strings which can have ambiguous derivations. This is one of the most interesting links that we found in the results. This link could not have been by co-occurrence based methods since there is no explicit statement about it in the textbook and the terms do not co-occur at all.

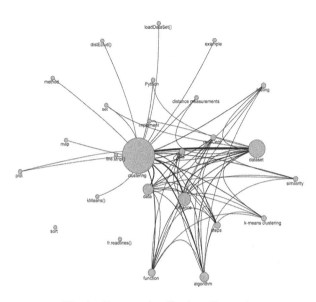

Fig. 4. Concept-visualization of a section

Figure 4 shows a typical visualization of a section. The figure depicts an introductory section on *Clustering and K-Means Clustering* as presented in the *Machine Learning* book. As described in the implementation section, the sizes of the nodes are determined by their centrality and the edge weights show the strength of their relationships. In the following section, we quantitatively evaluate the performance of our algorithm in generating concept visualizations of sections of a textbook.

6 Evaluation and Discussion

Having described the results qualitatively considering the concept-timelines of certain key terms, we now attempt to quantitatively evaluate the algorithm considering the concept visualization of an entire section. We compare the results obtained by our algorithm with the results of an algorithm that follows a windowed co-occurrence approach. A window size of n indicates that all key phrases in a window of n sentences are linked to each other and the strength of the link is defined by $L(W_1, W_2) = n(W_1) \times n(W_2)$, where W_1 and W_2 are key-phrases and $n(W_i) =$ number of occurrences of term W_i.

We randomly sampled 10 sections from each of the two textbooks for the purpose of this quantitative evaluation. In order to benchmark our algorithm with respect to the co-occurrence based algorithm, the following two sets of edges were considered:

1. Edges included in transitive closure and excluded (i.e., not found) by co-occurrence. In further discussions, this set is referred to as I. For the purposes of evaluation, the top ten edges of this set are considered.
2. The edges excluded in transitive closure and included by co-occurrence. In further discussions, this set is referred to as E. For the purposes of evaluation, the top ten edges of this set are considered.

The sampled sections and the outputs of both algorithms were presented to a set of subject experts who were asked to score each of the edges in both sets based on the rules defined in Table 2 and to assign an E-score and an I-score. E-score and I-score are defined as:

$$\text{E-score} = \frac{\sum_{e \in E} score(e)}{|E|} \text{ and I-score} = \frac{\sum_{e \in I} score(e)}{|I|},$$

where score is assigned according to the rules outlined in Table 2. In other words, we are computing a weighted precision@10 for sets I and E, which measure respectively, the precision of including and excluding edges with respect to our algorithm. Better the ability of our transitive closure-based algorithm to prefer other, indirect edges over those based only on co-occurrence, higher the E-score. Similarly, better its ability to find edges which cannot be found by co-occurrence, higher the I-score.

Table 2. Rules for scoring edges

	Edge makes sense given the context of the section	Edge makes sense given the context of the section and is one of the important takeaways of the section and should have been assigned significant score	Edge doesn't make sense given the context of the section	Edge doesn't make sense given the context of the section and the choice is not good
Included in transitive closure and excluded in co-occurrence	+1	+2	−1	−2
Included in co-occurrence and excluded in transitive closure	−1	−2	+1	+2

After obtaining results from the evaluators, we assigned an E-Score and an I-Score to each of the 10 sections by averaging them. Table 3 summarizes the results of the evaluation.

In both the textbooks, we observe that the E-score increases as we increase the co-occurrence window size, indicating that more and more co-occurrence relationships were added as the window size was increased whereas our algorithm excluded them in preference over others found through transitive closure. The values of E-score are consistently high. If we considered the overall mean of all 10 sections, we observe that the values are well above +1.0 and go up to +2. We see that the evaluators concur that the relationships chosen by co-occurrence do not generally make sense and increasing

Table 3. E-score and I-score values.

Algorithm used for comparison		Theory of Computation		Machine Learning	
		E-Score	I-Score	E-Score	I-Score
Co-occurrence with window size = 1	Mean of all 10 sections	+1.23	+0.954	+1.84	+0.91
	Mean of top 5 scores	+1.6	+1.68	+2	+1.53
Co-occurrence with window size = 2	Mean of all 10 sections	+1.52	+1.10	+1.95	+1.02
	Mean of top 5 scores	+1.96	+1.83	+2	+1.74
Co-occurrence with window size = 3	Mean of all 10 sections	+1.7	+1.07	+1.98	+1.2
	Mean of top 5 scores	+1.97	+1.79	+2	+1.6

window size makes matters worse as we see that the mean tends towards +2.0. This is further substantiated by the mean of top 5 scores. Further, we observe that the E-Scores for the Machine Learning textbook are consistently in the range of +1.9 to +2. On closer inspection, we found that most of the keywords in *Machine Learning Textbook* like *similarity, recommendation* are also part of common English vocabulary and the co-occurrence algorithm, which isn't capable of word sense disambiguation, couldn't reject such relationships. However, our algorithm with its limited sophistication was capable of rejecting such relationships.

The I-score, on the other hand, remains more or less centered around the same range irrespective of the algorithm against which it is compared. Interestingly, very similar trends are observed in both the textbooks. This shows that there are significant relationships that co-occurrence fails to capture despite a larger window size. The mean value of I-score of all ten sections is centered around +1.0 which can be taken to indicate that the included relationships make sense. Further, the means of the top 5 scores are considerably higher and are centered around +1.7 which says that the relationships make sense and it is important to include them to properly characterizing the section. We also observed that smaller sections (having one or two paragraphs) generally had lower I-scores when compared to larger ones, as a result of which, the average score is around +1.0. Thus, we see that across the board, there are no negative or very low E-scores or I-scores thereby indicating clearly that our algorithm is performing well in characterizing the concept-relations in sections of a textbook.

We noted earlier that co-occurrence based algorithms fail to capture certain relationships despite increasing the window size. Table 4 further substantiates this point. The percentage of edges in E is the percentage of additional edges that co-occurrence had captured but were excluded by our algorithm. Similarly, the percentage of edges in I represents additional edges that transitive closure captured. The table shows the mean percentage, where the mean is taken over all the 121 sections of the textbook.

Table 4. Percentage analysis of cardinalities of sets E and I.

Window size	Theory of computation			Machine learning		
	Mean Percentage of edges in E	Mean Percentage of edges in I	Mean Percentage of common edges	Mean Percentage of edges in E	Mean Percentage of edges in I	Mean Percentage of common edges
1	34.63	30.64	34.73	33.98	33.08	32.94
2	37.01	22.49	40.50	36.99	23.82	39.19
3	35.85	21.56	42.59	35.99	22.22	41.79
4	35.08	20.99	44.00	33.91	21.87	44.22
5	33.93	20.89	45.18	32.03	21.77	46.20

Interestingly, the results across two different textbooks are strikingly similar. We observe that the percentage of edges in E is around 35% even as the window size increases. This indicates that there are around 35% extra relationships (which are spurious, as shown by the scoring by the evaluators) that are to be excluded. We also observe that the percentage of edges in I saturates to around 21% even on increasing the window size. This shows that there are approximately 21% of the relationships (that make sense and are likely to be important in characterizing the section, as shown by the scoring by our evaluators) which co-occurrence cannot find even on increasing the window size. The percentage of common edges increases as the window size increases but continues to stay below the 50% mark. In summary, roughly a third (35%) of the relationships found by co-occurrence are not meaningful and need to be excluded while another fifth to a third (20%–30%) of other relationships need to be included and they can be discovered by our transitive closure algorithm.

7 Algorithm Extensions and Applications

In this section, we outline certain visual applications of our algorithm for characterizing a given unit (sub-section, section or chapter) of a textbook. First, we propose extensions to the algorithm to find concept-relationships in units larger than a section: chapters, entire books and the subjects they cover. We categorize the applications into three types in terms of the textbooks considered as input: Intra-Text, Intra-Subject and Inter-Subject. Intra-Text deals with applications which consider an entire textbook. Intra-Subject deals with applications which consider multiple textbooks on the same subject. Inter-Subject deals with applications which consider textbooks belonging to different but related subjects.

We used the following matrix for transitive closure for finding concept-concept relationships in a section:

$$A_{section} = \begin{bmatrix} Sent - Sent & Sent - Concept \\ Concept - Sent & Concept - Concept \end{bmatrix}.$$

We can extend this idea for capturing the relationships between concepts at a chapter level, subject level and course level as follows:

$$A_{chapter} = \begin{bmatrix} section - section & section - Concept \\ Concept - section & Concept - Concept \end{bmatrix}$$

$$A_{subject} = \begin{bmatrix} chapter - chapter & chapter - Concept \\ Concept - chapter & Concept - Concept \end{bmatrix}$$

$$A_{curriculum} = \begin{bmatrix} subject - subject & subject - Concept \\ Concept - subject & Concept - Concept \end{bmatrix}.$$

In each case, the Concept-Concept part of the matrix is initialized to zeros and the actual relationships are discovered by computing the transitive closure of the matrix. Similarity between sections, chapters and subjects can be measured using the concept relationships so discovered. Further, the importance of a concept in a section, chapter or subject can be measured using centrality measures. For example, the importance of a concept to a section can be determined by performing Eigenvector centrality on the concept-relationship graph of the section considered. Having thus extracted the relationships at various levels, we can provide a drill-down visualization showing the concept relationships at various levels of abstraction. For example, we can visualize the concepts linking various constituent sections of a chapter by considering the Concept-Section block and section-section block in conjunction with the Concept-Concept block. The visualization can be shown pictorially as in Fig. 5.

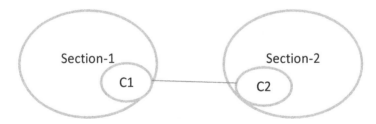

Fig. 5. Relating two sections belonging to same chapter

Figure 5 shows 2 sections section-1 and section-2 and a concept relationship between C1 of Section-1 and C2 of Section-2. It is important to note that C2 of Section-2 could be related with C1 of other sections as well. In order to determine which section's C1 is related and to what extent, it is necessary for us to use Concept-Section and Section-Section blocks along with the Concept-Concept block.

1. *Intra-Text Application: Reading Assistant-I*: An application that assists the reader by identifying potential sections to which he can refer when he feels that he needs to comprehend some concept in a section. Also, when a reader is attempting an exercise, it can help him trace back to identify sections and concept-relationships necessary for solving it.

2. *Intra-Subject Application: Reading Assistant-II:* We can create a reading assistant that links up different textbooks on the same subject. This idea has been explored by [8] which aims to create links across sections of different text books, primarily to enable readers to comprehend a section by referring to another textbook which could possibly explain the content better. The work uses LDA to characterize each section and uses a probabilistic model to link them up. However, we can get potentially better linking by considering the underlying concept-relationships characterizing each section in each book which are found by our transitive closure algorithm.
3. *Inter-Subject Application: Course visualization:* A big picture visualization of a curriculum or course which shows how various constituent subjects are related to one another through shared concepts. It can help a student visualize the entire breadth and depth of the curriculum allowing him to see how the various subjects are inter-connected, thereby helping him plan his studies, take necessary prerequisite courses, and make informed decisions about elective courses to take.

8 Conclusion

We proposed an elegant algorithm to extract concept-relationships characterizing the underlying text meanings in sections of a textbook. We have also shown the importance of extracting concept-relationships and proposed applications which can be developed into useful tools for students to enhance their learning by enabling them to make connections between different concepts within and across textbooks and subjects.

References

1. Erkan, G., Radev, G.R.: LexRank: graph-based lexical centrality as salience in text summarization. J. Artif. Intell. Res. **22**(1), 457–479 (2004)
2. Wan, X., Yang, J., Xiao, J.: Towards an iterative reinforcement of simultaneous document summarization and keyword extraction. In: Proceedings of ACL 2007, Prague, pp. 552–559 (2007)
3. Cao, C., et al.: Progress in the development of national knowledge infrastructure. J. Comput. Sci. Technol. **17**(5), 523–534 (2002)
4. Liu, H., Singh, P.: ConceptNet — a practical commonsense reasoning tool-kit. BT Technol. J. **22**(4), 211–226 (2004)
5. Cambria, E., Hussain, A., Havasi, C., Eckl, C., Munro, J.: Towards crowd validation of the UK national health service. In: Websci 2010 (2010)
6. Yuntao, Z., Ling, G., Yongcheng, W., Zhonghang, Y.: An effective concept extraction method for improving text classification performance. Geo-spatial Inf. Sci. **6**(4), 66–72 (2003)
7. Poria, S., Agarwal, B., Gelbukh, A., Hussain, A., Howard, N.: Dependency-based semantic parsing for concept-level text analysis. In: Gelbukh, A. (ed.) CICLing 2014. LNCS, vol. 8403, pp. 113–127. Springer, Heidelberg (2014). https://doi.org/10.1007/978-3-642-54906-9_10

8. Guerra, J., Sosnovsky, S., Brusilovsky, P.: When one textbook is not enough: linking multiple textbooks using probabilistic topic models. In: Hernández-Leo, D., Ley, T., Klamma, R., Harrer, A. (eds.) EC-TEL 2013. LNCS, vol. 8095, pp. 125–138. Springer, Heidelberg (2013). https://doi.org/10.1007/978-3-642-40814-4_11

9. Blei, D.M., Ng, A.Y., Jordan, M.I.: Latent Dirichlet allocation. J. Mach. Learn. Res. **3**, 993–1022 (2003)

10. Šimko, M., Bieliková, M.: Automatic concept relationships discovery for an adaptive E-course. In: International Conference on Educational Data Mining (2009)

11. Mahesh, K.: Theory of Computation: A Problem-Solving Approach. Wiley, New Delhi (2012)

12. Harrington, P.: Machine Learning in Action. Manning Publications Co., Greenwich (2012)

Matching, Re-Ranking and Scoring: Learning Textual Similarity by Incorporating Dependency Graph Alignment and Coverage Features

Sarah Kohail[(⊠)] and Chris Biemann

Language Technology Group, Computer Science Department,
Universität Hamburg, Hamburg, Germany
{kohail,biemann}@informatik.uni-hamburg.de

Abstract. In this work, we introduce a supervised model for learning textual similarity, which can identify and score similarity between a set of candidate texts and a given query text. By combining dependency graph similarity and coverage features with lexical similarity measures using neural networks, we show that most relevant documents to a given text can be more accurately ranked and scored than if the lexical similarity measures were used in isolation. Additionally, we introduce an approximate dependency subgraph alignment approach allowing node gaps and mismatch, where a certain word in one dependency graph cannot be mapped to any word in the other graph. We apply our model to two different applications, namely re-ranking for improving document retrieval precision on a new dataset, and automatic short answer scoring on a standard dataset. Experimental results indicate that our approach is easily adaptable to different tasks and languages, and works well for long texts as well as short texts.

1 Introduction

Semantic Textual Similarity (STS) measures the degree of semantic equivalence between a given pair of text. It also determines the notion that some texts are more similar than others. Measuring textual similarity, resulting from paraphrasing or summarization, may improve language understanding for many Natural Language Processing (NLP) applications, ranging from Information Retrieval (IR) [1] and machine reading comprehension [2,3] to question answering [4] and short answer scoring [5,6].

In this paper, we address the problem of assessing STS of alternative versions of candidate texts that have a varying degrees of similarity to a given query text. Specifically, we try to examine the impact of using dependency graph similarity and coverage features, and leverage supervised machine learning techniques in order to improve the relevancy identification and scoring. We also present an approximate subgraph alignment approach to find a subgraph in the candidate

© Springer Nature Switzerland AG 2018
A. Gelbukh (Ed.): CICLing 2017, LNCS 10761, pp. 377–390, 2018.
https://doi.org/10.1007/978-3-319-77113-7_30

text dependency graph that is similar to a given query text dependency graph, allowing for node gaps and mismatches, where a certain word in one dependency graph cannot be mapped to any word in the query text graph, as well as graph structural differences. We evaluate our method on two tasks: re-ranking for information retrieval and automatic short answer grading.

The remainder of this paper is organized as follows. In Sect. 2, we briefly discuss related work. Section 3 describes our textual similarity model and feature representation. In Sect. 4, we show the performance of our method and analyze results. Section 5 concludes and discusses further future work.

2   Related Work

Researchers have made substantial progress on STS motivated by the annual SemEval competitions [7–9]. Most of state of the art approaches often focus on training regression models on traditional lexical surface overlap features. Recently, deep learning models have achieved very promising results; the top three performing systems from SemEval STS 2016 used deep learning based models [10–12]. However, STS remains a hard problem when it comes to texts, which have both variable length and complex dependency structure [13].

To ensure that our method is generalizable over different languages and various text lengths, we evaluate our method using two different tasks, namely re-ranking for improving document retrieval, and automatic short answer grading.

Examining the effect of results ranking, Jansen and Spink [14] observed that most users do not browse results beyond the first page and the higher the document placement in the first page results, the more likely a user is to read that document. By minimizing the huge amount of relevant results to few highly relevant to the users' query, and re-ranking them to appear in the upper top rank, users are more likely to find their goal quickly and easily [15].

Since short texts might not contain sufficient statical information or syntax patterns, multiple evaluations have been proposed to operate for short texts separately [16–18], while Pilehvar and Navigli [19] investigate a unified approach to semantic similarity that operates at multiple levels.

The task of automatic short answer grading is to assess short natural language answers based on their similarity with expert-provided correct answers. Mohler et al. [6] train support vector machine (SVM) on a combination of graph-based alignment and lexical similarity measures to score short students answers using a 5-point scale. They find that the supervised model in this task outperforms the unsupervised model [5].

Numerous approaches have used the dataset of Mohler et al. [6] as a benchmark to evaluate their methods. We mention two recent comparable works, [20,21], which we use later in our comparative study. Ramachandran et al. [20] adopt a mechanism to automate the generation of regexp text patterns from the reference expert answer as well as top-scoring student answers, to capture the structural and semantic variations of good answers. Sultan et al. [21] train supervised model, namely a ridge regression model, on a set of similarity and word embeddings features for the task of short answer grading. They apply question demoting (QD) technique in an attempt to reduce the advantage of repeating words provided in the question by re-computing similarity features after removing these words from both the reference answer and the student response. Their ablation study shows that applying question demoting results in 0.021 correlation improvement and 0.016 reduction in Root Mean Square Error (RMSE).

3   Learning Textual Similarity

Given query text q and candidate text d, textual similarity captures the fact of how much a candidate text d conveys the same information as a query text q.

In this research, we employ two sets of features, similarity features and coverage features. The novel contribution of this work is constituted by three feature types: dependency structure features, expansion features and coverage features. In the following subsection, we describe each of these features in more details.

3.1   Similarity Features

Similarity is measured by the shared feature types between two texts a query text q and a candidate text d. For both texts d and q, we create a vector representation d and q for various feature types. Each entry in one vector corresponds to the existence/presence (i.e $d_i/q_i \in \{0,1\}$), frequency (i.e $d_i/q_i \in \mathbb{N}$) or Tf-Idf measure (i.e $d_i/q_i \in \mathbb{R}$) of a given feature type in a text. After removing stopwords, we consider the following feature types:

Bag of Words (BOW): We represent the content of each text by a bag of words. In this case, similarity is measured by the shared vocabulary between both d and q. We also employ a second version of this feature using stemmed words.

Topic Distribution: We also model each document as a vector of topics using Latent Dirichlet Allocation model (LDA) [22].

Dependency Structure: Another important similarity measure is dependency parse structure similarity. Based on the work by [23], we aggregate individual dependency relations obtained from a parser, weigh them with Tf-Idf and produce a graph which contains the highest-ranked content words and their dependency relations. For a text d, dependency graph $G_d = \{V_d, E_d\}$, where $V_d = \{w_1, ..., w_N\}$ represents the content words in a text, and E_d is an set of edges, where each edge e_{jk} represents a directed dependency relation between w_j and w_k. Written also as a list of triples as follows: "w_j" $\rightarrow$ "w_k"[$label = $"$e_{jk}$"].

A generated dependency graph is then filtered according to the following three conditions:

- Tf-Idf $(w_j, d) \geq \alpha$ or Tf-Idf $(w_k, d) \geq \alpha$
- Tf-Idf $(w_j\ w_k, d) \geq \beta$
- Tf-Idf $(e_{jk}\ w_j\ w_k, d) \geq \lambda$

where α, β and λ are ≥ 0. When α, β and $\lambda = 0$, no Tf-Idf filtering is applied. For q, dependency graphs are generated, and filtered in the same manner. Similarity is then measured on these three levels by representing each text as a vector of words, pairs and relations.

Named Entities: In this case, we measure similarity based on named entities terms only.

Expansion Features: Since the variability of language allows expressing the same concepts, entities and facts in different words, measuring similarity purely based on exact word matching does not fully capture conceptual matching. We expand content words, i.e. (common and proper) nouns, adjectives, verbs and adverbs in each text using the Distributional Thesaurus (DTs) from [24].

Once these vectors are constructed the similarity between each d and q texts pair can be measured using cosine similarity using the following equation:

$$cos(\boldsymbol{q}, \boldsymbol{d}) = \frac{\sum\limits_{i=1}^{N} q_i d_i}{\sqrt{\sum\limits_{i=1}^{N} q_i{}^2} \sqrt{\sum\limits_{i=1}^{N} d_i{}^2}} \tag{1}$$

where N is the dimension of q and d. Vectors are (length-) normalized by dividing by the L_2 norm $\sqrt{\sum\limits_{i=1}^{N} q_i{}^2}$ and $\sqrt{\sum\limits_{i=1}^{N} d_i{}^2}$ of q and d respectively.

3.2 Coverage Features

As a text gets longer, term frequency factors increase, and thus having a high similarity score is likelier for longer than for shorter texts. IR research has shown that document length normalization is important to guarantee that documents are retrieved with similar chances as their likelihood of relevance regardless of their length [25]. The same applies to answers grading: longer answers should not receive unjustly higher scores. Normalizing vectors using the cosine L_2 norm has proven to have several limitations due to the use of the individual terms weights for text length normalization [26,27]. This dependency is undesirable when the text includes infrequent terms with high Idf value, which can significantly increase the overall cosine normalization L_2 factor and cause inaccurate weighting for the other terms in the text. Accordingly, the new weights may not reflect the actual importance of the terms in content representation of the text. We try to solve this problem by incorporating a set of coverage features to measure the coverage of the query in the document.

Let $G_d = \{V_d, E_d\}$ and $G_q = \{V_q, E_q\}$ be the dependency graphs of d and q respectively. We measure coverage using the following equations:

Vocabulary Coverage: We calculate vocabulary coverage by computing the number of one-to-one nodes correspondence between both q and d dependency graphs divided by the overall number of nodes in the query text q dependency graph, as in the following equation:

$$\frac{|V_d \cap V_q|}{|V_q|} \tag{2}$$

Relation Coverage: We calculate relation coverage by computing the number of one-to-one edges (triple) correspondence between both q and d dependency graphs divided by the overall number of edges in the query text q dependency graph:

$$\frac{|E_d \cap E_q|}{|E_q|} \tag{3}$$

Pair Coverage: As in relation coverage, however in this case, we ignore the relation type and edge direction.

Graph Coverage: Before we present more details about how graph coverage features are measured, however, we first need to introduce our approximate dependency sub-graph alignment methodology.

The idea is to find a subgraph $G_s = \{V_s, E_s\}$, where $G_s \subseteq G_d$, that is approximately similar to a query text graph G_q. Algorithm 1 shows the pseudo-code of dependency sub-graph approximate matching algorithm.

Input: G_d, G_q, Threshold t
Output: G_s
$V_{intersection} \leftarrow \{V_d \cap V_q\}$;
for $j \leftarrow 1$ **to** $|V_{intersection}| - 1$ **do**
    for $k \leftarrow j + 1$ **to** $|V_{intersection}|$ **do**
        $Path \leftarrow dijkstra.getPath(G_d, w_j, w_k)$;
        if $Path \neq null$ **and** $Path.size(w_j, w_k) \leq t$ **then**
            | $G_s \leftarrow G_s \cup Path$;
        end
    end
end

Algorithm 1. Dependency sub-graph approximate alignment methodology.

First, we obtain the nodes intersection $V_{intersection}$ between both q and d dependency graphs. We then find the shortest path between every pair (w_j, w_k) of vertices belongs to the $V_{intersection}$ set in the candidate text dependency graph using Dijkstra's algorithm [28]. Each edge was given a weight of 1 and edges directions are ignored during the process of the algorithm. Due to linguistic variation, we may not find a sub-graph that match the exact query text graph, however, we might find a sub-graph that match the query text graph approximately. We define a threshold parameter t to allow node gaps and mismatch in the case where some nodes in the query text cannot be mapped to any nodes in the candidate text graph. If the shortest path size (i.e number of edges between w_j and w_k) is less than or equal t, the path will be added to the sub-graph G_s. By setting t to a value greater than 1, it is much more likely to capture syntactic variations. Figure 1 shows examples of sub-graph matching from the dataset of [6].

The resulting G_s is used to measure two graph coverage features as follows:

$$\frac{|E_s|}{|E_d|} \qquad (4)$$

$$\frac{|E_s|}{|E_q|} \qquad (5)$$

Since a much more relevant candidate text is much more likely to have a larger overlap with the query text than other less relevant candidates, this may well tend to improving relevant documents similarity assessment.

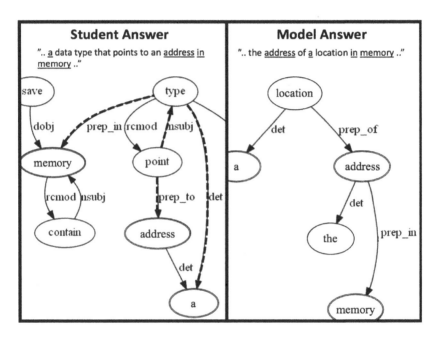

Fig. 1. Approximate sub-graph matching illustration. Example is taken from [6]. Given a model and a student candidate answer, double lined nodes represents the shared words between both answers and connections between words represents dependency relations. Direction and dependency types are ignored. Algorithm 1 uses the dependency structure similarity of local neighborhoods within Shortest Path (SP $\leq$ **t**), to find an approximate sub-graph that match the model answer. The selected subgraph is highlighted by bold dotted lines. In this example **t = 3**.

4   Applications and Analysis

We evaluate our method for two different tasks: re-ranking for article summary matching, and short answer grading.

4.1   Re-Ranking for Article-Summary Matching

Similar to [15], we utilize the ranking output of an IR system to do re-ranking. We choose to select the top n relevant documents ranked by Lucene[1] and incorporate our features to improve documents ranking precision.

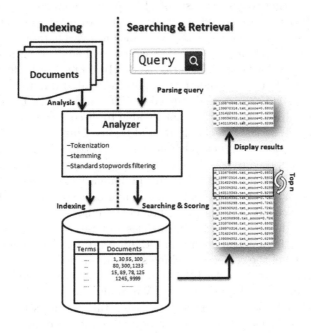

Fig. 2. Re-ranking top-n results of a retrieval system.

Lucene Ranking. Our re-ranking method is build on the top of Lucene. Lucene offers an open source information retrieval library, which provides IR-related tasks like indexing, querying, language analysis, results scoring and retrieval. Figure 2 shows how our re-ranking setup with Lucene works. Depending on the chosen language analyzer, the documents are internally tokenized, stemmed and filtered for stopwords. The analyzer preprocesses and extracts the terms on which the searching can be done. The terms are then indexed into a format that facilitates rapid searching. When a query is issued, it gets analyzed and relevant results are selected from the collection by matching the query against the index. Finally, relevant documents are scored according to the following equation:

$$Score(d, q) = \sum_{w \in q} tf(w, d) \times idf(w) \times norm_{length}(d) \tag{6}$$

where $tf(w, d)$ is the word w frequency in document d, $idf(w)$ is the inverse number of documents in which word w appears, and $norm_{length}(d)$ is the length normalization factor for document d.

---

[1]   https://lucene.apache.org/.

Dataset. We used a set of 37,164 German news articles collected from Spiegel Online over 2015[2]. It includes German online news articles from different genres like sports, politic, economics, health, entertainment, etc. We remove images/video only articles, filter out irrelevant information like: source agency, date and translator name, clean HTML tags and extract only articles with summaries. The resulting corpus consists of 1130 (article, summary) pairs. Each of these articles has one corresponding summary that was created manually by the article author. Summaries are abstractive, which involves paraphrasing the facts from the original article using new novel sentences. Length ranges are [45–950] and [1185–9560] characters, and [7–107] and [216–1390] words, for summaries and articles respectively. The news are highly correlated due to the short one-year interval. Similar events are discussed in different contexts, therefore simple features like word frequency would not be able to discern the correct summary from summaries of articles on closely related topics. Thus, this dataset is suited for testing the capability of methods that assess a certain semantic understanding of texts – as opposed to e.g. the DUC datasets[3], where we found string-based matching to yield almost perfect scores in a preliminary experiment. The remaining articles, which have no corresponding summaries, were used as a background corpus for Tf-Idf calculation and topic model training.

To create our dataset, we index the articles and use the summaries as a query to retrieve the articles. Then, we re-rank the n top-ranked documents returned by Lucene. We label the correct matching summary-article pairs as "1" and "0" otherwise. Overall, we have 5650 examples and 11300 examples, in the cases where $n = 5$ and $n = 10$ respectively. We apply the same process with the summaries indexed and the articles as queries. Table 1 shows Lucene retrieval results for both settings. Precision at n reports the fraction of documents ranked in the top n results that are labeled as relevant. Note that not all summary-article pairs could be correctly retrieved in all cases and the retrieval performance is not equally effective for retrieving texts and summaries, since summaries are much shorter and thus contain less distinctive words.

Experimental Setup. For computing features, we use the implementation provided by [29] for topic modeling, and [30] for named entities extraction. Dependency graphs for both queries and documents are generated using the German collapsed parser by [31], and filtered using Tf-Idf thresholds in three levels. By manual inspection, α, β and λ are set to 10, 5 and 2 respectively. For lexical expansions features, we obtain the top 10 DT expansions using the JoBimText API[4].

Once the similarity and coverage scores have been computed for each summary-article pairs, we use a cost-sensitive Multilayer Perceptron (MLP) neural network to handle the imbalance between positive and negative examples. False positives are assigned a larger cost than false negatives, so the classifier

---

[2] http://www.spiegel.de/.

[3] http://duc.nist.gov/.

[4] www.jobimtext.org/jobimviz-web-demo/api-and-demo-documentation/.

Table 1. Lucene retrieval performance. Retrieval is evaluated by P@n. The number in parenthesis shows the overall P@n for the entire dataset of 1130 article-summary pairs.

P@n	Sum/Text[a]	Text/Sum[b]
P@1	1029 (0.9106)	887 (0.7849)
P@3	1111 (0.9831)	1021(0.9035)
P@5	1122 (0.9929)	1052 (0.9309)
P@10	1128 (0.9982)	1090 (0.9646)
P@20	1128 (0.9982)	1114 (0.9858)
P@50	1129 (0.9991)	1123 (0.9938)
P@100	1129 (0.9991)	1129 (0.9991)

[a] summaries on index, articles for retrieval.
[b] articles on index, text for retrieval.

would not be biased toward negative instances. The cost of false negatives is fixed to 1. We explore different costs to find the best cost using the validation set for false negatives class. We have found that a cost of $n - 1$ performs the best across all the training/validation rounds.

We run different experiments with different MLP structures and learning parameters. For evaluation, we use 5-fold cross-validation. We choose the model which provides the best overall accuracy with a balanced classification error rate and stable performance scores between the two classes, which was determined on a smaller version of the dataset in preliminary experiments. The network structure includes 3 hidden layers, with $(f + c)/2$ neurons in each layer, where f is the number of input features and c is the number of classes (i.e $c = 2$). Training time is set to 1000 epochs.

Table 2. Binary relevancy classification results using MLP (n = 5). Results shows binary relevancy classification precision (P@1), recall, F-measure and true positives.

Features	Sum/Text				Text/Sum			
	Precision	Recall	F-Measure	TP	Precision	Recall	F-Measure	TP
All	**0.916**	**0.894**	**0.899**	**1026**	**0.915**	**0.889**	**0.896**	**960**
BOW	0.890	0.853	0.864	1013	0.893	0.850	0.862	929
Dependency	0.887	0.838	0.850	1003	0.890	0.843	0.856	921
Coverage	0.893	0.853	0.864	919	0.850	0.861	0.862	917
Cov+Dep	0.904	0.874	0.882	1011	0.904	0.870	0.880	942
All/(Cov+Dep)	0.902	0.865	0.874	1023	0.901	0.860	0.870	948

Table 3. Binary relevancy classification results using MLP (n = 10). Results shows binary relevancy classification precision (P@1), recall, F-measure and true positives.

Features	Sum/Text				Text/Sum			
	Precision	Recall	F-Measure	TP	Precision	Recall	F-Measure	TP
All	**0.951**	**0.929**	**0.936**	**1038**	0.949	0.923	0.931	994
BOW	0.938	0.894	0.908	1024	0.936	0.889	0.904	971
Dependency	0.933	0.879	0.895	1014	0.933	0.883	0.899	954
Coverage	0.930	0.881	0.896	972	0.935	0.888	0.903	962
Cov+Dep	0.945	0.918	0.926	1020	0.941	0.899	0.912	989
All/(Cov+Dep)	0.943	0.906	0.917	1035	0.942	0.903	0.915	989

To re-rank, MLP[5] is configured to return a probability distribution of each class label. We re-rank the relevancy according to the descending ordering of the probability distribution of the positive class.

Table 4. Re-ranking results using MLP probability distribution from different relevancy classification models when n = 5 and n = 10. The table shows the improvement in P@1 after the re-ranking. Numbers in brackets are the results of dividing by the number of cases, (1122, 1052), (1128, 1090), where the correct document is in the top 5 or top 10 Lucene results respectively, which form an upper bound.

Features	n = 5		n = 10	
	Sum/Text	Text/Sum	Sum/Text	Text/Sum
All	1080 (0.962)	996 (0.946)	1077 (0.954)	1018 (0.933)
BOW	1064 (0.948)	987 (0.938)	1048 (0.929)	994 (0.911)
Dependency	1050 (0.935)	980 (0.931)	1037 (0.919)	976 (0.895)
Coverage	1038 (0.925)	962 (0.914)	1031 (0.914)	973 (0.892)
Cov+Dep	1079 (0.961)	991 (0.942)	1072 (0.950)	1010 (0.926)
All/(Cov+Dep)	1064 (0.948)	986 (0.937)	1048 (0.929)	993 (0.911)

Results and Discussion. The best report relevancy classification results are reported in Tables 2 and 3. Since we are only aware of the correct article-summary pairs, we use P@1 only as a measure of performance. We also reported the recall, F-measure and true positives. We test the performance using different sets of features. From the results we observe the following: First, using all the features achieves the best performance over all measures in all cases. Second, using a combination of coverage and dependency features lead to the second best performance and play a role in providing comparable performance to that obtained using all the features with an unnoticeable drop in true positives. Third,

---

[5] Learning rate = 0.5, momentum = 0.2.

F-measures falls with average of 0.0215 (in four cases) when excluding these two features, and using each feature in isolation does not lead to any improvement, but achieves comparable results to the ones using BOW features. Forth, in most cases, we outperform Lucene P@1 in terms of true positives.

The improvement is most clearly seen when we use our neural networks models for re-ranking, see Table 4.

In manual error analysis, we generally observe limitations on very short summaries that have no intersection with the text – most extremely noticed for an article on fashion history (866 words) with the (translated) summary "The suit is the uniform of gentlemen". Other errors could be addressed by a German compound splitter, as there are frequently compounds where only the parts match, such as "Ratenkreditangebote" and "Ratenkredite", other examples include derivational matches like "vorweihnachtliche" and "Vorweihnachtszeit", which could be addressed by an improved morphology component that also includes compound analysis.

4.2 Automatic Short Answers Grading

We provide a second evaluation for our method on automatic short answers grading.

Dataset. We use the dataset by [6][6]. The dataset consists of 81 computer science questions on data structures course and 2273 student answers. The dataset was graded by two judges and normalized on a scale of 0.5 according to the extent to which the student answers are considered similar to the content of the correct answers. The reported inter-annotator agreement (IAA) between both judges is 0.586% (Pearson's ρ) and 0.659 Root Mean Square Error (RMSE).

Experimental Setup. Dependency graphs for both questions and answers are based on collapsed dependencies from the Stanford Parser[7]. By manual inspection, we set α, β and λ to 4, 2 and 1 respectively. As average text length in this case is shorter, we choose smaller values.

We used New York Times articles within the years 1998–2000 as a background corpus for Tf-Idf calculation, and our topic model was trained using set of 36 million sentences from the recent English Wikipedia dump.

We train a MLP with one hidden layer using default parameters[8] for 5000 training epochs to increase stability. Following [6], we apply a 12-fold cross validation over the entire dataset for evaluation.

Results and Discussion. Table 5 shows our results in comparison to previous approaches.

---

[6] http://web.eecs.umich.edu/~mihalcea/downloads.html#saga.

[7] http://nlp.stanford.edu/software/lex-parser.shtml.

[8] Learning rate = 0.3, momentum = 0.2.

Table 5. Comparing performance of different models trained on [6]. Comparison is based on Pearson's ρ correlation and Root Mean Square Error (RMSE). * results on a smaller test dataset, not directly comparable.

Features	ρ	RMSE
Tf-Idf	0.327	1.022
Lesk	0.462	1.050
Mohler et al. [6]	0.518	0.978
Inter-annotator Agreement (IAA)	0.586	0.659
Sultan et al. [21]	0.592	0.887
Sultan et al. [21] w/ Question Demoting	0.571	0.903
Ramachandran et al. [20]*	0.610	0.860
Our Method	**0.590**	**0.847**

Our approach exhibits superior performance over existing models when evaluating on RMSE except for IAA, and we perform quite well in comparison to IAA and [21] in terms of Pearson's correlation. Although [20] report better results; however, their evaluation is based on much smaller test data (453 examples) and they use in-domain model training. As can be seen as well, our coverage and alignment features proven to has a great effect on improving the performance than when only considering BOW or Tf-Idf features in isolation.

Further manual error analysis shows that a substantial portion of the errors are due to unstructured answers and misspelling. Again, a more lenient matching mechanism, e.g. using edit distance or automatic spelling correction, might alleviate these errors.

5   Conclusion

In this paper, we introduced a supervised approach for learning to rank and score a set of candidate texts that have varying degrees of similarity to a given query text. We showed that incorporating additional structural and content similarity features, coverage measures and lexical similarity from distributional thesaurus can produce better results than if each were used individually.

To enable interpretable similarity, we also developed an approximate dependency subgraph alignment algorithm. The idea is to find a subgraph in the candidate text dependency graph that is similar to a given query text dependency graph, allowing for syntactic variations.

To ensure that our method is generalizable over different languages and various text lengths, we evaluate our method using two different tasks, namely re-ranking for improving document retrieval, and short answer grading. Results indicate that our approach provide better or comparable performance to baseline and recent approaches.

In the future, we would like to improve the quality of our alignment algorithm by incorporating semantic similarity, which will help capturing synonyms and paraphrases. Further improvement would be to explore more lenient matching mechanisms to capture morphological variants and misspellings.

References

1. Amiri, H., Resnik, P., Boyd-Graber, J., Daumé III, H.: Learning text pair similarity with context-sensitive autoencoders. In: ACL, Berlin, Germany, pp. 1882–1892 (2016)
2. Chen, D., Bolton, J., Manning, C.D.: A thorough examination of the CNN/daily mail reading comprehension task. In: ACL, Berlin, Germany, pp. 2358–2367 (2016)
3. Hermann, K.M., et al.: Teaching machines to read and comprehend. In: Cortes, C., Lawrence, N.D., Lee, D.D., Sugiyama, M., Garnett, R. (eds.) Advances in Neural Information Processing Systems, vol. 28, Montreal, Canada, pp. 1693–1701 (2015)
4. Weston, J., Bordes, A., Chopra, S., Mikolov, T.: Towards AI-complete question answering: A set of prerequisite toy tasks. CoRR (2015)
5. Mohler, M., Mihalcea, R.: Text-to-text semantic similarity for automatic short answer grading. In: EACL 2009, Athens, Greece, pp. 567–575 (2009)
6. Mohler, M., Bunescu, R., Mihalcea, R.: Learning to grade short answer questions using semantic similarity measures and dependency graph alignments. In: ACL-HLT. HLT 2011, Portland, Oregon, USA, pp. 752–762 (2011)
7. Agirre, E., Diab, M., Cer, D., Gonzalez-Agirre, A.: Semeval-2012 task 6: A pilot on semantic textual similarity. In: SemEval, Montreal, Canada, pp. 385–393 (2012)
8. Agirre, E., et al.: Semeval-2016 task 1: Semantic textual similarity, monolingual and cross-lingual evaluation. In: SemEval, San Diego, California, pp. 497–511 (2016)
9. Agirre, E., Gonzalez-Agirre, A., Lopez-Gazpio, I., Maritxalar, M., Rigau, G., Uria, L.: Semeval-2016 task 2: Interpretable semantic textual similarity, California, San Diego, pp. 512–524 (2016)
10. Rychalska, B., Pakulska, K., Chodorowska, K., Walczak, W., Andruszkiewicz, P.: Samsung poland nlp team at semeval-2016 task 1: necessity for diversity; combining recursive autoencoders, wordnet and ensemble methods to measure semantic similarity. In: SemEval, San Diego, California, pp. 602–608 (2016)
11. Brychcín, T., Svoboda, L.: Uwb at semeval-2016 task 1: semantic textual similarity using lexical, syntactic, and semantic information. In: SemEval, San Diego, California, pp. 588–594 (2016)
12. Afzal, N., Wang, Y., Liu, H.: Mayonlp at semeval-2016 task 1: Semantic textual similarity based on lexical semantic net and deep learning semantic model. In: SemEval, San Diego, California, pp. 674–679 (2016)
13. Mueller, J., Thyagarajan, A.: Siamese recurrent architectures for learning sentence similarity. In: Thirtieth AAAI Conference on Artificial Intelligence, Phoenix, Arizona, USA (2016)
14. Jansen, B.J., Spink, A.H.: Investigating customer click through behaviour with integrated sponsored and nonsponsored results. Int. J. Internet Market. Adv. **5**, 74–94 (2009)
15. Hagen, M., Völske, M., Göring, S., Stein, B.: Axiomatic result re-ranking. In: CIKM 2016, Indianapolis, Indiana (2016)
16. Yang, S., Lu, W., Yang, D., Yao, L., Wei, B.: Short text understanding by leveraging knowledge into topic model. In: NAACL: HLT, pp. 1232–1237. Association for Computational Linguistics, Denver (2015)

17. Severyn, A., Moschitti, A.: Learning to rank short text pairs with convolutional deep neural networks. In: Proceedings of the 38th International ACM SIGIR Conference on Research and Development in Information Retrieval, SIGIR 2015, Santiago, Chile, pp. 373–382 (2015)
18. Gu, Y., Yang, Z., Zhou, J., Qu, W., Wei, J., Shi, X.: A fast approach for semantic similar short texts retrieval. In: ACL, Berlin, Germany, pp. 89–94 (2016)
19. Pilehvar, M.T., Navigli, R.: From senses to texts: an all-in-one graph-based approach for measuring semantic similarity. Artif. Intell. **228**, 95–128 (2015)
20. Ramachandran, L., Cheng, J., Foltz, P.: Identifying patterns for short answer scoring using graph-based lexico-semantic text matching. In: Proceedings of the Tenth Workshop on Innovative Use of NLP for Building Educational Applications, Denver, Colorado, USA, pp. 97–106 (2015)
21. Sultan, M.A., Salazar, C., Sumner, T.: Fast and easy short answer grading with high accuracy. In: NAACL: HLT, San Diego, California, pp. 1070–1075 (2016)
22. Blei, D.M., Ng, A.Y., Jordan, M.I.: Latent dirichlet allocation. J. Mach. Learn. Res. **3**, 993–1022 (2003)
23. Kohail, S.: Unsupervised topic-specific domain dependency graphs for aspect identification in sentiment analysis. In: Student Research Workshop Associated with RANLP 2015, Hissar, Bulgaria, pp. 16–23 (2015)
24. Biemann, C., Riedl, M.: Text: Now in 2D! a framework for lexical expansion with contextual similarity. J. Lang. Model. **1**, 55–95 (2013)
25. Albalate, A., Minker, W.: Semi-Supervised and Unervised Machine Learning: Novel Strategies. Wiley (2013)
26. Buckley, C., Singhal, A., Mitra, M., Salton, G.: New retrieval approaches using smart: Trec 4. In: TREC, Gaithersburg, Maryland, pp. 25–48 (1995)
27. Singhal, A., Salton, G., Buckley, C.: Length normalization in degraded text collections. In: Proceedings of Fifth Annual Symposium on Document Analysis and Information Retrieval, Las Vegas, Nevada, USA, pp. 15–17 (1996)
28. Dijkstra, E.W.: A note on two problems in connexion with graphs. Numerische Mathematik **1**, 269–271 (1959)
29. Phan, X.H., Nguyen, C.T.: GibbsLDA++: A C/C++ implementation of latent Dirichlet allocation (LDA) (2007). http://gibbslda.sourceforge.net
30. Benikova, D., Yimam, S.M., Santhanam, P., Biemann, C.: GermaNER: Free Open German Named Entity Recognition Tool. In: GSCL, Duisburg-Essen, Germany, pp. 31–38 (2015)
31. Ruppert, E., Klesy, J., Riedl, M., Biemann, C.: Rule-based Dependency Parse Collapsing and Propagation for German and English. In: GSCL, Duisburg-Essen, Germany, pp. 58–66 (2015)

Text Similarity Function Based on Word Embeddings for Short Text Analysis

Adrián Jiménez Pascual[1(✉)] and Sumio Fujita[2]

[1] The University of Tokyo, 3-8-1, Komaba, Meguro, Tokyo, Japan
adri@ms.u-tokyo.ac.jp
[2] Yahoo Japan Corporation, 1-3, Kioicho, Chiyoda-ku, Tokyo, Japan
sufujita@yahoo-corp.jp

Abstract. We present the Contextual Specificity Similarity (CSS) measure, a new document similarity measure based on word embeddings and inverse document frequency. The idea behind the CSS measure is to score higher the documents that include words with close embeddings and frequency of usage. This paper provides a comparison with several methods of text classification, which will evince the accuracy and utility of CSS in k-nearest neighbour classification tasks for short texts.

We experimentally confirmed that CSS performed excellent in the short text classification task as have been intended, outperforming traditional methods as well as WMD, the most recently proposed method.

1 Introduction

One of the most broadly used representations of text documents are bags of words (BOW) weighted by term frequency-inverse document frequency (TF-IDF). Nevertheless, some undesired results can arise specially when using these traditional representations to analyze short texts, as in the following example:

$$d_1 = The\,man\,walked\,into\,the\,bar\ ,$$
$$d_2 = He\,entered\,a\,pub\ . \tag{1}$$

These two sentences express the same action with almost synonymous words, yet when considering their BOW representation ρ under the basis [*the, man, walked, into, bar, he, entered, a, pub*], they become transversal (i.e., unrelated):

$$\rho(d_1) = [2, 1, 1, 1, 1, 0, 0, 0, 0]\ ,$$
$$\rho(d_2) = [0, 0, 0, 0, 0, 1, 1, 1, 1]\ . \tag{2}$$

This is an extreme case of the real problem which is the almost-perpendicularity of closely related short texts that use different terminology.

This work was carried out while the first author was in a research internship at Yahoo! JAPAN Research.

Many methods have been developed in order to tackle this problem [1,2]. Kusner et al. proposed an interesting distance in this direction called Word Mover's Distance (WMD) as analogy of the Earth Mover's Distance [3], translating an area transfer problem into a word transfer problem. The approach of WMD towards short texts distances served as inspiration for our work, which looks for a word transfer in a simpler and broader meaning than WMD, yet with better results in classification tasks – as will be seen in Sect. 4.

In this paper we provide a new document similarity measure based on the remarkable word embedding model by Mikolov et al. (2013) *word2vec* [4], which was proven by the authors to construct embedded word vectors that preserve semantic relationships when operated. For example, we could consider the following operation: $v(king)$ - $v(man)$ + $v(woman)$, which will result into a vector closest to $v(queen)$. We will represent text documents as arrays of their word vectors, and then make use of this property together with the document's words IDF to define our closeness measure Contextual Specificity Similarity (hereon *CSS*) between documents.

The rest of the paper is organized as follows: Sect. 2 overviews the previous studies of related domains and Sect. 3 explains our proposed similarity measure. In Sect. 4, we presented our evaluation experiments and analyzed the results. Finally, Sect. 5 concludes the paper.

2   Related Work

Computing textual similarity is of great interest not only for natural language processing society but for many related areas including document retrieval [5,6], text classification [7,8], news categorization and clustering [1,9], song identification [10], sentiment analysis [11], and multilingual document matching [12].

2.1   Text Similarity Measures in Information Retrieval

In information retrieval, the task consists of identifying relevant text documents of various length given the description of search requests typically in very short textual query such as TREC topic descriptions [13]. Given the representations of bags of words of both the query and documents, the vector space model computes the similarity between two vectors, each element of which is weighted by TF-IDF, local and global corpus statistics based on term frequencies [5]. More sophisticated text similarity measures based on bags of words include OKAPI BM25 TF [6], which approximates 2-poisson model term weighting and several language modeling approaches [14].

2.2   Context Vectors and Dimensional Reduction Approaches

The history of the discovery of word classes based on contextual information is as old as we may go back to the work of structural linguists in the middle of the 20th century [15]. The origins of several distributional word representations seem to be an inversion of bag of words representation of documents, where a word is represented by the centroid of vectors representing the textual contexts of the appearances [16]. On the other hands, statistical dimensional reduction approaches of document representations, initiated by *Latent semantic indexing* (LSA)[17], try to represent documents by a fewer dimension than the vocabulary size in order to solve word miss matching issues in several text matching applications. Recently, the most successful example is *Latent Dirichlet Allocation* (LDA) by Blei et al. [18], which learns *topic models* consisting of contextually related words by completely unsupervised manner.

2.3 Word Embedding

Finally, we adopted *word2vec*, continuous vector representations of words from very large corpora where the words occur, using a continuous skip-gram model. The neural network learning process is enabled by adopting the negative sampling method which approximately maximizes a softmax objective function of the probability of observing context words given the target word [4,19].

Kusner et al. proposed a document distance measure, WMD on the basis of word embedding representation of words and short texts [3], translating an area transfer problem into a word transfer problem.

3 Contextual Specificity Similarity Measure

Our purpose is to create a method that reckons meaning-related texts that use different terms (i.e., they are unrelated through BOW, as in (1)).

3.1 Background

As previously stated, Kusner et al.'s WMD idea was taken as a starting point, from which our development subsequently diverged.

Formally, WMD's original definition lies on an interpretation of the Earth Mover's Distance, transforming this earth moving problem into a word moving problem. On its basis, they assume that a sentence (area) can be thought of as a certain disposition of words (earth), and finding the similarity of two sentences would equal to minimizing the amount of work one has to do to transport all words in one sentence into the words of the second one.

Technically, we assume we have a word embedding matrix $X \in \mathbb{R}^{d \times n}$, where d is the dimensionality of the word vectors, and n the number of words in the vocabulary corpus. Each element $x_i \in \mathbb{R}^d$ is a vector that represents the ith word in the d-dimensional space. Let $c(i,j) = ||x_i - x_j||_2$ be the "cost" associated to travelling from the ith word to the jth word, and let $T \in \mathbb{R}^{n \times n}$ be a flow matrix whose ij-term represents the "amount of the ith word that travels to the jth word". Then, the problem of calculating the distance between two documents d_1 and d_2 is summed up in the formula:

$$\text{WMD}(d_1, d_2) = \min_{T \geq 0} \sum_{i,j=1}^{n} T_{ij} c(i,j) \,, \tag{3}$$

where certain constrains regarding the matrix T apply with respect to the words from the documents used.

Nevertheless, the WMD distance features a special assignment if the sentences to be compared have different number of words, in which case words are "weight-wise split" and one certain word may be divided into several portions, each of them being transformed into a different term in the comparing sentence. As the authors state, if we consider the sentences "*The President greets the press in Chicago*" and "*Obama speaks in Illinois*", we will get split associations such as "*Obama*" being transported to "*President*" and "*greets*", or "*speaks*" being moved to "*President*" and "*press*". This ambiguity in the assignment of target is what we tried to avoid in our method, which only assigns one word per word.

3.2   Definition

The idea behind our method also lies on a word-weight transfer. However, instead of performing a word-weight transfer from one word to all its "close" words, we only look for one closest word in terms of word embeddings and IDF. We achieve this by creating a word similarity matrix—whose entries will be defined as the product of the average IDF of the facing words and the *cosine similarity* of their vectors—and looking for the maximal values in it.

The reason for taking a matrix with such values as reference is because we consider that for two words to be similar they should have similar word embeddings (i.e., contextual similarity) and we would like higher IDF terms to contribute more in the weighting of the measure, since usually less frequency of occurrence is related to higher specificity of the terms, and this is specially valuable in short texts analysis. For this, we call our method *Contextual Specificity Similarity (CSS)*.

Thereby, despite *President, Prime Minister* and *Churchill* having almost equal word embeddings due to their appearance in similar contexts, the entry in the matrix corresponding to *Prime Minister* versus *Churchill* will show a bigger value thanks to their greater IDF, and therefore they would cast a bigger correspondence value than the pairs *President-Prime Minister* or *President-Churchill*. By doing this, we emphasize the focus of our similarity search on higher IDF words, as mentioned above.

Unlike most other methods, our method assigns higher values to closer words, potentially reaching a maximum when a word is compared to itself (proper distances would become 0 in this case). Therefore, instead of a minimizing function we require a maximizing one.

Similarly to WMD's technical definition, in our construction we assume we are given a word embedding matrix $V \in \mathbb{R}^{d \times n}$ (coming from *word2vec*), where n is the number of vectors (i.e., the number of unique words in our corpus) and d is their embedding dimension. As previously, we consider the vectors to be read in columns. Explicitly, the ith column of V, $v_i \in \mathbb{R}^d$, represents the d-dimensional embedding of the ith word in the corpus. Consider now $\sigma : D \to \mathbb{R}^d$ to be a function from a document D to the vector space $\mathbb{R}^d$ that assigns to every word $w \in D$ its correspondent vector in V, $v = \sigma(w) \in \mathbb{R}^d$.

Words Similarity Matrix. Let w_i and w_j be two words whose word embedding vectors are $v_i = \sigma(w_i)$ and $v_j = \sigma(w_j)$ respectively. We first define the following matrix:

$$M(w_i, w_j) = \frac{v_i \cdot v_j}{||v_i|| ||v_j||} \cdot \frac{IDF(w_i) + IDF(w_j)}{2} , \qquad (4)$$

where IDF is the *inverse document frequency* of the words through the all documents. Expressed in words, each entry of the matrix M is the *cosine similarity*[1] of the vectors associated to the words, weighted by the average of their IDF. This means that the closer two words are in the embedding, and the less frequent they are in appearances, the higher the assigned value will be. Please, observe that the diagonal of the matrix represents the IDF of all words: $M(w_i, w_i) = IDF(w_i)$.

Document Similarity Measure. Having created this word-similarity matrix M, we now define the similarity between documents as:

$$CSS(d_1, d_2) = \sum_{w_1 \in d_1} \max_{w_2 \in d_2} M(w_1, w_2) . \qquad (5)$$

By defining CSS in this manner, for every word w_1 in d_1, we look for the word in d_2 with the highest similarity to it. We do this for every word in d_1, therefore at the end we are adding the values of all most-similar words to the words of d_1 in d_2. This contrasts with the definition of WMD, with which one word can be transformed ("moved") into several words, while our method converts one word into the most similar it finds under these requisites.

It is important to note that, actually, the CSS measure is <u>not</u> a formal *distance*, since neither the properties $d(a, a) = 0$ or $d(a, b) = d(b, a)$ are satisfied in general. Nevertheless, this closer-contextuality-higher-value measure will be proven to be an effective measure for the problem that matters.

Example 1. Let us illustrate how CSS works with a basic example where traditional BOW based methods would fail to grasp texts similarities:

$$\begin{aligned} d_1 &= Child\,of\,mine , \\ d_2 &= Mother\,of\,his , \end{aligned} \qquad (6)$$

$$M = \begin{bmatrix} 2.2126 & 0.1777 & 1.4609 & \mathbf{1.9512} & 1.1209 \\ 0.1777 & 0.0333 & 0.2363 & \mathbf{0.2419} & -0.0629 \\ 1.4609 & \mathit{0.2363} & 3.0019 & 1.0636 & \mathbf{1.1829} \\ \mathit{1.9512} & 0.2419 & 1.0636 & 3.8312 & 1.6852 \\ 1.1209 & -0.0629 & \mathit{1.1829} & 1.6852 & 3.7512 \end{bmatrix} . \qquad (7)$$

---

[1] In practice we will implement the cosine similarity as the dot product without normalization, since the word vectors obtained from *word2vec* have a modulus close to 1, and making the whole calculation would increase the complexity to the algorithm while not improving the results.

The matrix M is expressed in the basis $[w_1, w_2, w_3, w_4, w_5] = [mine, of, Child, his, Mother]$. The similarity measure between d_1 and d_2 will therefore be:

$$
\begin{aligned}
CSS(d_1, d_2) &= M(w_1, w_4) + M(w_2, w_4) + M(w_3, w_5) \,, \\
CSS(d_1, d_2) &= 3.3760 \,,
\end{aligned}
\tag{8}
$$

where the maximal terms are **bold** in the matrix. Remember that this similarity measure is not symmetrical. As mentioned before, observe that:

$$
\begin{aligned}
CSS(d_2, d_1) &= M(w_4, w_1) + M(w_2, w_3) + M(w_5, w_3) \,, \\
CSS(d_2, d_1) &= 3.3704 \,,
\end{aligned}
\tag{9}
$$

which is different from $CSS(d_1, d_2)$. The maximal terms of M when calculating the similarity measure from d_2 towards d_1 are *italized* in the matrix.

Extending the example, if we added a third sentence $d_3 = Colors\ of\ signs$ to Eq. 6 (what gives us a bigger M), and calculated its similarity with the previous sentences, we would get:

$$
\begin{aligned}
CSS(d_1, d_2) &= 3.3760 \,, \\
CSS(d_1, d_3) &= 1.0927 \,, \\
CSS(d_2, d_3) &= 0.8199 \,.
\end{aligned}
\tag{10}
$$

These values go along with the idea of similarity with which we defined the measure.

3.3   Derived Document Similarity

In addition to the definition of CSS, we define yet another similarity measure based on it but with a slight change that improves its definition when targeted to long texts.

The basic approach remains the same: the similarity between words (i.e., the matrix M) is as previously defined. However, in this occasion instead of simply adding the similarity value of the word with most similar features for a given word, we will only count with the values of those words whose reciprocal corresponds to itself, and then take their average similarity value. In other words, if we have $w_1 \in d_1$ and $w_2 \in d_2$ such that w_2 is the most similar term to w_1 in d_2, we will count their (weighted) similarity value if and only if w_1 is the respective most similar term to w_2 among all the words in d_1.

We consider this new approach to emphasize the resemblance between sentences, since now only pairs of similar terms will contribute to the summation.

Definition. For the technical details of this similarity measure, consider M to be defined as in Eq. 4. Using the same notation as before, let us first define the following set:

$$
\mathcal{A}(w, d) := \{ w' \in d \,|\, M(w, w') = \max_{w_i \in d} M(w, w_i) \} \,.
\tag{11}
$$

$\mathcal{A}(w, d) \subset d$ is the set of words in the document d closest to the word w. This set would generally consist of one single word, and we shall assume so in the successive part. Consider now that the word w_1 belongs to the document d_1, and let d_2 be another document. Then, we define the following *delta* function:

$$\delta(w_1, d_1, d_2) := \begin{cases} 1 & \text{if } w_1 = \mathcal{A}(\mathcal{A}(w_1, d_2), d_1) \ , \\ 0 & \text{else} \ . \end{cases} \tag{12}$$

This function becomes 1 only when w_1 is the associated word corresponding to the associated word of itself. Equivalently, if $w_2 \in \mathcal{A}(w_1, d_2)$, then $\delta = 1$ only if $w_1 \in \mathcal{A}(w_2, d_1)$.

We now define the set of associated pairs between d_1 and d_2 as:

$$\mathcal{P}(d_1, d_2) := \{(w_1, \mathcal{A}(w_1, d_2)) \in d_1 \times d_2 | \delta(w_1, d_1, d_2) = 1\} \ . \tag{13}$$

Using this, we define our new measure:

$$CSS*(d_1, d_2) = \frac{1}{|\mathcal{P}(d_1, d_2)|} \sum_{(w_1, w_2) \in \mathcal{P}(d_1, d_2)} M(w_1, w_2) \ . \tag{14}$$

This new measure that we will call *CSS**, represents a weighted modification of CSS. The summation of CSS* partialy realizes the summation of CSS, since it only takes into account the summands when they are symmetrical in the sense of δ. The result is then averaged by the amount of terms that were actually summed, what gives us a mean value of the relevant word similarities.

This definition extracts the similarity between documents based on their reciprocal word similarity. The longer the documents are, the less likely it is to find a proper pair, yet the more precise the match when found.

Example 2. In *Example 1* we saw that the closest word to *"Child"* (w_1) was *"Mother"* (w_4), and the closest word to *"Mother"* was *"Child"*. So happened too with *"mine"* (w_3) and *"his"* (w_5). But we find that despite *"Mother"* (w_4) being the closest word to *"of"* (w_2), *"of"* is <u>not</u> the closest word to *"Mother"* (it is *"Child"*, as we already said). Therefore, in this scenario, we will only take into account the first couple of words, which are the "corresponded" ones, and the measure would result into:

$$\begin{aligned} CSS*(d_1, d_2) &= \frac{1}{2}\big(M(w_1, w_4) + M(w_3, w_5)\big) \ , \\ CSS*(d_2, d_1) &= \frac{1}{2}\big(M(w_4, w_1) + M(w_5, w_3)\big) \ . \end{aligned} \tag{15}$$

Please observe that this derived measure is actually symmetrical, since we only add the values if there is reciprocity in similarity terms.

3.4   Other Attempts

These two similarity measures (CSS and CSS*) were chosen after several attempts to design an adequate similarity measure for short text analysis. In particular, our mayor concern was to create a good distance matrix M (Eq. 4), since the definitions of CSS (Eq. 5) and CSS* (Eq. 14) arise quite logically considering what our goal is. Therefore we tried many variations of definition for such M. Specifically, we tested variations on the multiplicand in Eq. 4, since we thought the cosine similarity multiplier ought to remain unchanged to properly reflect a similarity feature between word embeddings.

Among the changes in definition that we performed and whose effect on the final result we compared, we tried taking the minima and maxima of the IDFs respectively instead of the finally chosen average expression:

$$M_1(w_i, w_j) = cos - sim(w_i, w_j) \cdot \min(IDF(w_i), IDF(w_j)) , \qquad (16)$$

$$M_2(w_i, w_j) = cos - sim(w_i, w_j) \cdot \max(IDF(w_i), IDF(w_j)) . \qquad (17)$$

None of these led to overall better results. Neither did considering other quantities such as the geometric mean:

$$M_3(w_i, w_j) = cos - sim(w_i, w_j) \cdot \sqrt{IDF(w_i) \cdot IDF(w_j)} , \qquad (18)$$

or the harmonic mean:

$$M_4(w_i, w_j) = cos - sim(w_i, w_j) \cdot 2 \cdot \frac{IDF(w_i) \cdot IDF(w_j)}{IDF(w_i) + IDF(w_j)} . \qquad (19)$$

We tried several other arrangements and formulae without further improvements. Nonetheless, we found a pattern which tends to improve the results for every modification that we tried: squaring the matrix (element-wise) –remember that for our usage of similarity, the bigger the value the higher the similarity.

$$M'_*(w_i, w_j) = M_*(w_i, w_j)^2 \qquad (20)$$

This would lead to a better performance than the non-squared case in most cases. However, we decided not to stick to this method due to our uncertainty of a plausible explanation for this effect.

4   Evaluations

4.1   Evaluation Environment

The test is run through two sets of $1,000$ and $10,000$ Japanese articles from Mainichi-shimbun documents in NTCIR-3 data [20] respectively classified with a section tag within the newspaper (*culture, sports, politics,* etc.). These articles are in turn split in their titles and bodies. Beside CSS and CSS*, we run the test using some classical retrieval methods (BM25 and TF-IDF) by Terrier

IR platform[2], a random classifier, and, for the sake of comparison, the WMD distance[3].

For all methods, the modus operandi is:

1. For every document (title or body) $d \in D$, calculate its distance to all remaining documents in the corpus $d_i \in D \setminus \{d\}$.
2. Rank d to all d_i's distances in closeness order (for CSS and CSS* closer are larger values).
3. Determine the class to which d should belong by utilizing the k-nearest neighbours (k-NN) method, applied with $k = 1$, $k = 5$ and $k = 15$.
4. Results are presented in terms of the macro-average of the *F-measure* of each section tag, where the F-measure is calculated as usual as the harmonic mean of the evaluation *precission* and *recall*:

$$F_1 = 2 \cdot \frac{precision \cdot recall}{precision + recall} \tag{21}$$

The word embedding used is the one provided by *word2vec*, where the vectors have been trained over more than 63 million words, spread through 220 thousand articles: 3+ million words in titles (15 words each in average), 60+ million words in bodies (275 words each in average).

4.2 Results

The results are presented in the four tables below, which are divided in the analysis of 1,000 documents on the left column and 10,000 documents on the right column, and row-wise the analysis of the body of the articles above and their title below. Maximum values are highlighted in **bold** (Tables 1, 2, 3, and 4).

Table 1. Body – 1,000 documents

Method	$k = 1$	$k = 5$	$k = 15$
BM25	**0.4399**	0.3495	0.2904
TF-IDF	0.4346	**0.3539**	0.2863
WMD	0.3151	0.2397	0.2182
Random	0.0447	0.0578	0.0478
CSS	0.3618	0.2598	0.1946
CSS*	0.4214	0.3382	**0.3026**

Table 2. Body – 10,000 documents

Method	$k = 1$	$k = 5$	$k = 15$
BM25	**0.5135**	**0.4815**	**0.4706**
TF-IDF	0.5107	0.4786	0.4675
WMD	—	—	—
Random	0.0638	0.0568	0.0486
CSS	0.3315	0.2783	0.2569
CSS*	0.3551	0.3035	0.2926

---

[2] http://terrier.org/.

[3] Due to calculations limits (memory error), the WMD distance was only calculated for the set of 1,000 articles.

Table 3. Title – 1,000 documents

Method	$k = 1$	$k = 5$	$k = 15$
BM25	0.4862	0.4591	0.3790
TF-IDF	0.4848	0.4537	0.3650
WMD	0.1322	0.1001	0.0871
Random	0.0558	0.0485	0.0472
CSS	**0.5332**	**0.4713**	**0.4151**
CSS*	0.4432	0.3829	0.3417

Table 4. Title – 10,000 documents

Method	$k = 1$	$k = 5$	$k = 15$
BM25	0.6396	0.6508	0.6127
TF-IDF	0.6433	0.6554	0.6167
WMD	—	—	—
Random	0.0566	0.0541	0.0471
CSS	**0.6523**	**0.6651**	**0.6326**
CSS*	0.4006	0.3947	0.3917

4.3   Discussions

The effectiveness of our method is clearly reflected in the two lower tables, which show how CSS outperforms any other contrasted method in short text classification tasks using the k-nearest neighbours method. Whilst on longer texts, both CSS and CSS* are overwhelmed by more broadly used methods such as BM25 or TF-IDF. Yet, as expected, CSS* shows a better performance in longer texts analysis than CSS, what supports the motivation behind CSS*.

As for WMD, which served as inspiration for developing our first method, it does not show a good performance specially in short texts analysis, what could be due to the lack of a broader background context that could help words find better pairings. Nonetheless, it performs at similar levels to CSS in long text classification tasks, as it can be seen in Table 1. Unlike the results that Kusner et al.'s paper [3] reported, BM25 as well as TF-IDF performed better in longer text as have been proven in the series of past evaluation forums in information retrieval [13]. One reason of such overwhelming performance of traditional approaches is that we used Terrier IR platform implementation for TF-IDF and BM25, which is properly configured at the out of box status. As these methods leverage local as well as global statistics, a carefully configured corpus setting and operational parameter setting are needed to be well performed; failing to do that leads to a very weak baseline performance.

In spite of such strong baselines, we can briefly summarize that CSS performs especially excellent in short text classification as have been intended, outperforming traditional methods such as TF-IDF and BM25 as well as WMD, the most recently proposed method. Although CSS* is fairly good in long text classification, traditional methods such as TF-IDF or BM25 performed much better when properly configured.

5   Conclusions

We proposed two text similarity measures, namely CSS and CSS*, among which CSS is intended to improve the effectiveness in short text matching where word miss matching is a crucial problem. According to our experiments described in Sect. 4, we can conclude that CSS and CSS* are powerful tools for short and long text classification tasks respectively, being CSS the best classifier among the compared methods for short texts. Especially CSS showed excellent performance in short text classification as have been intended, outperforming traditional methods such as TF-IDF and BM25 as well as WMD.

References

1. Greene, D., Cunningham, P.: Practical solutions to the problem of diagonal dominance in kernel document clustering. In: Machine Learning, Proceedings of the Twenty-Third International Conference (ICML 2006), Pittsburgh, Pennsylvania, USA, 25–29 June 2006, pp. 377–384 (2006)
2. Schölkopf, B., Weston, J., Eskin, E., Leslie, C., Noble, W.S.: A kernel approach for learning from almost orthogonal patterns. In: Elomaa, T., Mannila, H., Toivonen, H. (eds.) ECML 2002. LNCS (LNAI), vol. 2430, pp. 511–528. Springer, Heidelberg (2002). https://doi.org/10.1007/3-540-36755-1_44
3. Kusner, M.J., Sun, Y., Kolkin, N.I., Weinberger, K.Q.: From word embeddings to document distances. In: Proceedings of the 32nd International Conference on Machine Learning, ICML 2015, Lille, France, 6–11 July 2015, pp. 957–966 (2015)
4. Mikolov, T., Chen, K., Corrado, G., Dean, J.: Efficient estimation of word representations in vector space. CoRR abs/1301.3781 (2013)
5. Salton, G., Buckley, C.: Term-weighting approaches in automatic text retrieval. Inf. Process. Manage. **24**, 513–523 (1988)
6. Robertson, S.E., Walker, S.: Some simple effective approximations to the 2-poisson model for probabilistic weighted retrieval. In: Croft, B.W., van Rijsbergen, C.J. (eds.) SIGIR 1994, pp. 232–241. Springer, London (1994). https://doi.org/10.1007/978-1-4471-2099-5_24
7. Lewis, D.D.: An evaluation of phrasal and clustered representations on a text categorization task. In: Proceedings of the 15th Annual International ACM SIGIR Conference on Research and Development in Information Retrieval, SIGIR 1992, pp. 37–50. ACM, New York (1992)
8. Lewis, D.D.: Evaluating and optimizing autonomous text classification systems. In: Proceedings of the 18th Annual International ACM SIGIR Conference on Research and Development in Information Retrieval, SIGIR 1995, pp. 246–254. ACM, New York (1995)
9. Ontrup, J., Ritter, H.J.: Hyperbolic self-organizing maps for semantic navigation. In: Advances in Neural Information Processing Systems 14, Neural Information Processing Systems: Natural and Synthetic, NIPS 2001, 3–8 December 2001, Vancouver, British Columbia, Canada, pp. 1417–1424 (2001)
10. Brochu, E., de Freitas, N.: "Name that song!" A probabilistic approach to querying on music and text. In: Advances in Neural Information Processing Systems 15, Neural Information Processing Systems, NIPS 2002, 9–14 December 2002, Vancouver, British Columbia, Canada, pp. 1505–1512 (2002)
11. Pang, B., Lee, L.: Opinion mining and sentiment analysis. Found. Trends Inf. Retrieval **2**, 1–135 (2007)
12. Quadrianto, N., Smola, A.J., Song, L., Tuytelaars, T.: Kernelized sorting. IEEE Trans. Pattern Anal. Mach. Intell. **32**, 1809–1821 (2010)
13. Harman, D.: Overview of the first TREC conference. In: Proceedings of the 16th Annual International ACM SIGIR Conference on Research and Development in Information Retrieval, SIGIR 1993, pp. 36–47. ACM, New York (1993)

14. Zhai, C., Lafferty, J.: A study of smoothing methods for language models applied to information retrieval. ACM Trans. Inf. Syst. **22**, 179–214 (2004)
15. Harris, Z.: Structual Linguistics. University of Chicago Press, Chicago (1951)
16. Schütze, H.: Automatic word sense discrimination. Comput. Linguist. **24**, 97–123 (1998)
17. Deerwester, S., Dumais, S.T., Furnas, G.W., Landauer, T.K., Harshman, R.: Indexing by latent semantic analysis. J. Am. Soc. Inf. Sci. **41**, 391–407 (1990)
18. Blei, D.M., Ng, A.Y., Jordan, M.I.: Latent dirichlet allocation. J. Mach. Learn. Res. **3**, 993–1022 (2003)
19. Mikolov, T., Sutskever, I., Chen, K., Corrado, G.S., Dean, J.: Distributed representations of words and phrases and their compositionality. In: Advances in Neural Information Processing Systems 26: 27th Annual Conference on Neural Information Processing Systems 2013, Proceedings of a meeting held 5–8 December 2013, Lake Tahoe, Nevada, United States, pp. 3111–3119 (2013)
20. Chen, K., et al.: Overview of CLIR task at the third NTCIR workshop. In: Proceedings of the Third NTCIR Workshop on Research in Information Retrieval, Automatic Text Summarization and Question Answering, NTCIR-3, Tokyo, Japan, 8–10 October 2002 (2002)

Information Extraction

Domain Specific Features Driven Information Extraction from Web Pages of Scientific Conferences

Piotr Andruszkiewicz[✉] and Rafał Hazan

Institute of Computer Science, Warsaw University of Technology, Warsaw, Poland
P.Andruszkiewicz@ii.pw.edu.pl, R.Hazan@stud.elka.pw.edu.pl

Abstract. In this paper we describe information extraction from web pages of scientific conferences. We enrich already known features with our new features specific for this domain and show their importance in the process of extracting information. Moreover, we investigate various data representation models, e.g., based on single tokens or sequences, in order to find the best configuration for the task in question and set up a new baseline over publicly available corpus.

1 Introduction

Up-to-date information about conferences plays a vital role in scientific life. Therefore methods for automatic collection of data on conferences, e.g., homepages of a conference for the current and previous years, when and where a conference will be held, submission, notification, camera ready dates, etc., are important for scientific community.

In order to gather data about conferences, one may extract interesting information from relevant resources. It is easy to obtain data from structured services like WikiCFP. However, regarding data from this kind of sources, there might be the lack of information or outdated information. A service might not have information about conference we are looking for because it is field specific or covers only small part of all conferences in the field. Calls For Papers (CFPs) have limited range of information, e.g., usually there is no information about sponsors. Moreover, this kind of service provides CFPs that are not updated while changes are made, e.g., submission date extensions. Homepages of conferences provide updated information but in an unstructured way. Due to that fact, the methods of information extraction from unstructured text/web resources need to be employed. To this end, in most cases supervised methods are used. These methods need an annotated data set that will be used for training, optimisation and testing.

Bearing in mind drawbacks of CFPs as a data source, we deal with information extraction from conference web pages. Being more specific, we investigate the already known and new domain specific features for information extraction and check how different models handle extraction of specific entity types. In our

© Springer Nature Switzerland AG 2018
A. Gelbukh (Ed.): CICLing 2017, LNCS 10761, pp. 405–417, 2018.
https://doi.org/10.1007/978-3-319-77113-7_32

experiments we use Support Vector Machine (SVM) and Conditional Random Fields (CRF) and combine them with different data representation models. We verify our statements on publicly available corpus of scientific conferences web pages and make a new reproducible baseline for this corpus.

The remainder of this paper is organised as follows: Sect. 2 presents related work. In Sect. 3 we describe the corpus we use. In Sects. 4 the proposed features are presented. The experimental results are presented in Sect. 5. Finally, Sect. 6 summarises the conclusions of the study and outlines avenues to explore in the future.

2   Related Works

Previous works in the field of information extraction from scientific conferences focused mostly on information extraction from CFPs using different approaches. Extracting information from CFPs has already mentioned drawbacks. In [13] a rule based method was employed to extract date and country from a CFP. A linear CRF was used in [16] in order to extract seven attributes about conferences from CFPs with the use of layout features. However, in this approach only plain text of CFPs was used. We use HTML sourcecode of web pages, including formatting. As in [16] only plain text was used, layout features were based on lines of text, indicating, e.g., first token in line or first line in the text. We take into account, for instance, hyperlinks, blocks, and formating. Thus, our data has much richer layout. In [8] a general platform for performing and assessing information extraction from workshop CFPs was described. In [9] authors focused also on information extraction from CFPs, including those which come via e-mails. They used rule-based methods to extract information about conferences from conference services, like WikiCFP, and combined them in one system in order to facilitate the process of finding conferences that are of interest of a user. In contrast to aforementioned works [18] extracted information about conferences from web pages with Constrained Hierarchical Conditional Random Fields. However, the set of homepages used in experiments has not been published. Hence, we could not apply our approach to this set in order to compare the results. Furthermore, we could not recreate this system due to insufficient details in the paper.

In information extraction from documents of rich structure and plain text, many approaches have been proposed, regardless the domain of data. One of them is a rule-based method employed in [3,6]. A Support Vector Machines (SVM) classifier was also applied to extract information from web pages [1]. A variety of Conditional Random Fields (CRF) methods were widely used [1,17, 18]. Furthermore, Markov Logic Networks (MLNs) were used for information extraction from web pages [1].

In order to verify the necessity of domain specific features and set a new baseline for publicly available corpus we focused on information extraction from conference web pages.

3   The Corpus

The corpus we use is, to the best of our knowledge, the only one publicly available corpus of annotated scientific conferences homepages. It contains 943 annotated homepages of scientific conferences (14794 including subpages). The topics of conferences are equally distributed over five topics; namely, Artificial Intelligence, Natural Language Processing, computer science, telecommunication, and image processing. The following entities are annotated: *name* and *abbreviation* of the conference, *place*, *dates* of the conference, *submission, notification, final version due* dates. We call the last three entities *important dates*. In this paper all mentioned types of entities are considered to be extracted. This corpus is available in public and can be found on the website http://ii.pw.edu.pl/~pandrusz/ data/conferences.

4   Preprocessing and Features

Information extraction from web pages is a special case of information extraction, hence it requires specific techniques and approaches. We start the description of our approach from the preprocessing phase. Then we present group of features we developed. The described techniques and features are verified in models we build in order to find the best configuration for the given information extraction task.

4.1   Preprocessing

In the preprocessing phase we use Snowball stemmer [15] in order to reduce the number of features. Furthermore, we remove stopwords to reduce information noise in the data. We create our own stoplist by dividing words into two groups; namely, *far words* that are farther more than four words from the annotated entity in the data and *close words* that are closer than far words. We consider a word to be a stopword if it does not provide additional information and is in the far words group but not in the close group. The stoplist consists of 21095 words. Moreover, words which occur once or twice in the training set are also considered stopwords. This reduces words that come from wrongly parsed words or named entities that occur very rarely. This way of stoplist preparing reflects the specificity of the domain we are working with. Names of conferences often consist of words such as "the", "and", "on" that are commonly assumed to be stopwords. In this case we cannot remove them because we will not be able to extract a proper name of conferences.

Web pages often contain a lot of unnecessary information, e.g., advertisements, HTML code, menus, copyright notes, thus a specialised library can be used to clean an analysed web page. However, in the case of scientific conference web pages there are not many advertisements and unnecessary information. Hence, we use standard library, Boilerpipe [11], to extract a main article or paragraphs from a web page. We do not remove any other text from the web page to avoid removing important elements by mistake.

4.2   Features

In our approach we distinguish the following group of features: local, offset, layout, and dictionary features. Within these groups we enriched already known features with new features that to the best of our knowledge have not been used for information extraction before.

Local Features. Local features are calculated based on a current word we are analysing. The first and commonly used feature is a *word*. We do not create features for words from stoplist and those that contain nonalphabetic characters. Furthermore, we use part of speech (POS) tags for a current word provided by *Penn Pos Tagger* from *factorie* package [14]. Next feature is *short word* that is assigned with a value *true* if a word contains from 2 to 5 characters. This feature is designed for extraction of acronyms of conferences. 74% of conference's acronyms contain from 2 to 5 characters. *Shape of a word* is the next feature. The feature contains 'a' (for small letters), 'A' (for capital letters), and '1' (for numbers). If there are more than two the same characters in a row, the sequence is reduced to two the same characters. The example values for this feature are: AaaAA (WebET), Aaa (International), 1aa (5th), AA (NAACL), 11 (2016).

Last but not least is a *type of a word* feature. We distinguish eight types of words. *Date* represents whole dates that can be found on a web page. *Short phrase* is assigned to words that are part of sequence of length of one or two words (for more information about sequences please refer to Sect. 4.3, the example is the named entity with two words, for instance, Carl Brunto). *Long phrase* represents words of sequences that consist of at least three words. The reason behind the distinction between short and long phrases is that conference names are usually not short phrases but location of conferences usually are. Other types are: *Number* - assigned for numbers, e.g., 23, 3rd; *acronyms* are words of the following shapes: AA, AaaAaa AaaAA, AA1AA, AAaa, AaAA, AAa, AAaAA; *punctuation marks, special char* - all nonalphanumeric chars that are not punctuation marks, e.g., @, *.

All other words are of the type *standard word*. They represent words that probably do not contain interesting information we want to extract.

Table 1. The distribution of the interesting entities over blocks of a web page.

Entity	Name	Abbrev.	Place	Date	Submission	Notification	Final ver. due	Other
Head title	0.18	0.11	0.03	0.03	0.00	0.00	0.00	0.00
Title/subtitle	0.23	0.09	0.04	0.08	0.03	0.02	0.02	0.01
Paragraph	0.50	0.60	0.68	0.61	0.36	0.29	0.25	0.42
Table/list	0.05	0.14	0.16	0.21	0.44	0.51	0.51	0.14
Other	0.40	0.60	0.09	0.07	0.24	0.18	0.22	0.44

Offset Features. *Predecessor* represents features based on the word that precedes the current word. We take into account only one predecessor and *type of a word* feature. *Successor* is calculated for a word that follows the current word. We consider one word ahead and *type of a word* feature.

Sections with important dates of a conference are often organised with lists or tables. Though it is a convenient way for a human, machine learning algorithms poorly deal with learning patterns that occur on scientific conference web pages, because dates are placed on the right, left and even above and below the description of a date. In order to ease the process of learning, we bring into being *date surrounding words* features that extract the description of a given date in a way presented in Algorithm 1. Words returned by the aforementioned algorithm are used to create features for a current word, however, only for dates in order not to increase the number of features too much.

Algorithm 1. Extracting words surrounding a date

Data: a list item or a table cell // input text with a date
Result: a description of a date // words surrounding a date

if *a date is followed by a semicolon* **then**
| **return** *up to six words after a date*
end
if *a date is preceded by a semicolon* **then**
| **return** *up to six words before a date*
end
if *a date is in a short (less than 100 words) list item or a table cell* **then**
| **return** *up to six words before and up to six words after a date that are*
| *within a list item or a table cell*
end
else
| **return** *up to six words before a date*
end

General conditions that need to be met:
- returned words must come from the same sentence as a date,
- if a returned sequence of words contains a different date then choose a subsequence that starts from the first word and ends at the word before the first date in a sequence.

Layout Features. *Emphasised* feature indicates words that are modified by the following HTML tags: STRONG, B, U, and FONT which means that they are bold, underlined, or use different fonts. The underlined words are more often dates of a conference, however, names of conferences and abbreviations do not correlate with use of aforementioned HTML tags.

Hyperlink feature distinguishes words that are presented as links (A tag). Contrary to the first impression this feature is a good indicator of not being the important information to extract in our case; that is, correlation shows that hyperlinks more often lead to other conferences.

Block feature indicates a block a word belongs to. A separate value is assigned for each block. Considered blocks are head title, title and subtitle, paragraph, table, and list. Table 1 shows the distribution of the entities of our interest over blocks on a web page. Names and abbreviations of conferences, locations, and date occur mostly in paragraphs. Names and abbreviations are placed also in head title and title/subtitle. Dates of submission, notification and so on usually are provided in tables and lists, however, paragraphs also carry that information.

Paragraph number feature indicates the number of a paragraph a word belongs to. We count only the first 6 paragraphs as more than half of interesting entities are contained in these paragraphs according to the corpus. This feature helps in detection of conference names and abbreviations, dates and locations of conferences because as the corpus confirms these entities often occur at the beginning of a web page. The important dates usually occur further in a web page.

Entities we are looking for can be found on one of the subpages of the main conference web page. Thus, we add subpages to the training data, however, we restrict subpages to those that can be accessed through links with the following names: index, home, important dates, call for papers, registration. Furthermore, each word from subpage gets *subpage* feature that contains anchor text, e.g., SUB=home, SUB=index.

Dictionary Features. Detection that word(s) represent a location is helpful for conference location extraction. Hence, we used gazetter from ANNIE module of GATE [10] to add location names from the corpus. Each location found in a text generates a *location* feature, LOC=true. Moreover, each country gets feature COUNTRY=true and city CITY=true.

Out of dictionary feature indicates that a current word has not been found in our custom dictionary of English words that contains 112505 words. This feature is intended to help in abbreviations extraction as the percentage of words not found in the dictionary is the highest for conference abbreviations (0.89). The percentage for location (0.75) is also high, hence it is suggested to a model by this feature (for name it is only 0.23 and 0.14 for other words).

Promising surrounding words feature indicates whether there is at least one word from a given dictionary in a sentence a current word belongs to. We use dictionaries for the following types of entities: name and abbreviation of conference, place and date created based on the most frequent words that occur in sentences that contain an important entity. The dictionaries are not mutually exclusive, hence the *promising surrounding words* feature indicates whether it is an important entity rather than an entity is of a specific type.

4.3   Multi-token Sequences

While describing features for our model, we assume that a single token; that is, a word, a number, or a nonalphanumeric character, is considered a base object used by a model and assigned one of interesting entity types, including *other* that means an object is not of one of the interesting entity types. This leads to a case when a sequence of tokens may have different entity types assigned even if they are one entity of, e.g., conference name type. For instance, a sequence *International Conference on Artificial Intelligence & Applications* may have the following entity types assigned: *International* - conference name, *Conference* - conference name, *on* - other, *Artificial* - conference name, and so on. Therefore, we expand a base object of a model to be a sequence of tokens that groups words forming one instance of entity. While detection of dates is an easy task, finding sequences that represent other named entities is not a trivial one. Hence, we prepared a heuristic algorithm customised for finding token sequences on conference web pages that is based on the following rules: each sequence consists of words that begin with a capital letter; these words may be separated by one word that starts with small letter; sequences are found within a sentence; a sequence cannot be separated by any of the chars for this set: ',-:'. For example, words *International Conference on Advancements in Information Technology* is treated by this algorithm as one sequence.

For sequences with at least two words we need to calculate features in one of the following ways: (1) calculate features for the first word; (2) calculate features for each word separately and use all features; (3) combine features for

Table 2. The importance of features groups for types of entities extraction.

Features	Measure	Name	Abbrev.	Place	Date	Submission	Notification	Final ver. due
All	Precision	0.38	0.76	0.75	0.80	0.66	0.54	0.71
	Recall	0.34	0.75	0.60	0.80	0.54	0.40	0.59
	F1	0.36	0.76	0.67	0.80	0.60	0.46	0.65
No local features	Precision	0.10	0.51	0.72	0.58	0.64	0.47	0.67
	Recall	0.09	0.58	0.60	0.23	0.41	0.28	0.43
	F1	0.09	0.55	0.66	0.33	0.50	0.35	0.52
No offset features	Precision	0.36	0.73	0.69	0.67	-	-	-
	Recall	0.30	0.64	0.57	0.68	0.00	0.00	0.00
	F1	0.33	0.68	0.62	0.67	0.00	0.00	0.00
No layout features	Precision	0.33	0.63	0.68	0.79	0.62	0.53	0.67
	Recall	0.22	0.45	0.45	0.65	0.55	0.44	0.54
	F1	0.26	0.52	0.54	0.71	0.58	0.48	0.60
No dict. features	Precision	0.35	0.77	0.70	0.70	0.61	0.56	0.70
	Recall	0.32	0.72	0.46	0.67	0.52	0.44	0.57
	F1	0.33	0.74	0.55	0.69	0.00	0.00	0.58

all words into one feature. For example, feature *word* is calculated according to the second approach and, e.g., International Conference on Mechanics has the following features W=International, W=Conference, W=on, W=Mechanics. Third approach is used for POS features, e.g., 'Workshop on Applications of Software Agents' has a feature POS=INNNNNS.

5 Experiments

In our experiments we divide the corpus into training and test sets according to the proportion of 70/30. For the SVM model the training set is used to perform cross validation in order to find the best parameters, then the model is trained on the whole training set using these parameters.

For a web page, as an extracted entity we choose the only one instance of entity of a given type that has the highest score among those indicated by an algorithm. Only *location* entity may have two instances because usually a country and a city is provided on a web page as a *location* of a conference.

Table 3. The results of entities extraction with regard to different models (the best F1 results marked in bold).

Features	Measure	Name	Abbrev.	Place	Date	Submission	Notification	Final ver. due
Lin. SVM	Precision	0.14	0.79	0.74	0.72	0.41	-	0.32
	Recall	0.16	0.86	0.59	0.79	0.06	0.00	0.08
	F1	0.15	**0.82**	0.66	0.76	0.11	0.00	0.13
Lin. SVM seq.	Precision	0.38	0.76	0.75	0.80	0.66	0.54	0.71
	Recall	0.34	0.75	0.60	0.80	0.54	0.40	0.59
	F1	0.36	0.76	**0.67**	0.80	0.60	0.46	**0.65**
Lin. CRF	Precision	0.74	0.75	0.66	0.82	0.73	0.25	0.56
	Recall	0.47	0.82	0.53	0.69	0.09	0.01	0.14
	F1	**0.57**	0.78	0.59	0.75	0.17	0.02	0.22
Lin. CRF seq.	Precision	0.61	0.77	0.66	0.82	0.67	0.63	0.70
	Recall	0.40	0.84	0.56	0.82	0.57	0.40	0.50
	F1	0.48	0.80	0.61	**0.82**	**0.61**	**0.49**	0.58

5.1 Importance of Features

In our first group of experiments we verify how important the groups of features customised for information extraction from scientific conferences web pages are. We want to show how domain specific features influence the final results. As the groups of features contain sparse features, a model with only one group of features would obtain very low accuracy and the comparison of models built with only one group of features would not be reliable. Therefore we perform experiments with all groups of features but one. The results of the experiments with SVM (Table 2) show that the most important are *local* features. Lack of them causes the highest drop in accuracy of the results (more than 20 p.p. for *name* and *abbreviation*, almost 50 p.p. for *date* in F1). These features generate almost half of feature functions. This group contains *type of a word* feature and its absence makes extraction task harder for each type of interesting entities. Furthermore, lack of *shape of a word* and *short word* features decreases accuracy for abbreviation extraction. Only *place* noticed slightly drop of accuracy.

Lack of *offset* features reduces mostly the accuracy of conference *date*, about 13 p.p. in terms of F1, and *important dates* are not discovered at all. It is due to lack of *date surrounding words* features that characterise important dates well. This group generates high number of feature functions also.

Layout features help in extraction of *name* and *abbreviation* of a conference. They are also important for *place* and *date* of a conference, however, to lower extend. Within this group of features *block* and *paragraph number* features are the most important ones. These entities often occur in head title. They may be provided also in a title or a subtitle of a web page. If these entities are missing in aforementioned block, it is almost sure that they appear in the first or in a few first paragraphs of a web page. This information is carried over by mentioned features.

As we expected *dictionary* features play the most important role for *place* detection as a *location* feature is a key for this entity type.

To sum up, each group of features carries some information that is important (at least for one of) interesting entity types. Thus, we could say that it is crucial to prepare features that are specific for a given domain. As we have shown, lack of some features may reduce the accuracy for some entity types to zero, for instance, the lack of *offset* features for *important dates*. In the domain of web pages of scientific conferences *local* features identify more general objects, such as dates and named entities that contain desired information. *Offset* features describe surroundings of a word, its context, that is necessary for *important dates* extraction. *Layout* features generate important features functions that inform about a place within a web page a given word is located. They help in case when an entity is not placed in the main text of a web page. *Dictionary* features improve the results mostly by its *location* feature that indicates potential places where a conference is held.

5.2   Models Comparison

Having the influence of features verified, we investigate the applicability of different models with regard to variations of their basic objects used; namely, single tokens and sequences. In this set of experiments we use all mentioned groups of features and preprocessing described in Sect. 4.1.

SVM Model. As a base model we use Support Vector Machine (SVM) [4] with linear and radial basis function (RBF) kernel that is defined as follows:

$K(x,y) = e^{-\xi||x-y||^2}$. We use LibSVM implementation [2]. For multiclass classification we employ one versus the rest approach [5]. For SVM model we start with comparison of single tokens and sequences used as basic objects that the model is working with. The results for linear SVM classifier run on single tokens as basic objects[1] are shown in the first row of Table 3. The accuracy of the model, also linear SVM, that uses sequences as basic objects is presented in the second row in the same table. The single token SVM performs significantly poorer than sequence SVM for *name* of a conference and *important dates*. The reason behind is that the first model assigns a label to each single token independently and mentioned entities consists of several tokens. We try to help SVM with this task by incorporating *offset* features, however, it seems that it is not enough to help single token SVM with extraction of entities that consist of several consecutive words. By providing the SVM already extracted potential sequences we overcome this problem. For sequence SVM we observe also 6 p.p. decrease in F1 for *abbreviation* detection comparing to the single token SVM.

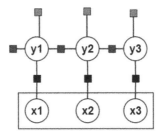

Fig. 1. Linear CRF structure.

We present only the results of linear SVM because the non-linear SVM with RBF kernel function has not obtained significantly better results. Therefore, we stay with linear one due to less complexity and shorter training time. Our model has a high number of features, hence there is no need to increase the dimensionality by applying a kernel function [7].

CRF Model. In the experiments, we also use Conditional Random Fields, CRF [12]. Figure 1 presents the structure of CRF model which is a linear one with three different templates of factors. First template connects factors with an input variable and an output variable. The second represents the relation between consecutive output variables. The third has only one argument that is an output variable. Equation 1 shows the formula of our CRF model, where $Z(\boldsymbol{x})$ is a normalisation factor.

---

[1] It means that the model assigns a label; that is, a type of entity, to a single token.

$$p(\boldsymbol{y}|\boldsymbol{x}) = \frac{1}{Z(\boldsymbol{x})} \exp\left(\sum_{j=1}^{n}\left(\sum_{i=1}^{m_1}\alpha_{1,i}f_{1,i}(y_j)+\right.\right.$$

$$\left.\left.\sum_{i=1}^{m_2}\alpha_{2,i}f_{2,i}(y_{j-1},y_j) + \sum_{i=1}^{m_3}\alpha_{3,i}f_{3,i}(y_j,\boldsymbol{x},j)\right)\right) \tag{1}$$

In our experiments we used CRF that operates on single tokens (Lin. CRF in Table 3) and sequences (Lin. CRF seq. in Table 3). Single tokens CRF significantly outperforms both SVM models in *name* extraction (0.57 versus 0.36 and 0.15 in F1) due to the fact that it models sequences of label (SVM lacks this feature). However, for entities that do not consist of several consecutive words we have not observed the improvement in the results; on contrary, we notice small decrease for *place* and *date*. Surprisingly, single token CRF cannot handle *important dates* extraction like in the case of single token SVM. However, sequence CRF discovers them on a comparable level to sequence SVM. Both models based on sequences handle *important dates* significantly better because the sequence discovery algorithm extracts potential entities, that may have different formats, very well. Moreover, sequences also help CRF in *date* extraction, like for SVM.

In case of *name* sequences discovery, which is not so perfect as for *important dates*, we observe 9 p.p. decrease in extraction of that entity for CRF based on sequences compared to the one based on single tokens. However, sequences slightly increase CRF results for *abbreviation* and *place*.

Summarising, dates are extracted better with models based on sequences than single tokens. For *place* the winner is SVM on both single tokens and sequences (only 1 p.p. difference), however, all other models are not worse than 8 p.p. in terms of F1. The single token models outperforms sequence models for *name* and *abbreviation*. The single token SVM obtains the best results for *abbreviation*, however, the sequence CRF is not far behind (0.82 vs. 0.80 in terms of F1). Furthermore, the results of all models in *abbreviation* extraction are within the difference of 6 p.p., hence, results from all models do not differ much. The reason behind may be that *abbreviation* is a single token entity and sequence models do not leverage their properties in this case. Surprisingly, *name* entity is handled the best with the single token CRF, despite having more than one token. This is probably due to lower accuracy of the algorithm that discovers conference name token sequences for the sequence models compared to date discovery (*dates* are extracted the best with sequence models).

Concluding the analysis of the obtained results, different models may be used for specific entity types in order to achieve the best cumulative results.

6   Conclusions and Future Works

In this paper we investigated information extraction from scientific conference web pages by verifying the applicability of different types of features and various models.

We designed different groups of features and verified their importance in this task. Based on the empirical results obtained on publicly available corpus we state that domain specific features are necessary for correct information extraction. Additionally entity type specific features are also necessary in order to obtain good results.

Despite having a broad range of features, the considered models (algorithms, representations of base objects for algorithms) achieve different results for different entity types. Thus, it is beneficial to apply specific models for specific entities.

Moreover, with help of our new features we set new baselines values of precision, recall, and F1 for information extraction from a publicly available corpus of scientific conference web pages.

In future work we plan to create a model for multi-token sequence detection and incorporate it in our models. We would also like to apply other models, e.g., MLNs, hierarchical CRF, to obtain better results.

References

1. Andruszkiewicz, P., Nachyla, B.: Automatic extraction of profiles from web pages. In: Bembenik, R., Skonieczny, L., Rybinski, H., Kryszkiewicz, M., Niezgodka, M. (eds.) Intelligent Tools for Building a Scientific Information Platform - Advanced Architectures and Solutions, pp. 415–431. Springer, Heidelberg (2013). http://dx. doi.org/10.1007/978-3-642-35647-6_25
2. Chang, C., Lin, C.: LIBSVM: a library for support vector machines. ACM TIST **2**(3), 27 (2011). http://doi.acm.org/10.1145/1961189.1961199
3. Ciravegna, F.: $(LP)^2$, an adaptive algorithm for information extraction from web-related texts. In: Proceedings of the IJCAI-2001 Workshop on Adaptive Text Extraction and Mining (2001). http://citeseer.ist.psu.edu/481342.html
4. Cortes, C., Vapnik, V.: Support-vector networks. Mach. Learn. **20**(3), 273–297 (1995)
5. Fan, R., Chang, K., Hsieh, C., Wang, X., Lin, C.: LIBLINEAR: a library for large linear classification. J. Mach. Learn. Res. **9**, 1871–1874 (2008). http://doi.acm.org/ 10.1145/1390681.1442794
6. Hazan, R., Andruszkiewicz, P.: Home pages identification and information extraction in researcher profiling. In: Intelligent Tools for Building a Scientific Information Platform - Advanced Architectures and Solutions, pp. 41–51 (2013). http:// dx.doi.org/10.1007/978-3-642-35647-6_4
7. Hsu, C.W., Chang, C.C., Lin, C.J., et al.: A practical guide to support vector classification (2003)
8. Ireson, N., Ciravegna, F., Califf, M.E., Freitag, D., Kushmerick, N., Lavelli, A.: Evaluating machine learning for information extraction. In: Raedt, L.D., Wrobel, S. (eds.) Machine Learning, Proceedings of the Twenty-Second International Conference (ICML 2005), Bonn, Germany, 7–11 August 2005. ACM International Conference Proceeding Series, vol. 119, pp. 345–352. ACM (2005). http://doi.acm.org/ 10.1145/1102351.1102395

9. Issertial, L., Tsuji, H.: Information extraction and ontology model for a 'call for paper' manager. In: Taniar, D., Pardede, E., Nguyen, H., Rahayu, J.W., Khalil, I. (eds.) iiWAS 2011 - The 13th International Conference on Information Integration and Web-based Applications and Services, 5–7 December 2011, Ho Chi Minh City, Vietnam, pp. 539–542. ACM (2011). http://doi.acm.org/10.1145/2095536.2095650

10. Kenter, T., Maynard, D.: Using GATE as an annotation tool, January 2005. http://gate.ac.uk/sale/am/annotationmanual.pdf

11. Kohlschütter, C., Fankhauser, P., Nejdl, W.: Boilerplate detection using shallow text features. In: Davison, B.D., Suel, T., Craswell, N., Liu, B. (eds.) Proceedings of the Third International Conference on Web Search and Web Data Mining, WSDM 2010, New York, NY, USA, 4–6 February 2010, pp. 441–450. ACM (2010). http://doi.acm.org/10.1145/1718487.1718542

12. Lafferty, J.D., McCallum, A., Pereira, F.C.N.: Conditional random fields: probabilistic models for segmenting and labeling sequence data. In: Brodley, C.E., Danyluk, A.P. (eds.) ICML, pp. 282–289. Morgan Kaufmann (2001)

13. Lazarinis, F.: Combining information retrieval with information extraction for efficient retrieval of calls for papers. In: 20th Annual BCS-IRSG Colloquium on IR, Autrans, France, 25–27 March 1998. Workshops in Computing, BCS (1998). http://ewic.bcs.org/content/ConWebDoc/4410

14. McCallum, A., Schultz, K., Singh, S.: FACTORIE: probabilistic programming via imperatively defined factor graphs. In: Bengio, Y., Schuurmans, D., Lafferty, J.D., Williams, C.K.I., Culotta, A. (eds.) Advances in Neural Information Processing Systems 22: 23rd Annual Conference on Neural Information Processing Systems 2009. Proceedings of a Meeting Held 7–10 December 2009, Vancouver, British Columbia, Canada, pp. 1249–1257. Curran Associates, Inc. (2009)

15. Porter, M.F.: Snowball: a language for stemming algorithms (2001)

16. Schneider, K.: Information extraction from calls for papers with conditional random fields and layout features. Artif. Intell. Rev. 25(1–2), 67–77 (2006). http://dx.doi.org/10.1007/s10462-007-9019-4

17. Tang, J., Zhang, J., Yao, L., Li, J., Zhang, L., Su, Z.: ArnetMiner: extraction and mining of academic social networks. In: Li, Y., Liu, B., Sarawagi, S. (eds.) KDD, pp. 990–998. ACM (2008)

18. Xin, X., Li, J., Tang, J., Luo, Q.: Academic conference homepage understanding using constrained hierarchical conditional random fields. In: Shanahan, J.G. et al. (eds.) Proceedings of the 17th ACM Conference on Information and Knowledge Management, CIKM 2008, Napa Valley, California, USA, 26–30 October 2008, pp. 1301–1310. ACM (2008). http://doi.acm.org/10.1145/1458082.1458254

Classifier-Based Pattern Selection Approach for Relation Instance Extraction

Angrosh Mandya[✉], Danushka Bollegala, Frans Coenen, and Katie Atkinson

Department of Computer Science, University of Liverpool, Liverpool, UK
{angrosh,danushka,coenen,katie}@liverpool.ac.uk

Abstract. A classifier-based pattern selection approach for relation instance extraction is proposed in this paper. The classifier-based pattern selection approach proposes to employ a binary classifier that filters patterns that extracts incorrect entities for a given relation, from pattern set obtained using global estimates such as high frequency. The proposed approach is evaluated using two large independent datasets. The results presented in this paper shows that the classifier-based approach provides a significant improvement in the task of relation extraction against standard methods of relation extraction, employing pattern sets based on high frequency. The higher performance is achieved through filtering out patterns that extract incorrect entities, which in turn improves the precision of applied patterns, resulting in significant improvement in the task of relation extraction.

1 Introduction

Pattern-based information extraction systems have focused on extracting entities for specific relations. For example, given a sentence "Mozart was born in 1756" the task for extracting entities for the relation PERSON-BIRTHYEAR is to extract predicates of the form PERSON-BIRTHYEAR *(Mozart, 1756)*. Similarly the triple COMPANY-CEO *(Google Inc., Sundar Pichai)* is extracted from the sentence "Sundar Pichai is the current CEO of Google Inc." for the relation COMPANY-CEO. Several studies have proposed various types of patterns for extracting entities related to such relations. For example, Ravichandran and Hovy (2002) [1] generated patterns using lexical terms between entities. Similarly, Wu and Weld (2010) [2] and Etzioni et al. (2011) [3] derived patterns by employing lexico-syntactic features such as Part-Of-Speech (POS) tags. Studies have also proposed dependency parse based syntactic features [2–4] and frame-based semantic features [5–8] for IE.

Equally important to the process of pattern learning and entity extraction is the creation of an optimum set of patterns to ensure extraction of correct entities for specific relations. The goodness measures commonly employed to create such optimum set of patterns considers measures such as frequency [9] or accuracy of patterns [10–12]. Filtering patterns employing such goodness measures results

ⓒ Springer Nature Switzerland AG 2018
A. Gelbukh (Ed.): CICLing 2017, LNCS 10761, pp. 418–434, 2018.
https://doi.org/10.1007/978-3-319-77113-7_33

in a fixed set of ranked patterns [13], which are then used to extract entities for specific relations. However such methods do not adjudge the quality of patterns with respect to the instances extracted by patterns. For example, while a pattern for a given relation, irrespective of its type i.e., whether it is lexical, syntactic or semantic extracts correct instances from sentence s, the same pattern may extract wrong instances from a different sentence s'. For instance consider the following example sentences:

1. The CEO of the company, **Steve Jobs** *announced the products of* **Apple** at WWDC.
2. Today, **Amazon** *announced the products of* **Apple** on their website.

In the example sentences above, while the lexical pattern "announced the products of" when applied on Sentence 1, extracts correct entities *(Steve Jobs, Apple)* for the relation CEO-COMPANY, the same pattern, when applied on Sentence 2, extracts incorrect entities for the relation CEO-COMPANY *(Amazon, Apple)*. Thus, it is difficult to adjudge patterns by simply considering the number of times the pattern extracts correct and incorrect instances. Further, the fixed set of patterns used for entity extraction is often created independent of the sentences on which the patterns are applied. This implies that none of the useful local information from the target sentence is considered before applying the pattern. For instance, in the example Sentence 1, the terms *CEO* and *WWDC* can serve as useful indicators to extract arguments for the relation CEO-COMPANY. However, such information is not considered before applying the pattern.

Thus, though the patterns obtained using goodness measures are useful, these patterns can still extract wrong entities from newer sentences, resulting in poor precision. To overcome this problem, a useful intermediate step before applying patterns would be to remove patterns that extract wrong entities. Against this notion, this paper presents a classification-based pattern selection approach for relation instance extraction. The key focus is to develop a classifier that filters patterns that extract incorrect entities from the large pool of patterns obtained using goodness measures such as frequency. The classification based approach is useful in selecting the subset of patterns that often extracts correct instances from sentences for particular relations, thereby improving the precision of the applied pattern set.

More specifically, the key contribution of this paper is a binary classifier that is trained to determine whether a pattern should be applied on test sentences. A seed set of relational instances is used to automatically generate positive and negative training instances for the classifier, thereby minimizing the manual effort required to build the classifier. The classifier is evaluated against employing fixed pattern sets created using global estimates such as frequency. Further, the experiments are conducted on two independent datasets: (a) Wikipedia dataset, developed following distant supervision assumption [14]; and (b) Riedel et al. (2010) dataset [15], which is developed by relaxing the distant supervision assumption.

The remainder of this paper is organised as follows. In Sect. 2, we describe the related work to this study. In Sect. 3 the proposed classification-based approach is presented. In Sect. 4, we describe the datasets and the evaluation metrics used in this study and also the results of this study. In Sect. 5, we conclude this paper.

2   Related Work

Riloff et al. (1996) [9] evaluated the relevance of a pattern for IE before application and employed a weighted conditional probability associating higher weight for high frequency to choose the best patterns. Brin (1998) [16] evaluated the derived patterns based on the specificity of the pattern, measuring the length of the middle context, prefix and suffix of the pattern. Patterns with low specificity were rejected to avoid overly general patterns. Thelen et al. (2002) [17] applied the ranking measure proposed by Riloff et al. (1996) [9] to learn semantic lexicons using extraction pattern contexts. Studies have also evaluated patterns based on their confidence by counting the number of positive and negative entities extracted by the pattern [10–12]. Agichtein et al. (2000) [10] also adopt the ranking measure of Riloff et al. (1996) [9] to consider the coverage of the pattern for evaluation.

Patwardhan et al. (2006) [18] computed "semantic affinity" as a ratio of the target semantic class extractions for each noun class over the total noun class extractions for a closed set of semantic categories. Patwardhan et al. (2007) [19] presented an IE system that decouples the tasks of finding relevant regions of text and applying extraction patterns; a sentence classifier was developed to identify relevant regions and employ semantic affinity measures to automatically learn domain-relevant extraction patterns. Alfonseca et al. (2012) [20] employed topic models to discriminate ambiguous patterns and learn more useful high-precision patterns. Goudong et al. (2005) [21] have shown that diverse lexical, syntactic and semantic knowledge are useful for relation extraction.

Thus, a significant number of studies have investigated various types of patterns, but have generally focused on global estimates such as frequency and accuracy to evaluate patterns. In comparison to the related studies, the focus of the study presented in this paper is not on creating a new type of pattern for entity extraction. However, unlike studies using global estimates, the goal of the study presented in this paper is to examine the set of available different types of patterns and identify the best patterns to apply in the context of a given sentence. This is achieved, as noted above, by developing a binary classifier that learns from features drawn from sentence-pattern pairings as explained further in the next section.

3   Classification Based Pattern Selection

The classification-based method to select a set of patterns per sentence for relation instance extraction is presented in this section. Given a tuple (R, s_j, p_i), consisting of a relation R, $s_j \in \mathcal{S}$ (set of sentences) and a pattern $p_i \in \mathcal{P}$ (set of different types of patterns such as lexical, syntactic etc.), a binary classifier h is trained to return the following prediction for a tuple (R, s_j, p_i):

$$h(R, s_j, p_i) = \begin{cases} +1 & \text{if } p_i \text{ correctly extracts both entities in } s_j \\ -1 & \text{otherwise} \end{cases} \tag{1}$$

Each sentence-pattern pair for a specific relation type is represented as a feature vector $\phi(R, s_j, p_i)$ comprising features {pattern_features, hybrid_features} and labels $\{1, -1\}$ to indicate whether pattern p_i correctly extracts arguments from sentence s_j. While any classifier such as Perceptron, logistic regression or Support Vector Machine (SVM) can be used to construct the above mapping function h, we used in this study an SVM [22] to construct the mapping function h.

3.1 Pattern Selection Using the Classifier

Given a test sentence s_j, the binary classifier selects an optimum set of patterns $\mathcal{P}(s_j)$ such that $\mathcal{P}(s_j) \in s_j = \{p_i : h(p_i, s_j) = 1\}$. The optimum set of patterns $\mathcal{P}(s_j) \in s_j$ consists patterns p_i that are classified as true by the classifier. Further the subset of patterns are ranked based on the confidence scores provided by the classifier. This is achieved by fitting a logistic regression model on top of the distance measure from the decision hyperplane in SVM. LIBSVM [23], a standard library for SVM is used to train the classifier and to fit the logistic regression model to derive confidence scores.

3.2 Features

The different features for the classifier examined in this study are focused on using n-grams from sentences and patterns and other information such as pattern type and length of the pattern. However, the features for the classifier need not be limited to these features alone and can include other features as well. For example, word embeddings can be used as features to represent the sentence-pattern pair. The different features used in this study are described below:

Sentence Features. The sentence features are designed to capture the context of the sentence and includes the following:

(a) *n-grams in sentence* - the unigram and the bigram terms in the sentence are used as feature terms.

(b) *sentence length* - information about the length of the sentence is provided as features to the classifier. Three features are defined to represent sentence length based on the number of tokens n in the sentence s and includes the following: (a) *sentence_length_small* if $n \leq 10$; *sentence_length_medium* if $n > 10$ *and* ≤ 30; and *sentence_length_long* if $n > 30$.

Pattern Features. The pattern features are designed based on the information obtained from the patterns and includes the following:

(a) *pattern type* - this feature distinguishes between different types of patterns i.e., whether a given pattern is a lexical or syntactic pattern based on words or grammatical relations or both, and semantic pattern. For example, a feature *pattern_type_lexical* is created to indicate a lexical pattern.

(b) *n-grams in pattern* - the unigrams in the pattern are used as feature terms.

(c) *length of patterns* - information about the length of the pattern is also used as a feature. Three features are defined to capture pattern length based on the number of tokens n in pattern p and includes the following: (a) *pattern_length_small* if $n \leq 10$; *pattern_length_medium* if $n > 10\ and\ \leq 30$; and *pattern_length_long* if $n > 30$. In addition to this information, the pattern type is also appended to pattern length to indicate the pattern type. For example, a lexical pattern with less than 10 tokens would be represented using the feature *lexical_pattern_length_small*.

(d) *position of patterns* - the position of the pattern in terms of its order of occurrence in the sentence is provided as a feature to the classifier.

3.3   Pattern Types

An important aspect of the proposed classification based approach for per sentence pattern selection presented in this paper is the ability for the classifier to select from the available different types of patterns for a given sentence. This study considered three different types of patterns for the classifier (a) lexical patterns; (b) syntactic patterns based on dependency parse; and (c) frame-based semantic patterns. It needs to be noted that the classifier is not confined to these three types of patterns. The classifier can be provided with other types of patterns that can be created for sentences. For example, patterns based on POS tags or named entity recognition can be used with the classifier. The process of deriving the pattern types considered in this study is explained below with the following example sentence:

1. company[Fenrir Inc] is based in the district of location[Osaka], location[Japan].

(a) **Lexical Patterns.** Following [1], regular expressions are used to define lexical patterns simply comprising lexical entries between the relevant entities as shown in List 1. The arguments for each pattern is shown in parenthesis following the pattern.

List 1 - Lexical patterns: (1) COMPANY is based in the district of LOCATION (Fenrir Inc, Osaka); (2) COMPANY is based in the district of LOCATION, LOCATION (Fenrir Inc, Japan)

(b) **Syntactic Patterns.** Syntactic patterns for sentences are defined using the shortest path in the dependency graph [2]. In this study, the following three variants of syntactic patterns are used: (a) patterns using words (List 2); (b) patterns using grammatical relations (grs) (List 3); and (c) patterns using both words and grs (List 4). The STANFORD PARSER [24] is used to obtain dependency parse for sentences in this study.

List 2 - Syntactic patterns using words in shortest path: (1) COMPANY is based district LOCATION LOCATION (Fenrir Inc, Osaka); (2) COMPANY is based district LOCATION (Fenrir Inc, Japan)

List 3 - Syntactic patterns using grammatical relations in shortest path: (1) COMPANY nsubj prep prep_in prep_of nn LOCATION (Fenrir Inc, Osaka); (2) COMPANY nsubj prep prep_in prep_of LOCATION (Japan, Osaka)

List 4: Syntactic patterns using words and grammatical relations in shortest path: (1) COMPANY nsubj_is_COMPANY based prep_based_is district prep_in_district prep_of_LOCATION nn_LOCATION_LOCATION LOCATION (Fenrir Inc, Osaka); (2) COMPANY nsubj_is_COMPANY based prep_based_is district prep_in_district prep_of_LOCATION LOCATION (Fenrir Inc, Japan)

(c) Frame-Based Semantic Patterns. The study also considers frame-based semantic patterns following the frame semantic framework [25]. Frame semantics assign *semantic frame elements* to words in a sentence [25] to provide a meaningful representations for lexical entries in the sentence. Semantic parsing tools such as SEMAFOR [26] is used to derive such semantic frames. The semantic parse obtained using SEMAFOR for the example sentence above is provided below and the semantic patterns obtained using the semantic parse is shown in List 5.

Frame based semantic parse for Sentence 1: Businesses[FENRIR INC] is based in Political_locales[DISTRICT] of Locale[OSAKA] Locale[JAPAN].

List 5: Frame-based semantic patterns: (1) COMPANY Businesses Political_locales LOCATION LOCATION (Fenrir Inc, Osaka); (2) COMPANY Businesses Political_locales LOCATION (Fenrir Inc, Japan)

4 Experiments

We explain in this section the two datasets used in this study in Sect. 4.1. We also describe the evaluation technique employed in this study in Sect. 4.2, where we explain the process of deriving negative samples for the classifier, the evaluation metrics and provide more details of the evaluation method followed in this study. Finally, in Sect. 4.3, we present the results obtained in this study.

4.1 Datasets

Wikipedia Dataset. The distant supervision method was followed to create the Wikipedia dataset. Specifically, we find all sentences that mentions a pair

Table 1. Relations in Wikipedia dataset. EP: Entity Pairs; TS: Total Sentences

Relation	EP	TS	Relation	EP	TS
ACTOR-MOVIE	1613	3147	COMPANY-FOUNDER	2062	14489
COMPANY-LOCATION	4550	6908	ALBUM-ARTIST	9474	20961
COMPANY-PRODUCT	2890	9122	BIRTHPLACE-PERSON	4114	21737
DIRECTOR-MOVIE	6537	10651	ALBUM-GENRE	8360	22934
AUTHOR-BOOKTITLE	5076	12245	COUNTRY-CITY	4647	45981
Total number of sentences: 168175					

of entities in the seed dataset, and consider those sentences as describing the semantic relationship between the two entities specified in the seed dataset. DBpedia [27] was used to obtain seed entity pairs for ten different relations, which were further used to obtain sentences from Wikipedia dump. Sentences with a mention of at least one entity pair were retained. The dataset for each relation (details are provided in Table 1) was randomly split in the ratio of 80:20 to create the training and the test set, respectively.

Riedel et al. (2010) [15] **Dataset.** The Riedel et al. (2010) [15] dataset was developed with a focus to relax the distant supervision assumption to extract relations from newswire instead of Wikipedia. However, Freebase was chosen as the knowledge base for obtaining relations and seed entities. Sentences containing two related entities were extracted from the New York Times data, resulting in a large dataset. In this study, we considered ten relations from this dataset (details are provided in Table 2) to evaluate the proposed classifier-based pattern selection method for relation extraction. Sentences for each of these relations

Table 2. Relations considered from Riedel et al. (2010) [15] dataset; TS: Total Sentences

	Relation	TS
REL_1	people_deceased_person_place_of_death	2541
REL_2	people_person_place_of_birth	4265
REL_3	business_person_company	7987
REL_4	location_administrative_division_country	8860
REL_5	location_country_administrative_divisions	8860
REL_6	location_neighborhood_neighborhood_of	9472
REL_7	people_person_place_lived	9829
REL_8	location_country_capital	11216
REL_9	people_person_nationality	11446
REL_10	location_location_contains	75969
Total number of sentences: 150445		

were randomly split in the ratio of 80:20 to create the training and test set, respectively.

4.2 Evaluation

Negative Samples for the Classifier. The process of developing the dataset based on the distant supervision method allows to create patterns that extract correct entities. However, the proposed method of developing the classifier for predicting patterns requires negative samples, i.e., patterns that extract wrong entities. The process of deriving negative samples for the classifier is explained below using the following example sentence:

> $_{COMPANY}$Interwoven was founded in $_{NUM}$1995 in $_{LOCATION}$California by Peng Tsin Ong of $_{LOCATION}$Singapore, who was also $_{COMPANY}$Interwoven's first CEO and chairman.

In the sentence above, the entity pair (*Interwoven, California*) is the correct argument for the COMPANY-LOCATION relation. Such entity pairs (seed instances) can be obtained from different knowledge sources such as DBpedia and Freebase. The distant supervision method allows us to derive the following two lexical patterns for extract the entities (Interwoven, California). The year information (1995) is changed to NUM and the word Singapore is changed to LOCATION to generalize the patterns.

Patterns extracting correct entities
1. <COMPANY> was founded in NUM in <LOCATION>
2. <LOCATION> by Peng Tsin Ong of LOCATION, who was also <COMPANY>

However, the presence of the LOCATION entity term 'Singapore' allows to derive the following lexical patterns that extract wrong entities (*Interwoven, Singapore*) for the COMPANY-LOCATION relation:

Patterns extracting wrong entities
1. <COMPANY> was founded in NUM in LOCATION by Peng Tsin Ong of <LOCATION>
2. <LOCATION>, who was also <COMPANY>

Thus, such patterns extracting wrong entities are used as negative samples for the classifier.

Evaluation Metrics. Further, given a pattern $l \in \mathcal{L}$, the pattern set obtained from train data, and a test sentence $s \in \mathcal{S}$, the following types of patterns are defined:

1. *matched pattern:* the pattern l is defined as a *matched pattern* for the test sentence s, iff (if and only if) the pattern l matches the test sentence s.

2. *correct pattern:* a pattern l is defined as a *correct pattern* for the test sentence s, iff the pattern l matches the test sentence s and correctly extracts the two arguments (e_1, e_2) for a given relation r.

The precision of a pattern l is defined as the ratio of number of times the pattern l is seen as a *correct pattern* to the number of times it is seen as a matched pattern on the test set $\mathcal{S}$. Thus, the precision of a pattern l on the test set $\mathcal{S}$ is given by:

$$\text{Precision } (l) = \frac{\#\ \text{pattern } l \text{ is a } correct\ pattern \text{ in } \mathcal{S}}{\#\ \text{pattern } l \text{ is a } matched\ pattern \text{ in } \mathcal{S}} \tag{2}$$

The overall precision P of the pattern set is obtained by:

$$P = \frac{1}{|\mathcal{L}|} \sum_{l \in \mathcal{L}} \text{Precision } (l) \tag{3}$$

where $|\mathcal{L}|$ is the total number of patterns in the pattern set.

The recall of a pattern set is measured in terms of its effectiveness or coverage in applying correct patterns on the test set and is defined as the ratio of the total number of test sentences on which *correct patterns* are applied to the total number of test sentences. Thus, the recall of a pattern set is given by:

$$R = \frac{\#\ \text{of test sentences with } correct\ patterns}{\#\ \text{of test sentences}} \tag{4}$$

Given Precision P and Recall R, the F-score of a pattern set is obtained by:

$$\text{F-Score} = \frac{2 \times PR}{P + R} \tag{5}$$

The classification accuracy of the classifier on the test set is reported on two different set of features: (a) pattern features; (b) hybrid features - combining features obtained from sentences and patterns. If p_c is the total number of correctly classified patterns and p_t is the total number of patterns in test set, the accuracy of the classifier Accuracy(c) is obtained using the equation:

$$\text{Accuracy(c)} = \frac{p_c}{p_t} \tag{6}$$

Model selection for optimal parameter estimation was performed as a grid search through cross-validation on the development set [28].

Evaluation Method. In a regular setting, the large pattern set obtained from the training set are applied on the test set for relation extraction. At this stage, the patterns from the train set that match the patterns in the test set are applied for relation extraction. However, there could be many patterns in the matched pattern set that extract wrong entities for the targeted relation. The proposed per-sentence classifier approach for relation extraction is employed to filter such patterns from the matched pattern set that extract wrong entities. This filtering process can help in achieving higher precision, without losing on recall. Thus, in this study, the matched pattern set from the training set is evaluated against the filtered pattern set obtained using the classifier for relation extraction. The evaluation method employed in this study is shown in Fig. 1.

As shown in Fig. 1, the following three pattern sets based on their size are evaluated: (a) 10% of most frequent patterns; (b) 50% of most frequent patterns; and (c) full pattern set obtained from the train set. Further, as seen in Fig. 1, with regard to applying different sets of frequent patterns, the patterns are initially matched against the test set, following which the *matched patterns* are applied to identify *correct patterns*. However, with regard to the classifier, the *matched pattern* set is examined against the classifier to obtain a *filtered pattern* set, comprising only those patterns that are positively classified by the classifier. The *filtered pattern* set is now applied on the test set for relation extraction. The precision, recall and F-score for the applied pattern sets in both cases are recorded for different relations.

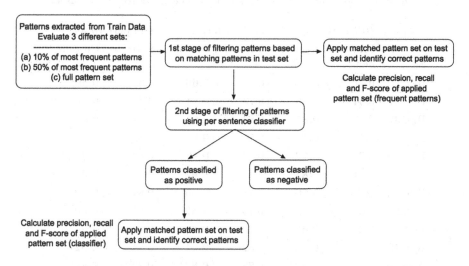

Fig. 1. Evaluation methodology for evaluating pattern set obtained using frequent patterns and classifier.

4.3   Results

The evaluation results of applying the classifier approach proposed in this study against the regular use of frequency based pattern set for relation extraction is presented in this section.

Classifier vs. High Frequency Patterns. As seen in Tables 3 and 4, the per sentence classifier-based approach for relation extraction, achieves statistically significant average F-scores ($p \leq 0.05$; Wilcoxon Signed-Rank Test) for

Table 3. F-score values for relations in Wikipedia dataset.

Relation	10% patterns		50% patterns		Full pattern set	
	Frequent patterns	Per sentence classifier	Frequent patterns	Per sentence classifier	Pattern set	Per sentence classifier
ACTOR-MOVIE	0.72	**0.79**	0.73	**0.82**	0.74	**0.84**
COMPANY-LOCATION	0.72	**0.80**	0.73	**0.82**	0.73	**0.84**
COMPANY-PRODUCT	0.76	**0.82**	0.77	**0.83**	0.77	**0.84**
DIRECTOR-MOVIE	0.70	**0.84**	0.70	**0.86**	0.68	**0.87**
AUTHOR-BOOKTITLE	0.75	**0.83**	0.74	**0.84**	0.74	**0.85**
COMPANY-FOUNDER	0.82	**0.85**	0.83	**0.87**	0.84	**0.88**
ALBUM-ARTIST	0.69	**0.80**	0.69	**0.83**	0.69	**0.84**
BIRTHPLACE-PERSON	0.64	**0.75**	0.64	**0.78**	0.64	**0.80**
ALBUM-GENRE	0.69	**0.81**	0.69	**0.83**	0.69	**0.85**
COUNTRY-CITY	0.75	**0.82**	0.76	**0.85**	0.77	**0.86**
Average	0.72	**0.81**[a]	0.73	**0.83**[a,b]	0.70	**0.85**[a,b]

[a]statistically significant against applying patterns based on frequency and full pattern set.
[b]statistically significant than using the previous pattern set size.

Table 4. F-score values for relations in Riedel et al. (2010) [15] dataset.

Relation	10% patterns		50% patterns		Full pattern set	
	Patterns only	Classifier	Patterns only	Classifier	Patterns only	Classifier
REL_1	0.57	**0.57**	0.65	**0.66**	0.68	**0.69**
REL_2	0.64	**0.68**	0.73	**0.79**	0.79	**0.90**
REL_3	0.85	**0.87**	0.86	**0.91**	0.84	**0.92**
REL_4	0.70	**0.72**	0.79	**0.84**	0.82	**0.89**
REL_5	0.71	**0.73**	0.79	**0.83**	0.82	**0.87**
REL_6	0.81	**0.87**	0.84	**0.92**	0.83	**0.94**
REL_7	0.74	**0.80**	0.78	**0.86**	0.80	**0.89**
REL_8	0.68	**0.71**	0.75	**0.82**	0.81	**0.90**
REL_9	0.60	**0.63**	0.80	**0.89**	0.79	**0.90**
REL_10	0.63	**0.66**	0.82	**0.91**	0.84	**0.92**
Average	0.69	**0.72**[a]	0.78[b]	**0.84**[a,b]	0.80[b]	**0.88**[a,b]

[a]statistically significant against applying patterns based on frequency and full pattern set.
[b]statistically significant than using the previous pattern set size.

relations both in Wikipedia dataset and Riedel et al. (2010) [15] dataset for pattern sets varying in different sizes. For example, for relations in Wikipedia dataset (Table 3) the classifier achieves statistically significant average F-scores of 0.81, 0.83 and 0.85 ($p \leq 0.05$; Wilcoxon Signed-Rank Test) against the average F-score of 0.72, 0.73 and 0.70 achieved by using 10% and 50% of high frequency patterns, and the full pattern set obtained from the training set, respectively. A similar performance is seen for relations in Riedel et al. (2010) [15] dataset

(Table 4), with the classifier achieving statistically significant average F-scores of 0.76, 0.84 and 0.88 ($p \leq 0.05$; Wilcoxon Signed-Rank Test) against the average F-score values of 0.69, 0.78 and 0.80 achieved for 10% and 50% of high frequency patterns, and the full pattern set obtained from the training set, respectively. These results indicate that classifier-based approach is significantly useful for the task of relation extraction.

Further as seen in Tables 3 and 4, the increase in the size of the applied pattern set on test sentence, results in a significant increase in the performance of the classifier. For example, with the Wikipedia dataset, while the classifier achieves an average F-score of 0.81 with 10% of high frequency patterns, the classifier achieves a statistically significant higher average F-score of 0.83 using 50% of high frequency patterns ($p \leq 0.05$; Wilcoxon Signed-Rank Test). The classifier achieves a further higher average F-score of 0.85 with the use of full-pattern set obtained from the training data. A similar performance is also seen for relations in Riedel et al. (2010) [15] dataset as shown in Table 4. While the classifier achieves an average F-scores of 0.76 with the use of 10% of high frequency patterns, a statistically significant higher average F-score o 0.84 is achieved using 50% of high frequency patterns ($p \leq 0.05$; Wilcoxon Signed-Rank Test. A higher performance is achieved with the use of full pattern set, with the classifier achieving a higher average F-score of 0.88.

The precision values scored for relations in Wikipedia dataset and Riedel et al. (2010) [15] dataset is shown in Figs. 2 and 3. The recall values are not reported here, since both the classifier and high frequency patterns achieve the same recall for all relations for both the datasets. As seen in Figs. 2 and 3, the increase in the applied pattern set size results in a decrease in precision values for majority of relations in both the datasets. This can be the reason for the decrease in the performance of high frequency patterns, when larger set of patterns are used. However, on the other hand, the precision score improves with the increase in the applied pattern set size for all relations. These results further prove the usefulness of classifier-based approach for the purpose of relation extraction.

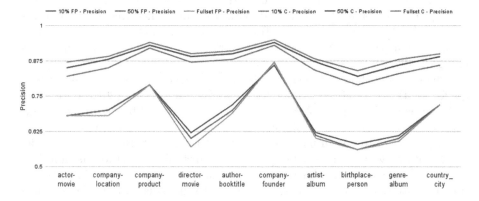

Fig. 2. Precision values for relations in Wikipedia dataset.

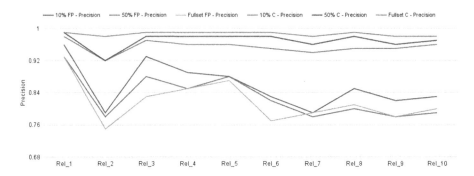

Fig. 3. Precision values for relations in Riedel et al. (2010) [15] dataset.

Wikipedia vs. Riedel et al. (2010) [15] **Dataset.** The proposed classifier-based pattern selection approach was evaluated on relations drawn from two different datasets: (a) Wikipedia dataset; and (b) Riedel et al. (2010) [15] dataset. Interestingly, the increase in the size of high frequency patterns lowers the performance of relation extraction, particularly for the Wikipedia dataset. As seen in Table 3, while an average F-score of 0.72 is achieved with 10% of higher frequency patterns, a slightly higher F-score of 0.73 is achieved with using 50% of high frequency patterns. However, with the use of the full pattern set obtained from the training data, a further lower average F-score of 0.70 is achieved. The decrease in the performance with the increase in the size of the pattern set is not statistically significant. This indicates that a significantly large proportion of patterns seen in the test sentences are covered in the top 10% of high frequency patterns, indicating the usefulness of higher frequency patterns.

However in the case of Riedel et al. (2010) [15] dataset, the increase in the pattern set size of high frequency patterns does not lower the performance of relation extraction. The performance obtained for larger pattern set of high frequency patterns is statistically significant, with an average F-score of 0.78 being achieved with 50% of high frequency patterns, while a lower average F-score of 0.69 is obtained with 10% of higher frequency patterns ($p \leq 0.05$; Wilcoxon Signed-Rank Test). The average F-score (0.80) obtained using the full pattern set is further higher than employing smaller proportions of high frequency patterns.

These results show that using high frequency patterns and the full pattern set for relation extraction is more beneficial for Riedel et al. (2010) dataset. It needs to be noted that the Wikipedia dataset is developed based on the distant supervision method where training knowledge base derived from the training text is employed to obtain patterns from the training text. However, the Riedel et al. dataset was developed by relaxing the distant supervision assumption, where an external training knowledge base, not derived from the training text is used for pattern extraction. Thus, in the case of Wikipedia dataset, the obtained pattern set from training data suffers from poor precision i.e., extract wrong entities in spite of matching the test sentences, resulting in the poor performance for relation extraction. However, the patterns obtained in Riedel et al. (2010) dataset are more precise, extracting correct entities given match on test sentences. These results show that high frequency patterns are more useful for datasets where distant supervision assumption is relaxed. However, the classifier-based pattern selection approach surpasses the performance obtained using high frequency patterns and the complete pattern set, indicating the usefulness of classifier-based approach on both types of datasets.

Classification Accuracy of the Classifier. As mentioned previously in Sect. 3 SVM was adopted as a binary classifier for this study. The following two types of feature sets were examined: (a) pattern features - features obtained from pattern alone; and (b) hybrid features - combining pattern features along with features obtained from the sentence. The pattern and sentence features were previously discussed in Sect. 3.

To choose the best kernel for SVM, the classification accuracy of the classifier for various kernels was examined for the development set of COMPANY-LOCATION as shown in Table 5. As seen in Table 5, the Radial Basis Function (RBF) kernel using the optimal parameters from grid search, achieved the best performance scoring an accuracy of 69.35 for hybrid features (HF).

Table 5. Classification accuracy (c) of the classifier for various kernels for COMPANY-LOCATION dataset (development set). PF - pattern features, HF - hybrid features

Function	PF	HF
Linear	63.42	63.85
Polynomial (degree 2)	64.71	68.96
Polynomial (degree 3)	62.08	67.05
Radial basis	65.21	69.35

The results in Table 5 indicates that the classification accuracy for all the kernels improves with the use of hybrid features (HF), compared to using pattern features (PF) alone. To further confirm this, we evaluated the classifier (choosing RBF kernel) for different relations both in the Wikipedia dataset and Riedel et al. (2010) [15] dataset, as shown in Table 6. As seen in Table 6, the classifier achieves statistically significant performance using hybrid features (HF) in comparison to using pattern features (PF) alone for both the datasets. For example, the hybrid features scores statistically significant average of 72.28 and 85.10 against lower average scores of 68.19 and 76.59 for Wikipedia dataset and Riedel et al. (2010) [15], respectively ($p \leq 0.05$; Wilcoxon Signed-Rank Test). Thus, these results further strengthen our proposed argument that features drawn from both sentences and patterns are useful for relation extraction. Based on these results, the classifier using RBF kernel and hybrid features as chosen as the classification model.

Table 6. Classifier results using RBF kernel for different relations in the Wikipedia and Riedel et al. (2010) datasets. PF - pattern features, HF - hybrid features, *performance obtained using hybrid features is statistically significant than using pattern features alone.

Wikipedia dataset			Riedel et al. (2010) Dataset [15]		
Relation	PF	HF	Relation	PF	HF
ACTOR-MOVIE	67.24	68.57	REL_1	70.79	88.81
COMPANY-LOCATION	65.21	69.35	REL_2	74.67	79.55
COMPANY-PRODUCT	68.62	75.55	REL_3	80.44	88.74
DIRECTOR-MOVIE	76.20	77.04	REL_4	75.44	87.26
AUTHOR-BOOKTITLE	74.24	74.34	REL_5	76.11	87.65
COMPANY-FOUNDER	68.48	82.68	REL_6	82.15	90.84
ALBUM-ARTIST	69.33	71.41	REL_7	79.61	86.17
BIRTHPLACE-PERSON	60.84	61.52	REL_8	72.95	81.60
ALBUM-GENRE	67.68	71.21	REL_9	74.51	82.66
COUNTRY-CITY	64.06	71.14	REL_10	73.45	81.45
AVERAGE	68.19	**72.28***		76.59	**85.10***

5   Conclusion

We presented in this paper a classifier-based pattern selection approach for relation instance extraction. The classifier-based approach for relation extraction was evaluated against using different proportions of high frequency patterns and also employing the full pattern set for relation extraction. This paper showed that employing a classifier to remove patterns that extract wrong entities for a given relation before applying on test sentences, helps in improving precision without compromising on recall, which in turn facilitate significant improvement in the relation extraction task. The results show that an increase in the applied high frequency patterns results in lowering the performance for relation extraction, particularly on datasets developed based on distant supervision method. The results further show that the classifier-based pattern selection approach is useful for relation extraction on different types of datasets that are developed following distant supervision and also where the distant supervision assumption is relaxed.

References

1. Ravichandran, D., Hovy, E.: Learning surface text patterns for a question answering system. In: Proceedings of the COLING, pp. 41–47. Association for Computational Linguistics (2002)
2. Wu, F., Weld, D.S.: Open information extraction using Wikipedia. In: Proceedings of COLING, pp. 118–127. Association for Computational Linguistics (2010)

3. Etzioni, O., Fader, A., Christensen, J., Soderland, S., Mausam, M.: Open information extraction: the second generation. Proc. IJCAI. **11**, 3–10 (2011)
4. Gamallo, P., Garcia, M., Fernández-Lanza, S.: Dependency-based open information extraction. In: Proceedings of the Joint Workshop on Unsupervised and Semi-supervised Learning in NLP, pp. 10–18. Association for Computational Linguistics (2012)
5. Kim, J.T., Moldovan, D.: Acquisition of linguistic patterns for knowledge-based information extraction. IEEE Trans. Knowl. Data Eng. **7**, 713–724 (1995)
6. Moschitti, A., Morarescu, P., Harabagiu, S.M.: Open domain information extraction via automatic semantic labeling. In: Proceedings of FLAIRS, pp. 397–401 (2003)
7. Shen, D., Lapata, M.: Using semantic roles to improve question answering. In: Proceedings of EMNLP-CoNLL, pp. 12–21 (2007)
8. Søgaard, A., Plank, B., Alonso, H.M.: Using frame semantics for knowledge extraction from twitter. In: Proceedings of AAAI (2015)
9. Riloff, E.: Automatically generating extraction patterns from untagged text. In: Proceedings of the National Conference on Artificial Intelligence, pp. 1044–1049 (1996)
10. Agichtein, E., Gravano, L.: Snowball: extracting relations from large plain-text collections. In: Proceedings of the Fifth ACM Conference on Digital Libraries, pp. 85–94. ACM (2000)
11. Yangarber, R., Lin, W., Grishman, R.: Unsupervised learning of generalized names. In: Proceedings of COLING, pp. 1–7. Association for Computational Linguistics (2002)
12. Lin, W., Yangarber, R., Grishman, R.: Bootstrapped learning of semantic classes from positive and negative examples. In: Proceedings of ICML-2003 Workshop on the Continuum from Labeled to Unlabeled Data, vol. 1, p. 21 (2003)
13. Gupta, S., Manning, C.D.: Improved pattern learning for bootstrapped entity extraction. In: CoNLL, pp. 98–108 (2014)
14. Mintz, M., Bills, S., Snow, R., Jurafsky, D.: Distant supervision for relation extraction without labeled data. In: Proceedings of the Joint Conference of the 47th Annual Meeting of the ACL and the 4th International Joint Conference on Natural Language Processing of the AFNLP: Volume 2, vol. 2, pp. 1003–1011. Association for Computational Linguistics (2009)
15. Riedel, S., Yao, L., McCallum, A.: Modeling relations and their mentions without labeled text. In: Balcázar, J.L., Bonchi, F., Gionis, A., Sebag, M. (eds.) ECML PKDD 2010. LNCS (LNAI), vol. 6323, pp. 148–163. Springer, Heidelberg (2010). https://doi.org/10.1007/978-3-642-15939-8_10
16. Brin, S.: Extracting patterns and relations from the world wide web. In: Atzeni, P., Mendelzon, A., Mecca, G. (eds.) WebDB 1998. LNCS, vol. 1590, pp. 172–183. Springer, Heidelberg (1999). https://doi.org/10.1007/10704656_11
17. Thelen, M., Riloff, E.: A bootstrapping method for learning semantic lexicons using extraction pattern contexts. In: Proceedings of EMNLP, pp. 214–221. Association for Computational Linguistics (2002)
18. Patwardhan, S., Riloff, E.: Learning domain-specific information extraction patterns from the web. In: Proceedings of the Workshop on Information Extraction beyond the Document, pp. 66–73. Association for Computational Linguistics (2006)
19. Patwardhan, S., Riloff, E.: Effective information extraction with semantic affinity patterns and relevant regions. Proc. EMNLP-CoNLL **7**, 717–727 (2007)

20. Alfonseca, E., Filippova, K., Delort, J.Y., Garrido, G.: Pattern learning for relation extraction with a hierarchical topic model. In: Proceedings of the 50th Annual Meeting of the ACL, pp. 54–59. Association for Computational Linguistics (2012)
21. GuoDong, Z., Jian, S., Jie, Z., Min, Z.: Exploring various knowledge in relation extraction. In: Proceedings of COLING, pp. 427–434. Association for Computational Linguistics (2005)
22. Cortes, C., Vapnik, V.: Support-vector networks. Mach. Learn. **20**, 273–297 (1995)
23. Chang, C.C., Lin, C.J.: LIBSVM: a library for support vector machines. ACM Trans. Intell. Syst. Technol. **2**, 27:1–27:27 (2011)
24. De Marneffe, M.C., MacCartney, B., Manning, C.D.: Generating typed dependency parses from phrase structure parses. Proc. LREC **6**, 449–454 (2006)
25. Fillmore, C.: Frame semantics. In: Linguistics in the Morning Calm, pp. 111–137 (1982)
26. Das, D., Chen, D., Martins, A.F., Schneider, N., Smith, N.A.: Frame-semantic parsing. Comput. Linguist. **40**, 9–56 (2014)
27. Auer, S., Bizer, C., Kobilarov, G., Lehmann, J., Ives, Z.: DBpedia: a nucleus for a web of open data. In: 6th International Semantic Web Conference, Busan, Korea (2007)
28. Hsu, C.W., Chang, C.C., Lin, C.J., et al.: A practical guide to support vector classification (2003)

An Ensemble Architecture for Linked Data Lexicalization

Rivindu Perera$^{(\boxtimes)}$ and Parma Nand

Auckland University of Technology, Auckland, New Zealand
{rivindu.perera,parma.nand}@aut.ac.nz

Abstract. Linked Data has revamped the representation of knowledge by introducing the triple data structure which can encode knowledge with the associated semantics including the context by interlinking with external resources across documents. Although Linked Data is an attractive and effective mechanism to represent knowledge as created and consumed by humans in the form of a natural language, it still has a dimension of separation from natural language. Hence, in recent times, there has been an increase interest in transforming Linked Data into natural language in order to harness the benefits of Linked Data in applications interacting with natural language. This paper presents a framework that lexicalizes the Linked Data triples into natural language using an ensemble architecture. The proposed architecture is comprised of four different pattern based modules which lexicalize triples by analysing the triple features. The four pattern mining modules are based on occupational metonyms, Context Free Grammar (CFG), relation extraction using Open Information Extraction (OpenIE), and triple properties. The framework was evaluated using a two-fold evaluation process consisting of linguistic accuracy analysis and human evaluation for a test sample. The linguistic accuracy evaluation showed that the framework can produce 283 accurate lexicalization patterns for a set of 25 ontology classes resulting in a 70.75% accuracy, which is an approximately 91% increase compared to the existing state-of-the-art model.

Keywords: Lexicalization · Natural Language Generation
Linked Data · DBpedia

1 Introduction

The past decade has seen a rapid development of Linked Data [1] in many application areas [2]. In essence, Linked Data is a knowledge representation model which uses a triple data structure which is in the form of ⟨*subject, predicate, object*⟩$_T$. The benefit of this representation is that it supports interlinking information by connecting subject or object of the triple using Uniform Resource Identifiers (URIs). Figure 1 depicts an example scenario of information linking for two events. Although not shown in this figure, when developing the Linked Data resource, the subjects are always represented as URIs which support the

© Springer Nature Switzerland AG 2018
A. Gelbukh (Ed.): CICLing 2017, LNCS 10761, pp. 435–449, 2018.
https://doi.org/10.1007/978-3-319-77113-7_34

linking of information. However, the current specification does not enforce using URIs for objects, hence they can be represented as URIs, literals, or even as date-time values. The main benefit of this knowledge representation in the web is that machines can now consume this structured data unambiguously with less effort compared to the unstructured text representation.

Fig. 1. A portion of the Linked Data graph. The graph shows how two entities, Margaret Thatcher and Denis Thatcher, are linked based on the common attributes.

Since the objective of Linked Data is to make the web information machine understandable, it does not specifically focus on the human friendliness. In essence, structured knowledge representation in the form of triples is not obviously an effective method for knowledge acquisition and manipulation for humans. This creates the need to transform the Linked Data triples into natural language which is the dominant and the most preferred form of knowledge manipulation for humans. Existence of such a transformation framework would have applications in a wide range of domains. For example, a museum kiosk can store information in Linked Data form, and which can generate the textual description of artefacts based on the user requests. The importance of such a service is that users can specify which information they need, and application would be able to generate a customized description for the user which suits his/her information need. This humanizes the Linked Data by presenting it as a natural language, hence would somewhat lessen the feel of interacting with a machine. In order to implement an application of this type, we need to focus on techniques that would be able to transform the individual triples into a natural language text.

This paper presents RealText$_{lex}$ lexicalization framework which focuses on transforming triples to their syntactically and semantically correct natural language form. The RealText$_{lex}$ lexicalization framework[1] is a part of the RealText project [3–5] which is designed to generate descriptive answers in Question Answering over Linked Data (QALD). The framework is comprised of four sub-frameworks, namely, lexicalization, aggregation (using clustering and rule based approach), Referring Expression Generation (using ontology classes and

---

[1] http://www.rivinduperera.com/information/realtextlex.html.

personal pronouns), and structure realization. However, in this paper we limit the scope to the lexicalization sub-framework. Table 1 depicts a sample set of lexicalization patterns that we expect to build with corresponding triples. The notation $\langle X_1, X_2, X_3 \rangle_T$ denotes a triple where X_1, X_2, and X_3 are phrases. Similarly, $\langle X_1, X_2, X_3 \rangle_L$ and $\langle X_1, X_2, X_3 \rangle_R$ represent a lexicalization pattern and a relation respectively.

Table 1. Sample set of lexicalization patterns that can be used to transform triples (in the right) to natural language sentences. Subject and Object are denoted as $S?$ and $O?$ respectively.

Triple (subject, predicate, object)	Lexicalization pattern
$\langle Margaret\ Thatcher,\ birthDate,\ 1925\text{-}10\text{-}13 \rangle_T$	$\langle S?,\ was\ born\ on,\ O? \rangle_L$
$\langle Margaret\ Thatcher,\ birthPlace,\ Grantham \rangle_T$	$\langle S?,\ was\ born\ in,\ O? \rangle_L$
$\langle Margaret\ Thatcher,\ spouse,\ Denis\ Thatcher \rangle_T$	$\langle S?,\ married,\ O? \rangle_L$
	$\langle S?,\ is\ the\ wife\ of,\ O? \rangle_L$
	$\langle O?,\ is\ the\ husband\ of,\ S? \rangle_L$
$\langle Margaret\ Thatcher,\ successor,\ John\ Major \rangle_T$	$\langle O?,\ succeeded,\ S? \rangle_L$

The framework uses DBpedia [6–8] as the Linked Data resource and the reasons behind this selection is explained in detail in Sect. 2. The lexicalization is accomplished using four pattern mining modules implemented in an ensemble architecture. Each module is equipped with a selection function which selects the triples to lexicalize.

The rest of the paper is structured as follows. Section 2 provide a discussion on DBpedia, which is used as the Linked Data resource. The section also describes the reasons for selecting DBpedia as the principle source of information. Section 3 discusses the framework including the algorithmic details of the four pattern extraction modules. Section 4 focuses on the evaluation of the framework. We conclude the paper with future directions in Sect. 6.

2   DBpedia as a Linked Data Resource

A Linked Data resource collects and organizes triples in one of the following ways; through information extraction, community effort, or by a combination of the above two. Currently there are hundreds of Linked Data resources providing structured data on different domains.

Among all these LOD resources, DBpedia plays a significant role in Linked Open Data cloud. DBpedia extracts information from the Wikipedia and makes them available as structured data under an extended ontology. It is often referred to as the crystallization point and nucleus for the web of open data due to the wide coverage and linkage to other Linked Data resources.

Table 2 depicts the growth rate of DBpedia for last 7 releases. This shows that it has evolved rapidly in terms of the number of triples, ontology classes and the entities.

Table 2. DBpedia growth rate in last 7 releases. Only the number of entities, triples and ontology classes are considered in the English edition of DBpedia.

Release version	Entities (millions)	Triples (billions)	Ontology classes
2015(b)	6.2	4.3	739
2015(a)	5.9	3.13	735
2014	4.58	3	685
3.9	4.26	2.46	529
3.8	3.77	1.89	359
3.7	3.64	1	320
3.6	3.5	0.672	272

3   RealText$_{lex}$ Lexicalization Framework

The framework is comprised of four lexicalization pattern extraction modules which are implemented in an ensemble architecture. The occupational metonym patterns has the highest priority and it is based on a lexicon which uses the metonyms to derive lexicalization patterns. The Context Free Grammar (CFG) patterns use the CFG language generation rules to determine lexicalization patterns. The relational pattern extraction module uses the unstructured text to extract lexicalization patterns by aligning triples with extracted relations from the unstructured text. The lowest priority is given for the property patterns which is based on a predetermined set of templates. This is given lowest priority because high coverage in template based approach can reduce the language variety. Furthermore, if one of the module finds a lexicalization pattern for a triple, then the following modules are not executed.

Fig. 2. Schematic representation of the ensemble architecture for lexicalization pattern extraction. This depicts the pipeline architecture of pattern extraction. However, presence of a lexicalization pattern extracted from a previous module leads other modules to be skipped.

The high-level architecture is shown in Fig. 2. The framework takes a collection of triples and associated with metadata as input. This metadata represents some of the information which is derived computationally and utilizing databases. The metadata includes the verbalizations of the triple, ontology class hierarchy of the subject, information of predicate (whether it requires a measured number, a date value, or a numerical value as the object), grammatical gender of the subject, and whether predicate can contain multiple object values. The triple and metadata collection is then passed to the pattern extraction modules. The modules are prioritized in the same order shown in the Fig. 2, however, whether or not to associate a given triple with a lexicalization pattern in decided by the modules. Furthermore, once a triple is associated with a pattern, the rest of the modules will not be executed.

3.1 Occupational Metonym Patterns

Occupational metonym [9] can be considered as a semantic representation of occupation. Some of the examples of occupational metonyms are teacher (used to represent a person who teaches something), and producer (someone who produces something).

The commonly used method to form a occupational metonym is the -er nominalization, where a base verb is transformed to a noun by adding the -er suffix. The base verb which is nominalized is the verb related to the occupation and thus it provides a semantic representation. Although, -er nominalization can be generalized to a certain extent, there are some exceptions. For instance, *scratcher* and *broiler* are -er nominalized nouns which are used to denote a ticket which is scratched and a broiled chicken respectively.

In this research we use these -er nominalized occupational metonyms to derive patterns for lexicalization. The resulting lexicon is built by identification of -er nominalized nouns and then manual validation is done to confirm that they represent a real-world occupational metonym. The patterns are derived based on the base verb that is provided in the occupational metonym.

Table 3 lists a sample set of occupational metonyms derived from the DBpedia with the respective lexicalization patterns.

Table 3. Lexicalization patterns based on occupational metonyms

Metonym	Subject ontology class	Lexicalization pattern
Publisher	Book	$\langle S?, \text{ is published by, } O? \rangle_L$
Developer	Software	$\langle S?, \text{ is developed by, } O? \rangle_L$
Designer	Software	$\langle S?, \text{ is designed by, } O? \rangle_L$

3.2 Context Free Grammar Patterns

Context Free Grammar (CFG) [10] can be used to parse the language as well as to generate the language. To utilize CFG in language generation tasks, a sample corpus needs to be parsed and all known CFG rules need to be extracted. In some scenarios, more representative CFG rules are given the priority to boost the accuracy. These rules can be then reapplied to generate language which will exhibit the same syntactic structure of the source rule. It is also possible to compile CFG rules manually.

Although CFG based language production is quite straightforward, it cannot cover all the possible scenarios in a natural language which is full of idiosyncratic rules. In this research, we utilize CFG as one of the supporting method and its functionality is limited to a certain pattern type. Essentially, only the triples which contain the predicate as a verb (either past tense of past participle) are allocated for CFG based lexicalization patterns. The only exception is that some of the predicates which are formed using a passive form are also considered as triples which can be lexicalized using CFG rules.

The limited grammar formalism ($\mathcal{G}$) that is employed in this module can be expressed as below.

$$
\begin{aligned}
S &\Rightarrow NP \ \ VP \\
NP &\Rightarrow NNP \dots NNP \\
VP &\Rightarrow VBD \ \ NP
\end{aligned}
$$

where, S denotes a sentence which is formed using a NP, a noun phrase and a VP, a verb phrase. The NP can be further modularized as a collection of proper nouns. Similarly, VP can be modularized into a past form of a verb and a noun phrase.

3.3   Relational Patterns

The relational pattern processing module takes the unstructured text collection as the input and extract lexicalization patterns from the text using Open Information Extraction (OpenIE) based methodology. Figure 3 depicts the overview of the module. The module takes the triple and associated metadata collection as the input and then proceeds to the following sequence of steps. The first step pre-processes web based text related to the triples and then extracts relations from the text collection. The extracted relations are then aligned with the triples to derive lexicalization patterns. The following sections discuss the process in detail.

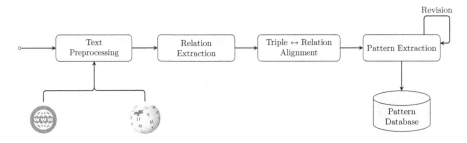

Fig. 3. Overview of the relational Pattern extraction process

3.4   Text Pre-processing

In the text pre-processing stage, we extract text related to the triples from
Wikipedia and other web based text resources which provides information related
to the subject of the triple. Wikipedia is considered as a main resource as DBpedia (the Linked Data resource used) is based on the Wikipedia infoboxes. Therefore, there is a high probability that Wikipedia text contains the textual representation for the DBpedia triples. However, Wikipedia itself is not sufficient for
pattern extraction. This is because in some scenarios textual representation is
presented in a form that cannot be extracted as a relation. For example, birth
dates in Wikipedia is provided in brackets next to the first occurrence of the
person name (e.g., Margaret Thatcher (October 13, 1925–April 8, 2013)). To be
able to include the textual representation of such information we collect information from other factsheets provided on the internet including biography.com,
imdb.com, and other relevant web sites. The web based text is generally wrapped
in the HTML boilerplates which present another burden in extracting them. To
avoid this we employ the shallow text feature based boilerplate removal algorithm, Boilerpipe [11].

The extracted text is then passed to a co-reference resolution step and resolve
all identified co-references with the name of the subject. This step is designed to
support the alignment phase (see Sect. 3.6) where we attempt to align the triple
with relation arguments derived from the extracted text. In essence, it will not
be possible to align a triple components with referring expressions. Therefore, we
use the Stanford CoreNLP toolkit [12] to substitute anaphora with antecedent.
For example, a paragraph such as,

"Barack Obama is an American politician who served as the 44th President
of the United States. He is the first African American to have served as president.
He previously served in the U.S. Senate representing Illinois from 2005 to 2008."

Will be converted to,

"Barack Obama is an American politician who served as the 44th President
of the United States. *Barack Obama* is the first African American to have served
as president. *Barack Obama* previously served in the U.S. Senate representing
Illinois from 2005 to 2008."

3.5   Extracting Relations

The relation extraction from free text is a widely studied phenomena in NLP.
Although, several relation extraction approaches exist, they can be categorised
into two main camps, namely, Closed Information Extraction (ClosedIE) and
Open Information Extraction (OpenIE) [13].

ClosedIE which is the traditional relation extraction approach is based on rule
based, kernel, or sequence labelling methods. These methods induce several key
pitfalls in the ClosedIE based applications. These are the need of hand-crafted
rules and hand-tagged data where both require expensive human involvement.
In addition to the resource expensiveness, the labelling process need to make
sure to maintain a high inter-annotator agreement. Another inherent key issue

with such tagging is that they are domain sensitive which in turn necessitates development of multiple closed domain relation extractors.

On the other hand, OpenIE addresses the issues highlighted in the ClosedIE process by extracting relations based on relational phrases. A relational phrase is a natural language phrase that represents a relation. This approach makes the relational extraction scalable and as well as generalizable for different languages. On the other hand as relational phrases are based only on the linguistic knowledge (not on the domain knowledge), it can be applied over multiple domains. Thus this makes it more suitable for a relation extraction process which utilizes the web based unstructured text.

In particular, we utilize the Ollie for the relation extraction phase. Ollie presents some key advantages over other OpenIE frameworks, such as Reverb and ClauseIE. Both Ollie and Reverb use seed relations to initiate the process which is not seen in ClasueIE [14] which is naively employs the dependency parsing to identify the relations. And Ollie is the successor of Reverb and contains more seed relations compared to the Revereb.

3.6   Aligning Triples with Relations

Once the relations are extracted using OpenIE, we have to then align them with the triples to extract potential lexicalization patterns. The alignment is made only for the subject and the object. The predicate does not need to be aligned as we are looking for the natural language representation of the predicate.

The alignment is calculated using the Phrasal Overlap Measure (POM) as below.

$$sim_{pom}(s_1, s_2) = tanh\left(\frac{overlap\,(s_1, s_2)}{|s_1| + |s_2|}\right) \tag{1}$$

where, s_1 and s_2 are two text strings and $overlap_{phrase}\,(s_1, s_2)$ is calculated using (2).

$$overlap(s_1, s_2) = \sum_{i=1}^{n}\sum_{m} i^2 \tag{2}$$

where, m is a number of i-word phrases that appear in text string pairs.

There are occasions where triple objects can be represented in multiple ways. For example, a date object can be represented in multiple formats (e.g., 2012-06-18 $\Rightarrow$ {18 June 2012; June 18, 2012}), measured number can be represented in different units (e.g., 100 m $\Rightarrow$ {100 m; 100 m; 10000 cm; 0.1 km}), numerical values can be verbalized to different formats (e.g., $2.3 \times 10^8 \Rightarrow$ {2.3 billion, 2300 million}).

Table 4 explains the three data formats that are further verbalized associating them different formats.

Since, we have no prior knowledge whether arguments in relation is aligned with the triple subject and object, an initial calculation is performed to find out the best matching order. In essence, given relation $\langle arg_1, rel, arg_2 \rangle_R$ and the triple $\langle subject, predicate, object \rangle_T$, subject and object align with arg_1 and

Table 4. Verbalization process for selected data formats

Data format	Verbalization process
Date	The dates are transformed to most commonly used 7 different formats. Technically this is accomplished by Java date formatting functions
Measured Values	We provided coverage for 9 different measurements including meter, dollar, celsius, and kilogram. The measurement information associated with the triple as metadata is used for the conversion process with hand-coded functions for each measurement unit
Normal numbers	Normal numbers are verbalized for 5 scales and if they contain decimal values they are formatted for four different decimal formats

arg_2 respectively as determined by the first record in Table 5. The inverse of this function is used to determine whether *subject* and *object* align with arg_2 and arg_1 respectively as shown in the second record in Table 5. The pattern alignment phase generates alignments with different POM values. Therefore, a threshold value must be defined to select the best alignment from the candidates to extract the pattern. This threshold value is set to the alignment of a triple to a relation were in both subject and objects have a single token ($\langle Steve, founder, Apple \rangle_T \Rightarrow \langle Steve, is\ the\ founder\ of,\ Apple \rangle_R$). This resulted in the value of 0.21 as the POM measure to select the alignment to extract lexicalization patterns.

Table 5. Conditions required to decide the alignment of the triple with relation

Required Conditions	Alignment result
$(sim_{pom}(subject, arg_1) > sim_{pom}(subject, arg_2))$ $\wedge\ (sim_{pom}(object, arg_2) > sim_{pom}(object, arg_1))$	$(subject \xrightarrow{align} arg_1) \wedge$ $(object \xrightarrow{align} arg_2)$
$(sim_{pom}(subject, arg_2) > sim_{pom}(subject, arg_1))$ $\wedge\ (sim_{pom}(object, arg_1) > sim_{pom}(object, arg_2))$	$(subject \xrightarrow{align} arg_2) \wedge$ $(object \xrightarrow{align} arg_1)$

3.7   Extracting Lexicalization Patterns

Aligned patterns from the previous module are used to extract lexicalization patterns by substituting the subject and object with expressions, $S?$ and $O?$ respectively. However, the pattern extraction process does not naively extract relation patterns from the relations, instead following two steps (described in Sects. 3.7 and 3.8) are executed to generate generalizable and cohesive patterns.

Search for the Best Matching Verbalization. Instead searching for the object in the relation to substitute it with expression, the best matching verbalization of the object is used. This is if the triple object is used then a partial match may cause a information loss or existence of unnecessary information.

Mapping Triple Object to Compound Tokens in Relation Arguments. In some scenarios, although a triple object cannot be directly aligned with the relation argument, it can be mapped to a compound noun phrase in relation argument. For example, the relation ⟨*Bill Clinton, is married to, Hillary Diane Rodham Clinton*⟩ $_R$ and the triple ⟨*Bill Clinton, spouse, Hilary Clinton*⟩ $_T$. Although both are aligned, the pattern extraction process cannot simply substitute the arg_2 with object expression because arg_2 is a compound phrase although it has the same meaning as triple object.

To address this scenario, we introduced the compound phrase substitution with dependency tree based compound noun identification. We first aggregate the relation elements and then do a dependency parse of the sentence. Figure 4 depicts the dependency parse of the aforementioned relation with respective POS tags. The typed dependencies with compound relations are extracted and transformed to phrases. These phrases are then analysed to check whether triple object values exist in the compound phrase. If such an existence is found then it is substituted with pattern object expression.

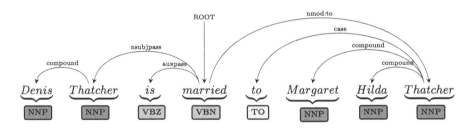

Fig. 4. Compound nouns with the compound typed dependency relation

3.8   Property Patterns

The property patterns are predefined pattern templates which are based on the predicates. The module contains 5 such patterns which can be applied only for particular predicates. Table 6 depicts the property patterns with example predicates, triples, and resulting lexicalizations.

Table 6. Property patterns with example triples

Pattern	Predicates	Triples	Resulting lexicalization
$\langle S?\text{'s } P?, \text{ is, } O? \rangle_L$	birthName	$\langle$Margaret Thatcher, birthName, Margaret Hilda Thatcher$\rangle_T$	Margaret Thatcher's birth name is Margaret Hilda Thatcher
$\langle S?, \text{ has, } O? \text{ } P? \rangle_L$	Championships	$\langle$Michael Schumacher, championships, 7$\rangle_T$	Michael Schumacher has 7 championships
$\langle S?, \text{ is, } O? \rangle_L$	Type	$\langle$Manhattan Bridge, type, Suspension bridge$\rangle_T$	Manhattan Bridge is a suspension bridge
$\langle P? \text{ in } S?, \text{ is, } O? \rangle_L$	numberOfPages	$\langle$The Lost World, numberOfPages, 816$\rangle_T$	Number of pages in The Lost World is 282
$\langle S?, \text{ } P?, \text{ } O? \rangle_L$	isPartOf	$\langle$Scotland, isPartOf, United Kingdom$\rangle_T$	Scotland is part of United Kingdom

4   Evaluation and Results

The evaluation of the framework focused on both linguistic correctness and a human evaluation which focused on readability and the accuracy for a sub-sample of the generated lexicalization patterns. The second test is limited to a random sub-sample due to the resource limit which is pertinent to any human evaluation process. The complete test set contained 28 entities randomly selected from DBpedia which represented 25 ontology classes. To support the relational patterns, text was extracted from the web and pre-processed for each of the entity presented in the collection.

4.1   Linguistic Accuracy Evaluation

This evaluation focused on evaluating the lexicalization patterns extracted for each of the entity for their linguistic accuracy. Figure 5 shows the result of this evaluation phase. The figure shows all triples for a particular entity, triples associated with correct lexicalization patterns, and triples which are not associated with a lexicalization pattern.

4.2   Human Evaluation

Five postgraduate students evaluated a random sample of 40 lexicalization patterns for the accuracy and readability in 5-point Likert scales. All the participants rated the questions for accuracy and readability with inter-rater agreement measured using Cronbach's alpha with values 0.866 and 0.807 respectively.

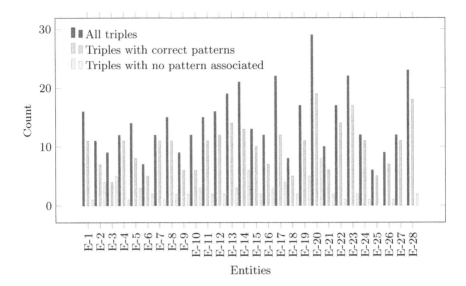

Fig. 5. Lexicalization pattern evaluation for linguistic accuracy. Entities are denoted as E-1 to E-28.

Table 7 shows further statistics related to the inter-rater agreement It is clear according to the table that participants have a good agreements as dropping of any participant's rating does not significantly affect the overall alpha value.

Table 7. Summary of the statistics related to inter-rater agreement in evaluation

Participant	Accuracy			Readability		
	Cronbach's α	Item-total correlation	α if Item Deleted	Cronbach's α	Item-total correlation	α if Item Deleted
P1	0.866	0.685	0.841	0.807	0.591	0.773
P2		0.672	0.843		0.683	0.740
P3		0.667	0.844		0.593	0.771
P4		0.705	0.834		0.561	0.781
P5		0.723	0.829		0.551	0.783

Table 8. Categorized weighted average of ratings for accuracy and readability

Weighted Average of ratings	Accuracy	Readability
1.0–2.0	1	1
2.1–3.0	0	0
3.1–4.0	11	10
4.1–5.0	28	29

Table 8 shows the lexicalization patterns categorized based on the weighted average of human ratings for both accuracy and readability.

4.3 Discussion

The linguistic accuracy evaluation showed that the framework has generated 283 accurate lexicalization patterns for a 400 triples. This shows that framework has achieved 70.75% accuracy level. Compared to a similar existing system, the Lemon model [15] which achieved 37% accuracy level in full automatic mode, our model performed with a much higher level of accuracy. The weighted rating for pattern accuracy and readability as rated by human evaluators showed a correlation for our surface level analysis. To analyse this further we performed a two-tailed Spearman correlation analysis which resulted in 0.828 correlation coefficient ($p < 0.01$). This shows that accuracy of a lexicalization pattern has an effect on the readability. The weighted averages from the human evaluation (refer Table 8) shows that majority of the participants have rated the lexicalization patterns in the range of 4.1–5.0 which shows that generated patterns are in good level of accuracy and with a readability level.

5   Related Work

Duma and Klein [16] introduce template based approach to lexicalize triples. The approach is mainly focused on triple graph lexicalization instead of individual triple lexicalization. The proposed approach by Duma and Klein [16] uses a triple collection and a text collection which contains the natural language representation of the triples. The model then attempts to extract candidate sentences that contain the triple subjects and objects. Next these sentences can be used as templates by substituting the subjects and objects with new triple components. One of the main drawback in this model is that it uses a very naive approach towards the template generation as it does not consider removing adjectives or adverbs in the candidate sentence as such additional tokens can decrease the generalizability of a template. On the other hand, a text collection with multiple sentences will often contain higher number of referring expressions and it will not be possible to match the exact subject or object text to the ones that are in the sentences. Therefore, a co-reference resolution step is necessary to carry out before the template extraction. However, the proposed approach in this paper introduces such additional functions to generate more accurate lexicalization patterns for a given triple.

The corpus based triple lexicalization approach presented by Walter et al. [15], is the only triple lexicalization approach that closely resembles with our objective which is lexicalizing individual triples. Walter et al. [15] use the dependency paths between subject and object mentioned in a candidate sentence to extract lexicalization patterns. However, our initial experiments showed that this approach fails completely when provided with sentences with grammatical conjunctions. Specially, when there exist complex dependency paths between the subject and object, the model is not able to find a cohesive pattern.

Ell and Harth [17] present a framework that generates RDF verbalization templates for a triple graph. The templates are extracted using the maximal sub-graph pattern extraction. However, the proposed approach in this paper uses four different pattern extraction methods which collectively contribute for the triple lexicalization. On the other hand, as mentioned in the beginning of this section, our focus in this research is to generate lexicalization patterns for individual triples while Ell and Harth focus on lexicalizing a triple graph.

6   Conclusion and Future Directions

This paper introduced RealText$_{lex}$ lexicalization framework which is designed as an ensemble architecture of four different pattern mining modules. The framework was evaluated in two forms and both showed that framework is able to generate syntactically correct lexicalization patterns without compromising the semantics associated with the Linked Data triples. The framework also achieved higher accuracy compared to the approach presented by Walter et al. [15] which works in a very similar context to the our research.

The future work of this research will focus on improving the pattern mining modules further to generate high quality lexicalization patterns. In addition, the lexicalization framework will be used in Question Answering (QA) systems [18–20] to generate descriptive answers using lexicalized triples. To accomplish this aim, the framework proposed in this paper will be integrated into some of the other research frameworks [21,22] that we have built in the QA domain which focus on answer presentation.

References

1. Berners-Lee, T.: Linked Data Design Issues. Technical report, World Wide Web Consortium (W3C) (2006)
2. Ngomo, A., Auer, S., Lehmann, J., Zaveri, A.: Introduction to linked data and its lifecycle on the web. In: 7th International Conference on Reasoning Web: Semantic Technologies for the Web of Data. ACM (2014)
3. Perera, R., Nand, P., Klette, G.: Realtext-lex: a lexicalization framework for RDF triples. Prague Bull. Math. Linguist. **106**(1), 45–68 (2016)
4. Perera, R., Nand, P.: RealText asg : a model to present answers utilizing the linguistic structure of source question. In: 29th Pacific Asia Conference on Language, Information and Computation (PACLIC). Association for Computational Linguistics (2015)
5. Perera, R., Nand, P.: Answer presentation in question answering over linked data using typed dependency subtree patterns. In: Open Knowledge Base and Question Answering Workshop collocated with 26th International Conference on Computational Linguistics (COLING), p. 44 (2016)
6. Bizer, C., Lehmann, J., Kobilarov, G.: DBpedia-a crystallization point for the Web of Data. Web Semant. **7**(3), 154–165 (2009)
7. Auer, S., Bizer, C., Kobilarov, G., Lehmann, J., Cyganiak, R., Ives, Z.: DBpedia: a nucleus for a web of open data. In: Aberer, K. (ed.) ASWC/ISWC -2007. LNCS, vol. 4825, pp. 722–735. Springer, Heidelberg (2007). https://doi.org/10.1007/978-3-540-76298-0_52

8. Kobilarov, G., Bizer, C., Auer, S., Lehmann, J.: DBpedia - a linked data hub and data source for web and enterprise applications. Int. World Wide Web Conf. **18**, 1–3 (2009)
9. Panther, K., Thornburg, L.: A conceptual analysis of English-er nominals. Appl. Cogn. Linguist. **1**, 149–200 (2002)
10. Jurafsky, D., Martin, J.H.: Speech and Language Processing: An Introduction to Natural Language Processing, Computational Linguistics, and Speech Recognition, 1st edn. Prentice Hall PTR, Upper Saddle River (2000)
11. Kohlschütter, C., Fankhauser, P., Nejdl, W.: Boilerplate Detection using Shallow Text Features. In: ACM International Conference on Web Search and Data Mining, pp. 441–450 (2010)
12. Lee, H., Peirsman, Y., Chang, A., Chambers, N., Surdeanu, M., Jurafsky, D.: Stanford's multi-pass sieve coreference resolution system at the CoNLL-2011 shared task. In: Conference on Natural Language Learning, Portland. Association for Computational Linguistics (2011)
13. Schmitz, M., Bart, R., Soderland, S., Etzioni, O.: Open language learning for information extraction. In: Joint Conference on Empirical Methods in Natural Language Processing and Computational Natural Language Learning, Jeju Island, pp. 523–534. ACL, July 2012
14. Del Corro, L., Gemulla, R.: ClausIE: clause-based open information extraction, pp. 355–366, May 2013
15. Walter, S., Unger, C., Cimiano, P.: A corpus-based approach for the induction of ontology lexica. In: Métais, E., Meziane, F., Saraee, M., Sugumaran, V., Vadera, S. (eds.) NLDB 2013. LNCS, vol. 7934, pp. 102–113. Springer, Heidelberg (2013). https://doi.org/10.1007/978-3-642-38824-8_9
16. Duma, D., Klein, E.: Generating natural language from linked data: unsupervised template extraction. In: 10th International Conference on Computational Semantics (IWCS 2013), Potsdam. ACL (2013)
17. Ell, B., Harth, A.: A language-independent method for the extraction of RDF verbalization templates. In: 8th International Natural Language Generation Conference, Philadelphia. ACL (2014)
18. Perera, R., Nand, P.: Interaction history based answer formulation for question answering. Commun. Comput. Inf. Sci. **468**, 128–139 (2014)
19. Perera, R.: Scholar - cognitive computing approach for question answering. Ph.D. thesis, University of Westminster (2012)
20. Perera, R.: IPedagogy: question answering system based on web information clustering. In: Proceedings - 2012 IEEE 4th International Conference on Technology for Education, T4E 2012, Hyderabad, pp. 245–246. IEEE Press (2012)
21. Perera, R., Nand, P., Naeem, A.: Utilizing typed dependency subtree patterns for answer sentence generation in question answering systems. Prog. Artif. Intell. **6**(2), 1–15 (2017)
22. Perera, R., Nand, P.: Generating lexicalization patterns for linked open data. In: NLP&LOD2 Collocated with 10th Recent Advances in Natural Language Processing (RANLP), Hissar, Bulgaria. Association for Computational Linguistics (2015)

A Hybrid Approach for Biomedical Relation Extraction Using Finite State Automata and Random Forest-Weighted Fusion

Thanassis Mavropoulos[✉], Dimitris Liparas, Spyridon Symeonidis, Stefanos Vrochidis, and Ioannis Kompatsiaris

Information Technologies Institute, Centre for Research and Technology Hellas, Thermi-Thessaloniki, Greece
{mavrathan,dliparas,spyridons,stefanos,ikom}@iti.gr

Abstract. The automatic extraction of relations between medical entities found in related texts is considered to be a very important task, due to the multitude of applications that it can support, from question answering systems to the development of medical ontologies. Many different methodologies have been presented and applied to this task over the years. Of particular interest are hybrid approaches, in which different techniques are combined in order to improve the individual performance of either one of them. In this study, we extend a previously established hybrid framework for medical relation extraction, which we modify by enhancing the pattern-based part of the framework and by applying a more sophisticated weighting method. Most notably, we replace the use of regular expressions with finite state automata for the pattern-building part, while the fusion part is replaced by a weighting strategy that is based on the operational capabilities of the Random Forests algorithm. The experimental results indicate the superiority of the proposed approach against the aforementioned well-established hybrid methodology and other state-of-the-art approaches.

Keywords: Natural language processing · Relation extraction
Supervised learning · Support Vector Machines · Random Forests
Weighted fusion

1 Introduction

The onset of the digital era and notably the advent of the internet have not only changed the way people communicate and entertain themselves but have also altered fundamentally their working practices and needs. The medical domain has been on the forefront of these changes, as medical professionals have been exploiting the latest advancements of research and technology in order to improve their services since the very beginning. But this wealth of information is sometimes overwhelming and difficult to tackle manually. A certain level of automation in information extraction is imperative, especially when non-medical practitioners, like patients or their families, are involved. In most cases these people do not possess the ability to fully understand the language used by the professionals since there is a great knowledge gap between the two groups. The rich in terminology patient history reports is one such area,

© Springer Nature Switzerland AG 2018
A. Gelbukh (Ed.): CICLing 2017, LNCS 10761, pp. 450–462, 2018.
https://doi.org/10.1007/978-3-319-77113-7_35

especially when these are riddled with acronyms tailored to the medical domain. The same holds for online resources, like dedicated medical sites and forums, which users often consider when soliciting for information on drugs, diseases or treatments.

Medical concept relation extraction deals with the automatic extraction of relations that exist between entity types relevant to this domain, such as treatment, test or disease, among others. This task has been the focal point for a lot of researchers, due to many applications that it can support, such as the creation of medical ontologies and content representation that could serve as basis for medical content retrieval and question answering systems, as well as decision support services for doctors. According to [1], *"identifying relations between medical entities in clinical data can help in stratifying patients by disease susceptibility and response to therapy, reducing the size, duration, and cost of clinical trials, leading to the development of new treatments, diagnostics, and prevention therapies"*.

Traditionally, studies on medical relation extraction have relied on rule/pattern-based linguistic approaches, machine learning ones and also on hybrid systems that combine linguistic templates and machine learning in order to improve their results. An example of a hybrid framework for medical relation extraction is the approach introduced in [2] and further evaluated in [3], which relied on two different methodologies: (a) relation patterns defined by human experts via regular expressions and (b) Support Vector Machine (SVM)-based classification based on three types of extracted features, namely lexical, morphosyntactic and semantic features. Fusion of the results from these two methodologies was achieved by means of a strategy, which relied on the training examples of a given dataset, giving more influence to the relation patterns when few training examples were available for a certain relation type and more influence to the machine learning approach when enough examples were provided.

In this paper, the focus is shifted towards the relation extraction task of the 2010 i2b2/VA challenge, which required the extraction of eight types of semantic relationships found between the medical concepts of the given dataset. The other parts of the contest involved the extraction of the medical concepts themselves and also the annotation of the assertions made about these concepts. We are inspired by the hybrid approach described above and we extend it with an innovative pattern-construction method, based on finite state automata, and a novel weighted fusion strategy. More specifically, we approach the creation of linguistic patterns not via the use of regular expressions, as in the case of [2], but by using node-based finite state automata, which can include information like the part of speech (POS) and the inflection of a lexical unit or even contain whole gazetteers of words inside a node.

As an additional novelty, we introduce the use of a Random Forests (RF) classification model, which provides the weighted fusion values for the pattern-based and machine learning modules of the relation extraction framework based on its operational performance on the training set, with the use of the out-of-bag (OOB) error estimate [4]. It should be noted that we keep the use of the SVM classifier for the machine learning module of our framework, due to its demonstrated superiority in many natural language processing (NLP)-related classification tasks. Our hybrid framework is applied to the currently available partial version of the 2010 i2b2/VA challenge dataset [5] and the experimental results demonstrate its superior performance, compared to a number of considered approaches.

The rest of this paper is organised as follows: In Sect. 2, the theoretical background and an outline of the relevant literature are provided. In Sect. 3 the proposed hybrid relation extraction approach is described, while Sect. 4 provides the experimental framework of our study. In Sect. 5, the results of the experiments are presented and discussed. Finally, Sect. 6 concludes the paper.

2   Related Work and Theoretical Background

In this section, since the biomedical domain constitutes the point of interest of the current study, we report previous work on relation extraction in this field. In addition, we provide information on the theoretical background, as well as the related work for the Random Forests (RF) and Support Vector Machines (SVMs) machine learning methods.

As already mentioned in Sect. 1, three main types of methodologies have been proposed over the years for concept relation extraction: the rule/pattern-based linguistic approaches, the statistical/machine learning approaches and the hybrid ones, which combine both approaches.

Pattern-based systems have been used in the biomedical domain since the early 2000s and have mainly approached the problem as a text classification one. [6] tried to extract and structure information related to molecular pathways with their GeneWays system. A year later, [7] attempted to extract similar relationships between genes, proteins, drugs and diseases.

However, the term "relation extraction" is only part of the problem called "relation classification", which was first introduced in [8] and entails the extraction of the semantic roles and the recognition of the relationship that holds between them. It was a very influential study that explored five generative graphical models and a neural network to identify seven different relationships that can be found between "treatment" and "disease" entities. The corpus that was used in their work originates from "The BioText Project", is known as the "MEDLINE 2001" corpus and has since been widely used in relation extraction tasks. In [9], a Conditional Random Fields (CRF) classifier was used because of the need to detect the medical entities and at the same time, the relations between them. The semantic relations between diseases and treatments, as well as between genes and treatments were targeted, which were classified into seven and five predefined types respectively. All experiments were conducted on the "MEDLINE 2001" corpus. Relation extraction between entities in literature text (Medline abstracts) was conducted by [10], via the use of kernel-based learning methods. The method involved a customization of the standard tree kernel function "by incorporating a trace kernel to capture richer contextual information" and resulted in outperforming word and sequence kernels.

The framework that currently claims the best results between treatments and diseases on the MEDLINE 2001 corpus is the one presented in [11], which uses a hybrid feature set for the classification of relations. The major differentiation is in the semantic feature set, where verb phrases are ranked using the Unified Medical Language System (UMLS), while the relations are classified by SVM and Naïve Bayes models.

2010 was a year that marked a great insurgence of research in the medical concept extraction domain and this was due in no small part to the respective i2b2 Shared-Task and Workshop. The contest gave the research community the incentive by supplying a pre-annotated corpus with concepts, relations and assertions. Since then, the contest's best ranking systems are considered as the reference, against which all new ones are benchmarked.

The research, which is underway in the extraction of biomedical relationships, has also been receiving growing attention, "with numerous biological and clinical applications including those in pharmacogenomics, clinical trial screening and adverse drug reaction detection", as [12] are outlining in great detail. In addition, there have been some recent approaches based solely on Convolutional Neural Network (CNN) models. For instance, in [13], a CNN-based model is implemented in order to extract the semantic relations found between medical concepts and with the goal "to learn features automatically and thus reduce the dependency on manual feature engineering". The method is applied to the currently available partial version of the 2010 i2b2/VA challenge dataset with promising results.

Random Forests (RF) is a well-known machine learning method [4], used with great success in many applications. Its basic idea is the construction of a multitude of decision trees, which can be used for classification and regression purposes. There is randomness in the operational procedures of RF in two different ways: 1) Each decision tree is constructed on a different group of data, sampled randomly with replacement (bootstrap) from the training set, and 2) During the construction of each decision tree, the best split at each node is determined based on a randomly selected subset of the variable set. An estimation of the generalisation error of RF can be provided by means of an inherent method called out-of-bag (OOB) error. In a nutshell, only approximately 2/3 of the original data examples are used in a specific bootstrap sample during the construction of a decision tree. The rest of the original data examples (approximately 1/3), called OOB data, are used for testing the performance of the constructed decision tree. The OOB error is the averaged prediction error for each training case, using only the decision trees that do not have that training case in their bootstrap sample. As already mentioned, RF has been successfully applied to many disciplines. Specifically in the biomedical domain, there have been applications of RF for automated diagnosis of diseases [14], electromyography (EMG) signal classification [15], or in the context of brain-computer interfaces (BCI) [16], among others. Finally, the use of late fusion strategies based on RF's operational capabilities in the context of multimodal news articles classification has been investigated in [17].

Support Vector Machines (SVMs) [18] are supervised learning methods used for solving pattern recognition problems. Their basic notion lies in hyperplanes, which are used to separate sets of data points with different class memberships in multidimensional spaces. The effectiveness of SVMs in NLP classification tasks and more specifically, for relation extraction, can be highlighted by the fact that the highest performance for the relation extraction task in the 2010 i2b2/VA challenge was achieved by [19] with their supervised approach. This approach employed an SVM classifier to identify relations, which was informed by several resources such as Wikipedia, WordNet, General Inquirer and a relation similarity metric. Furthermore, the only hybrid system participating in the challenge, employing an SVM classifier together with manually constructed linguistic patterns was developed by [20]. Finally, [1] used an SVM classifier with a combination of lexical, syntactic and semantic features, terms extracted from a vector-space model created using a random projection algorithm, as well as additional contextual information extracted at sentence-level to detect relations.

3 Hybrid Relation Extraction Approach

In this section we present the proposed framework for the medical relation extraction problem, which is illustrated in Fig. 1. It consists of two main modules for relation extraction (a pattern-based and a machine learning one) and a weighting module for the fusion of the results provided by each module.

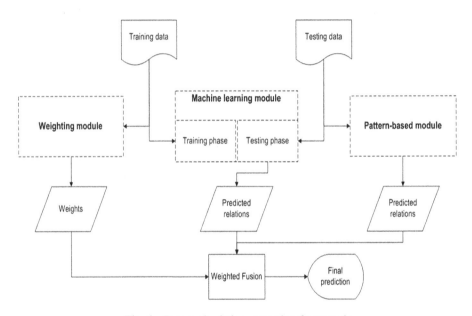

Fig. 1. Proposed relation extraction framework

Pattern-Based Module. While developing a pattern based method one has to consider the many forms that are often utilised in natural language to express the same thing. These variations need to be taken into consideration when devising the manually constructed rules and patterns, in order for the system to deliver the optimal results. This exact fact is also what makes pattern based methods complex and time consuming to develop. The procedure is not without limitations, with the most important being that the patterns need to be exhaustive enough to cover all possible language variations but manage it without overlapping each other. Therefore, the set of patterns that is created must be finely tuned in order to avoid conflicts that can invalidate the extraction results. Another limitation is that the pattern creation is largely dependant on the corpus, which dictates a certain degree of flexibility and adaptation when there is the need to apply the method to a new corpus.

The method of choice revolves around finite state automata, which, while being the simplest level of grammar and well understood by users who write rules, is also a technique versatile enough to enable detailed description of complex linguistic phenomena as well as permit the generation of output files rich in linguistic information.

Thus, for the semantic relation extraction task, a set of patterns is constructed for each target relation after examining the structure of certain natural language expressions and detecting common forms in them. This is usually possible with the use of regular expressions and by exploiting keywords usually found in clinical texts, like *cure, treat, drug* and *side effect*. It is the most commonly used method and the one employed by [3] in their MEANS system. However, the current paper adopts an approach which is based on the exploitation of finite state automata (or graphs) via the use of the corpus processing suite *Unitex* [21], in order to overcome any limitations that are encountered when utilising regular expressions. The pattern-building procedure is done through a powerful interface that enables the manipulation of interconnecting nodes, in order for the user to achieve the most descriptive pattern possible. These nodes may contain a POS, a regular expression, a multitude of linguistic filters (e.g. the feminine plural forms of an adjective) or even whole graphs. A major differentiation compared to simple regular expressions, which ultimately plays a pivotal role in the effectiveness of a Unitex-made graph, is the ability to exploit the rich in linguistic information incorporated dictionaries. These have been manually created and contain the grammatical attributes, such as POS or inflection, for the whole of the English vocabulary. In addition to the default integrated dictionaries, *Unitex* also supports the creation of custom ones which can be populated with specialised entries such as disease or treatment terminology.

Each relation targeted by the pattern-based module is being represented by a number of dedicated, manually constructed patterns that locate medical entities/concepts, which appear in pairs in a sentence. A weighted label of specificity is allocated to each pattern in order to solve ambiguous matches, since different relations can be expressed in similar manners (for each pattern, the more detailed the representation of the lexical context, the more specific the weight that gets allocated). The pattern weights that correspond to the assigned labels take the values of 1 for the most specific relation type pattern, 0.75 for a fairly specific one and 0.50 for low specificity patterns (i.e. R1 = 1, R2 = 0.75, R3 = 0.50, with R1 being the most specific relation (R)). When the entity pair meets the criteria laid out by one of these patterns, the respective label is assigned. To be more precise via an example, the phrase "He had been noting night sweats, increasing fatigue, anorexia, and dyspnea, which were not particularly improved by increased transfusions or alterations of hydroxy urea." can be represented with the automaton of Fig. 2, while one of the possible output sentences is represented as (E1 = entity1 and E2 = entity2): *He had been noting night sweats, increasing fatigue, anorexia, and <E2> dyspnea </E2> which were not particularly <TrWP2> improved by </TrWP2> <E1> increased transfusions </E1> or alterations of hydroxy urea.*

All grey boxes invoke secondary graphs with similar formalism to this one, which contain relevant information to their title. The nodes "disease/signORsymptom" and

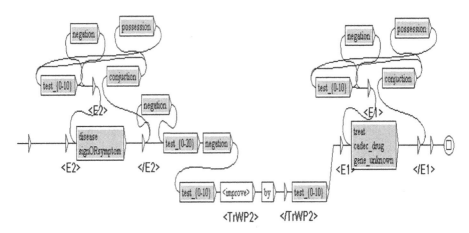

Fig. 2. Finite state automaton representing the "TrWP" relation type.

"treat/cadec_drug/gene_unknown" enclose the relevant dictionaries, while the nodes "negation", "possession", "conjunction" describe the respective syntactic functions. Lastly, the white node, which is the only one not evoking another graph, is determining the output of the box, which in this case is the relation type <TrWP2> (Treatment Worsens Problem with level 2 specificity). In total, around 350 patterns were created, a number that also includes assistive graphs, like the ones used to handle lexical units of trivial importance found between or around the target entities (test_{10}/test_{20}).

Machine Learning Module. In the training phase, a linear SVM classifier is trained on features extracted from a given dataset in order to describe each example. The extracted features fall into three types: lexical, morphosyntactic and semantic features.

The lexical features include the entities' position in the phrase, the words that form each entity and their immediate context; the words before, after and between them. Also of importance are their lemmas. The morphosyntactic features include the POS (extracted by the Stanford CoreNLP suite [22]) of the lexical units in question, the number of words that form each entity, the verbs before, after and between the entity pairs. Finally, the semantic features refer to the concepts associated to the target entities, as well as those found in their close vicinity; before, after and between them. They are all derived from the online resource UMLS [23], which is a software suite that encompasses various health-related vocabularies and standards to allow for interactions between computer systems. Another type of feature, which carries semantic information and is provided in the dataset, is the concept type of each entity. However, it was decided that, while such a feature is positively helpful and already available in the given dataset, it wouldn't be included in the feature set of the used classifier. The reason behind this decision lies in the non-existent availability of a reliable resource/procedure that can provide equivalent values in a real–life, non-laboratory scenario.

In the testing phase, for any instance where its relation type is considered to be unknown, the trained SVM model outputs a prediction of the relation type in the form of probability scores.

Weighting Module. The probability scores from the pattern-based and machine learning modules are combined using weighted fusion. Different weights are assigned to each module and for each class (relation type). In order to output the final probability that a case is relevant to a class R, the predicted scores P_{pb} (from the pattern-based module) and P_{ml} (from the machine learning module) are first multiplied by their corresponding weights W_{pb} and W_{ml} and are then summed, as in Eq. (1). The relation type with the highest fused probability score is assigned to each test set instance.

$$P_{\text{fused}}(R) = \big(W_{pb}(R) * P_{pb}(R)\big) + \big(W_{ml}(R) * P_{ml}(R)\big) \tag{1}$$

In this study, we propose a weighting method, which relies on a different classifier than the one used in the machine learning module. Specifically, an RF model is trained on the training examples in order to leverage an operational capability exclusive to this algorithm. This capability is the out-of-bag (OOB) error, which provides an estimation of the generalisation error of RF. During the training of the RF model, a portion of the original data examples (approximately 1/3), called OOB data, are used for testing the performance of each constructed decision tree. The accuracy of the trained RF model on the OOB data is calculated for each class (relation type) separately and the corresponding scores are assigned as weight values to the machine learning module. The sum of the weights for the two modules must be strictly equal to 1. This means that the pattern-based weight for a relation R is the complement of the corresponding machine learning weight, $W_{pb}(R) + W_{ml}(R) = 1$.

4 Experimental Framework

Dataset. The proposed approach was evaluated on the relation extraction task of the 2010 i2b2/VA challenge, which has been the reference for nearly every competing system working on medical relation extraction. The task's focus was on eight relation categories, as it can be seen in Table 1. The eight relationships can be further classified into three sub-groups of the treatment-problem (TrIP, TrWP, TrCP, TrAP, TrNAP), test-problem (TeRP, TeCP) and problem-problem (PIP) variety. The vast majority of training examples that can be found in the dataset belongs to the TrAP, PIP and TeRP relations, with 885, 755 and 992 examples respectively. This number amounts to 84.39% of the dataset examples, which is a problem in itself as the remaining 15.61% that represents the five less populated classes is not enough to effectively feed the training procedure of the classifier in order to produce acceptable results. This fact alone renders the presence of a pattern-based module imperative, which not only rectifies the problem of the sub-populated classes, but also aids in the amelioration of the final results in their entirety.

Table 1. Details of the dataset.

Relation type		Examples
TrIP	Treatment improves medical problem relations	51
TrWP	Treatment worsens medical problem relations	24
TrCP	Treatment causes medical problem relations	184
TrAP	Treatment is administered for medical problem relations	885
TrNAP	Treatment is not administered because of medical problem relations	62
PIP	Medical problem indicates medical problem relations	755
TeRP	Test reveals medical problem relations	992
TeCP	Test conducted to investigate medical problem relations	166

The original dataset consisted of 394 training reports, 477 test reports, and 877 unannotated reports, while currently, the dataset is only partially available for research, due to IRB limitations, with 170 training and 256 test reports, respectively.

Experimental Setup. The LibSVM [24] wrapper class contained in the Weka machine learning software was used to train the linear SVM models of the machine learning module. The main SVM parameters C and gamma, received values of 1 and 0, respectively. In the training procedure one binary classifier (mono-class) was trained for each relation type. For weight assignment, two different strategies were tested. In the first strategy (proposed in [2]), the weight values are directly analogous to the frequency of each relation type in the training set examples. The second strategy is the one we propose for our hybrid approach, based on the RF OOB error estimate. The RF classification model was trained using the scikit-learn python library. The number of trees in the trained RF model was set to 1000. Finally, for the evaluation of the performance of all configurations the micro-averaged values for the precision, recall and F-score measures were computed.

5 Experimental Results

The test set results from the experiments conducted in this study are compared in Table 2 with state-of-the-art systems. Rows 2 and 3 of Table 2 contain the results from our system and from the one we use as a baseline approach. It should be noted that all experiments for these two hybrid systems were conducted with the use of our own patterns, as it is not possible to recreate the exact patterns used in [2]. We observe a 2.6% relative improvement (in terms of micro-averaged F-score) in the performance of our system, when compared to the baseline system. This improvement is satisfactory, considering that only the weighting strategy changes are taken into account. No reliable comparison can be made on a pattern level, until the two systems are compared on the same dataset. We would like to note that in general, the finite state automata of our pattern-based module were specific in their performance, with a high micro-averaged precision value (0.772) and a low micro-averaged recall value (0.292). In row 4 of Table 2, [13] trained a convolutional neural network on the exact same limited I2b2 dataset that we also used in our experiments. Rows 5–8 of Table 2 present the performance and type of the relation extraction systems that scored the highest in the I2b2/VA challenge (they used the full dataset, so the machine learning part was trained with more data). We notice that our proposed system outperforms all considered state-of-the-art approaches, to a lesser or greater extent. Most notably, there is an approximate 7% relative improvement in our system's performance, compared to the best I2b2 hybrid system [20].

Table 2. Performance evaluation of the proposed hybrid system vs. the baseline system and state-of-the-art approaches.

System	Approach	F-score
Our method	Hybrid	0.758
Abacha and Zweigenbaum [2]	Hybrid	0.739
Sahu et al. [13]	Semi-supervised	0.712
Roberts et al. [25]	Supervised	0.737
DeBruijn et al. [26]	Semi-supervised	0.731
Grouin et al. [20]	Hybrid	0.709
Patrick et al. [27]	Supervised	0.702

Furthermore, Table 3 presents the added value that the integration of the pattern-based module brings to our hybrid system, compared to the use of the machine learning module only. We notice an overall improvement in the F-score values for the different relation types of the dataset. The biggest gains are observed in the TrNAP and TeCP relation types, with a relative improvement of 320.6% and 133.6%, respectively. If we specifically observe the results for the relation types, for which there are fewer training examples available in the dataset (compared to the TrAP, PIP and TeRP relations), namely TrIP, TrWP, TrCP, TrNAP and TeCP, it is evident that there are noticeable improvements from the integration of the pattern-based module, not only in relative terms, but also in terms of absolute values. For instance, for the TrIP relation we have a 3.9% (absolute) improvement, for the TrWP relation the F-score of the hybrid system is computed only on the performance of the pattern-based module and for the TrCP relation, there is a 6% (absolute) improvement. Hence, it can be said that the afore-mentioned performance improvements warrant the manual effort needed for the construction of our hybrid system's pattern-based module. Finally, we used the nonparametric Wilcoxon signed ranks test for comparing the 8 F-score values in Table 3 from the supervised (machine learning) module with the corresponding values from the hybrid system. The p-value is 0.008, showing that there is evidence for statistically significant difference between the performance of the machine learning module and that of the hybrid system.

Table 3. Performance difference (in terms of F-score) from the integration of the pattern-based module into the proposed system.

Relation type	Supervised	Hybrid	Relative difference
TrIP	0.240	0.279	+16.2%
TrWP	0.0	0.275	N/A
TrCP	0.456	0.516	+13.2%
TrAP	0.749	0.782	+4.4%
TrNAP	0.068	0.286	+320.6%
PIP	0.792	0.823	+3.9%
TeRP	0.817	0.829	+1.5%
TeCP	0.125	0.292	+133.6%

6 Conclusions and Future Work

In this study, we have proposed a novel medical concept relation extraction framework by extending [2] with the use of a more sophisticated pattern-constructing method and a weighting strategy, which leverages an inherent operational feature of the RF algorithm. Based on experiments conducted on a well-known dataset for relation extraction, we have demonstrated that our methodology outperforms a number of state-of-the-art approaches. It should be noted that in [2] the evaluation is conducted on the MEDLINE 2001 corpus and the patterns of the corresponding module are constructed in a different way. In the future, we plan to fully compare our approach with the latter on the MEDLINE 2001 corpus, as well as investigate the use of alternative weighting strategies for our framework.

Acknowledgments. This work was supported by the project KRISTINA (H2020-645012), funded by the European Commission. Deidentified clinical records used in this research were provided by the i2b2 National Center for Biomedical Computing funded by U54LM008748 and were originally prepared for the Shared Tasks for Challenges in NLP for Clinical Data organized by Dr. Ozlem Uzuner, i2b2 and SUNY.

References

1. Frunza, O., Inkpen, D.: Extracting relations between diseases, treatments, and tests from clinical data. In: Butz, C., Lingras, P. (eds.) AI 2011. LNCS, vol. 6657, pp. 140–145. Springer, Heidelberg (2011). https://doi.org/10.1007/978-3-642-21043-3_17
2. Ben Abacha, A., Zweigenbaum, P.: A hybrid approach for the extraction of semantic relations from MEDLINE Abstracts. In: Gelbukh, A. (ed.) CICLing 2011. LNCS, vol. 6609, pp. 139–150. Springer, Heidelberg (2011). https://doi.org/10.1007/978-3-642-19437-5_11
3. Ben Abacha, A., Zweigenbaum, P.: Means: A medical question-answering system combining nlp techniques and semantic web technologies. Inf. Process. Manage. **51**(5), 570–594 (2015)
4. Breiman, L.: Random forests. Machine learning **45**(1), 5–32 (2001)

5. Uzuner, Ö., South, B.R., Shen, S., DuVall, S.L.: 2010 i2b2/va challenge on concepts, assertions, and relations in clinical text. J. Am. Med. Inform. Assoc. **18**(5), 552–556 (2011)
6. Friedman, C., Kra, P., Yu, H., Krauthammer, M., Rzhetsky, A.: Genies: a natural-language processing system for the extraction of molecular pathways from journal articles. Bioinformatics **17**(suppl 1), S74–S82 (2001)
7. Feldman, R., Regev, Y., Finkelstein-Landau, M., Hurvitz, E., Kogan, B.: Mining biomedical literature using information extraction. Current Drug Discov. **2**(10), 19–23 (2002)
8. Rosario, B., Hearst, M.A.: Classifying semantic relations in bioscience texts. In: Proceedings of the 42nd Annual Meeting on Association for Computational Linguistics, p. 430. Association for Computational Linguistics, July 2004
9. Bundschus, M., Dejori, M., Stetter, M., Tresp, V., Kriegel, H.P.: Extraction of semantic biomedical relations from text using conditional random fields. BMC Bioinform. **9**(1), 1 (2008)
10. Li, J., Zhang, Z., Li, X., Chen, H.: Kernel-based learning for biomedical relation extraction. J. Am. Soc. Inf. Sci. Technol. **59**(5), 756–769 (2008)
11. Muzaffar, A.W., Azam, F., Qamar, U.: A relation extraction framework for biomedical text using hybrid feature set. In: Computational and Mathematical Methods in Medicine (2015)
12. Luo, Y., Uzuner, Ö., Szolovits, P.: Bridging semantics and syntax with graph algorithms state-of-the-art of extracting biomedical relations. Briefings in Bioinformatics (2016)
13. Sahu, S.K., Anand, A., Oruganty, K., Gattu, M.: Relation extraction from clinical texts using domain invariant convolutional neural network. arXiv preprint arXiv:1606.09370 (2016)
14. Tripoliti, E.E., Fotiadis, D.I., Manis, G.: Automated diagnosis of diseases based on classification: dynamic determination of the number of trees in random forests algorithm. IEEE Trans. Inf Technol. Biomed. **16**(4), 615–622 (2012)
15. Gokgoz, E., Subasi, A.: Comparison of decision tree algorithms for EMG signal classification using DWT. Biomed. Signal Process. Control **18**, 138–144 (2015)
16. Steyrl, D., Scherer, R., Faller, J., Müller-Putz, G.R.: Random forests in non-invasive sensorimotor rhythm brain-computer interfaces: a practical and convenient non-linear classifier. Biomed. Eng./Biomedizinische Technik **61**(1), 77–86 (2016)
17. Liparas, D., HaCohen-Kerner, Y., Moumtzidou, A., Vrochidis, S., Kompatsiaris, I.: News articles classification using random forests and weighted multimodal features. In: Lamas, D., Buitelaar, P. (eds.) IRFC 2014. LNCS, vol. 8849, pp. 63–75. Springer, Cham (2014). https://doi.org/10.1007/978-3-319-12979-2_6
18. Vapnik, V.N.: The Nature of Statistical Learning Theory (1995)
19. Rink, B., Harabagiu, S., Roberts, K.: Automatic extraction of relations between medical concepts in clinical texts. J. Am. Med. Inf. Assoc. **18**(5), 594–600 (2011)
20. Grouin, C., et al.: CARAMBA: concept, assertion, and relation annotation using machine-learning based approaches. In: i2b2 Medication Extraction Challenge Workshop, November 2010
21. Paumier, S., Nagel, J.S.: UNITEX 3.1BETA. User Manual (2013)
22. Manning, C.D., Surdeanu, M., Bauer, J., Finkel, J.R., Bethard, S., McClosky, D.: The stanford corenlp natural language processing toolkit. In: ACL (System Demonstrations), pp. 55–60, June 2014
23. Lindberg, D.A., Humphreys, B.L., McCray, A.T.: The unified medical language system. In: IMIA Yearbook, pp. 41–51 (1993)
24. Chang, C.C., Lin, C.J.: LIBSVM: a library for support vector machines. ACM Trans. Intell. Syst. Technol. (TIST) **2**(3), 27 (2011)

25. Roberts, K., Rink, B., Harabagiu, S.: Extraction of medical concepts, assertions, and relations from discharge summaries for the fourth i2b2/VA shared task. In: Proceedings of the 2010 i2b2/VA Workshop on Challenges in Natural Language Processing for Clinical Data. i2b2, Boston (2010)
26. de Bruijn, B., Cherry, C., Kiritchenko, S., Martin, J., Zhu, X.: NRC at i2b2: one challenge, three practical tasks, nine statistical systems, hundreds of clinical records, millions of useful features. In: Proceedings of the 2010 i2b2/VA Workshop on Challenges in Natural Language Processing for Clinical Data. i2b2, Boston (2010)
27. Patrick, J.D., Nguyen, D.H.M., Wang, Y.: I2b2 challenges in clinical natural language processing 2010. In: Proceedings of the 2010 i2b2/VA Workshop on Challenges in Natural Language Processing for Clinical Data. i2b2, Boston (2010)

Exploring Linguistic and Graph Based Features for the Automatic Classification and Extraction of Adverse Drug Effects

Tirthankar Dasgupta[1,2(✉)], Abir Naskar[1,2], and Lipika Dey[1,2]

[1] TCS Innovation Lab, New Delhi, India
[2] TCS Innovation Lab, Kolkata, India
{dasgupta.tirthankar,abir.naskar,lipika.dey}@tcs.com

Abstract. Adverse drug effects (ADEs) are known to be one of the leading causes of post-therapeutic death. Thus, their identification constitutes an important challenge as the effects of ADEs are often underreported. However, the recent popularity of different social media sources has make it a promising source for ADE extraction. In this paper, we have explored different linguistic and graph topological features to automatically classify short sentences or tweets into ADEs or Non-ADEs. We have further represented the ADE knowledge base into an bipartite network structure of drugs and their side effects to model drug-side effect relationships. The proposed model can also be used to discover implicit ADEs that are not represented in the source data. We have evaluated our proposed models with two openly available ADE dataset. Our evaluation results shows that the proposed model have surpasses the performance of the existing baseline systems.

Keywords: Adverse Drug Reactions · Graph topology
Semantic context · Classification · Information extraction

1 Introduction

An Adverse Drug Events/Effects (ADEs) or Adverse Drug Reactions (ADRs) can be defined as "any undesirable effect of a drug beyond its anticipated therapeutic effects occurring during clinical use" [15]. ADEs are known to be potentially harmful to patients and are amongst the top causes of morbidity and mortality [19]. It has been estimated that around 4.7% of hospitalizations each year and 6.7% incidence rate among already hospitalized patients are caused due to ADEs. ADEs are between the fourth and sixth leading cause of death in hospitals [19]. Many ADEs are difficult to diagnosed due to the fact that drug effects may prone to age, gender, geographic locations and other conditions that may take a long time to expose. Consequently, ADEs have led to the withdrawal of several drugs whose long term risks were identified later [14].

The extensive proliferation of social media platforms has provided unprecedented access to large streams of information related to drug effects. Such data

© Springer Nature Switzerland AG 2018
A. Gelbukh (Ed.): CICLing 2017, LNCS 10761, pp. 463–474, 2018.
https://doi.org/10.1007/978-3-319-77113-7_36

can provide significant insights about drugs and their possible effects. Consider for example this (real) Twitter message: "Cipro destroyed my entire body, from head to toe. Bed ridden since 2009. Be VERY careful." Evidently, the tweet contains significant drug effect and adverse side-effect information. The content of such streams is widely available and may be used for pharmacovigilance and/or for collecting large-scale information on drugs and their side effects.

Recently computational analysis of unstructured data such as medical reports [11,12] or social network data [10,18] have been used to detect content that contains ADRs. A large number of tweets or Facebook status contains ADRs may expose serious or unknown consequences. Typically, different machine learning techniques are used to detect ADEs. Most of these models uses traditional linguistic or statistical features [20,21,24]. These approaches suffer from the fact that their models do not take in account long distance dependencies between terms and their orders.

In this paper, we have explored computational linguistic and graph topological features to automatically classify unstructured texts into ADEs and Non-ADEs. The primary contributions of the paper are enumerated below:

1. A combined computational linguistic and graph based techniques to automatically classify unstructured social media texts into ADEs.
2. Representing the ADE knowledge base into an bipartite network structure to model drug and side effect relationships.
3. The proposed model can also be used to discover implicit ADEs that are not represented in the source data.

The rest of the paper is organized as follows. Section 2 discusses related work on ADR detection from text. Section 3 presents the classification framework including the pre-processing units and the feature extraction unit. Section 4 presents the technique to extract the adverse effects. Section 5 presents the Experimental setup and results. Finally, Sect. 6 concludes the paper.

2   Related Works

A plethora of works have been done in automatic classification of ADEs from different unstructured texts that includes twitter, or other biomedical news artiles. Traditionally, ADEs have been identified through expert review of the event reports. A technique called Gamma Poisson Shrinking (GPS) and Multi-Item Gamma Poisson Shrinking (MGPS) algorithms is proposed in the literature [22]. The World Health Organization has used Bayesian Confidence Propagation Neural Networks (BCPNN) at its Upsalla Monitoring Center [6]. Moreover, empirical Bayes screening [1], and odds ratios [7], have also been proposed. These techniques have also been used to identify novel drug-drug interactions through regression modeling. Apart from this, analysis of some structured data has also been performed like EMRs contain both structured data (for example ICD9 codes) and unstructured data (clinical free text), both of which have been targeted for ADE discovery [16]. Information Extraction techniques from the field of Natural Language Processing have been used to identify drug-effect pairs in patient discharge summaries [23] to identify ADEs. Topic modeling techniques have been applied to find common pharmacological topics or distributions of effects amongst the drug labels [2]. Works have also been done to detect off target drug interactions were predicted using the molecular similarity between drugs and known ligands. These predicted off-target interactions were shown to explain several known ADEs.

3   The ADE Classification Framework

The overall architecture of the ADE classification and analysis framework has four primary modules: (a) The *Linguistic pre-processing unit* (b) *Feature Extraction Unit* (c) *classifier unit* and (d) the *analysis unit*. Each of the modules are briefly discussed in the following subsections.

3.1   The Dataset

We use two datasets for training and testing of our proposed classification and extraction models. The first one is a Twitter dataset [21] published for a shared task in Pacific Symposium on Biocomputing, Hawaii, 2016. The tweets associated with the data were collected using generic and brand names of the drugs. The tweets were annotated for presence of ADRs. We have downloaded around 3,100 tweets as many tweets links are no longer accessible.

The second dataset, the ADE (adverse drug effect) corpus, was created by [12] by sampling from MEDLINE case reports. Each case report provides important information about symptoms, signs, diagnosis, treatment and follow-up of individual patients. The ADE corpus contains 4,272 sentences are annotated with names and relationships between drugs, adverse effects and dosages.

3.2   The Linguistic Pre-processing Unit

Here, the input text is first passed to the preprocessing unit that removes html tags, and foreign language characters from the text. The preprocessed text is then passed to the Stanford parts-of-speech (POS) tagger and parser [5] for syntactic processing. The POS tagger tokenizes the text and labels each word with their corresponding POS. From the output of the POS tagger, root verbs are extracted and passed to an English morphological analyzer to identify the tense, aspect and modality of the root verb. The Stanford parser is used to extract different dependency relations within the sentences.

3.3   Feature Extraction

The syntactically analyzed text is then passed to the feature extraction unit. In the present work we have considered the following features:

N-gram (N): We used counts of bigram features to train our model. However, instead of considering all the bigrams in a text, we have computed Pointwise mutual information between all possible bigram pairs and considers only those bigrams as features whose PMI score exceeds the *average + stddev* threshold.

Term Relevance (TR): We have considered all the unigram terms that are relevant to a given class. We have computed the term relevance of a word following the technique as described in the literature [4].

Dependency features (D): We took the count of all Stanford typed dependency features that includes adverbial clause modifier, auxiliary, negation modifier, marker, referent, open clausal complement, clausal complement, expletive, coordination, passive auxiliary, nominal subject, direct object, copula, conjunct as the most important features. It is worth mentioning here that the collected twitter dataset was extremely noisy. Therefore, dependency features can only be applied to the ADE/MEDLINE data set.

Sentiment features (S): We also took normalized sentiment scores of the drug descriptors tweets as features. Sentiment scores are calculated using qdap package in R using polarity function.

3.4   Graph Topological Features

Apart from the above traditional linguistic features, we have also explored the language independent graph topological features as discussed in [8,9]. The graph topological features comprises of construction of two different type of graphs, (a) the Word Distance Graph (WD) and (b) the Drug Adverse Effect Graph (DAE). A WD graph is constructed based on the principle that each of the word in a sentence has got some relationship to all other words in that sentence and the degree of relationship between two words is proportional to the distance between them in the sentence. Therefore, The closer two words are to each other, the stronger their connection tends to be. Accordingly, we have constructed the WD graph by considering each of the word pair (W_1, W_2) in a sentence and assigning an edge between them if W_1 is in the neighborhood[1] of W_2.

Apart from word distance, we have also used pointwise mutual information (PMI) between two words/vertices to calculate the edge weight. Co-occurrence is counted at the sentence level, i.e. $P(i, j)$ is estimated by the number of sentences that contain both terms W_i and W_j, and $P(i)$ and $P(j)$ are estimated by counting the total sentences containing W_i and W_j, respectively. As the set of seen sentences grows and co-occurrence between words becomes more prevalent, PMI becomes more influential on edge weights, strengthening edges between words that have high PMI. Formally, the weight wt for each edge in the graph is defined below, where $d_{i,j}$ is the distance between words W_i and W_j and $PMI(i, j)$ is the pointwise mutual information between words W_i and W_j, given the sentences seen so far.

---

[1] We imposed a "window size" of 8 as a limit on the maximum distance between two words to enter an edge relationship.

$$PMI(W_i, W_j) = log_2(\frac{P(W_i, W_j)}{P(W_i)P(W_j)}) \qquad (1)$$

$$Wt(W_i, W_j) = \frac{1 + PMI(W_i, W_j)}{d^2(W_i, W_j)} \qquad (2)$$

Based on the generated graph discussed above, we have applied the TextRank metric, as described in [17]. TextRank is inspired by the PageRank metric which is used for web page ranking. Here, weight of each vertex is computed as:

$$TR(V_i) = (1 - d) + d * (\sum_{V_j \in NB(V_i)} \frac{Wt_{ji}}{\sum_{V_k \in NB(V_j)}(Wt_{jk})} * TR(V_j)) \qquad (3)$$

where $TR(V_i)$ is the TextRank score for vertex i, $NB(V_i)$ is the set of neighbors of V_i, i.e. the set of nodes connected to V_i by a single edge, Wt_{xy} is the weight of the edge between vertex x and vertex y, and d is a constant dampening factor, set at 0.85. All vertices were initially assigned a score of 1 and the above equation is applied iteratively until the difference in scores between iterations falls below a threshold of 0.0001 for all vertices.

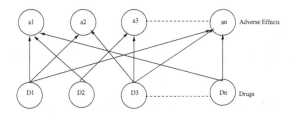

Fig. 1. Illustration of the Drug-Adverse effect bipartite graph.

Constructing the DAE Graph. Apart from the WD graph, we have observed that the specific drugs and their adverse effects also exhibits a relationship that can be represented via a bipartite graph (see Fig. 1) [8]. This motivated us to construct a separate graph using only the drugs and their adverse effects. Our objective is to explore the topological properties of such a drug-adverse effect bipartite graph. With respect to this, the first feature that we consider in classifying an edge (d, e) is the number of edges incident to each node, or the edge degrees. We have also considered the number of common neighbors for each edge. As discussed in [8], the common neighborhood of an edge (d, e) defines a connected sub-graph of the DAE graph centered around (d, e). Here, the maximum distance between any two nodes from the same set (i.e., two drug nodes) is two. The maximum distance between nodes of a different set is three. The intuition behind this measure is that the size of this sub-graph indicates a level of relatedness. The larger the neighborhood is, the more drug and effect neighbors the two nodes have in common.

Again, following the work of [8], we have computed the similarity of the two neighborhood nodes by normalizing the common neighborhood size by the size of the combined neighborhoods. Thus, it is apparent that the neighborhood of a node is also influenced by the individual connectedness of the neighborhood member nodes.

3.5 SVM Based ADE Classification Model

The ADE classification unit classifies unknown tweets into "ADE" or "Not an ADE" class. Once the textual feedback is provided as an input to the system, it is first preprocessed and analyzed by the linguistic processing unit. The analyzed text is then passed to the classification module. have primarily employed support vector machines (SVM) to develop a binary classifier. Given a training set instance-class pairs $(x_i, y_i), i = 1, 2, ..., l$, where $x_i \in R^n$ is the feature space and $y \in 1, -1^l$ is the label space, the general equation of a SVM is:

$$\frac{1}{2}\boldsymbol{W}^T\boldsymbol{W} + C\sum_i^l (\zeta_i) \; is \; minimized \tag{4}$$

$\boldsymbol{W}$ = weight vector;
C = regularization term.

$$y_i(\boldsymbol{W}^T + \phi(\boldsymbol{x_i}) + b) \geq 1 - \zeta_i, \zeta_i(slack variable) \geq 0 \tag{5}$$

4 Extraction of New Adverse Effects

Apart from classification of tweets into ADEs we have also tried to extract the specific adverse effects from the data streams that may not be explicitly mentioned in the sentence. We hypothesized that in most cases Adverse Effects contains a noun or a noun phrase. For example, *Hemolytic uremic syndrome following the infusion of oxaliplatin* is an example of an adverse effect. In order to extract the adverse events we have followed the following steps:

Step-1: Each of the drug description sentences are passed to the stanford parser where all the noun phrases are identified and extracted.

Step-2: We have extracted all the words that have a first level or second level dependency relation with the noun phrase. For example, consider the dependency relation of the sentence as depicted in Fig. 2. Here, the words like, "developed" and "Symptomatic" belongs to the first level dependency with the noun "hyponatremia" and "after" and "She" belongs to the second level dependency.

Step-3: We have considered all those words as a feature to learn a linear SVM classifier.

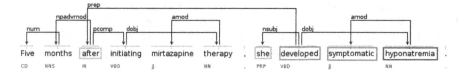

Fig. 2. Illustration of the dependency structure of a drug description.

5   Experimentation and Evaluation

We have evaluated the performance of the proposed ADE classification models by comparing its output with that of the expert annotations. As discussed earlier, we have used different features like, Point wise mutual information (PMI), Term Relevance (TR), Dependency features, Sentiment features, graph based features and feature combinations like, $PMI + Dependency$, $PMI + Sentiment$, $Graph + Dependency$ and $Graph + Dependency + Sentiment$ to study the model accuracies. We quantify the performance score in terms of the precision, recall and F-measure values. We have also compared the performance of our proposed model with the some of the state of the art baseline models. Presently, we have considered the term-matching classifier based on an ADR lexicon, maximum entropy with n-grams and TFIDF weightings or NB log-count ratio [20,24] and the deep learning based convolution neural network (CNN) model [13] as baseline models.

As discussed earlier, the SVM classifier was implemented using the LIBSVM software [3]. We have tested four type of kernels namely linear, polynomial, radial basis and sigmoid on the data. However, we have presented results against only the polynomial kernels as the other functions were found to be significantly poor performers. Moreover, different parameter configurations (regularization variable, slack variable value) of the SVM were evaluated and only the optimum configuration has been presented. To evaluate the quality of the classifications for SVM, multiple correlations (R) has been used. R denotes the extent to which the predictions are close to the actual classes and its square (R^2) indicates the percentage of dependent variable variation that can be explained by the model. Therefore, while percentage accuracy is an indicator to how well the model has performed to classify, R indicates the extent of explanatory power it posses. A better fit will have large R-value as well as Accuracy. Table 1 depicts the result of our evaluation of the ADE classification model in terms of the parameters and features discussed above. We have presented the classification results for the Twitter dataset and MEDLINE dataset separately. Overall, We found that taking the graph based features into account the model significantly reduces the false negative score and achieved a high true positive score thereby achieving a high precision and recall. In case of the MEDLINE dataset taking the graph based, sentiment and the dependency features into consideration $(graph + dependency + sentiment)$, we have achieved an highest accuracy of 94% with a precision of 87% and recall of 98% which is best as compared to the other baseline methods. Around 12% of the cases were found where our

Table 1. Evaluation results of the ADE classifier for Twitter and MEDLINE Dataset. Table 2(a) is the results as reported by the existing baseline systems and Table 2(b) reports the results of our proposed models. Note that "ALL" implies combination of graph, dependency and sentiment features

Method	Twitter			MEDLINE		
	P	**R**	**F1**	**P**	**R**	**F1**
TM	13	89	23	30	99	46
TFIDF	33	70	45	74	86	80
NBLCR	79	14	23	91	79	84
CNN	47	57	51	85	89	87

(a)

Method	Twitter				MEDLINE			
	P	**R**	**F1**	R^2	**P**	**R**	**F1**	R^2
PMI	57	66	61	.47	51	63	56	.43
Dep.	27	78	22	.43	69	78	73	.47
TR	65	41	50	.43	54	47	50	.41
Sen	65	41	50	.43	54	47	50	.41
G	75	81	78	.53	84	87	85	.57
ALL	**80**	**88**	**84**	**.61**	86	90	88	.63

(b)

system failed to classify the ADEs correctly. Out of this, in around 11% cases the system incorrectly predicted a drug description as valid ADE which is actually not, whereas only 1% of the drug descriptions were incorrectly identified as "Not an ADE" despite being marked as "ADE" by the experts. On the other hand, maximum accuracy of 88% was obtained for the twitter dataset taking only the graph based features into consideration. Here, we have observed that the *Graph + dependency + sentiment* features are not performing well. The primary reason behind this is due to fact that most of the collected tweets are extremely noisy, as a result of which the dependency parser fails to parse the texts properly and thus returning incorrect dependency relations.

We have observed that a large number of drug descriptions are written in languages other than English. As a result of this, the classifier failed to correctly predict. Another source of error is the occurrence of incomplete sentences in the drug descriptions that restricts the classification engine to correctly label the descriptions.

We have also evaluated the performance of the extraction of adverse effects from the MEDLINE dataset. We have achieved an F-measure of 78% with

Table 2. Illustration of some of the tweets with drugs and their adverse effects.

A 44-year-old man taking **naproxen** for chronic **low back pain**
We describe the side effects of **5-FU** in a colon cancer patient who suffered severe, **mucositis desquamating dermatitis, prolonged myelosuppression, and** neurologic toxicity that required admission to the intensive care unit
Lupus-like syndrome caused by **5-aminosalicylic acid** in patients with inflammatory bowel disease
Methemoglobinemia after axillary block with **bupivacaine** and additional injection of **lidocaine** in the operative field

precision of 77% and recall of 79% respectively. Table 2 illustrates the extraction of adverse effects from some of the tweets. We found most of the errors are due to the errors in dependency parsing. Presently, the system is incapable of handling such errors.

During our extraction of adverse effects, we have observed that a lot of adverse are of similar types which none of the present models could handle efficiently. However, due to slight change in their name or spelling they are considered as different effects. In order to overcome this, we have applied K-Means algorithm to cluster out similar adverse effects together. Similarly we have created clusters of drugs corresponding to similarities in adverse effects. Table 3 depicts some of the common clusters of drugs and adverse effects.

Table 3. Illustration of clustering of drugs and their adverse effects.

Cluster	Frequent adverse effects	Drug names
C1	Cough, headache, dizziness, tired feeling	LOSARTAN, ENALAPRIL, MANNITOL, ANASTROZOLE
C2	Renal failure	T-PA, CEFOXITIN, DEXTRAN-40, AMPICILLIN, RIFAMPICIN
C3	Dizziness, Stevens-Johnson, syndrome, hypoglycaemia	TRIMETHOPRIM-SULFAMETHOXAZOLE, PREDNISOLONE, NEVIRAPINE

Figure 3(a) and (b) depicts the overall distribution of drugs in terms of their number of adverse effects for both Twitter and MEDLINE datasets respectively. We can observe that in Twitter dataset drugs like, VYVANSE, SEROQUEL, EFFEXOR and FLUOXETINE have got the maximum adverse effects whereas METHOTREXATE, LITHIUM, FLUOXETINE and OLANZAPINE have the maximum side effects in MEDLINE dataset (Table 4).

Table 4. List of top 5 drugs and their respective AEs.

Drug	AEs
METHOTREXATE	Abdominal pain, accelerate HIV disease, Accelerated nodulosis
LITHIUM	Akathisia, angry outbursts, ARDS, asterixis, hyperparathyroidism
FLUOXETINE	Renal failure, akathisia, angry outbursts, ARDS, asterixis,
LANZAPINE	Intoxication, agitation, agranulocytosis, akathisia, anxiety, camptocormia
VYVANSE	Addicted, adhd, All nighter, anxiety, can't go back to sleep, chest hurt

Figure 3(c) and (d) depicts the distribution of adverse effects associated with number of drugs for both Twitter and MEDLINE datasets respectively. We can observe that in Twitter dataset effects like, TIRED, NAUSEA, WEIGHT GAIN and RASH have got the maximum number of associated drugs whereas FEVER, SEIZURES, HEPATOTOXICITY, and ACUTE RENAL FAILURE have the maximum associated drugs in the MEDLINE dataset.

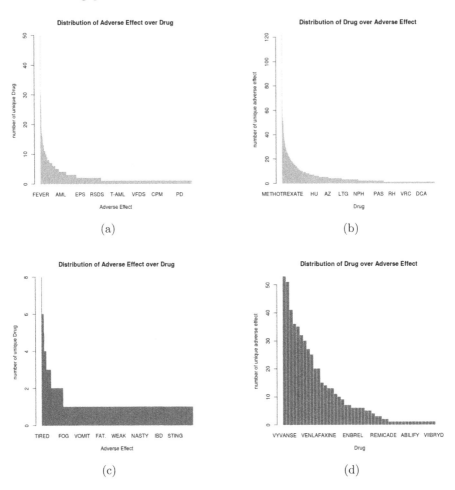

Fig. 3. Distribution of Adverse effects and Drugs across the MEDLINE Dataset.

6   Conclusion

In this paper we have explored different linguistic and graph topological features to automatically classify short sentences or tweets into ADEs or Non-ADEs. We have further represented the ADE knowledge base into an bipartite network structure of drugs and their side effects to model drug-side effect relationships. We have also proposed models to discover implicit ADEs that are not represented in the source data. We have evaluated our proposed models with two openly available ADE dataset namely the twitter dataset and the MEDLINE dataset. Our evaluation results shows that the proposed model have surpasses the performance of the existing baseline systems. Overall, We found that taking the graph based features into account the model significantly achieves a high precision and recall. In case of the MEDLINE dataset taking the graph based, sentiment and the dependency features into consideration, we have achieved an highest accuracy of 94%. On the other hand, due to the noisiness in data none of the linguistic features could be applied to the Twitter dataset. Consequently we have managed an accuracy of 88% taking only the graph based features into account.

References

1. Bate, A., et al.: A Bayesian neural network method for adverse drug reaction signal generation. Eur. J. Clin. Pharmacol. **54**(4), 315–321 (1998)
2. Bisgin, H., Liu, Z., Fang, H., Xu, X., Tong, W.: Mining FDA drug labels using an unsupervised learning technique-topic modeling. BMC Bioinform. **12**(10), S11 (2011)
3. Chang, C.-C., Lin, C.-J.: LIBSVM: a library for support vector machines. ACM Trans. Intell. Syst. Technol. (TIST) **2**(3), 27 (2011)
4. Chirawichitchai, N., Sa-nguansat, P., Meesad, P.: Developing an effective Thai document categorization framework base on term relevance frequency weighting. In: 2010 8th International Conference on ICT Knowledge Engineering, pp. 19–23. IEEE (2010)
5. De Marneffe, M.-C., Manning, C.D.: The Stanford typed dependencies representation. In: Coling 2008: Proceedings of the Workshop on Cross-Framework and Cross-Domain Parser Evaluation, pp. 1–8. Association for Computational Linguistics (2008)
6. Duan, K.-B., Rajapakse, J.C., Wang, H., Azuaje, F.: Multiple SVM-RFE for gene selection in cancer classification with expression data. IEEE Trans. Nanobiosci. **4**(3), 228–234 (2005)
7. DuMouchel, W.: Bayesian data mining in large frequency tables, with an application to the FDA spontaneous reporting system. Am. Stat. **53**(3), 177–190 (1999)
8. Eshleman, R., Singh, R.: Leveraging graph topology and semantic context for pharmacovigilance through Twitter-streams. BMC Bioinform. **17**(13), 335 (2016)
9. Gamon, M.: Graph-based text representation for novelty detection. In: Proceedings of the First Workshop on Graph Based Methods for Natural Language Processing, pp. 17–24. Association for Computational Linguistics (2006)
10. Ginn, R., et al.: Mining Twitter for adverse drug reaction mentions: a corpus and classification benchmark. In: Proceedings of the Fourth Workshop on Building and Evaluating Resources for Health and Biomedical Text Processing. Citeseer (2014)
11. Gurulingappa, H., Mateen-Rajpu, A., Toldo, L.: Extraction of potential adverse drug events from medical case reports. J. Biomed. Semant. **3**(1), 15 (2012)
12. Gurulingappa, H., et al.: Development of a benchmark corpus to support the automatic extraction of drug-related adverse effects from medical case reports. J. Biomed. Inform. **45**(5), 885–892 (2012)
13. Huynh, T., He, Y., Willis, A., Rüger, S.: Adverse drug reaction classification with deep neural networks (2016)
14. Lazarou, J., Pomeranz, B.H., Corey, P.N.: Incidence of adverse drug reactions in hospitalized patients: a meta-analysis of prospective studies. JAMA **279**(15), 1200–1205 (1998)
15. Lindquist, M., Edwards, I.R., Bate, A., Fucik, H., Nunes, A.M., Ståhl, M.: From association to alert-a revised approach to international signal analysis. Pharmacoepidemiol. Drug Saf. **8**(S1), S15–S25 (1999)

16. Liu, M., et al.: Comparative analysis of pharmacovigilance methods in the detection of adverse drug reactions using electronic medical records. J. Am. Med. Inform. Assoc. **20**(3), 420–426 (2013)
17. Mihalcea, R., Tarau, P.: TextRank: bringing order into texts. Association for Computational Linguistics (2004)
18. Nikfarjam, A., Sarker, A., O'Connor, K., Ginn, R., Gonzalez, G.: Pharmacovigilance from social media: mining adverse drug reaction mentions using sequence labeling with word embedding cluster features. J. Am. Med. Inform. Assoc. (2015). https://doi.org/10.1093/jamia/ocu041
19. Pirmohamed, M., et al.: Adverse drug reactions as cause of admission to hospital: prospective analysis of 18 820 patients. BMJ **329**(7456), 15–19 (2004)
20. Rastegar-Mojarad, M., Komandur Elayavilli, R., Yu, Y., Hiu, H.: Detecting signals in noisy data-can ensemble classifiers help identify adverse drug reaction in tweets. In: Proceedings of the Social Media Mining Shared Task Workshop at the Pacific Symposium on Biocomputing (2016)
21. Sarker, A., Nikfarjam, A., Gonzalez, G.: Social media mining shared task workshop. In: Proceedings of the Pacific Symposium on Biocomputing, pp. 581–592 (2016)
22. Szarfman, A., Machado, S.G., O'Neill, R.T.: Use of screening algorithms and computer systems to efficiently signal higher-than-expected combinations of drugs and events in the us FDA's spontaneous reports database. Drug Saf. **25**(6), 381–392 (2002)
23. Wang, X., Hripcsak, G., Markatou, M., Friedman, C.: Active computerized pharmacovigilance using natural language processing, statistics, and electronic health records: a feasibility study. J. Am. Med. Inform. Assoc. **16**(3), 328–337 (2009)
24. Zhang, Z., Nie, J., Zhang, X.: An ensemble method for binary classification of adverse drug reactions from social media. In: Proceedings of the Social Media Mining Shared Task Workshop at the Pacific Symposium on Biocomputing (2016)

Extraction of Semantic Relation Between Arabic Named Entities Using Different Kinds of Transducer Cascades

Fatma Ben Mesmia[1]([⊠]), Kaouther Bouabidi[2], Kais Haddar[2],
Nathalie Friburger[3], and Denis Maurel[3]

[1] Laboratory MIRACL, Multimedia InfoRmation Systems
and Advanced Computing Laboratory,
University of Tunis El Manar, Tunis, Tunisia
`fatmabm@ymail.com`
[2] Laboratory MIRACL, Multimedia InfoRmation Systems
and Advanced Computing Laboratory, University of Sfax, Sfax, Tunisia
`bouabidikaouther@gmail.com, Kais.Haddar@fss.rnu.tn`
[3] LI, Computer Laboratory, University François Rabelais of Tours,
Tours, France
`{nathalie.friburger,denis.maurel}@univ-tours.fr`

Abstract. The extraction of Semantic Relationship (SR) is an important task allowing the identification of relevant semantic information in the annotated textual resources. Besides, extracting SR between Named Entities (NE) is a process, which consists in guessing the significant semantic links related to them. This process is very useful to enhance the NLP-application performance, such as Question Answering systems. In this paper, we propose a rule-based method to extract and annotate SR between Arabic NEs (ANE) using an annotated Arabic Wikipedia corpus. In fact, our proposed method is composed of two main cascades regrouping respectively analysis and synthesis transducers. The analysis transducer cascade is dedicated to extract five SR types, which are synonymy, meronymy, accessibility, functional and proximity. However, synthesis one is devoted to normalize the SR and NE annotation using the TEI (*Text Encoding Initiative*) recommendation. Furthermore, the established transducer cascades are implemented and generated using the CasSys tool available under Unitex linguistic platform. Finally, the obtained results showed by the calculated measure values are encouraging.

Keywords: Semantic Relation · Arabic Named Entity · Arabic Wikipedia Transducer cascade · TEI

1 Introduction

The extraction of SR, a sub-task of the Information Extraction (IE) task, consists in identifying semantic information existing in the annotated textual resources. This sub-task as a new research line is evolving to realize many objectives, namely enhancing the NLP-application performance. Therefore, extracting SRs between NEs offers an

© Springer Nature Switzerland AG 2018
A. Gelbukh (Ed.): CICLing 2017, LNCS 10761, pp. 475–487, 2018.
https://doi.org/10.1007/978-3-319-77113-7_37

opportunity to enrich and enhance NE electronic dictionaries and to improve disambiguation of Entity Linking (EL) task. Besides, the SR extraction participates in a large part to enrich corpora, in the semantic level, that can index search engines. Furthermore, the extracted SR can increase the efficiency of Question Answering systems in order to provide relevant responses. The SR extraction between NEs is a process highly linked to the recognition one. Generally, the NER is a preprocessing step concentrated on the textual resource analysis in order to recognize NEs and classify them in predefined categories. The classification process is always referring to the previous categorizations. However, the extraction of SRs between NEs can encounter many challenges. The first one is related to the NE representation. If the NE is not recognized and annotated, it must envisage a recognition process. The second challenge concerns the prediction of the context describing a SR. Predicting the context means that we must delimit it by respecting the SR and the treated language limits. The third challenge consists in a deep linguistic conducting study to guess the SR types. This linguistic study should ensure a good semantic linking. The other challenge that can be envisaged that is the richness of the necessary resources used to identify the SRs. Among the envisaged challenges, we can also cite the identified SR annotation, which is not an easy task. The annotation needs to be based on standards, like XML (*eXtensible Markup Language*) and TEI. These standards facilitate the detailing of both SR and NE components. The last challenge consists of the coupling of the SR extraction and the annotation processes. This coupling requires a powerful formalism to reduce the execution time.

In this context, we begin by making a linguistic study based on an annotated Arabic Wikipedia corpus. The annotated corpus contains selected articles that were collected from different Arabic countries. A symbolic NER transducer cascade ensures the recognition and the annotation of the ANEs existing in this annotated corpus. To identify the SR types, we use an ANE hierarchy associated with the annotated corpus describing the different categories and sub-categories. In addition, we propose a rule-based method to extract and annotate SR between ANE. In fact, our proposed method is composed of two main cascades regrouping respectively the analysis and synthesis transducers. The analysis transducer cascade is dedicated to extract SRs. However, synthesis one is devoted to the normalization of the SR and NE annotation using the TEI recommendation.

The work originality consists in exploiting the finite-state transducer formalism, which can couple the SR extraction and the annotation processes. In fact, transducers are very efficient, robust and speedy. Besides, we take full advantage of transducer cascade to organize their passage order. The organized passage order helps the regrouped transducers to provide a degree of certitude. In addition, we refer to the TEI recommendation to normalize the annotation for both SR and ANE components. The TEI offers several tag types and attributes providing structured corpora that can be easily integrated in the semantic Web.

The present paper is composed of five sections. Section 2 presents the related work for the SR extraction using the three main approaches. Section 3 consists in exploring and analyzing an annotated corpus extracted from Arabic Wikipedia by describing the identified SR between ANE. Section 4 details the proposed method to extract and annotate SR. The implementation and evaluation of our proposed method are presented in Sect. 5. Finally, we give a conclusion and some perspectives.

2 Related Work

The extraction of SR between NE is highly connected to the NER task. In fact, extracting a SR depends on the NE definition determining its limits. The SR extraction also needs the NE categorization that facilitates the prediction of the SR types. Generally, the systems extracting SR are based on the output of the NER one. The two systems have common points as the establishment of each system is based on the three main approaches (symbolic, statistic and hybrid) [2, 3, 5, 11]. The choice of an annotation norm, which can represent and detail both SR and NE components, is also a common point. In the current section, we give an overview on the related work.

The symbolic approach is based on formal local grammars that are described by rules. These local grammars can be modeled using the adequate formalisms (i.e. regular expression, transducer) in order to facilitate their implementation and management. Among the systems based on this approach, [9] proposed a semi-automatic method to obtain local grammars extracting relations between NEs in corpora. The NE detection is done using Arisem analyzer based on the symbolic approach. The used analyzer annotates the text with different labeling levels (morphological, syntactic and semantic) for the relevant words and phrases. Finite-state transducer ensured the SR extraction. The proposed method extracts one relation, which is "contact". This extracted relation is annotated using XML (*eXtensible Markup Language*). To evaluate the produced grammar, the author used a sample of the newspaper "Le Monde" for the year 2007. In [6], the authors proposed a rule-based system to extract relations between ANEs. They focused on Functional relations (Director, Responsible and president) that link ENAMEX and ORG categories describing an ANE. The proposed system consisted on a process composed of three steps. The first step is dedicated to the NE recognition, especially ENAMEX and ORG ANEs. The second one is devoted to identify functional relations between the recognized ANE. The last step is the generation of the predicate forms representing the identified relation. The proposed system enables the translation of the Arabic functional relations that are identified. The target language is French. The system implementation was done using Nooj linguistic platform taken from journalistic and Wikipedia corpora. Nevertheless, the annotation of the extracted relations does not respect an annotation norm. For the medical domain, [7] proposed a method to extract SR between medical entities. This proposed method relies on a preprocessing step to recognize medical entities from sentences and determine their categories. The SRs are extracted from every couple of medical entities and their semantic types are determined using the UMLS (*Unified Medical Language System*) semantic network. Based on the UMLS, the elaborated method helps to extract 6 relations, which are causes, diagnoses, treats, prevents, complicates, sign or symptom of. The authors implemented their method using MeTAE (*Medical Texts Annotation and Exploration*) platform. In fact, the annotation process of both medical entities and relations was done through RDF (*Resource Description Framework*) annotation. The corpora that were used in the proposed method were collected from PCM (*PubMed Central*).

The Machine Learning (ML) approach takes advantage of ML-algorithms to learn SR tagging decisions from annotated corpora. This approach requires the availability of large annotated data. The most used ML-techniques are the Supervised Learning (SL) techniques such as Vector Machines (SVM) and Decision Trees. Using this approach, we quote the semi-supervised system elaborated by [1]. This system consists in extracting binary relations from Arabic texts collected using open source Web contents. The proposed system is generic, which can be used in different domains. The relation extraction is described as an iterative process where each iteration contains two main phases, which are the pattern and the instances extraction. The new instances are filtered to avoid noisy in each new iteration. The proposed system helps to extract four types of relations, which are author-of (person, book) relation, president-of (person, country) relation, play-in (person, club) relation and CEO-of or chairman (person, company) relation. Nevertheless, the SR annotation does not respect an annotation norm.

The hybrid approach is the fusion of the already mentioned approaches, which are complementary. This approach uses manually written rules and those extracted from data through learning algorithms and decision trees. Among the systems based on this approach, we can mention the system developed by [8] to extract relations between ANEs. The developed system used linguistic modules employed as post-processing to ameliorate the obtained results. Initially, these results were obtained from a machine learning-based method. This system extracts the SRs, which are complicated or expressed through more than one word and annotates them using a defined markup. An NER processing preceded the SR extraction and annotation processes.

The previous work to extract SR between NEs using different formalisms and techniques. All the illustrated systems use a NER process as a preprocessing step. There are systems creating their own NER module as transducer establishment and others calling the existing tools such as Arisem and MetaMap. Nevertheless, the recognized NEs were assigned to a few number of categories so that the detected SR types will be reduced. The reduced number of the SR types can be due to the exploited corpora having restraint domains. We remark also that the annotation process is usually using a specific defined Markup associated with the elaborated system. Moreover, there are systems that use an annotation norm as XML and RDF. Generally, the elaborated systems extracting SRs are based on annotated textual resources that are not always exhaustive. Moreover, the free resources can be a solution for this difficulty. The NER process is highly coupled with the SR extraction. In this case, a deep categorization increasing the granularity level is required. This categorization can, on the one hand, enhance the SR detection and, on the other hand, guess their adequate types. For this reason, the NER must offer an NE hierarchy containing refined categories and subcategories. In the previous work, an SR was always detected as a binary relation between NEs. However, they adopted no formal definition determining the SR limits. In the following section, we will explain our linguistic study made on an annotated corpus extracted from Arabic Wikipedia.

3 Identification of SR Between ANEs

To identify SR between ANE, we explore an annotated study corpus, which is the NER process output. Initially, this corpus is extracted from Arabic Wikipedia. Our linguistic study is divided in two phases. In the first phase, we schematize an ANE hierarchy including all the appearing categories and sub-categories. In the second one, we identify the different SR types related to ANEs and their associated forms. During the exploration of our study corpus, we remark that the ANEs are assigned to several categories and sub-categories. For this reason, we regroup them in an ANE hierarchy. The ANE hierarchy helps, on the one hand, describe the deep categorization made through the NER process and, on the other hand, facilitate to guess the SR types in the next phase. In fact, we identified five main categories that appear Date, Person name, Location, Event and Organization.

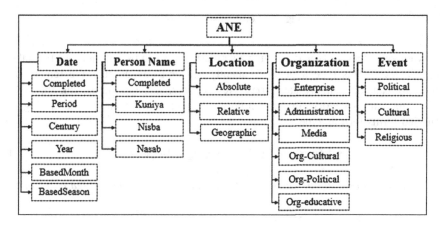

Fig. 1. ANE hierarchy identified from the annotated study corpus

Figure 1 describes the ANE hierarchy associated with our annotated study corpus. All the categories are refined to be composed of sub-categories. The illustrated categories are extended to increase the granularity level. This extension depends, in a large part, on the appearing ANE forms. The category Location is refined to form a sub-hierarchy described in the following figure.

Figure 2 shows the sub-categorization made to extend the category Location. From the illustrated sub-hierarchy, we remark that Relative Location ANEs are the most appearing in our study corpus. Hence, 16 sub-categories are identified where each one has the appropriate forms.

Exploring and analyzing our annotated corpus is an important task. This task enables us to identify each SR between ANEs using the already mentioned categories. In fact, we identified five SR types, which are synonymy, meronymy, accessibility, proximity and functional. For each SR type, we detected several alternative forms. Furthermore, we consider an SR as a semantic link relating two ANEs in our annotated corpus.

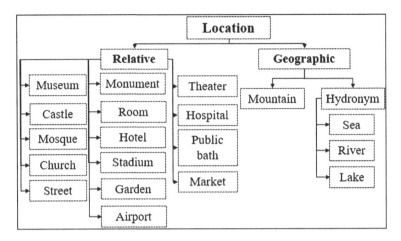

Fig. 2. Location sub-categories

Synonymy is an SR type describing a similarity between two or more ANEs in their semantic level. This type appears frequently in our study corpus. We remark that synonymy can touch many categories. Nevertheless, this type must link the ANEs having the same category. In what follow, we give some examples describing the identified forms for this type.

$$الثورة التونسية التي تعرف أيضا بـثورة الحرية والكرامة \qquad (1)$$

The Tunisian revolution, which is also known as the freedom and dignity revolution

$$القديس سمعان المعروف بـسمعان العامودي \qquad (2)$$

St. Simeon known as Simeon Stylites

In (1), synonymy links two ANEs having the same category that is Event especially the sub-category Political Event. This type is deduced through the expression "تعرف أيضًا التي / *which is also known*". This expression contains significant elements, such as a verb and an adverb describing synonymy in their meaning. Nevertheless, the second ANE is introduced using the preposition "بـ / *as*" to indicate that it is the synonym of the first ANE. In (2), synonymy concerns the category Person name. Consequently, the linked ANEs have the same meaning but different forms. The first ANE is composed of a religious function "القديس / *St.*" followed by a first name "سمعان / *Simeon*". However, the second one is a succession of a first name and a surname. The synonym ANEs are linked through the expression "المعروف بـ / *known as*". In our study corpus, we identified other expressions replacing it as "المسمى / *named*" or "المكنى / *nicknamed*"..

Meronymy is an SR type describing an inclusion relation. This type refers to a part of a whole. If two ANEs are linked with a meronymy, this means that the first ANE is a part of the second. In our study corpus, this current type is relating ANE having different categories. However, we remark that the second ANE has always the category Location especially Absolute one.

كمال الملاخ من القاهرة   (3)

Kamal Al-Mallakh from Cairo

وزارة الدفاع في مصر   (4)

Ministry of Defense in Egypt

In (3) and (4), the ANEs connected with a meronymy using prepositions. The preposition " من/ *from*" relates a Person name category to an Absolute Location sub-category. However, the preposition "من/ *from*" links an Organization name and an Absolute Location one. From the identified forms, we notice that a Relative Location can be used instead of the Absolute one.

Accessibility is an SR type of related place names to express a kind of inclusion. This type means that it is possible to access one place from another. Based on our study corpus, we remark that accessibility is related to the NEs having the category Location and its sub-categories Absolute, Relative and Geographic.

محافظة إدلب في سوريا   (5)

Idleb province in Syria

جبل سيدي عبد الرحمان بولاية نابل   (6)

Mount Sidi Abderrahmane in governorate of Nabeul

From (5) and (6), we notice that accessibility links the ANEs having the category Location through specific prepositions expressing a kind of inclusion. The semantic link can be between two Absolute Location, such as the city " إدلب/ *Idleb*" and the country " سوريا/ *Syria*" or a Geographic Location, such as جبل سيدي عبد الرحمان/ *Mount Sidi Abderrahmane*" that is a mountain name and an Absolute one, such as " نابل/ *Nabeul*". The ANEs having the category Absolute Location are associated to a set of trigger words that facilitate their recognition.

We also notice the existence of a new semantic link between ANE having the same category, such as Location or Organization. This link expresses an SR defined in its sense by an approximation of places. We find several forms describing this SR. For this reason, we decide to identify its different forms and call it Proximity.

مؤسسة الحسن حذو وزارة الثقافة   (7)

Al Hassan Foundation next to The Ministry of Culture

جبل سيدي علي بن عون قرب مدينة بئر الحفي   (8)

Mount Sidi Ali Ben Aoun near the city of Bir El Hafey

Proximity in (7) and (8) is expressed through the two nouns playing the role of trigger words. The first one named " حذو/ *next to*" links two ANEs having the category

Organization. However, the second noun "قرب/ *near*" relates two ANEs having the same category Location but a different sub-category, which are respectively Geographic and Absolute. There are other trigger words that can describe proximity as "بجانب/*near*" and "يقرب/*near*". These trigger words are synonym. Recall that the noun "بقرب/*near*" is a typographic variant of "قرب/*near*".

Functional is an SR type describing a functional role between ANEs. This type can link NEs having different categories, such as Person name related to an Organization. From our study corpus, we identified several forms expressing a function role that helped us to determine and detect this type.

$$بشار الأسد رئيس سوريا \qquad (9)$$

Bashar al-Assad, President of Syria

$$(10) \quad مهرجان جده الغنائي هو مهرجان موسيقي افتتح على يد صاحب السمو الملكي الأمير مشعل بن ماجد$$

Jeddah musical festival is a musical festival inaugurated by the Royal Highness Prince Mishaal bin Majed

In (9) and (10), the functional type is detected on a functional role or an expression containing different components. The functional role " رئيس / *President* " is a trigger word belonging to a class named political function identified by the NER process. Other classes helped us to detect the meaning of the trigger words for the current SR type such as peerage function. The " الأمير / *Prince*" belongs to the already mentioned class. In some cases, we found a conjunction of trigger word, such as "مدير ومؤسس/*director and funder*". We must treat this agglutination to recognize the second noun. Expressions identifying the functional type generally include a significant verb, such as the verb "افتتح/ *inaugurated*".

4 Proposed Method

Our proposed method for the extraction and annotation of SRs between ANEs is composed of the following steps: the exploration of our Arabic Wikipedia corpus to construct dictionaries and extraction rules, the transformation of these extraction rules into regular expressions and the establishment of two sets of transducers. The first set is called analysis transducer ensuring the extraction and annotation of the five identified SR types. The second one is called synthesis dedicated, on the one hand, to normalizing the ANE using the TEI recommendation and, on the other hand, to conforming their annotation to the SR one. The established transducers will be regrouped in two kinds of cascades, the analysis and synthesis cascades.

The resource identification is an important step that enables a good extraction and annotation. In fact, we identified 141 extraction rules distributed as follows; 11 for synonymy, 18 for meronymy, 92 for accessibility, 12 for functional and 8 for proximity. Besides, the identified extraction rules are represented as regular expressions. In what follows, we will give some examples from our established regular expressions related to some SRs.

- **Synonymy regular expression:**
  <V+synonymy>+placeName+type="city">+<Adv>+<Prps>+<Nc>+<placeName +type="city">
- **Meronymy regular expression:**
  <persName>+<Adj+meronymy>+<Nc>+<placeName+type="city">
- **Accessibility regular expression:**
  <V+placeName>+<Nc>+<placeName+type="city">+<Prps+accessibility>+<pla-aceName+type="country">

The analysis phase combines both SR extraction and annotation processes ensured by analysis transducers. Recall that the transducers are a semi-formal specification of the established regular expressions. In the elaborated transducers, we find alternative paths with the correspondent SR annotation. The annotation in conformity with the TEI recommendation. We choose this recommendation because it provides elements that can describe the identified SR components. Nevertheless, we also chose TEI because we noticed that all the ANEs appearing in our corpus were annotated using TEI. As a result, we use the following form to annotate all the detected SRs:

<Relation type="Name of SR" Entity1="ANE" Entity2="ANE"/>

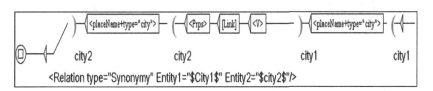

Fig. 3. Transducer recognizing the synonymy SR between two ANEs having the sub-category Absolute Location

Figure 3 defines the transducer extracting an SR expressing synonymy between two ANEs. The linked ANEs have the same sub-category, which is the Absolute Location. The feature describing this sub-category is not related to a dictionary as in the second box containing <V>. In reality, this feature is related to the NER process that was made on our corpus. In fact, the NER process used the { } markers, which is not an arbitrary choice. These markers make the recognized ANE a polylexical word that cannot be detected by another transducer and a feature like the dictionary one. The boxes containing the important features are surrounded by variables having as values symbolic names (city1 and city2). The symbolic names are called in the output node to organize the tag form respecting the TEI recommendation. Calling variables consist in writing them using the $ symbol.

Figure 4 describes the form of a transducer extracting an SR expressing meronymy. In this case, the meronymy links two ANEs having respectively the category Person name and the sub-category Absolute Location more precisely country names. Respecting the same principle of the previous figure, the current transducer uses variables and a negative context helps us avoid linguistic phenomena related to the Arabic language such as anaphora and ellipsis.

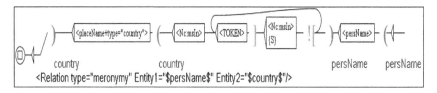

Fig. 4. Transducer recognizing meronymy between a person name and an Absolute Location

The synthesis phase consists in normalizing the ANE annotation using the TEI recommendation [10, 12]. Indeed, we are based on the TEI syntax, which is defined as follows: an opening tag describing such category, like <persName> and a closing tag </persName> will surround the ANE. The tag <persName> can include in its turn other tags as <forename>, <surname> and <roleName> to annotate respectively a first name, a last name and the trigger word that can precede a person name. In the tag < roleName>, we can specify the type of this role name, such as profession and political function.

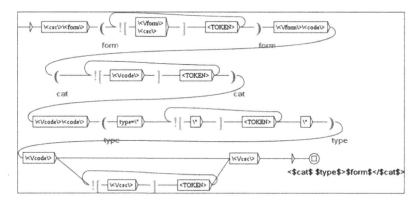

Fig. 5. Transducer transforming the ANE annotation into TEI

Figure 5 describes the transducer normalizing the ANEs annotation into TEI. The illustrated transducer treats the case when a tag has an element "type" to specify the sub-category. In fact, we propose a path taking as input the annotation format generated by the tool utilized to generate the transducer cascade. In the proposed path, we use the concept of negative context through ![,] markers and variables using (,) markers. The output is organized in the last box by the variables "form", "cat" and "type" respecting the already mentioned TEI syntax.

5 Implementation and Evaluation

To experiment our proposed method, we exploit a test corpus comprising annotated ANEs. The test corpus allows not only the application of our proposed transducer cascade for both analysis and synthesis phases in order to annotate it but also to testing the performance of our elaborated system. The test corpus contains text files for a cumulative 95 373 tokens. The study corpus containing text files for a cumulative 246 001 tokens permits us to create new dictionaries and update dictionaries available under Unitex linguistic platform.

The elaborated analysis transducer cascade calls five main graphs. Nevertheless, we elaborated 73 graphs in total. Additionally, it should be recalled that a transducer cascade regroups the main graphs respecting a certain order of passage that enables us to avoid ambiguities [4].

#	Disabled	Name	Merge
1	☐	Accessibility.fst2	✔
2	☐	Meronymy.fst2	✔
3	☐	Fonctionnal.fst2	✔
4	☐	Synonymy.fst2	✔
5	☐	VicinityPlaceName.fst2	✔

Fig. 6. Transducers cascade extracting SR betwen ANE

Figure 6 shows the form of our transducer cascade for the analysis phase implemented in Unitex linguistic platform. In fact, each graph adds its appropriate annotations to the text using the mode "Merge". This mode provides, as output, a recognized SR defined in tags having the forms defined in analysis graphs.

The synthesis transducer cascade is dedicated to normalizing ANEs using TEI recommendation and providing a structure output. Moreover, this synthesis cascade normalizes the ANE annotation based on the CasSys tags. The transducers inside the synthesis cascade also needs an order of passage and differs from the analysis one in the mode of passage and in the format of the input file.

#	Disabled	Name	Merge	Replace	Until Fix Point	Generic
1	☐	balisageType.fst2	☐	✔	✔	☐
2	☐	balisage.fst2	☐	✔	✔	☐

Fig. 7. Transducer cascade for the synthesis phase

Figure 7 shows the form of the transducer cascade for the synthesis phase. The two graphs are passed in mode "replace" in order to replace the old tags with the new one and in the mode "Until fix point" to treat the occurrence of the ANEs and the imbrication inside them. The first graphs must be passed before the second to treat the tags having the element "type".

Table 1. Table summarizing the measure values

Precision	Recall	F-score
0.96	0.77	0.84

After applying the analysis and the synthesis transducer cascade, we obtain a structured test corpus (extracted from Arabic Wikipedia). Furthermore, the evaluation of our proposed system is an important task that proved its performance. For this reason, we evaluate it using the measure values.

Table 1 demonstrates that our proposed analysis transducer cascade extracting SRs between ANEs shows a precision of 96%, a recall of 77% and an f-score of 84%. Therefore, we find that the obtained results are encouraging. The number of SRs detected in error causes the obtained recall value. The errors presented in this extraction process due to dictionaries' coverage that must be improved especially those storing verbs indicating the SR type. These errors can be caused by the structure of Arabic Wikipedia's articles. For example, there are prepositions playing the role of an indicator of an SR existence written in error, such as "قي" instead of " في/ in" in the following sentence "مهرجان القطن يقام سنوياً قي مدينة حلب/Cotton Festival takes place annually in Aleppo city".

6 Conclusion

In the present paper, we developed a system extracting and annotating SRs between ANEs. The treated ANEs exist in an annotated corpus extracted from Arabic Wikipedia. Our developed system was represented through two kinds of transducer cascades. The first transducer cascade was dedicated to the analysis phase consisting in the SR extraction. The second one was devoted to the synthesis phase to normalize the ANE annotation using TEI recommendation. Finite-state transducers converting the identified regular expressions ensured both the analysis and synthesis phases. The evaluation of our proposed system performance was made through the measure values. The obtained results are satisfactory. In the future work, we will improve the size of our adopted corpus in order to discover new types describing an SR. We will also exploit the adopted corpus to construct an electronic dictionary storing ANE and SR between them annotated in TEI. This electronic dictionary will be an initiation to realize EL task. This task aims at linking the ANE to free resources, such as Dbpedia and Geonames.

References

1. Abd El-salam, Sh.M., El Houby, E.M.F., Al sammak, A.K., El-shishtawy, T.A.: Extracting Arabic relations from the web. Int. J. Comput. Sci. Inf. Technol. (IJCSIT) **8**(1), 85–102 (2016)
2. Ben Mesmia, F., Friburger, N., Haddar, K., Maurel, D.: Construction d'une cascade de transducteurs pour la reconnaissance des dates à partir d'un corpus Wikipédia. Colloque pour les Étudiants Chercheurs en Traitement Automatique du Langage naturel et ses applications, pp. 8–11, Sousse, Tunisie (2015)
3. Ben Mesmia, F., Friburger, N., Haddar, K., Maurel, D.: Arabic named entity recognition process using transducer cascade and Arabic Wikipedia. In: Proceedings of Recent Advances in Natural Language Processing, pp. 48–54, Hissar, Bulgaria (2015)
4. Ben Mesmia, F., Friburger, N., Haddar, K., Maurel, D.: Transducer cascade for an automatic recognition of Arabic Named Entities in order to establish links to free resources. In: First International Conference on Arabic Computational Linguistics (ACLing), pp. 61–67 (2015)
5. Ben Mesmia, F., Friburger, N., Haddar, K., Maurel, D.: Recognition and TEI annotation of Arabic event using transducers. Will appear in IEEE Proceedings of CiLing 2016 (2016)
6. Ben Hamadou, A., Piton, O., Fehri, H.: Multilingual extraction of functional relations between Arabic named entities using NooJ platform. In: Proceedings of the Nooj 2010 International Conference and Workshop, pp. 192–202 (2011)
7. Ben Abacha, A., Zweigenbaum, P.: Automatic extraction of semantic relations between medical entities: a rule based approach. J. Biomed. Semant. **2**(Suppl 5), S4 (2011). PMC. Web, January 2017
8. Boujelben, I., Jamoussi, S., Hamadou, A.B.: A hybrid method for extracting relations between Arabic named entities. J. King Saud Univ. Comput. Inf. Sci. **26**(4), 425–440 (2014)
9. Ezzat, M.: Acquisition de grammaire locale pour l'extraction de relations entre entités nommées. TALN (RECITAL), Montréal (2010)
10. Maraoui, H., Haddar, K.: Automatisation de l'encodage des lexiques arabes. Colloque pour les Étudiants Chercheurs en Traitement Automatique du Langage naturel et ses applications, pp. 74–82, Sousse, Tunisie (2015)
11. Sarhan, I., El-Sonbaty, Y., Abou El-Nasr, M.: Arabic relation extraction: a survey. Int. J. Comput. Inf. Technol. **5**(5), 430–437 (2016). ISSN: 2279–0764
12. Text Encoding Initiative Consortium 2016. TEI P5: Guidelines for Electronic Text Encoding and Interchange. In: Sperberg-McQueen, C.M., Burnard, L. (eds.) ACH-ALLC-ACL. Version 3.1.0, 1887 pages (2016)

Semi-supervised Relation Extraction from Monolingual Dictionary for Russian WordNet

Daniil Alexeyevsky[(✉)]

School of Linguistics, Faculty of Humanities, National Research University Higher
School of Economics, Moscow, Russia
dalexeyevsky@hse.ru

Abstract. Monolingual dictionaries are a voluminous loosely structured source of lexical and ontological information. Numerous attempts were made to extract WordNet or ontology relations from monolingual dictionaries with varying success. Most such attempts are based on morphosyntactic rules. Difficulty of the information extraction task greatly depends on discipline of dictionary editors. Despite frequently being excellent for the human reader the discipline is rarely strict enough to allow effortless data mining on dictionaries.

Here an improvement to rule-based approach to relation extraction is put forward. The improved approach is to automatically cluster similar definitions, then manually create either one or two relation extraction rules per cluster. This helps to reduce amount of annotator work, to increase quality of rule application and to pay more attention to some of rare cases. To group definitions with similar structure mixed n-gram features were employed, their usefulness is discussed.

The work is performed on Big Explanatory Dictionary of Russian language. Definitions are grouped to 100 clusters, annotated and correctness assessed. The average accuracy is 86% for hypernym extraction, which is high for works of the same scope.

Keywords: Relation extraction · Monolingual dictionary
Machine-readable dictionary · WordNet

1 Introduction

Some NLP tasks demand high quality set of semantic relations that represent common knowledge. Such sets of relations are needed to build thesauri, ontologies, for sentiment analysis etc. Methods of building such sets range from based on lexicographers' labor [6] to mostly automated approaches. There are two basic kinds of automated approaches: to translate relations from existing structured resource or to extract relations from existing unstructured resources in the same language. To rely upon translated relations one needs some proof that taxonomic structure of languages involved in translation is similar for a selected purpose. Whether this is true for lexical taxonomies is questionable.

© Springer Nature Switzerland AG 2018
A. Gelbukh (Ed.): CICLing 2017, LNCS 10761, pp. 488–499, 2018.
https://doi.org/10.1007/978-3-319-77113-7_38

Monolingual dictionary has some traits of good language resource for relation mining: it is a small corpus with limited syntactic variability that describes large part of the common knowledge and explicitly separates word senses. Many research has been applied to extracting relations from monolingual dictionaries. One of the best approaches is to parse a dictionary using a specialized grammar with annotation of relations within grammar rules.

This work is focused on extracting noun semantic relations from monolingual dictionaries. We put forward an approach that requires less work than creating a full specialized grammar and provides excellent accuracy.

2 Related Work

The most precise way to create a thesaurus to day remains manual creation by lexicographers. This is the way Princeton WordNet was created. Another example of manual approach is FinnWordNet [10] created as a translation of Princeton WordNet by a group of professional translators. It is also the most expensive approach. Many approaches were applied to automate some or all of the work at the possible cost of some loss of precision or coverage.

Perhaps, the most precise way to automatically obtain a network of semantic relations is to extract the network from some other manually created resource with well-defined semantic relations. This is the case, for example, with DanNet [14] which reuses genus term annotation built into Den Danske Ordbog dictionary by the same team.

One of the most popular approaches to automate hyponym extraction was initially demonstrated by Hearst [7]: he manually developed a set of lexico-syntactic rules for matching definitions in a corpus. Although he proposed bootstrapping approach to extend the set of rules, in many cases translations of the same 6 rules provide reasonable performance on different languages, including Russian [16]. Navigli applied machine learning to automatically create new patterns during bootstrap [11] to increase extraction recall.

For monolingual dictionaries it is often feasible to create a special-purpose grammar. An elaborate example with analysis of dictionary structures can be found in [2]. Oliveira et al. used such grammar to convert a general dictionary of Portugese into a set of semantic relations PAPEL [13].

Another popular approach is to automatically translate an existing Word-Net. An example of this approach is Catalan WordNet [3]. Large part of the EuroWordNet project [19] is based upon machine translation.

Distributional semantics promises to open new possibilities to automatically extend taxonomies. SemEval2015 task 17 [4] brought attention to this group of apporaches. Generally vector semantic approach is known to be able to extract pairs of semantically related words, but is not able to distinguish kinds of relations. Weeds et al. [21] applied SVM classifier to tell apart the different relations with moderate success. Yamane et al. [22] applied a specialized model in spirit of bootstrapping to simultaneously create a distance metric for hyponymy and organize a set of word pairs into taxonomy.

In WordNet synonymy is different from relations. Many works focus specifically on extracting groups of synonyms from texts. Wang compares [20] three approaches to extract synonyms from dictionary definitions. The first approach, called inverse index approach, is based on similarity of definitions in which a word is mentioned. The second approach uses machine learning to classify each word in a definition as being either synonym or not. The third approach is to manually create a set of lexico-syntactic rules. Wang shows comparably high performance in terms of both precision and recall with either of the three approaches. Rule-based approach to extract synsets from dictionary definitions was also used to extend Onto.PT [12].

3   Approach Overview

Our approach to relation extraction requires a corpus of dictionary definitions and some basic NLP tools detailed below in Sect. 5. The steps of relation extraction are:

– cluster word sense definitions automatically,
– manually annotate each cluster with relation extraction rule,
– run the relation extractor and evaluate the results.

The first step is clustering. The aim of this step is to reduce the amount of work for human annotators. This means that the dictionary should be separated into as few clusters as possible. On the other hand definitions in each cluster should have as similar structure as possible.

The second step is annotation. On this step an annotator is given a few definitions from a cluster. The annotator has to:

– determine if most of the definitions contain a word that can serve as hypernym, hyponym, meronym, holonym, or synonym to the defined word sense,
– write a morpho-syntactic pattern that extracts the word.

The third step is relation extraction and evaluation. To evaluate the rule quality the expert examines the result of rule application for 25 cases in each cluster (or the whole cluster, if the cluster is smaller than 25 word senses) and annotates result of rule application to examined sentence as either correct or incorrect. Summary of such annotations is given below in Sect. 6

The annotator guidelines strongly suggests to the expert to annotate exactly one rule per each cluster. This means that it is more dangerous to merge unrelated clusters in clustering step than to split similar definitions into several clusters.

4   Definition Features for Clustering

The aim of clustering in the work is to group together definitions that have the same phrase structure and are likely be parsed using the same morpho-syntactic rule. To construct definition features that extract structure similarity we employed the following assumptions about definition structure:

- lexicographers creating a dictionary restrict themselves to just a few phrase structures,
- looking at a few first words is usually enough to guess the phrase structure,
- phrase structure manifests itself in syntax or specific terms and their coordination structure.

Based on these assumptions, we attempt to find manifestation of phrase structure by considering the following features of definitions:

- lexical unigrams: `word_form`, `lemma`
- morphological unigrams: `pos`, every morphological feature as a tag (`gr_atoms`)
- compound morphological unigrams: full morphological parse (`gr`), immutable morphological features (`immutable_gr`, e.g. part of speech, gender and animacy for nouns), mutable morphological features (`mutable_gr`, e.g. case and number for nouns)
- mixed trigrams [5] with templates:
  (`lemmas, immutable_gr, immutable_gr`),
  (`immutable_gr, lemmas, immutable_gr`),
  (`immutable_gr, immutable_gr, lemmas`)

Each individual feature assigns some numerical value to a definition, e.g. number of occurrences of a given unigram. Using every possible of the listed features implies that a definition is characterized with a feature vector of very large dimension, for example there are 25508 unique lemmas in the corpus of definitions.

The assumptions above allow us to expect that for each wording style there exist one or several features listed above that are shared among all definitions that have the given wording style.

Each set of linguistic features is restricted to 200 most frequent (e.g. 200 most frequent lemmas out of 25508). This helps to both reduce the dimensionality of feature space and to increase the relevance of the features.

Vector representation f_i of definition features is produced for each position number i in the definition. For the purposes of clustering definition is represented using feature vector f_c defined as $f_c = f_1 \| f_2 \| f_3 \| \frac{1}{N} \sum_{i=1}^{N} f_i$, where N is number of words in the definition and $\|$ denotes vector concatenation.

4.1 Mixed N-Grams

Noun definitions in monolingual dictionaries are an example of restricted language. To illustrate, in the Big Explanatory Dictionary of Russian language approximately 91% of the definitions consist entirely of noun phrases (according to our unpublished research). General-purpose syntactic parsers have prohibitively-high error rate on restricted languages [8]. This prevents us from directly using syntactic features for clustering. In this work we used mixed n-grams as a simpler replacement.

Let us give a definiton of mixed n-gram by comparison with a definiton of n-gram. Given a sequence of L tokens $T_L = t_i$, let us say that n-gram is a n-tuple of sequential elements from a list starting at some position i: $(t_i, t_{i+1}, \ldots, t_{i+n-1})$.

Similarly, given n different lists of L tokens each $T_L^n = t_i^k$ let us define mixed n-gram as a n-tuple of elements, where each next element in the tuple corresponds both to the next list and to the next position in the list: $(t_i^1, t_{i+1}^2, \ldots, t_{i+n-1}^n)$.

Please refer to Table 1 for an illustration. In the linguistic domain let us call the set of token lists used *n-gram template*.

Table 1. Example phrase and it's (POS, word) mixed bigrams

	word	Отколовшийся	от	чего-л.	кусок.
		Chipped-off	of	smth.	piece.
		A piece that chipped off of something.			
	POS	V	PR	SPRO	S
(POS, word) bigram		(V	от)		
			(PR	чего-л.)	
				(SPRO	кусок.)

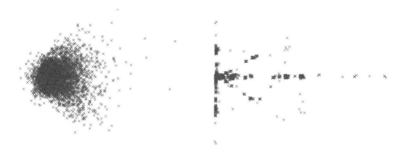

Fig. 1. Dictionary definitions converted to feature space then number of dimensions reduced to 2 using PCA. Each point is a word sense definition. Axes are PC1 and PC2 (essentially meaningless). Left: all features except mixed n-grams. No structure is visible. Right: all features including mixed n-grams. Some clusters are already visible

4.2 Do Mixed N-Grams Help?

To have a rough visualization of usefulness of feature sets the whole dictionary was converted to vector representation according to the selected feature set. Next, number of dimensions was reduced to 2 using principal component analysis and the data set visualized (see Fig. 1). We employ a fuzzy criteria for feature set usefulness: if the image is uniform, it is unlikely that any clustering algorithm can extract meaningful clusters from the data with the given features; if the image has visible irregularities, some suitable clustering algorithm might yield interesting results. While this criteria is vague, in the current case the interpretation of the plots is quite clear: mixed n-grams capture some structure in dictionary definitions that ngrams are incapable of. Whether the structure they capture is of any use for the task of relation extraction is a question of the rest of the paper.

5   Materials and Methods

The experiments were performed on Big Explanatory Dictionary of Russian language (BTS)[9]. We use a simplified representation of the dictionary as a table with four columns: headword, raw definition of one sense, full definition, references. Raw definition is textual definition of the headword. Full definition contains the raw definition, and also includes examples, labels, and possibly adjacent definitions if they contribute to understanding of the given definition.

For tokenization and morphology information we used mystem [17] version 3.1. mystem was invoked with -c flag to retain punctuation as separate tokens. Whitespace tokens were removed from mystem output.

Headwords were attributed with POS tags using mystem. To verify POS attribution quality on the specific material 1000 random headwords were annotated by an expert, of those 424 as nouns. The test displayed 97.9% recall and 89% accuracy of noun attribution by mystem. Accuracy results in dictionary test were expected to be lower than in corpus test, since headwords have no suitable context and uniform distribution, while Zipf's distribution in corpus test allows to obtain higher accuracy by correctly handling the most frequent words. Another problem of POS detection is that without context suitable for the language model of mystem it is impossible to handle words that have variable part of speech. This turned out to be a negligible problem for Russian language, responsible for 3/424 false negatives and 1/424 false positives in the selected subset.

The dictionary was filtered to contain only noun headwords. This resulted in 33683 words with 58621 word sense definitions.

Clustering and PCA analysis was performed using algorithms implemented in Scikit-learn [15] package for Python 3 [18] programming language. Visualization performed using numpy, scipy, matplotlib packages of Python 3.

Series of preliminary experiments with DBSCAN algorithm with varying ϵ parameter were performed in order to estimate the optimal number of clusters. Both cluster contents and distribution were assessed by expert to determine the necessary cluster count. The final clustering was performed with k-means algorithm to form 100 clusters.

Full pipeline for parsing the dictionary, preprocessing, building feature vectors and clustering was developed as a set of Python 3 scripts[1].

---

[1] Available at https://bitbucket.org/dendik/russian-wordnet-rules, http://www.cicling.org/2016/data/311.

6   Results

This section answers two questions. The most important question is how good is the suggested approach for extracting each relation. The second question is how good is each of the pattern-matching rules defined by annotator.

Table 2. Estimate on number of extracted relations and extraction accuracy as compared to PAPEL.

Relation	Russian.WN		PAPEL	
	Amount	Accuracy	Amount	Accuracy
Hypernym	53246	85.54%	29,563	59.10%
Synonym	10044	75.69%	11,862	86.10%
Junk	7175	100.00%		
Hypernym synonyms	4160	76.11%		
Hyponym	2761	53.71%		
Part of	1017	100.00%	1,287	52.60%
Domain	495	51.72%		
Instance of	253	61.26%		
Hypernym hypernym	125	100.00%		
Has part	105	92.38%		
Dictionary	58621	83.93%	37898	76.64%

To answer the first question we grouped clusters by relation that can be extracted from each cluster. For each group of clusters we counted number of definitions in the group and estimated overall accuracy of relation extraction. It is difficult to estimate either precision or recall for loosely-defined thesauri like WordNet. However, for a specific case of hypernyms it is fair to assume that every node in WordNet has roughly one hypernym (the top node, entity, has none, and very few of the lower nodes have two or more hypernyms). Thus, fraction of definitions that have hypernyms might be thought of as hypernym extraction recall. Accuracy of relation extraction recall was estimated by a human judge by reviewing random 25 examples from each of the 100 clusters.

To our best knowledge there exist no published work on relation extraction from Russian language dictionaries with which we can compare our results. Thus to provide some sort of baseline we refer to PAPEL [13] as one of the latest comparable works among other languages. Table 2 shows the results.

From the point of view of accuracy, hypernym and meronym extraction gains significant advantage from preliminary clustering, while synonym extraction still requires more work. One notable kind of "relation" is "junk", which is a group of clusters that contain either headwords of wrong POS or somehow distorted by dictionary parser.

Most notably the hypernym extraction accuracy exceeds the baseline by approximately 25%. Some part of the difference might be explained by different methods of accuracy estimation, different language and source data. We assume that some part of the improvement is due to different extraction approach employed. By applying a rule to a cluster of similar definitions we essentially prevent the rule from being applied applying to definitions with unsuitable syntactic structure.

Table 3. Pattern usefulness. For each pattern we show the number of definitions for which the pattern is applied and accuracy of relation extraction using the pattern.

pattern	definitions	accuracy
1st nominative noun	49348	85.15%
1nd nominative noun	9804	70.06%
junk	7175	100.00%
1st genitive noun	4165	62.95%
"человек" ("human being")	2988	100.00%
1st prepositional noun	2487	77.04%
1st noun after "часть" ("part")	1017	100.00%
1st genitive noun after "один" ("one")	726	100.00%
3rd nominative noun	556	80.04%
1st dative noun	382	57.59%
1st instrumental noun	242	64.05%
"люди" ("people")	218	100.00%
1st instrumental noun after "с" ("with")	105	92.38%
"растение" ("plant")	86	100.00%
"название" ("name")	82	80.49%

The second question this section needs to answer is how good is each of the patterns created by the annotator. We estimate the usefulness of each pattern likewise: first, estimate accuracy of each relation extraction pattern by reviewing a small subset of each cluster, then combine clusters that were annotated with the same pattern for relation extraction and estimate accuracy for the combined cluster.

For each relation extraction pattern employed in the work Table 3 lists the number of definitions in the combined cluster annotated with the pattern and the estimated extraction accuracy. The pattern definitions are converted to human-readable form. The quoted patterns (e.g. "human") match any definition and assign the result (e.g. value of hypernym) to every word sense in the cluster. The rest of the patterns name some word in the definition of word sense. The result (e.g. value of hypernym) of pattern application is the lemma of the matched word.

Since the result of a pattern is a word and the goal of relation extraction is to establish relations between word senses, the relation extraction should be followed by some disambiguation, e.g. [1]. This is out of the scope of the current work.

7 Discussion

The estimates are mostly accuracy for the given dictionary with the given annotator effort applied. We expect to see similar performance on other dictionaries,
but this needs to be checked. Of more interest, however, is to learn to infer
cluster types and relevant rules automatically, leading to a fully automated relations extraction pipeline. This might enable us to compare hypernym hierarchies
based on different monolingual dictionaries.

In this paper we somewhat incorrectly use the term "semi-supervised learning". Semi-supervised learning is a class of machine-learning approaches that
extend the training set by similarity of data points. We argue, however, that
clustering followed by annotation is similar: here we swap steps of annotating
training set and grouping together the similar data points.

The difference between straightforward semi-supervised learning and the presented approach is in the number of cases an annotator has to consider. If annotation is the first step and the annotators pick samples randomly, then they
inevitably examine samples of the same class multiple times. On the other hand,
if annotation is the second step and clustering performs well, the annotators
have no chance of observing samples of the same class: such samples would fall
into the same cluster. This essentially means that "cluster then annotate" approach allows the annotators to observe more different rare cases, and this does
not impair annotation of the most frequent cases.

As an illustration of the approach let us review fragments from some of the
clusters. Cluster 0 comprises definitions starting with "тот, кто" ("the one who"):

(1) ВИЗАВИ: Тот, кто находится напротив
     VIS-A-VIS: the one who is situated across

(2) ВИЗИТЁР: Тот, кто пришёл с визитом.
     VISITER: the one who came to visit

(3) ВИНОВНИК: Тот, кто является причиной, источником чего-л.
     CURLPIT: the one who is cause, source of something

This is a clearly well-detected cluster. Definitions within the cluster do not
mention hypernym of word senses, and direct hypernym is difficult to establish.
Indirect hypernym is trivial in this case: "человек" (human).

Cluster 18 consists of definitions starting with "О" ("about") followed by
adjective and a comma:

(4) ГОРГОНА: О злой, сварливой женщине.
     GORGON: About evil, gnarly woman.

(5) ГРЕНАДЕР: О рослом, плечистом, сильном человеке.
     GRENADIER: About tall, broad-shouldered man.

(6) ДЕБИЛ: О тупом, несообразительном человеке.
     DEBILE: About dull, unapprehensive person.

This cluster has a well-defined meaning: all definitions within the cluster
are metaphoric. The use of "about" to denote metaphoric senses seems very

strict, but is not mentioned in the dictionary user manual. Existence of this cluster helps the researcher to either set aside metaphoric meanings, or to write a morpho-syntactic rule: in all definitions within the cluster the hypernym is marked with prepositional case.

Cluster 46 is an example of bad cluster. The distinctive features of the cluster are that first word is nominative case noun and that comma is present among the first few words:

(7)   ДВУЛЕТНИК: Растение, живущее два года и дающее семена на второй год после посева.
      BIENNIAL: Plant, living for two years and bearing seeds on the second year since planting.

(8)   ДЕКЛИНАЦИЯ: Отклонение, уклонение.
      DECLINATION: diversion, deflection.

(9)   ДЕКОМПОЗИЦИЯ: Разделение, расчленение какой-л. сложной структуры на составляющие части.
      DECOMPOSITION: Separation, partition of some complex structure into composing parts.

From the point of view of relation extraction the here senses fall into major categories. Definitions like (8) consist entirely of nouns: such definitions along with the head word form a synset. Definitions like (7) contain specification clause along with conjunction of nouns: in this case nominative case nouns define hypernym synset of the given sense of head word. Some definitions in the cluster do not fall into either of the two categories, as is the case with (9).

8   Conclusion

Here we presented a semi-supervised approach to relation extraction. The novelty of the approach as compared to traditional rule-based approaches is preceding the rule-writing with clustering.

This little change appears very beneficial in many aspects:

1. The approach allows to extract a majority of definitions with limited and very predictable amount of work.
2. As compared to similar works, the approach seems to considerably increase accuracy of rule extraction without noticeable loss in amount of extracted relations, although precise comparison is not possible.
3. The approach is driven by the corpus. Along with the obvious frequent cases the annotator is put into position to pay attention to some of the rare definition styles if they are well-defined.

Mixed n-gram features used for the clustering deserve special mention. The feature set is very trivial to construct and requires minimal resources. The features proved suitable for the task of detecting small similar clauses. This is an important tool for the cases where quality syntax is unavailable, such as corpora with restricted language.

Combined with word sense disambiguation this approach allows to create prototype WordNets for low-resource languages bypassing the commonly-used translation approach. The prototyping requires some limited expert effort and very primitive linguistic resources: a machine-readable monolingual dictionary, a POS tagger and a word sense disambiguation tool.

References

1. Alexeyevsky, D., Temchenko, A.V.: WSD in monolingual dictionaries for Russian WordNet. In: Fellbaum, C., Forǎscu, C., Mititelu, V., Vossen, P. (eds.) Proceedings of the Eighth Global WordNet Conference, pp. 10–15. Bucharest, Romania, January 2016
2. Barnbrook, G., Sinclair, J.: Specialised corpus, local and functional grammars. Small Corpus Stud. ELT: Theor. Pract. **5**, 237 (2001)
3. Benitez, L., Cervell, S., Escudero, G., Lopez, M., Rigau, G., Taulé, M.: Methods and Tools for Building the Catalan WordNet. CoRR cmp-lg/9806009 (1998). http://arxiv.org/abs/cmp-lg/9806009
4. Bordea, G., Buitelaar, P., Faralli, S., Navigli, R.: Semeval-2015 task 17: Taxonomy Extraction Evaluation (TExEval). In: Proceedings of the 9th International Workshop on Semantic Evaluation. Association for Computational Linguistics (2015)
5. Bramsen, P., Escobar-Molano, M., Patel, A., Alonso, R.: Extracting social power relationships from natural language. pp. 773–782. Association for Computational Linguistics (2011)
6. Fellbaum, C.: WordNet. Wiley Online Library (1998)
7. Hearst, M.A.: Automatic acquisition of hyponyms from large text corpora, pp. 539–545. Association for Computational Linguistics (1992)
8. Kummerfeld, J.K., Hall, D., Curran, J.R., Klein, D.: Parser showdown at the wall street corral: an empirical investigation of error types in parser output, pp. 1048–1059. Association for Computational Linguistics (2012)
9. Kuznetsov, S.A.: The Newest Big Explanatory Dictionary of Russian Language. RIPOL-Norint, St.Petersburg (2008)
10. Lindén, K., Niemi, J.: Is it possible to create a very large wordnet in 100 days? An evaluation. Lang. Resour. Eval. **48**(2), 191–201 (2014)
11. Navigli, R., Velardi, P.: Learning word-class lattices for definition and hypernym extraction, pp. 1318–1327. Association for Computational Linguistics (2010)
12. Oliveira, H.G., Gomes, P.: Automatic Discovery of Fuzzy Synsets from Dictionary Definitions, pp. 1801–1806 (2011)
13. Oliveira, H.G., Santos, D., Gomes, P.: Relations extracted from a portuguese dictionary: results and first evaluation, pp. 541–552 (2009)
14. Pedersen, B.S., Nimb, S., Asmussen, J., Sørensen, N.H., Trap-Jensen, L., Lorentzen, H.: DanNet: the challenge of compiling a wordnet for Danish by reusing a monolingual dictionary. Lang. Resour. Eval. **43**(3), 269–299 (2009). https://doi.org/10.1007/s10579-009-9092-1
15. Pedregosa, F., et al.: Scikit-learn: machine learning in Python. J. Mach. Learn. Res. **12**, 2825–2830 (2011)

16. Sabirova, K., Lukanin, A.: Automatic extraction of hypernyms and hyponyms from russian texts. In: Ignatov, D.I., Khachay, M.Y., Panchenko, A., Konstantinova, N., Yavorsky, R., Ustalov, D. (eds.) Supplementary Proceedings of the 3rd International Conference on Analysis of Images, Social Networks and Texts (AIST 2014), vol. 1197, pp. 35–40. Citeseer (2014)

17. Segalovich, I.: A Fast Morphological Algorithm with Unknown Word Guessing Induced by a Dictionary for a Web Search Engine, pp. 273–280. Citeseer (2003). https://tech.yandex.ru/mystem/

18. Van Rossum, G.: Python Programming Language, vol. 41 (2007)

19. Vossen, P.: A Multilingual Database with Lexical Semantic Networks. Springer, Dordrecht (1998). https://doi.org/10.1007/978-94-017-1491-4

20. Wang, T., Hirst, G.: Extracting Synonyms from Dictionary Definitions, pp. 471–477 (2009)

21. Weeds, J., Clarke, D., Reffin, J., Weir, D., Keller, B.: Learning to distinguish hypernyms and co-hyponyms, pp. 2249–2259. Dublin City University and Association for Computational Linguistics (2014)

22. Yamane, J., Takatani, T., Yamada, H., Miwa, M., Sasaki, Y.: Distributional Hypernym Generation by Jointly Learning Clusters and Projections (2016)

Speech Recognition

ASR Hypothesis Reranking Using Prior-Informed Restricted Boltzmann Machine

Yukun Ma[1,2]([✉]), Erik Cambria[1], and Benjamin Bigot[2]

[1] School of Computer Science and Engineering, Nanyang Technological University, Singapore, Singapore
cambria@ntu.edu.sg
[2] Rolls-Royce@NTU Corporate Lab, Nanyang Technological University, Singapore, Singapore
mayu0010@e.ntu.edu.sg, bbigot@ntu.edu.sg

Abstract. Discriminative language models (DLMs) have been widely used for reranking competing hypotheses produced by an Automatic Speech Recognition (ASR) system. While existing DLMs suffer from limited generalization power, we propose a novel DLM based on a discriminatively trained Restricted Boltzmann Machine (RBM). The hidden layer of the RBM improves generalization and allows for employing additional prior knowledge, including pre-trained parameters and entity-related prior. Our approach outperforms the single-layer-perceptron (SLP) reranking model, and fusing our approach with SLP achieves up to 1.3% absolute Word Error Rate (WER) reduction and a relative 180% improvement in terms of WER reduction over the SLP reranker. In particular, it shows that proposed prior informed RBM reranker achieves largest ASR error reduction (3.1% absolute WER) on content words.

1 Introduction

Reranking models have been shown effective for reducing errors in a variety of Natural Language Processing tasks such as Named Entity Recognition [1, 2], Syntactic Parsing [3,4] and Statistical Machine Translation [5]. A reranking model typically treat the baseline system as a black box and is trained to rank the competing hypotheses based on more complex or global information.

In Automatic Speech Recognition (ASR), discriminative language model (DLM) was first introduced by Roark et al. [6] for reranking ASR hypotheses. They adopt a single perceptron to modify the confidence scores of hypotheses generated by a baseline ASR system. By using only n-gram features, their reranking model was shown capable of reducing the Word Error Rate (WER) of an ASR system. His work is followed by several variants with a variety of feature choices such as syntactic features [7,8], which try to capture correlation between simple features on the feature level. However, existing DLMs still suffer from poor generalization power and are vulnerable to shortage of training data,

© Springer Nature Switzerland AG 2018
A. Gelbukh (Ed.): CICLing 2017, LNCS 10761, pp. 503–514, 2018.
https://doi.org/10.1007/978-3-319-77113-7_39

because most of them rely on linear or log linear models that fail to take into consideration the correlation of input features on the model level. Apart from feature engineering, using hidden variables encoding semantic information helps improving the generalization power.

Koo et al. [4] proposes a hidden-variable model to rerank syntactic parsing trees. By linking input features to hidden states corresponding to word senses or classes, they achieves improved accuracy over a linear baseline. Inspired by the success of Koo et al., we propose to use the computational structure of Restricted Boltzmann Machine (RBM) [9] for the task of ASR hypotheses reranking. RBM is a neural network composed of one hidden layer and one input layer. The hidden layer of RBM has been shown capable of capturing high-order correlation and semantic information in the context of language modeling [10,11]. These approaches model the probability of a fixed length of word sequences, i.e., N-grams, using only local information, and are trained with a generative objective function. However, RBM cannot be directly used for ASR reranking due to its generative training manner.

We propose two modifications to train RBM in a more task-specific way. We modify the energy function of RBM to incorporate the ASR confidence score, which has been proved critical for reranking by previous DLMs [6–8]. We then propose a novel discriminative objective function for training RBM with N-best lists of ASR hypotheses. Our method differs from existing RBM-based language models [10,11] in two major aspects. Firstly, the proposed RBM reranker is trained discriminatively. Secondly, RBM in our method represents sentences of variable length as global feature vectors. Another attractive property of RBM is that the computational structure is flexible enough to incorporate various sources of prior knowledge [12]. As function words have little meanings and are less important for language understanding [13], we decide to focus more on content words, e.g., named entities. We hence further integrate to our model two types of prior knowledge: named entity related prior and a pre-trained hidden layer.

To our knowledge, this paper is the first to consider using hidden layer and prior knowledge in the context of ASR hypotheses reranking. The remainder of this paper is structured as follows: Sect. 2 describes in detail the proposed work; Sect. 3 shows the empirical results as well as analyses; finally, Sect. 4 concludes this paper and discusses about future work.

2   Training RBM for ASR Reranking

2.1   Restricted Boltzmann Machine

A Restricted Boltzmann Machine [9] (see Fig. 1) is a neural network composed of: one n-dimension input feature layer $\phi(\mathbf{t}) = [\phi_1(t), \phi_2(t), \cdots, \phi_n(t)]$, which is a global feature vector extracted for a raw input t, and one d-dimension binary hidden layer $\mathbf{h} = [h_1, h_2, \cdots, h_d]$. The joint probability $P_{\text{RBM}}(t, h)$ of hidden variables and raw input is defined as

$$P_{\text{RBM}}(t,h) = \frac{e^{-E_{\text{RBM}}(t,h)}}{\sum_{t,h} e^{-E_{\text{RBM}}(t,h)}}$$

$$E_{\text{RBM}}(t,h) = -\phi(t)^T W h - b^T \phi(t) - c^T h,$$

where $W \in R^{n \times d}$ is the matrix specifying the weights of connections between hidden and input layer, and $b \in R^n$ and $c \in R^d$ are the bias vectors of the two layers. $E_{\text{RBM}}(t,h)$ is called the energy function of RBM. The probability of a raw input t is then defined as the marginal probability of t

$$P_{\text{RBM}}(t) = \sum_h P_{\text{RBM}}(t,h),$$

and the training objective is to maximize the log likelihood of training data D

$$\sum_{t \in D} \ln P_{\text{RBM}}(t)$$

2.2 Maximum Margin Training for RBM-based Reranker

The goal of generative training of RBM is to learn a probability distribution, which is not necessary for choosing correct ASR hypotheses. Instead, the discriminative training allows the model to explicitly select ASR hypotheses containing fewer errors. In this section, we describe our discriminatively trained RBM-based reranking model, denoted as **dRBM**.

Before introducing the training objective function, we first introduce the energy function of RBM model. ASR posterior probabilities produced by the baseline ASR system have been shown useful for reranking in previous works on DLM [6–8]. We hence add ASR posterior to the energy function of RBM. The modified energy function is expressed as

$$E_{\text{dRBM}}(t,h) = E_{\text{RBM}}(t,h) + E_{asr}(t)$$

$$E_{asr}(t) = -w_0 \ln(P(t|a)),$$

where $P(t|a)$ is the posterior probability of a given ASR hypothesis t given the acoustic input a, and w_0 is the weight of ASR confidence score fixed during training. We represent each hypothesis as a global feature vector $\phi(t)$ using a predefined set of feature functions. In this paper, we mainly consider using unigram features, yet using more complicated features does not need to change the model.

Inspired by the maximum margin training for Bayesian Networks [14], we adopt a discriminative objective function L using likelihood ratio,

$$L = \frac{1}{|D|} \sum_{a \in D} \sum_{t' \in \text{GEN}(a)} \max(1 - \ln \frac{P_{\text{dRBM}}(\hat{t})}{P_{\text{dRBM}}(t')}, 0),$$

where D is the training set for the discriminative training of RBM and $|D|$ denotes the number of utterances in training set. $\text{GEN}(a)$ refers to the list of

N-best hypotheses generated by the baseline ASR system for the acoustic input a, while $\hat{t}$ is the oracle-best in the N-best list of t. Intuitively, the learning process finds the parameter setting maximizing the margin between the oracle-best hypotheses and other hypotheses in the N-best list. The subgradient of the objective function is

$$\frac{\partial L}{\partial \theta} = \sum_{a \in D} \sum_{t' \in \text{GEN}(a)} \mathbb{I}(\mathcal{F}(\hat{t}) - \mathcal{F}(t') < 1)\left(\frac{\partial \mathcal{F}(t')}{\partial \theta} - \frac{\partial \mathcal{F}(\hat{t})}{\partial \theta}\right),$$

where $\mathcal{F}(\cdot)$ is the free energy of RBM defined as

$$\mathcal{F}(t) = -\ln \sum_h e^{-E_{\text{dRBM}}(t,h)}$$

Algorithm 1. Discriminative training for RBM

Input:
D: the training data set
$\text{GEN}(a)$: N-best list for an utterance t in the reference
λ: learning rate
for $k{=}1{:}K$ **do**
    for $a \in D$ **do**
        Positive:
        $\hat{t} = \text{argmin}_{t' \in \text{GEN}(a)} \text{WER}(t')$
        Negative:
        $T^- = \{t'|1 + Score(t') > Score(\hat{t})\}, t' \in \text{GEN}(t)\}$
        for $t' \in T^-$ **do**
            $\theta \leftarrow \theta + \lambda\frac{\partial -\mathcal{F}(\hat{t})}{\partial \theta}$
            $\theta \leftarrow \theta - \lambda\frac{\partial -\mathcal{F}(t')}{\partial \theta}$

The training algorithm is described in Algorithm 1. For each acoustic input in training set, we select a set of hypotheses T^-, which are ranked higher than the oracle best hypothesis. Based on our analysis of the loss function, we boost the score of oracle best with its derivate of negative free energy and penalize hypothesis in T^- with their derivates of negative free energy. Note that, as compared with standard standard RBM training [15], which iterates over input space or samples of input space, our discriminative training needs only to iterate over the N-best list which grows linearly with the size of training data and N-best list.

2.3   Training with Prior Knowledge

The binary hidden layer of RBM allows for easily incorporating prior knowledge into the reranking model. We consider using two types of prior knowledge:

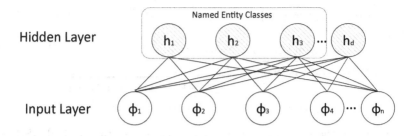

Fig. 1. Structure of RBM with entity-related prior

named entity labels and pretrained latent layer from texts. Firstly, to improve the capability of recognizing content words, we capture prior of a special class of content words – named entities. As entity related prior also encodes information about word classes, it helps improving the generalization power of language models [16] as well.

Specifically, we extract pairs of named entity words and their classes from texts using a named entity tagger, which annotates the text with 3 widely-adopted named entity classes, i.e., LOCATION, ORGANIZATION and PERSON. As show in Fig. 1, 3 variables in the hidden layer of RBM are used to represent named entity classes. For purpose of reducing ambiguity, we remove words belonging to multiple entity classes. We denote the list of entity-class pairs as $G = \{w, e\}$, where w is an index of the unigram feature in the input layer, and e an index of entity-class variable in the hidden layer. The objective function is then augmented with an entity-related regularizer,

$$L - \lambda \ln \prod_{w, e \in G} \prod (P(h_e = 1 | \phi_w) - 1)^2,$$

$$P(h_e | \phi_w) = \sigma(c_e + W_{e,w} \phi_w).$$

As introduced in Wang et al. [12], $P(h_e | \phi_w)$ denotes the probability of a hidden variable h_e being activated by a given input feature ϕ_w.

To handle the data sparsity, we initiate connection matrix W of RBM with values pretrained using a large text corpora and the generative training. The pretraining captures the distributional semantics of input features [17].

2.4   Scoring ASR Hypotheses

To score a given hypothesis, we propose two scoring functions using our RBM-based reranker and its combination with SLP. First of all, the RBM-based reranking score S_{RBM} is defined as the logarithm of the unnormalized probability $\tilde{P}_{\text{dRBM}}(t)$ assigned by the RBM-based reranker solely,

$$S_{\mathrm{RBM}}(t) = \ln \tilde{P}_{\mathrm{dRBM}}(t)$$

$$= \underbrace{w_0 \ln(P(t|a))}_{\text{ASR posterior}} + \underbrace{\sum_i^n b_i \phi_i(t)}_{\text{linear part}} + \underbrace{\sum_j^d \ln(1 + e^{(c_i + W_i \phi(t))})}_{\text{hidden variable part}}.$$

As shown above, the re-scoring function is composed of the original ASR posterior, a linear bias, and the hidden variable component. In addition, SLP and RBM are likely to have encoded information complementary to each other due to their different structures and training methods. Therefore, we propose a late fusion of the two methods, which combines their confidence scores in the testing phase. The combined reranking score is

$$S(t) = S_{\mathrm{RBM}}(t) + \alpha S_{\mathrm{SLP}}(t),$$

where $S_{\mathrm{SLP}}(t)$ is the single perceptron based confidence score weighted by α.

3   Related Work

DLM has been first introduced by Roark et al. in [6], where simple features like N-gram was shown able to effectively reduce WER. This previous work is using a Single Layer Perceptron (SLP) to modify the original posterior probabilities of the outputs of a baseline ASR system using a linear function,

$$\log P(t|a) + \sum_i w_i f_i(t),$$

where $\log P(t|a)$ is the log probability of a word sequence t given the acoustic signal a, and $\{f_i(\cdot)\}$ are the set of feature functions of an utterance weighted by $\{w_i\}$. Different types of features extracted from syntactic trees [7] and dependency trees [8] have also been used to enrich the feature set.

Apart from feature engineering and using linear combination of feature functions, inferring hidden variables from the observed input captures semantic information related to word classes and word senses. Our work is closely related with Koo et al. [4] who proposed a hidden-variable model to rerank syntactic parsing trees. For the tractability of their model, they put constrains on the connections between latent variables and visible variables (i.e., input layer) by splitting features into two sets. However, the way they divide features is specific to syntactic parsing, and thus is not applicable to our task. Our model differs from Koo et al. [4] in the sense that the connection is not constrained by their feature type, but instead relying on the structure of RBM to build connections between input and hidden layer.

RBM-based models [10,11] have been explored for language modeling. Both approaches model the probability of a fixed length of word sequences, i.e., N-grams, and trained with a generative objective function. Our method differs from these methods in two major aspects: the training of proposed RBM reranker is discriminative, and it represents sentences of variable length as global feature vectors.

4 Experiment

4.1 Dataset

We evaluate our work on the latest release of TedLium Corpus [18] which is a set of audio and manually transcribed texts of Ted talks. As shown in Table 1. We split the training set of TedLium Version 2 into two parts: former Tedlium Training set Version 1 and the rest. The Version 1 part is a set of 774 Ted talks consisting of 56,800 utterances and more than 1.7 million words, while the remaining of the TedLium training set contains another 718 talks. The evaluation set of our experiment is the testing set of TedLium corpus, which is composed of 11 talks. Our text corpus is the ukWaC corpus [19], which is a collection of texts containing about 1.8 billion words.

Table 1. Characteristics of the data sets used in experiments

	Utterances	Talks	Words
ASR AM train	56.8K	774	1.7M
Reranking train (speech)	36.2K	718	0.9M
Reranking train (text)	24M	-	1.8B
Reranking test	1.15K	11	29K

4.2 Baseline

The baseline ASR system is based on KALDI[1] toolkit [20] including a DNN-based acoustic model. It uses a pre-trained language model[2], which is released as part of Sphinx project [21] and has achieved a perplexity of 158.3 on a corpus of Ted Talks. The acoustic model is trained using the training set of TedLium Version 1. The rest of training set of TedLium Version 2 is used for training reranking models. The baseline SLP reranker is trained by following the work of Lambert et al. [8] that randomly selects K pairs of hypotheses from the N-best lists. Specifically, we randomly select 100 pairs from the 100-best list. Learning process is ran for 10 iterations, as we cannot observe further WER reduction with more iterations.

---

[1] http://kaldi.sourceforge.net/.

[2] http://cmusphinx.sourceforge.net/2013/01/a-new-english-language-model-release/.

4.3   RBM Setup

We refer to the system integrating prior knowledge with dRBM as p-dRBM. Both dRBM and p-dRBM use 200 hidden units and are trained using the same data set and 100-best hypotheses as SLP reranker. Since our focus is not on feature engineering, and for simplicity of interpreting our experiment results, we use only unigram features (i.e., single words) in our experiments, which have also been shown as the most effective features by previous works [8]. For training p-dRBM, we first crawled down a set of text summaries of ted talks from Ted website. We then create a list of 20 words for each entity category by tagging the collected text summaries with Stanford named entity recognition tool [22] following description in Sect. 2.3. A basic RBM is trained using ukWaC corpus and used for initiating the connection matrix W of p-dRBM. The late fusion of SLP and our proposed methods are denoted as SLP + dRBM and SLP + p-dRBM. We use $\lambda = 0.01$ and $\alpha = 1.0$ as weights of entity-related regularizer and SLP scores in late fusion.

4.4   Evaluation

First of all, we analyze the behavior of p-dRBM by computing the most-activated words by p-dRBM as shown in Table 2. As shown in previous section, the scoring function used by our method is a combination of ASR confidence score, a linear component (denoted as p-dRBM-L) and a hidden-variable component (denoted as p-dRBM-H). p-dRBM then takes as input one-hot vectors of words to compute their reranking scores. It shows that the linear component is mainly accounting for the function words, while the hidden component favors content words that are mostly nouns and adjectives. The final scoring function is a trade-off between function words and content words through a combination of the two components.

Table 2. Most activated word by p-dRBM

p-dRBM-H	p-dRBM-L	p-dRBM
integrated	of	integrated
demeanor	and	demeanor
disgust	the	disgust
tattoo	to	tattoo
formula	a	formula

To investigate on what is captured by RBM and potentially effective for improving the Word Error Rate (WER), we represent each hidden variable as a vector of words. These vectors represent how much a word is activated by a given hidden variable. Table 3 shows a selected set of hidden variables that can be seen as a set of topics. We found that RBM can capture meaningful topics by

Table 3. Example topics learned by p-dRBM

Working	Media	Higher education	Entertainment
Security	News	Cambridge	Scene
Services	Forum	Mary	Story
Office	Business	Professor	Tv
Home	Press	William	Songs
For	New	Royal	Moving

using only sentence-level co-occurrence. We then represent each word as a vector of hidden variables by taking the rows of the matrix $W \in R^{|V| \times d}$ of p-dRBM.

We rank words based on their cosine similarity with the queries and select the top 5 words for four query words. As shown in Table 4, the top ranked words all seem very relevant to the query words. Since our RBM is trained with sentence-level concurrence, which is different from the window-based methods, the 'similarity' looks more like a topical relatedness rather than syntactical similarity. In general, we can conclude that the resultant RBM-based reranking model to some extent captures the distributional semantics related to the topics of words.

Table 4. Most-similar words for queries using p-dRBM word embeddings

Japan	Film	Bible	Computer
India	Story	Greatest	Software
Italy	Music	Holy	Database
Asia	Beautiful	Truth	Digital
Germany	Famous	Gospel	User
China	Classic	Spirit	Server

As shown in Table 5, we evaluate WER of reranking systems. It shows that the proposed discriminatively trained RBM produces greater WER reduction than baseline SLP rerankers. Effectiveness of using prior knowledge is validated by further improving WER over dRBM. The greatest absolute WER reduction (1.3%) is achieved by the late fusion of SLP and p-dRBM, which confirms that our reranker captures information complementary to SLP.

Since the latent layer of p-dRBM incorporates prior knowledge related to content words (e.g., named entities), it is desirable that the proposed method can better recognize content words, which are more critical for downstream applications such as spoken language understanding. To evaluate the performance of our proposed methods on recognizing content words in a more general way, we words that have higher TF-IDF scores are more likely to be content words.

Table 5. Performance of reranking model on TedLium corpus

	WER	WER (TF-IDF$\geq$ 3)
ASR 1-best	18.23	46.9
Oracle 1-best	11.42	36.1
SLP	17.76	46.3
dRBM	17.51	44.6
p-dRBM	17.36	**43.8**
SLP + dRBM	17.11	45.2
SLP + p-dRBM	**16.91**	44.2

We hence assign more weight to errors involving a set of keywords with high TF-IDF instead of treating all words equally. Specifically, the list of keywords are chosen based on TF-IDF scores ($\geq$ 3.0) computed from whole TedLium corpus. We use the weighted-word-scoring implementation in NIST SCLITE tool[3] by aggressively assigning weight 1.0 to words on the list and 0.0 to the rest.

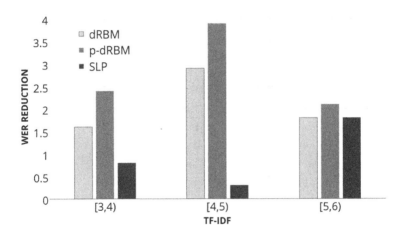

Fig. 2. WER reduction for words versus TF-IDF scores

Table 5 clearly shows that baseline reranking systems (SLP) fail to reduce much WER for selected keywords. In comparison, proposed RBM rerankers, especially p-dRBM, have reduced more errors on chosen keywords without sacrificing overall performance. We further break down the TF-IDF scores into 3 bins. Figure 2 shows the WER reduction by all three approaches. Thanks to hidden variables, our methods are capable of better capturing the discriminative information for most content words. In particular, p-dRBM is shown working

---
[3] http://www1.icsi.berkeley.edu/Speech/docs/sctk-1.2/sclite.htm.

significantly better than other methods on words with medium TF-IDF scores (<5), which is a result of injecting named entity words, e.g., washington (TF-IDF= 3.87), that mostly have TF-IDF between 3.0 and 5.0.

5   Conclusion

In this paper, we proposed an RBM-based language model that is discriminatively trained for reranking ASR hypotheses. In comparison with single perceptron based reranker, our proposed approach reduces more word errors. The success of fusing single perceptron and RBM-based reranker suggests that two models actually capture complementary information useful for selecting less erroneous ASR hypotheses. In addition, we found that introducing prior knowledge to RBM-based reranker results in a better recognition of content words. In the future, we would like to explore the use of lexical knowledge obtained from different resources, e.g., WordNet [23] or SenticNet [24], as additional prior knowledge for the proposed model.

Acknowledgements. This work was conducted within the Rolls-Royce@NTU Corp Lab with support from the National Research Foundation Singapore under the Corp Lab@University Scheme.

References

1. Collins, M.: Ranking algorithms for named-entity extraction: boosting and the voted perceptron. In: Proceedings of the 40th Annual Meeting on Association for Computational Linguistics, pp. 489–496 (2002)
2. Ma, Y., Cambria, E., Gao, S.: Label embedding for zero-shot fine-grained named entity typing. In: COLING, Osaka, pp. 171–180 (2016)
3. Collins, M., Koo, T.: Discriminative reranking for natural language parsing. Comput. Linguist. **31**, 25–70 (2005)
4. Koo, T., Collins, M.: Hidden-variable models for discriminative reranking. In: Proceedings of Human Language Technology Conference and Conference on Empirical Methods in Natural Language Processing, Vancouver, British Columbia, Canada, pp. 507–514 (2005)
5. Li, Z., Khudanpur, S.: Large-scale discriminative n-gram language models for statistical machine translation. In: Proceedings of AMTA (2009)
6. Roark, B., Saraclar, M., Collins, M., Johnson, M.: Discriminative language modeling with conditional random fields and the perceptron algorithm. In: Proceedings of the 42nd Meeting of the Association for Computational Linguistics (ACL 2004), Main Volume, Barcelona, Spain, pp. 47–54 (2004)
7. Collins, M., Roark, B., Saraclar, M.: Discriminative syntactic language modeling for speech recognition. In: Proceedings of the 43rd Annual Meeting on Association for Computational Linguistics, pp. 507–514 (2005)
8. Lambert, B., Raj, B., Singh, R.: Discriminatively trained dependency language modeling for conversational speech recognition. In: INTERSPEECH, pp. 3414–3418 (2013)

9. Smolensky, P.: Information processing in dynamical systems: foundations of harmony theory. In: Parallel Distributed Processing: Explorations in the Microstructure of Cognition, vol. 1, pp. 194–281 (1986)

10. Niehues, J., Waibel, A.: Continuous space language models using restricted boltzmann machines. In: IWSLT, pp. 164–170 (2012)

11. Dahl, G.E., Adams, R.P., Larochelle, H.: Training restricted boltzmann machines on word observations. arXiv preprint arXiv:1202.5695 (2012)

12. Wang, L., Liu, K., Cao, Z., Zhao, J., de Melo, G.: Sentiment-aspect extraction based on restricted boltzmann machines. In: Proceedings of the 53rd Annual Meeting of the Association for Computational Linguistics and the 7th International Joint Conference on Natural Language Processing, vol. 1: Long Papers. Association for Computational Linguistics, Beijing (2015)

13. Fries, C.C.: The Structure of English. Harcourt Brace, New York (1952)

14. Pernkopf, F., Wohlmayr, M., Tschiatschek, S.: Maximum margin Bayesian network classifiers. IEEE Trans. Pattern Anal. Mach. Intell. **34**, 521–532 (2012)

15. Hinton, G.E.: Training products of experts by minimizing contrastive divergence. Neural Comput. **14**, 1771–1800 (2002)

16. Levit, M., Parthasarathy, S., Chang, S.: Word-phrase-entity language models: Getting more mileage out of n-grams. In: Proceedings of Interspeech, Singapore, ISCA - International Speech Communication Association, pp. 666–670 (2014)

17. Salakhutdinov, R., Hinton, G.E.: Replicated softmax: an undirected topic model. In: NIPS, vol. 22, pp. 1607–1614 (2009)

18. Rousseau, A., Deléglise, P., Estève, Y.: Enhancing the TED-LIUM corpus with selected data for language modeling and more ted talks. In: Proceedings of the Ninth International Conference on Language Resources and Evaluation (LREC 2014), pp. 3935–3939 (2014)

19. Ferraresi, A., Zanchetta, E., Baroni, M., Bernardini, S.: Introducing and evaluating ukWaC, a very large web-derived corpus of English. In: Proceedings of WAC-4 (2008)

20. Povey, D., et al.: The kaldi speech recognition toolkit. In: IEEE 2011 Workshop on Automatic Speech Recognition and Understanding. IEEE Signal Processing Society (2011)

21. Walker, W., et al.: Sphinx-4: a flexible open source framework for speech recognition. Technical report, Mountain View, CA, USA (2004)

22. Finkel, J.R., Grenager, T., Manning, C.: Incorporating non-local information into information extraction systems by gibbs sampling. In: Proceedings of the 43rd Annual Meeting on Association for Computational Linguistics, pp. 363–370. Association for Computational Linguistics (2005)

23. Fellbaum, C.: WordNet: An Electronic Lexical Database (Language, Speech, and Communication). The MIT Press, Cambridge (1998)

24. Cambria, E., Poria, S., Bajpai, R., Schuller, B.: SenticNet 4: a semantic resource for sentiment analysis based on conceptual primitives. In: COLING, pp. 2666–2677 (2016)

A Comparative Analysis of Speech Recognition Systems for the Tatar Language

Aidar Khusainov[(✉)]

Institute of Applied Semiotics of the Tatarstan Academy of Sciences,
Kazan Federal University, Kazan, Russia
khusainov.aidar@gmail.com

Abstract. This paper presents a comparative study of several different approaches to speech recognition for the Tatar language. All the compared systems use a corpus-based approach, so recent results in speech and text corpora creation are also shown. The recognition systems differ in acoustic modelling algorithms, basic acoustic units, and language modelling techniques. The DNN-based system shows the best recognition result obtained on the test part of speech corpus.

Keywords: Speech recognition · Acoustic modelling · Language modelling
Tatar language

1 Introduction

The conventional way of building large vocabulary speech recognition systems is to obtain required acoustic models, a pronunciation dictionary, a language model, and use some of the decoders. The situation can be worse whenever you have to recognize the speech of an under-resourced language. In that case some (or all) of the required resources and algorithms may not exist. In this article we present our recent results in creating a very large vocabulary continuous speech recognition system for the Tatar language. Tatar is spoken by several million people but still is an under-resourced language. Therefore, we have to determine an acoustic alphabet, to record and annotate speech corpora, to build models with different existing approaches and to evaluate the recognition quality of combinations of the system's parts.

The paper is organized in the following way. In Sect. 2, the main acoustic and linguistic features of the Tatar language are presented. Section 3 gives an overview of our speech corpus and the text corpus of the Tatar language that are used to train and test acoustic and language models. Section 4 shows different types of developed Tatar speech recognition systems. Section 5 discusses the experiment results of proposed recognition systems. Last section concludes the paper.

2 The Tatar Language

Tatar is the second spoken language in Russia. There are 4.2 million of speakers in Russia and near 5.2 million of speakers in the world [1].

© Springer Nature Switzerland AG 2018
A. Gelbukh (Ed.): CICLing 2017, LNCS 10761, pp. 515–523, 2018.
https://doi.org/10.1007/978-3-319-77113-7_40

The Cyrillic alphabet (unified in 1939) consists of 39 characters. There are 12 vowel and 28 consonant sounds. Different dialects of Tatar can be identified: Western, Kazan (Middle) and Eastern. Based on the existing language classification [2, 3], in 2013 it was assigned to the under-resourced language class [4]. However, recent results in machine translation [5–7], speech analysis and synthesis [8, 9] fields can change this situation.

3   Data

The size and the quality of training data play an essential role in modern recognition approaches. To train speech recognizers for the Tatar language we have created a continuous speech data set representing different types of speakers. The data set consists of read speech mostly spoken by native speakers with a common Kazan dialect. As for training language models we have used the preprocessed Tatar National Corpus [10].

3.1   Speech Corpus

The most modern systems use speech corpora with a total duration of hundreds and thousands hours to create robust acoustic models. This amount of training data gives a possibility to create robust recognizers. The robustness in that case means relatively equal recognition accuracy for male and female speakers, speakers of different sex and age, noise conditions, etc.

Building and annotating the multi-speaker speech corpus for the Tatar language is currently in progress. Now it consists of two main parts. The first part – "Core part" – has been created to cover all the Tatar phonemes pronounced by the large number of speakers. Due to this goal, each speaker has been asked to utter approximately 2 min: 11 sentences from literature texts, 13 separate words, and 7 sentences from news. Each of the resulting set of 31 text fragments has been adjusted to contain all the Tatar phonemes and the maximum number of phonemes' contexts (left and right phonemes) based on 2- and 3-grams number. The "Core part" is now used in several algorithms in their first stages to create initial monophone models (see Sect. 4 for details).

The second part of the corpus is "Read part". We asked people to read randomly selected texts for 30-min. The source of used texts is the Tatar National Corpus described in 3.2. The only text adaptation was transcribing of all abbreviations and numbers.

Both "Core" and "Read" parts have been manually annotated, for now the speech corpus contains speech files, corresponding text and phonetic transcriptions. The corpus also contains additional meta-information about speakers (sex, age, mother tongue) and an expert's score of speakers' proficiency in Tatar. In addition, we plan to continue recording and annotating the "Spontaneous" and start collecting/transcribing "Radio and TV" parts of this corpus.

The main characteristics of the speech corpus are presented in Table 1.

Table 1. The characteristics of multi-speaker speech corpus for the Tatar language

Parameter	Value
Number of speakers	**377**
Male speakers	109
Female speakers	268
Duration	**57:55:09**
Number of speakers in training subcorpus	361
Duration in training subcorpus	52:50:15
Number of speakers in test subcorpus	16
Duration in test subcorpus	5:04:54
Number of speakers in "Core" part	251
Duration of "Core" part	8:12:16
Average duration per speaker in "Core" part	1:58
Number of speakers in "Reading" part	126
Duration of "Reading" part	49:42:53
Average duration per speaker in "Reading" part	23:40

3.2   Text Corpus

Texts that have been used to create language models are from the Tatar National Corpus. Some preprocessing steps have been implemented to prepare texts for language modelling:

1. Duplicate fragments have been removed;
2. All texts have been lower-cased;
3. Abbreviations, numbers, dates have been transcribed;
4. Texts have been split into separate sentences.

The main characteristics of the text corpus after the processing steps are presented in Table 2.

Table 2. The characteristics of the text corpus

Parameter	Value
Number of files	217 294
Number of words	69 810 033
Number of words in learning part	64 629 794
Number of words in test part	5 180 239
Number of syllables	186 014 478 (2,66 per word)
Number of morphemes	110 280 448 (1,58 per word)
Number of letters	434 636 548 (6,23 per word)

4 Systems Description

4.1 General Overview

We have built and evaluated several recognition systems that differ in an acoustic modelling unit, the size of used training data, and in the algorithms used for models' creation and decoding phases. All the training and evaluation have been done using the Kaldi toolkit [11].

We experimented with two different types of acoustic units: monophones and triphones. As we have already mentioned in Sect. 3.1, we used short utterances from the "Core" part of the speech corpus to create acoustic models in the initial training stages (see Training audio data column in Table 3).

Table 3. The overview of built Tatar speech recognition systems

System	Acoustic unit	Training audio data	Features	Language models
Mono	monophone	separate words	MFCCs	small 3-gram
Tri1	triphone	separate words	+delta, delta-delta	+3-gram full
Tri2	triphone	"Core" part	+LDA/MLLT	as above
Tri3	triphone	"Core" part	+fMLLR	as above
Tri4	triphone	full training corpus	as above	+4-gram
NN	triphone	full training corpus	as above	as above

The basic speech feature we used is mel-frequency cepstral coefficients (MFCCs), but also delta and delta-delta coefficients have been used to form 39-dimension feature vector. In more advanced systems (Tri2, Tri3, Tri4, NN) we have used LDA/MLLT feature transformation algorithm, SAT and fMLLR speaker adaptation techniques [12].

LDA-MLLT stands for Linear Discriminant Analysis – Maximum Likelihood Linear Transform. LDA reduces feature space for all data, whilst MLLT takes this reduced feature space from LDA and calculates a transformation for each speaker to implement speaker normalization.

SAT stands for Speaker Adaptive Training and performs speaker and noise normalization. After SAT training, the acoustic models are trained on speaker-normalized features. fMLLR stands for Feature Space Maximum Likelihood Linear Regression. The inverse of the fMLLR matrix is used to remove the speaker identity from the original features.

Three language models have been created: pruned and full 3-gram and 4-gram models.

We have used step by step training process so each next system should improve recognition quality by using more training data, advanced adaptation techniques and larger language models.

The resulting set of speech recognition systems is presented in Table 3.

The latter system from listed in Table 3 is DNN-based recognizer. This neural network uses a p-norm activation function and predicts the posterior probabilities of context-dependent states [13]. It has been trained on the training corpus data aligned by Tri4 recognition system.

All the systems use the same 200 k words vocabulary consisting of most frequent words.

4.2 Acoustic Models

In this work we have created acoustic models for rather good quality recordings: 16 bits, 22 kHz. We could use them to recognize speech in offices, in front of home PC, to analyze speech in not very noisy conditions. In future, we plan to create separate acoustic models for speech transmitted over telephone, TV and radio channels.

The simplest monophone and triphone acoustic models are created for 32 non-silence and 2 silence phones. Context dependency is introduced with left and right adjacent phonemes. Therefore, each context-dependent (CD) phoneme is presented with a triple of context-independent phonemes designated as a − b + c where b is a central phoneme name, a and c are names for left and right context phonemes respectively.

Phonemes names are taken from the basic phoneme alphabet for Tatar (a, ae, b, ch, d, dzh, e, f, g, h, i, j, k, kh, l, m, n, ng, o, oe, p, r, s, sh, t, ts, u, ue, v, y, z, zh) accomplished with a silence sil and short-pause sp, which makes total 34 items.

4.3 Language Models

Language model creation task arises in many applications from spellchecking to machine translation systems. In all the cases, language model has to describe language grammar rules and has the ability to estimate the probabilities of word sequence in a specified language.

We have built the language model for the Tatar language using the SRILM toolkit (Speech Technology and Research (STAR) Laboratory) [14]. This tool has the functionality to create n-gram models, can interpolate different models and estimate the quality of built models. The common way to use SRILM is as follows:

1. Executing "ngram-count" function to calculate the count of n-grams.
2. Executing "ngram-count" function to build the language model based on the results of the first step. A smoothing algorithm has to be specified.
   (a) Executing "ngram" function with –prune option with threshold as a value.
3. Model quality estimation using "ngram" function with 'ppl' parameter.

In addition to conventional 3- and 4-gram models, we have built a pruned 3-gram model. Language model pruning can help in dealing with the limited amount of memory in computing device. Used algorithm prune n-gram probabilities if their removal causes perplexity of the model to increase by less than threshold value (0.0000003 in our experiment) [15].

According to the limit of the corpus size, the developed language models cannot be complete. Thus, there will be unseen n-grams with a zero probability. As the probability of the entire speech utterance is calculated as the multiplication of separate n-grams, this can lead to the situation, in which even one unseen n-gram zeroes out the total utterance probability. To overcome this drawback, we used Kneser-Ney smoothing algorithm [16].

The Tatar language belongs to the agglutinative language family. Thus, its main characteristic is rich morphology. In case of word-based models, the only approach to better cover the entire lexicon is to use large vocabulary. Our experiments have shown that 20 k words vocabulary gives 17% OOV rate, 50 k – 10%, even 200 k vocabulary shows 4.4% OOV on test data set.

For these experiments we have chosen 200 k words vocabulary, because using bigger vocabulary will reduce the speed of the recognition system, while reducing the size of the vocabulary will increase the number of OOV words to an unsatisfactory level.

The quality of built models have been evaluated on the following parameters: memory usage, perplexity (model confidence level in analysis of the test data set), Table 4.

Table 4. Language models' features

Language models	Memory	Perplexity	OOV
pruned 3-gram	25 MB	1600.9	4.4%
3-gram	152 MB	299.5	4.4%
4-gram	189 MB	422.8	4.4%

4.4   Evaluation Method

The most common performance metric in speech recognition is word error rate (WER), that is computed as follows:

$$WER = \frac{I + D + S}{N} * 100\%,$$

where I is the number of insertion errors, D – deletion errors, S – substitution errors, N – the total number of words in uttered text.

The agglutinative nature of the Tatar language can lead to such a situation where one incorrectly recognized affix will be counted as a whole word error in WER. For example, in the third record from the test subcorpus word "калтырадым" have been recognized as "калтырады", and the WER statistics can't show real quality of this recognition result since it observes only the substitution error. To give a different source of evaluation information we have computed an additional metric: syllable error rate (SER).

One of the applications of speech recognition systems is the dictation system. In this type of programs users estimate the recognition quality mostly by the number of corrections they have to make in recognized texts. Therefore, such characteristic as character error rate (CER) can be representative.

5   Results

Table 5 shows the performance of the different recognizers on the 5-h test subcorpus.

Table 5. Evaluation results for Tatar speech recognition systems

System	Language models	WER	SER	CER
Mono	pruned 3-gram	52,06	39,65	28,70
Tri1	pruned 3-gram	28,80	18,32	12,54
Tri1	3-gram	22,59	14,09	9,78
Tri2	pruned 3-gram	24,14	13,95	8,69
Tri2	3-gram	19,08	10,86	6,91
Tri3	pruned 3-gram	21,16	11,35	6,67
Tri3	3-gram	17,21	9,04	5,37
Tri4	pruned 3-gram	18,57	9,29	5,24
Tri4	3-gram	15,19	7,46	4,18
Tri4	4-gram	15,10	7,41	4,15
NN	pruned 3-gram	16,47	8,27	4,94
NN	3-gram	12,99	6,44	3,86
NN	**4-gram**	**12,89**	**6,38**	**3,83**

The analysis of the WER values shows the main component of the errors: the substitution errors. For the conducted experiments the number of substitution errors is from 5 to 10 times bigger than the insertion or deletion errors. For example, for the best NN 4-gram LM system error numbers are as follows: 496 insertions, 395 deletions, 2362 substitutions.

We have found two possible reasons for these results. The first one is the OOV rate, and the second one is the rich morphological structure of the Tatar words. The number of OOV words in speech test corpus is near 1% (213 from 25240 words). The influence of rich morphological structure can be seen in SER and WER difference: syllable-based error rates are nearly two times less than word-based.

As it can be seen from Table 1, the number of male speakers is only one third from the total number of speakers in the corpus. Therefore, we have calculated the WER for NN with 4-gram LM system in per speaker manner to see if this difference in the amount of training data for male and female speakers causes recognition quality difference. The results of this experiment are given in Table 6.

A similar result is obtained on the test data set for male and female speakers (12.5% and 13.1% WER, respectively). This fact supports that our speech corpus can be used for training acoustic models for speaker-independent systems despite of two times less audio data available for male voices.

Table 6. Speech recognition word error rates for male and female speakers

Speaker	D	S	I	N	WER
Male	*168*	*792*	*167*	*9016*	*12,50%*
388	19	134	35	1886	9,97%
390	35	189	44	1834	14,61%
392	59	192	24	2049	13,42%
393	38	159	39	1711	13,79%
394	17	118	25	1536	10,42%
Female	*227*	*1570*	*329*	*16224*	*13,10%*
379	31	245	38	2009	15,63%
380	16	97	52	1215	13,58%
381	3	63	14	697	11,48%
382	41	223	48	2236	13,95%
383	30	169	40	2091	11,43%
384	28	217	28	1750	15,60%
385	23	243	31	1668	17,81%
386	1	19	7	213	12,68%
387	11	70	20	1152	8,77%
389	27	153	42	2179	10,19%
391	16	71	9	1014	9,47%

6   Conclusions

In this paper we present a comparative study of several different approaches to create speech recognition systems for the Tatar language. First multi-speaker speech corpus has been created and used to model Tatar acoustic units (monophones and triphones). The Tatar National Corpus has been used to build three language models.

The best result obtained on the test part of speech corpus by DNN-based system is more than 87% word recognition accuracy on 200 k words vocabulary. The resulting recognition system also showed robustness to speaker's sex.

For the future work we plan to implement sub-word units' recognition, that can give an opportunity to reduce vocabulary size and OOV rate, making a step to unlimited vocabulary recognition system for the Tatar language.

References

1. Lewis, M.P., Simons, G.F., Fennig, C.D. (eds.). Ethnologue: Languages of the World, 9th (edn.). SIL International, Dallas (2016). http://www.ethnologue.com
2. Berment, V.: "Me'thodes pour informatiser des langues et des groups de langues peu dotées", Ph.D. thesis, J. Fourier University, Grenoble I (2004)
3. Krauwer, S.: The basic language resource kit (BLARK) as the first milestone for the language resources roadmap. In: Proceedings of International Workshop Speech and Computer SPEECOM, Moscow, Russia, pp. 8–15 (2003)

4. Khusainov, A.: Tekhnologiya avtomatizatsii sozdaniya I otsenki kachestva programmnikh sredstv analiza rechi c uchetom osobennostey maloresursnykh yazikov, Ph.D. thesis, Kazan, 162 p (2014)
5. Salimzyanov, I., Washington, J., Tyers, F.: A free/open-source Kazakh-Tatar machine translation system. In: Proceedings of the Machine Translation Summit XIV, Nice, France (2013)
6. Yandex Translate. https://translate.yandex.com/translator/Russian-Tatar
7. Suleymnov, D., Gatiatullin, A., Gilmullin, R.: Lexicograficheskaya baza dannykh dlya system mashinnogo perevoda blizkorodstvennykh yazykov. In: Proceedings of Third International Conference «Informatizatciya obschestva», Astana, Kazakhstan, pp. 585–587 (2012)
8. Khusainov, A., Khusainova, A.: Speech human-machine interface for the Tatar language. In: Artificial Intelligence and Natural Language Conference, FRUCT Oy, Helsinki, pp. 60–65 (2016)
9. Khusainov, A., Suleymanov, D.: Language identification system for the tatar language. In: Železný, M., Habernal, I., Ronzhin, A. (eds.) SPECOM 2013. LNCS (LNAI), vol. 8113, pp. 203–210. Springer, Cham (2013). https://doi.org/10.1007/978-3-319-01931-4_27
10. Suleymanov, Dz., Nevzorova, O.A., Khakimov, B.: National corpus of the tatar language "Tugan Tel": structure and features of grammatical annotation. In: Proceedings of International Conference Georgian Language and modern Technology, Tbilisi, pp. 107–108 (2013)
11. Povey, D., et al.: The kaldi speech recognition toolkit. In: Proceedings of ASRU, pp. 1–4 (2011)
12. Rath, S.P., Povey, D., Vesely, K., Cernocky, J.H.: Improved feature processing for deep neural networks. In: Proceedings of InterSpeech (2013)
13. Zhang, X., Trmal, J., Povey, D., Khudanpur, S.: Improving deep neural network acoustic models using generalized maxout networks. In: 2014 IEEE International Conference on Acoustics, Speech and Signal Processing (ICASSP), May 2014, pp. 215–219. IEEE (2014)
14. Stolcke, A.: SRILM – an extensible language modeling toolkit. In: Proceedings of International Conference on Spoken Language Processing, vol. 2, Denver, pp. 901–904 (2002)
15. Stolcke, A.: Entropy-based pruning of backoff language models. In: Proceedings of DARPA Broadcast News Transcription and Understanding Workshop, Lansdowne, pp. 270–274 (1998)
16. Kneser, R., Ney, H.: Improved backingoff for m-gram language modeling. In: Proceedings of the IEEE International Conference on Acoustics, Speech and Signal Processing, vol. 1 (1995)

Applications to Linguistics and the Humanities

Interactive Data Analytics
for the Humanities

Iryna Gurevych[(✉)], Christian M. Meyer[(✉)], Carsten Binnig,
Johannes Fürnkranz, Kristian Kersting, Stefan Roth, and Edwin Simpson

Department of Computer Science, Technische Universität Darmstadt,
Darmstadt, Germany
{gurevych,meyer}@ukp.informatik.tu-darmstadt.de
http://www.informatik.tu-darmstadt.de

Abstract. In this vision paper, we argue that current solutions to data analytics are not suitable for complex tasks from the humanities, as they are agnostic of the user and focused on static, predefined tasks with large-scale benchmarks. Instead, we believe that the human must be put into the loop to address small data scenarios that require expert domain knowledge and fluid, incrementally defined tasks, which are common for many humanities use cases. Besides the main challenges, we discuss existing and urgently required solutions to interactive data acquisition, model development, model interpretation, and system support for interactive data analytics. In the envisioned interactive systems, human users not only provide annotations to a machine learner, but train a model by using the system and demonstrating the task. The learning system will actively query the user for feedback, refine its model in real-time, and is able to explain its decisions. Our vision links natural language processing research with recent advances in machine learning, computer vision, and data management systems, as realizing this vision relies on combining expertise from all of these scientific fields.

1   Challenges in Analyzing Humanities Data

Automated data analytics, aka. data mining and machine learning, is a key technology for enriching and interpreting data, making informed decisions, and developing new data-driven scientific methods across many disciplines in industry and academia. Although the potential of interactive problem solving was recognized early on [16], this field has not progressed very far beyond the initial work. In particular, interactive machine learning and data analytics have only recently received increased attention [98]. Current data analytics solutions focus predominantly on well-defined tasks that can be solved by processing large, homogeneous datasets available in a structured form. Consider for example recommender systems, which suggest new products based on the product's properties, the products that the customer has previously bought, and the collective behavior of the customer database [43]. The state-of-the-art relies on huge amounts of data—over one billion pairs of users and news items passively gathered—to train a

© Springer Nature Switzerland AG 2018
A. Gelbukh (Ed.): CICLing 2017, LNCS 10761, pp. 527–549, 2018.
https://doi.org/10.1007/978-3-319-77113-7_41

deep neural network [28]. This may explain, why data analytics is conceived in a rather impersonal way, with algorithms working autonomously on passively collected data, although practice is quite the opposite. Most of the influence practitioners have, comes through interacting with data, including crafting the data and examining results.

In the late 1990s, digitized data became widely available in the humanities as well. Since then, there has been a clear demand for data analytics approaches to tap into these textual and visual data, including cultural heritage collections. The research questions and strategies in the humanities are, however, radically different from data analytics tasks in other disciplines.

First, despite the large amount of digitized data, there is typically only a tiny fraction that qualifies as training data for machine learning systems, because most of the data lacks cleaning, preprocessing, and gold standard labels. Data preparation tasks are often highly complex in the humanities. For text, they range from transcribing Gothic script or handwriting through the labeling of references to persons and their actions to a manual analysis of the text's argumentative structure. For images and video, e.g., we need to correct distortions, annotate gestures, or manually describe scenes. Rather than depending on big input data, future data analytics methods for the humanities must therefore be able to cope with *small data scenarios*, generalize from few input signals, and at the same time avoid overfitting to the idiosyncrasies of the dataset.

Second, the analysis of humanities data requires highly specific *expert knowledge*. This may include historical and legal facts, understanding ancient and special languages, or recognizing gestures or architectural properties in images and video. Relying on expert knowledge further limits our possibilities to manually label data, as common annotation procedures, such as crowdsourcing [54] or gamification [2], can only be used for certain subproblems or must be customized for laypeople. An even more severe problem is, however, interpreting the output of a data analytics system, which is only possible with expert domain knowledge. So far, training such a system requires vast machine learning expertise, preventing domain experts from directly participating in the development process. Inspecting and refining a model is particularly challenging in neural network architectures, as there is still little insight into the internal operation and behavior of complex models [109]. Future methods need to communicate directly with domain experts and allow them to steer the data analytics process.

Third, most research questions in the humanities are not clearly defined in advance, but developed over time as the research hypothesis evolves. We therefore need data analytics methods that allow for *fluid problem definitions*. This is particularly true for subjective tasks, for which multiple, partially contradicting theories co-exist. Examples are different schools and traditions in philosophy as well as disparate sources and opinions in history or law. Rather than aiming at a single, universal problem definition, we thus need methods that adapt to particular users or theories and recognize shifting goals.

Although some of these challenges are relevant for data science tasks in general (e.g., the small data scenario in the biomedical domain [93]), fluid problem definitions are prototypical for the humanities, as researchers have to pursue and develop competing theories and standpoints before judging them according to their merits. The humanities therefore need specific solutions for future data analytics. This requires a close cooperation of natural language processing and computer vision with machine learning and data management systems research.

2   Interactive Data Analytics

In this paper, we advocate research on *interactive machine learning* approaches for *data analytics tasks in the humanities*. Interactive machine learning is characterized by incremental model updates based on a user's actions and feedback, yielding a system that is simultaneously developed and used. Rather than teaching a machine learning system with a predefined set of training instances, as is the most common practice today, we envision an intelligent system that a user teaches by *using* it. This is triggered by the insight that a user will not necessarily start with a pre-defined concept that must be modeled as accurately as possible (as is often assumed in machine learning); the concept sought after is likely to evolve during the discovery process and, hence, during the process of selecting data and training a machine learning system.

Indeed, this is akin to *active learning*—the system may ask the user to label a certain instance while learning—but transgresses it by removing the strong focus on data labeling. Active learning removes the passivity of the learning system which, in the classical setting, only receives data, and allows it to actively pose questions on the data. However, the teacher (i.e., the human expert) is passive in the sense that she has no direct influence on the models that the learner induces from the data. Her only way of influencing the results is via the provided data or labels. For that reason Shivaswamy and Joachims [94] extended this towards *coactive learning*, where the teacher can also correct the learner during learning if necessary, providing a slightly improved but not necessarily optimal example as feedback.

In *interactive learning*, we envision a process where the teacher and the learner not only interact at the data and example level, but also at the model level itself. The user should be enabled to directly interact with the model, to provide feedback on the model that influences the learner, or to even directly modify parts of the learner. This way, learning becomes a fully co-adaptive process, in which a human is changing computer behavior, but the human also adapts to use machine learning more effectively and adjusts his or her data and goals in response to what is learned. This requires on the one hand ways for communicating information or feedback about the models to the learner, and, on the other hand, relies on innovative methods for communicating learned models to a domain expert who is typically inexperienced in machine learning. Thus, we envision future interactive data analytics to essentially consist of four components:

Interactive Data Acquisition: The domain expert and the learning system need to interact to acquire the appropriate data as well as for annotating and labeling the data.

Interactive Model Development: Besides influencing the learning process by providing suitable training data, the domain expert can interact with the learning algorithm during the model's construction and use this to continually alter and refine the model.

Interactive Model Interpretation: The learned model is not passive and intransparent, but can be actively understood and explored by the domain expert.

Interactive System Support: To support the iterative learning process and the effective human–computer interaction under real-time constraints, it is essential to link interactive machine learning with data management systems.

All four components have, to some extent, been explored in the literature before, but for interactive data analytics it is essential that all four are realized and tightly integrated so that their synthesis facilitates the interaction between the domain expert and the analytics tool at multiple levels. Figure 1 shows how the four components enrich the traditional data analytics process based on explicit feedback in the form of labeled data. By putting the human in the loop, we can approach data analytics methods also in small data scenarios requiring expert knowledge, as it is the case in the humanities. The interactive learning paradigm does not only allow a user to steer the learning process, but also to simultaneously develop the actual task and learning goal.

It should be noted that the general idea of interactive machine learning is anything but new. De Raedt and Bruynooghe [80] used the term already in 1992 for a logical rule learning system that interactively queries the user whether a newly learned rule is considered correct. Interactive machine learning gained also some attention during the *Intelligent User Interfaces* conference series, since Fails and Olsen [29] introduced an interactive approach for image segmentation in 2003. Unlike in previous works, the users of Fails and Olsen's system are more than oracles for assessing the learning process. Instead, the users roughly crayon the outlines of an object and iteratively refine their input as the system updates its prediction. Or consider the GrabCut system due to Rother et al. [86]. It takes this further and considers different interaction modes to make the communication between the user and the learner more efficient. This is what Amershi et al. [4] later entitled as "power to the people": the users teach a machine learning system by demonstrating how it should behave, rather than just providing a (large) number of hand-labeled training instances.

This learning paradigm is known as *imitation learning* or *learning from demonstration* [5,6,77,88]. Though it is an active research topic at the *Neural Information Processing Systems* (NIPS) conferences, research continues to focus mostly on teaching robots. To facilitate data analytics for the humanities, however, we need to leverage methods for text and visual data, which to date have only been cursorily researched. Despite image segmentation [29,37,86], there is recent work on natural language generation [55] and natural language

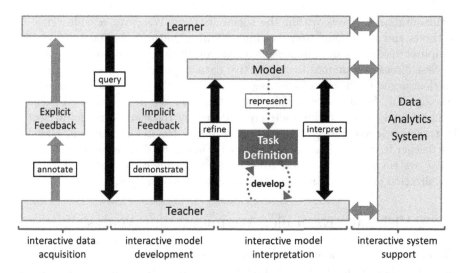

Fig. 1. Overview of our interactive data analytics vision: Besides the traditional approach (gray arrows) of training a learner by explicit feedback in the form of labeled data, the learner should be enabled to actively pose questions to the teacher, integrate implicit feedback and direct changes to the underlying model as well as foster interpretation of the learned model yielding a fluid task definition. The machine learning setup is backed by interactive system support to ensure the learner's response in real-time.

understanding [104]. The advantages of learning from demonstration are backed up by the user study by Cakmak et al. [15], who found that a standard approach to active learning is not perceived as a "real interaction" and some users complained about an imbalanced and badly structured stream of inquiries that hindered effective teaching to some extent. This was different in a setup that allowed users to ask the robot themselves, as the learning felt more natural.

Moving on with interactive approaches to data analytics for text and visual data is relevant for a large number of tasks in the humanities. A prototypical application is argumentation analysis. This is relevant for communication science (e.g., the analysis of political speeches), philosophy and ethics (e.g., controversial standpoints about cloning), history (e.g., changes in the public debate about corruption), and journalism (e.g., evidence retrieval and fact checking in news items). Typically, these use cases have a high impact on our society, but no clear, predefined task definition. There are, for instance, multiple competing theories of argumentative structures or how to define a fact. Instead, researchers approach a task from different perspectives with research questions that evolve while working with the data. Most tasks would benefit from a multimodal analysis, as text, images, and video (e.g., of political debates or to generate evidence for fact checking) contain complementary information.

In the remainder of the paper, we describe in detail our vision of interactive data analytics for the humanities and how we need to go beyond previous work. In Sect. 3, we introduce an integrated example for a fluid task in argumentation analysis within the social sciences. We cover a wide range of methods and techniques from machine learning, natural language processing, computer vision, and data management systems. Following our four components, we first discuss techniques for interactive data acquisition in Sect. 4. In Sect. 5, we then argue for means of allowing users to directly participate in the model development, for instance, by demonstrating how the learner should behave. This is closely linked to interactively interpreting the learned model in Sect. 6 by tracing a learner's decision and understanding the model internals, allowing the teacher to effectively guide the learning. Finally, we discuss methods of interactive system support in Sect. 7 to deal with real-time constraints and effective visualization of the results. In Sect. 8, we conclude the paper.

3   A Visionary Example: Argumentation Analysis

To demonstrate the potential of the interactive data analytics paradigm, we consider the following integrated example targeted towards argumentation analysis in the humanities. Imagine a social scientist (S) investigating a controversial research question such as "Should Europe accept more refugees?" by using the envisioned interactive data analytics system (D) with the goal of compiling a customized summary of the relevant standpoints and arguments present in arbitrary web sources.

In the first step, S acquires data by advising D to crawl relevant documents and video clips from a newspaper, YouTube, and a number of online forums dedicated to discussing the European Union. Instead of implementing this step as a preprocessing step tweaked by an information retrieval specialist, D would come up quickly with a few first results and asks S to select the ones that best fit her needs. D would pick up this feedback and iteratively refine both the crawling and the relevance ranking of the results. Already while the corpus is growing, S and her team would annotate some of the retrieved documents and individual scenes in the video clips, e.g., for claims and premises. From these annotations, D would develop a machine learning model that cannot only improve the information retrieval and exploration, but also pre-annotate the crawled documents to allow S to correct the system responses rather than having to create all annotations from scratch. Since S is not satisfied with the initial quality of the model, she first asks her colleague for a large annotated dataset of claims and premises and guides the system in transferring knowledge about how to effectively detect claims and premises although the newly crawled data spans totally different genres and domains. In the interplay of responding to system queries and annotating new data, S explores the dataset while it is developing. By observing which documents S prefers and that she tends to skip the first two minutes of a video, D gets a better notion of what is important for the task.

As the data grows and S feels that the claim and premise annotations work much better, she starts to label argumentative fallacies, such as the *shifting the burden of proof* fallacy as in: "There must be thousands of terrorists immigrating to our country. I challenge you to prove me wrong!" To this end, S starts with rather broad fallacy categories and iteratively refines them. D needs to adapt the learned model on-the-fly and assist S, for example, by suggesting a decision boundary between overly large, inhomogeneous groups of fallacious arguments. At several points, S is puzzled why D suggests a certain fallacy type, so she asks the system for explanation. D might respond that the decision is largely influenced by certain parts of a neural network, for which D shows a visualization. S soon finds that the voice recognition component keeps conflating *refugee* and *refuse*. Therefore, she draws an improved decision boundary into the visualization, which yields a strong constraint during retraining of the model.

Having analyzed much of the crawled data, S gets interested in ordering the arguments in a timeline. She asks D to do this by marking where the document or video creation time can be found. She also demonstrates that D should select the most important arguments and place the oldest argument at the top and the newest argument at the bottom. While doing so, she decides that it would make more sense to mark related arguments, which is why she introduces a separate column per argumentative strand.

This integrated example demonstrates that concepts sought after are likely to evolve during the discovery process and, hence, when selecting data and training a system. S starts with practically no data, but generates everything while she develops the research question and the result format. The envisioned system has to be highly flexible and interactive so that S can make all these inputs herself while D responds in real-time—even though the processing of the entire dataset might still be running in the background.

4   Interactive Data Acquisition

In a traditional data analytics setup, a machine learning algorithm is trained with massive amounts of data. This is particularly true in recent approaches making heavy use of deep neural networks [57]. Acquiring such large amounts of data is, however, a key problem in the humanities, where annotations typically depend on expert knowledge. We are thus facing small data scenarios, in which the learner often has no initial data for the current domain or task at all—which is generally known as the *cold-start problem*

To address this problem, new methods are needed to make better use of existing data and obtain new annotations for learning as efficiently as possible. Interactive annotation processes offer the opportunity for learners to request feedback when they are uncertain (active learning) and for human teachers to gradually refine the model while they are annotating (online learning) or intervene when they encounter mistakes. With interactive data acquisition, learning and annotation should become a single intertwined process that is guided by the teacher to rapidly learn a good model.

Research in *active learning* provides a first step into interactive data analytics by transgressing the conventional model of machine learning [20, 91]. Active learning has already found frequent use in natural language processing (e.g., [3, 33, 70]), in particular also for annotating texts [103]. For example, the open-source annotation tool WEBANNO has recently added active learning techniques for suggesting potential annotations to the annotator [108]. In computer vision, active learning has been investigated for object categorization [45], where a human teacher interactively labels images with the corresponding object categories, or for object attributes [74]. This paradigm has also been applied to domains where only experts are able to provide the appropriate fine-grained category information [12].

Typically, active learning techniques focus on identifying examples for which the currently learned hypothesis is most uncertain in its prediction. Using the most unreliable matches of the current hypothesis in the text to query the annotator for more information is the key idea of uncertainty sampling [58], a variant of which has, e.g., been applied to learning statistical grammars [8]. Bayesian active learning is a commonly used technique to globally optimize uncertainty [44]. Many classifiers not only yield a prediction but also a confidence score or probability value indicating the reliability of the prediction. Alternatively, uncertainty can be measured using the disagreement in a committee of diverse classifiers [92]. It may also be good to select batches of examples instead of single ones [13].

The active learning paradigm is particularly suited for dealing with the cold-start problem in the humanities, as it yields very steep learning curves [42]. To date, however, the available active learning methods are severely limited in the types of annotation and learning tasks they may be applied to. For instance, the active deep learning networks proposed by Zhou et al. [110], focus on atomic user annotations (e.g., sentiment labels). In contrast, our humanities setting requires a suitable representation of complex analysis units composed of multiple variables of different types, such as events, claims, or gestures. The representation also has to reduce the burden of the domain experts to express their expertise to the learning system. Experts may have many years of experience, and simply using only data and labels ignores all of the valuable insights that they could offer. As an example, the expert may say that if the author of a short story is "Edgar Allan Poe" then the preferred genre of the story is "mystery". Thus, users may provide programs that label some subset of the data as proposed by Ratner et al. [82]. This *data programming* is an instance of *statistical relational learning* [26], which learns models in domains such as the humanities with both complex relational structure—variable number of objects of different types with relations among them—and rich probabilistic structure. This generalizes weakly-supervised learning [107] and allows for a seamless integration of different learning systems and knowledge bases.

However, while users are domain experts they are not machine learning experts. Thus, deciding what knowledge to provide to the learner apriori is a difficult problem. Even if users were able to intuitively offer their knowledge, it is impractical for them to completely summarize years of experience before the learning starts. Hence, the learning algorithm should *actively seek advice* from the user as proposed by Odom and Natarajan [69]. For instance, the learner may ask the user, "What is your choice of label if a student and a professor are co-authors?" The expert replies saying, "I prefer the student to be advised by the professor". This preference is then explicitly weighed against the data while continuing to learn the model.

Making use of unlabeled data for training is generally an attractive way of addressing the cold-start problem and small data scenarios. Unsupervised (as well as semi- and weakly-supervised) learning methods incorporate general knowledge in the model design and use this to *extract latent structure* from unlabeled data. This latent structure simplifies the learning problem without relying on annotated training data. Early work aimed at giving recommendations based on sparse data [89]. Similar techniques have been transferred to a number of natural language processing (e.g., semantic analysis [10,102]) and computer vision [73,84] tasks. Unsupervised methods can further accelerate the learning process by identifying structure in the data before training data is available. Bayesian and approximate Bayesian methods, such as variational auto-encoders [52], provide novel techniques for handling this uncertainty within a deep model. The challenge is to identify such suitable models that are general enough to suit the variety of learning tasks that the system must adapt to.

We can also leverage data that has been labeled for different, but related annotation schemes or tasks using *transfer learning* [72]. Daumé III [25] introduced a simple method for domain adaptation using an augmented feature space, and Kim et al. [51] suggest using label embeddings for cases in which the annotation schemes vary substantially. Recent work has introduced new methods for transferring information from different hidden layers in deep neural networks for image representations [62,71]. The potential for transfer learning in expert-based data analytics tasks has not yet been fully explored and typically has not considered the transfer of personalized models between similar people. However, this will be necessary for subjective tasks in which humanities researchers follow alternative hypotheses or analysis strategies (e.g., different argumentation theories). Though these researchers aim at developing a personalized, user-centered model, they can benefit from integrating general latent properties of the task that hold across multiple strategies, or that are highly similar to other users.

Crowdsourcing has been successfully used to generate large amounts of labeled data in both natural language processing and computer vision. Major challenges to applying crowdsourcing for humanities tasks are fluid task definitions and the need for expert knowledge. There have been several attempts at modeling complex annotation tasks as games with a purpose [2] (e.g., for predicting protein structures [21]), but so far there is little work for humanities data. As crowdsourcing is limited to clearly defined tasks [14], it will be necessary to interactively translate a vaguely defined task to a clear-cut description that is comprehensible for lay workers – or identify subproblems for which this is possible. Recent work approached the task of intelligently selecting a worker's task to optimize annotation cost and quality [47,95,105]. While this improves learning rates, it is also a way to iteratively match a worker's skills with the (latent) demands of the task. Other works have focused on machine learning methods that are suitable for learning in the presence of the high numbers of comparably unreliable labels that often result from crowdsourcing annotations [41].

5 Interactive Model Development

Humanities experts each have their own personal working style and complex, changing goals, yet existing tools typically assume a static model that cannot adapt on-the-fly to the user's needs. These tools do not account for the different steps the user may take to complete a task, and do not learn how best to present information to assist the user at each stage. Furthermore, model development depends entirely on explicit annotations from the user. A new, dynamic approach for interactive model development is needed to adapt to such fluid problem definitions through both explicit and implicit user feedback.

Explicit annotation is time-consuming and typically constrains the user to a single, narrow way for passing information to the model—most often by labeling instances with one of multiple predefined classes. Annotation costs can often be considerably reduced by learning from multiple types of user feedback, including the implicit information in user navigation patterns recorded as mouse clicks. This feedback may not be in the form of class labels that can directly be used to train a model but may instead represent a choice for one action over another. For example, a user clicking on an item in a list may be interpreted as a preference for that item over the other items in the list [79]. Developing a ranking model from such pairwise comparisons is the goal of *preference learning* [31]. For example, Dzyuba et al. [27] infer a general ranking function for patterns from user-provided feedback over a small set of patterns. Training preferences for such models can be implicitly inferred from the user's behavior [79].

A further complication is the inconsistency of such implicit feedback signals, which are likely to have varying levels of noise or bias over time. Bayesian techniques have been successfully used to handle such unreliability when learning from pairwise preferences [18] or combining crowdsourced classifications with labelers whose behavior changes over time [96]. Such techniques could be used to train models for analyzing language or image data with multiple types of user feedback, and can be integrated using variational inference [7], which allows composition of models in a modular fashion. Recent works on deep exponential families [81] and variational auto encoders [52] show how this idea can be executed to create deep models using approximate Bayesian methods for complex modeling tasks.

As well as tailoring models of data to individual users, an interactive approach to humanities tasks could adapt the way that models are used to select and present information to assist users with different steps in a complex analysis task. Depending on the end user's perspective, the way that model outputs are presented may have very different costs or benefits. For example, omitting a crucial piece of information from a summary of an argument may have higher cost than including redundant text. The learner may also request feedback explicitly, but the future benefits of learning from this information must be traded off against the time the user takes to provide it and the need to provide her with immediate benefits. This balance is known as the *exploration vs. exploitation trade-off*, and can be optimized using reinforcement learning (RL) techniques [99]. To apply RL, we view our interactive scenario as a partially observable Markov decision process (POMDP), in which the agent aims to maximize a cumulative future reward by choosing an action given the current state of its environment. The state includes the available text and visual data, latent structures inferred from that data, such as arguments, as well as user behavior data (e.g., a record of clicks) and latent variables representing the user's preferences and task. The agent can perform different actions such as choosing information to present to the user, how this should be presented (e.g., the order of a list), and explicitly requesting feedback. The reward indicates the value of the new state to the user but may not be provided explicitly, so may need to be inferred from implicit feedback. The task of the learning system is to learn a so-called policy that lets it choose its actions in a way that maximizes the expected reward. A successful approach for complex tasks with large state spaces are relational RL [56] and deep RL [66], but this may require a large number of steps to train. In practice, the agent may encounter previously unseen states, which can be handled effectively using Bayesian RL to account for the uncertainty in the best course of action [34].

A crucial issue for interactive model development is that the learning system develops a model of the user and her task, so that techniques such as RL can effectively reduce the amount of interaction required. In machine learning, several techniques have recently been developed to facilitate what is also known as *apprenticeship learning*. Such techniques can be employed to learn human skills that the experts cannot directly communicate, or to personalize interaction processes.

Most notably, variants of RL have been developed that do not aim at optimizing a system's behavior by trial and error given a numeric feedback signal, but instead try to mimic observed behavior. The corresponding field—*learning from demonstration*—has become particularly successful in robotics [5,6,88] but has not yet been popularized in applications in the humanities. *Inverse reinforcement learning* [1,67,68,77] is such a technique, where the goal is to learn a hidden reward function, which may guide the teacher's observed behavior.

When working with a fluid problem definition, the relevance of previously accumulated data varies over time depending on the user's current task. Models must adapt to this *concept drift*, for example, by employing Bayesian techniques to handle the uncertainty caused by changing user behavior [96]. The long-term value of data must also be taken into account when acquiring explicit feedback from a user. Techniques for *lifelong machine learning* [19] could be brought to bear on this problem by providing mechanisms for balancing long-term value against the cost of interrupting a user [46].

A necessity for an interactive collaboration between a data analytics system and a human domain expert is that the machine learning algorithm does not need to be configured by a data science expert. This includes problems such as the automated selection of an appropriate algorithm [60] and tuning its hyper-parameters [40]. For example, AUTO-WEKA is an extension of the WEKA data mining and machine learning library, which can automatically find an appropriate configuration for a given learning task [101]. Humanities experts could also be empowered to directly modify the model, for example, by defining logical rules that alter neural networks models [39]. Another relatively unexplored approach would be to allow users to provide feedback to an *attention mechanism*, which directs a neural network to focus on the relevant parts of an image [24] or piece of text [63]. Given suitable user interfaces, humanities users could also create new features on the fly that provide useful abstractions from raw data. Intuitive latent features, such as topic clusters, could be modified directly by users, for example, by moving items between clusters. However, extensive modification of the internal components of a model depends on suitable techniques for interpreting models, which we discuss in the next section.

6   Interactive Model Interpretation

Most existing machine learning algorithms are integrated into applications as a black box that presents users with one type of output, such as classifications, without exposing them to the underlying workings of the method. However, sophisticated models for vision or language understanding often have multiple components, such as the different layers in a deep network, or multiple algorithms used together in a pipeline. While users should not be expected to understand the details of how a method works, complete opaqueness can undermine a user's confidence in an algorithm because its mistakes become unpredictable, meaning that the user may spend more time checking the automated method's work. An algorithm can indicate confidence in its decisions through probabilities, but these do not provide an explanation of the decision and therefore do not solve the problem. Understanding the model or important parts of it can be crucial when the algorithm encounters a new domain and must transition from a state of ignorance to earn the user's trust in carrying out its intended task. Moreover, in many applications the goal is not so much to maximize predictive performance, but to gain insight into the data. For this reason, one commonly distinguishes between *predictive* and *descriptive data mining*.

Algorithms for descriptive data mining typically rely on a rule-based representation of the results because rules offer the best trade-off between human and machine understandability [30]. Nevertheless, the aspect of interpretability still needs to be further explored. For example, it is conventional wisdom in machine learning that shorter explanations are better. Occam's Razor, *"Entia non sunt multiplicanda sine necessitate"*,[1] is often cited as support for this principle. Typically, it is understood as *"given two explanations of the data, all other things being equal, the simpler explanation is preferable"*. However, there are a few rule learning algorithms that explicitly aim for longer rules, and it is not clear that

---

[1] Entities should not be multiplied beyond necessity.

shorter rules are indeed more comprehensible for human experts [97]. Other criteria, such as semantic coherence of the conditions of a rule, should thus be considered in the learning process [32].

For other types of learning algorithms, it is harder but often nevertheless crucial to be able to explain and justify the outputs of the learned model to the user. For example, the strength of many recent learning algorithms, most notably deep learning [57,90], word embeddings [64] or topic modeling [11], is that latent variables are formed during the learning process. Understanding the meaning of these hidden variables is crucial for transparent and justifiable decisions. Consequently, visualization of such model components has recently received some attention [17,85,109]. Several works addressed the visualization of a network's long short-term memory (LSTM) and attention mechanisms, e.g., in machine translation [87]. Hendricks et al. [38] identify discriminating features in a deep visual classification task and learn to associate explanations in natural language with such features. Learning human-readable explanations at an appropriate level of abstraction is a core open research question. In any case, the need for learning interpretable models has been identified in several disciplines, and, not surprisingly, workshops at various conferences have been devoted to this topic [35,49,106].

A common restriction for most of the above methods is that even though the explanation quality is lifted from a technical to a semantic level, the user is still a comparably passive consumer of the presented models. A key step forward would be if users could directly interact with the provided models, visualize them from different (semantic) angles, pose multiple question types to the data, and, eventually, even correct the models. Systems like MININGZINC [36] and relational mathematical programming frameworks [48,65], which allow users to declaratively define a data analytics problem with a high-level constraint-based language, are currently being developed.

A further step ahead would be to allow the user to directly modify parts of the model in interaction with the learner. For example, Beckerle [9] explored an interactive rule learning process where learned rules could be directly modified by the user, thereby causing the learner to re-learn subsequently learned rules. Alternatively, one can imagine a user who is able to directly interact with other types of model components, such as hidden layers in a neural network. For example, Hu et al. [39] proposed a combination of deep neural models with structured logic rules to foster the interpretation of the model and allow users to (indirectly) steer the learning process. Such an interactive approach is perhaps closer to the way that people train each other – by explaining how they make decisions as well as providing examples. This approach could therefore improve both the user's trust in the model and increase training speed for new tasks and domains by reducing the need to provide numerous examples before the important features are identified. Research towards such truly interactive machine learning systems has just started.

7   Interactive System Support

Many systems and tools exist already to support developers when curating complex machine learning models for text and image data (e.g., R and Spark MLLib or TensorFlow). However, these tools are limited in a number of fundamental ways in their support for a human-in-the-loop when curating or developing models. First, existing tools require well-trained data scientists to select the appropriate techniques and adjust the hyperparameters to build models and to evaluate their outcomes. Second, even when working on small labeled datasets, many of these tools still require large amounts of data as background knowledge (e.g., large knowledge bases, corpora, pre-trained embeddings) and thus are often too slow to provide interactive feedback to domain experts in the model development process. Third, many of the machine learning techniques require heavy data pre-processing steps before text and visual data can be used for the actual analytics task. This can further limit the overall interactivity of the system.

In this section, we discuss how existing machine learning tools have to change to better support interactive data analytics from a systems perspective. This systems perspective complements the interactive data acquisition, model development, and model interpretation components described in the previous sections.

Interactive Data Acquisition: The first step in data acquisition is typically the pre-processing of the text and visual data used as input or background knowledge. One important step when pre-processing text documents is, for example, to retrieve the embeddings for each word from an existing corpus of pre-trained word vectors and then apply the pre-trained model (e.g., for classifying arguments as supportive or not). However, corpora of word embeddings can be huge (e.g., billions of vectors in the case of [76]). Therefore, we require efficient techniques for storing and retrieving embeddings or similar input data. Furthermore, pre-processing steps must be able to be applied incrementally to new data sources to support a progressive execution of the upstream machine learning pipeline. In the computational argumentation example, the user does not first want to pre-process the complete set of documents before applying it to the classification model to find out which arguments support her hypothesis. Instead, pre-processing and classification should be intertwined to provide progressive answers to the user while streaming over the text of the document.

The already mentioned cold-start problem is another challenge for system support to data acquisition. While there exist many techniques in active learning, none of these techniques focuses on how to suggest new examples at interactive speeds. Instead, their main objective is to find examples where the current model is most uncertain in its prediction, since labeling these items promises the biggest benefit. However, finding the most uncertain examples can be extremely expensive if large amounts of data are involved. One idea to achieve interactivity in this learning process is to use ideas from neighbor-sensitive hashing [75] to quickly find the k-nearest unlabeled neighbors of already labeled examples that are close to the decision boundary (i.e., where the model is the most uncertain).

For solutions based on transfer learning, we require systems that are able to store a large number of datasets and models and provide efficient search capabilities that allow users to interactively retrieve related data or pre-trained models that are best suited for their task and data at hand.

Interactive Model Development: Throughout the overall model training and development process, data analytics tools must consistently provide response times low enough to guarantee fluid user interactions and integrate user feedback. In fact, a recent study [61] has shown that even small delays of more than 500 ms significantly decrease a user's activity level, dataset coverage, and insight discovery rate. However, none of the existing tools can guarantee interactive latencies [53]. Previous work by Crotty et al. [23] approached this problem for structured data, but we are not aware of any work focusing on unstructured data or use cases in the humanities.

One important challenge for incorporating user feedback is to enable the retraining of models in real-time when new user input is available either through explicit labeling or through implicit feedback. While there has been significant work in online machine learning that allows models to be progressively updated, existing techniques often cannot be applied directly, as the incremental retraining might not be able to keep up with the high update rate resulting from implicit feedback (e.g., clickstreams). Another issue is that models may become too complex and thus updating the model incrementally might exceed the interactive threshold. There are multiple ways to tackle this issue: adaptively batching updates based on the incoming update rate, employing parallelization [83] or online learning algorithms such as MIRA [22], and avoiding that the model forgets already learned concepts, e.g., by applying experience replay [59] to speed up an RL-based learner.

Since implicit feedback with its heavily varying quality is of a different nature to explicit feedback, it might turn out useful to adaptively drop low quality updates by applying techniques known in the systems community for load shedding [100]. However, these techniques must be adapted to perform load shedding based on the quality of the user input. Another interesting direction to address this problem is to approximate the model (e.g., by representing weights in a neural network in an approximate manner). Finding the right approximation to achieve the best model quality under a fixed time budget to apply the update appears a promising avenue of research.

To make interactive learning tools accessible to humanities experts, we discussed automatically selecting machine learning algorithms and their hyperparameters in Sect. 5. However, this is a challenging task due to the large search space, which means that naïve grid search approaches, for instance, do not allow us to compute a good set of hyperparameters in real time. Recent techniques such as scalable kernel composition [50] provide viable solutions for certain types of machine learning models, but further research is required to enable automated model selection in real time.

Interactive Model Interpretation: Interactive systems support is also required for many of the model interpretation tasks outlined in Sect. 6. The reason is that models, such as neural networks, can get large and thus exploring the model and summarizing important aspects at interactive speeds is a challenge on its own. Furthermore, as a result of model interpretation, users might want to manually adjust the model internals. These techniques are sometimes referred to as Specialized Programming [78]. That is, the user "programs" the machine learning model, for instance, by directly modifying internal layers of a neural network. The corresponding challenge from the systems perspective is again to retrain the model based on the user modifications and interactively allow the user to inspect the model quality after updating the model.

Generally, if interactive systems are to have truly broad impact, building and maintaining them needs to become substantially easier. They should support the rapid combination, deployment, and maintenance of existing data analytics algorithms and domain knowledge. For that, one should identify and validate programming and data abstractions as building blocks. Identifying, optimizing, and supporting abstractions as primitives could make systems for interactive data analytics substantially easier to setup, to understand the user, and to scale. This can bring us a step closer to unleashing the full potential of data analytics in various domains, even beyond humanities. To ensure that such a platform is accessible to many users, the programming interface must be small, clean, and composable to enhance productivity and enable users to try and accommodate many data analytics algorithms; the ability to integrate diverse data sources and types requires the data model of the programming interface to be versatile. A combination of the relational data model and a statistical relational language such as Markov logic and relational mathematical programs [26] satisfies these criteria. In combination with imperative languages this seems to be a promising direction but further research is required to realize interactiveness.

8   Conclusions

Current data analytics is mostly limited to tasks for which large-scale homogeneous benchmarks exist. However, real-world use cases typically have different properties: they are highly heterogeneous and involve infrequent and complex phenomena, for which only small-scale datasets are available. The need for new data analytics approaches is particularly pressing for the humanities, as the use cases in these disciplines do not only require background expert knowledge for developing a system, labeling data, and interpreting the results, but are also only vaguely defined in advance. Such a fluid problem definition – which a human expert develops while doing the actual task – calls for a totally different approach to data analytics.

In this paper, we have laid out our vision towards future data analytics in the humanities based on interactive machine learning. The main idea is to put the human in the loop and iteratively refine the model based on the user's feedback. By focusing on the (expert) user and her task, we need to think beyond natural language processing and closely cooperate with computer vision to enable multimodal systems to learn jointly from text and visual data and mutually benefit from recent advances in the research of suitable (deep) machine learning architectures. Interactive data analytics also requires core research in machine learning, since existing techniques almost exclusively learn from indirect input in the form of labeled examples or an algorithm's parameter settings. Instead, our vision is that a human expert can steer the learning process by using the system for her task, demonstrating how the system should behave and interpret the learned model in order to identify specific patterns or errors. Even though there is little labeled data, we will have to include large amounts of background knowledge and computationally heavy learning algorithms, which must be able to return their estimations in real-time. Research into efficient data management and systems engineering is therefore the fourth major pillar of our vision.

In all four fields of study, there is already a vast body of existing work with which we can fulfill parts of our vision. However, there is a clear demand for future efforts to close the gaps in interactive model development and interpretation as well as systems supporting this approach. If natural language processing joins forces with computer vision, machine learning, and data management systems, we can make a great leap forward.

References

1. Abbeel, P., Ng, A.Y.: Apprenticeship learning via inverse reinforcement learning. In: 21st International Conference on Machine learning (ICML), ACM, New York (2004)
2. von Ahn, L.: Games with a purpose. Computer **39**(6), 96–98 (2006)
3. Ambati, V., Vogel, S., Carbonell, J.G.: Active learning-based elicitation for semi-supervised word alignment. In: 48th Annual Meeting of the Association for Computational Linguistics (ACL), pp. 365–370. ACL, Stroudsburg (2010)
4. Amershi, S., Cakmak, M., Knox, W.B., Kulesza, T.: Power to the people: the role of humans in interactive machine learning. AI Mag. **35**(4), 105–120 (2014)
5. Argall, B.D., Chernova, S., Veloso, M., Browning, B.: A survey of robot learning from demonstration. Robot. Auton. Syst. **57**(5), 469–483 (2009)
6. Atkeson, C.G., Schaal, S.: Robot learning from demonstration. In: 14th International Conference on Machine Learning (ICML), pp. 12–20. Morgan Kaufmann, San Francisco (1997)
7. Attias, H.: A variational Bayesian framework for graphical models. In: Advances in Neural Information Processing Systems 12 (NIPS), pp. 209–215. MIT Press, Cambridge (2000)
8. Becker, M., Osborne, M.: A two-stage method for active learning of statistical grammars. In: 19th International Joint Conference on Artificial Intelligence (IJCAI), pp. 991–996. Morgan Kaufmann, San Francisco (2005)
9. Beckerle, M.: Interaktives Regellernen. Diploma thesis, Technische Universität Darmstadt (2009). [in German]
10. Bejan, C.A., Harabagiu, S.: Unsupervised event coreference resolution. Comput. Linguist. **40**(2), 311–347 (2014)
11. Blei, D.M.: Probabilistic topic models. Commun. ACM **55**(4), 77–84 (2012)

12. Branson, S., et al.: Visual recognition with humans in the loop. In: Daniilidis, K., Maragos, P., Paragios, N. (eds.) ECCV 2010. LNCS, vol. 6314, pp. 438–451. Springer, Heidelberg (2010). https://doi.org/10.1007/978-3-642-15561-1_32
13. Brinker, K.: Active learning of label ranking functions. In: 21st International Conference on Machine Learning (ICML), pp. 129–136. ACM, New York (2004)
14. Burger-Helmchen, T., Pénin, J.: The limits of crowdsourcing inventive activities: what do transaction cost theory and the evolutionary theories of the firm teach us? In: Proceedings of the Workshop on Open Source Innovation, Strasbourg, France, pp. 1–26 (2010)
15. Cakmak, M., Chao, C., Thomaz, A.L.: Designing interactions for robot active learners. IEEE Trans. Auton. Ment. Dev. **2**(2), 108–118 (2010)
16. Chambers, R.A., Michie, D.: Man-machine co-operation on a learning task. In: Parslow, R.D., Prowse, R., Elliott-Green, R. (eds.) Computer Graphics: Techniques and Applications, pp. 179–185. Plenum, London (1969)
17. Chaney, A.J., Blei, D.M.: Visualizing topic models. In: 6th International Conference on Weblogs and Social Media (ICWSM). AAAI Press, Palo Alto (2012)
18. Chen, X., Bennett, P.N., Collins-Thompson, K., Horvitz, E.: Pairwise ranking aggregation in a crowdsourced setting. In: 6th ACM International Conference on Web Search and Data Mining (WSDM), pp. 193–202. ACM, New York (2013)
19. Chen, Z., Liu, B.: Lifelong Machine Learning. Synthesis Lectures on Artificial Intelligence and Machine Learning. Morgan & Claypool, San Rafael (2016)
20. Cohn, D.A., Atlas, L., Ladner, R.: Improving generalization with active learning. Mach. Learn. **15**(2), 201–221 (1994)
21. Cooper, S., Foldit Players, et al.: predicting protein structures with a multiplayer online game. Nature **466**, 756–760 (2010)
22. Crammer, K., Singer, Y.: Ultraconservative online algorithms for multiclass problems. J. Mach. Learn. Res. **3**, 951–991 (2003)
23. Crotty, A., Galakatos, A., Zgraggen, E., Binnig, C., Kraska, T.: The case for interactive data exploration accelerators (IDEAs). In: Workshop on Human-In-the-Loop Data Analytics (HILDA@SIGMOD), p. 11. ACM, New York (2016)
24. Das, A., Agrawal, H., Zitnick, L., Parikh, D., Batra, D.: Human attention in visual question answering: do humans and deep networks look at the same regions? In: 2016 Conference on Empirical Methods in Natural Language Processing (EMNLP), pp. 932–937. ACL, Stroudsburg (2016)
25. Daumé III, H.: Frustratingly easy domain adaptation. In: 45th Annual Meeting of the Association of Computational Linguistics (ACL), pp. 256–263. ACL, Stroudsburg (2007)
26. De Raedt, L., Kersting, K., Natarajan, S., Poole, D.: Statistical Relational Artificial Intelligence: Logic, Probability, and Computation. Synthesis Lectures on Artificial Intelligence and Machine Learning. Morgan & Claypool, San Rafael (2016)
27. Dzyuba, V., van Leeuwen, M., Nijssen, S., De Raedt, L.: Interactive learning of pattern rankings. Int. J. Artif. Intell. Tools **23**(6), 1460026 (2014). https://doi.org/10.1142/S0218213014600264
28. Elkahky, A.M., Song, Y., He, X.: A multi-view deep learning approach for cross domain user modeling in recommendation systems. In: 24th International Conference on World Wide Web (WWW), pp. 278–288. International World Wide Web Conferences Steering Committee, Geneva (2015)
29. Fails, J.A., Olsen, Jr., D.R.: Interactive machine learning. In: 8th International Conference on Intelligent User Interfaces (IUI), pp. 39–45. ACM, New York (2003)

30. Fürnkranz, J., Gamberger, D., Lavrač, N.: Foundations of Rule Learning. Springer, Heidelberg (2012). https://doi.org/10.1007/978-3-540-75197-7
31. Fürnkranz, J., Hüllermeier, E. (eds.): Preference Learning. Springer, Heidelberg (2010). https://doi.org/10.1007/978-3-642-14125-6
32. Gabriel, A., Paulheim, H., Janssen, F.: Learning semantically coherent rules. In: Cellier, P., Charnois, T., Hotho, A., Matwin, S., Moens, M.F., Toussaint, Y. (eds.) 1st International Workshop on Interactions between Data Mining and Natural Language Processing. CEUR Workshop Proceedings, vol. 1202, pp. 49–63 (2014)
33. Gambäck, B., Olsson, F., Täckström, O.: Active learning for dialogue act classification. In: 12th Annual Conference of the International Speech Communication Association (INTERSPEECH), pp. 1329–1332. International Speech Communication Association, Baixas (2011)
34. Ghavamzadeh, M., Engel, Y., Valko, M.: Bayesian policy gradient and actor-critic algorithms. J. Mach. Learn. Res. **17**, 1–53 (2016)
35. Gillies, M., et al.: Human-centered machine learning. In: CHI Conference on Human Factors in Computing Systems, pp. 3558–3565. ACM, New York (2016)
36. Guns, T., Dries, A., Nijssen, S., Tack, G., De Raedt, L.: MiningZinc: a declarative framework for constraint-based mining. Artif. Intell. **244**, 6–29 (2017)
37. He, H., Daumé III, H., Eisner, J.: Imitation learning by coaching. In: Pereira, F., Burges, C.J.C., Bottou, L., Weinberger, K.Q. (eds.) Advances in Neural Information Processing Systems 25 (NIPS), pp. 3149–3157. Curran Associates, Red Hook (2012)
38. Hendricks, L.A., et al.: Generating visual explanations. In: Leibe, B., Matas, J., Sebe, N., Welling, M. (eds.) ECCV 2016. LNCS, vol. 9908, pp. 3–19. Springer, Cham (2016). https://doi.org/10.1007/978-3-319-46493-0_1
39. Hu, Z., Ma, X., Liu, Z., Hovy, E., Xing, E.: Harnessing deep neural networks with logic rules. In: 54th Annual Meeting of the Association for Computational Linguistics (ACL), pp. 2410–2420. ACL, Stroudsburg (2016)
40. Hutter, F., Lücke, J., Schmidt-Thieme, L.: Beyond manual tuning of hyperparameters. Künstl Intell. **29**(4), 329–337 (2015)
41. Ipeirotis, P.G., Provost, F.J., Sheng, V.S., Wang, J.: Repeated labeling using multiple noisy labelers. Data Min. Knowl. Disc. **28**(2), 402–441 (2014)
42. Jamieson, K.G., Jain, L., Fernandez, C., Glattard, N.J., Nowak, R.: NEXT: a system for real-world development, evaluation, and application of active learning. In: Cortes, C., Lawrence, N., Lee, D., Sugiyama, M., Garnett, R. (eds.) Advances in Neural Information Processing Systems 28 (NIPS), pp. 2638–2646 (2015)
43. Jannach, D., Zanker, M., Felfernig, A., Friedrich, G.: Recommender Systems: An Introduction. Cambridge University Press, Cambridge (2010)
44. Kandasamy, K., Schneider, J., Poczos, B.: Bayesian active learning for posterior estimation. In: 24th International Joint Conference on Artificial Intelligence (IJCAI), pp. 3605–3611. AAAI Press, Menlo Park (2015)
45. Kapoor, A., Grauman, K., Urtasun, R., Darrell, T.: Active learning with Gaussian processes for object categorization. In: 11th International Conference on Computer Vision (ICCV), pp. 1–8. IEEE, New York (2007)
46. Kapoor, A., Horvitz, E.: Principles of lifelong learning for predictive user modeling. In: Conati, C., McCoy, K., Paliouras, G. (eds.) UM 2007. LNCS (LNAI), vol. 4511, pp. 37–46. Springer, Heidelberg (2007). https://doi.org/10.1007/978-3-540-73078-1_7
47. Karger, D.R., Oh, S., Shah, D.: Iterative learning for reliable crowdsourcing systems. In: Shawe-Taylor, J., Zemel, R.S., Bartlett, P.L., Pereira, F., Weinberger,

K.Q. (eds.) Advances in Neural Information Processing Systems 24 (NIPS), pp. 1953–1961. Curran Associates, Red Hook (2011)

48. Kersting, K., Mladenov, M., Tokmakov, P.: Relational linear programming. Artif. Intell. **244**, 188–216 (2017)

49. Kim, B., Malioutov, D., Varshney, K. (eds.): Proceedings of the ICML 2016 Workshop on Human Interpretability in Machine Learning, New York (2016) https://sites.google.com/site/2016whi/

50. Kim, H., Teh, Y.W.: Scalable structure discovery in regression using Gaussian processes. In: Hutter, F., Kotthoff, L., Vanschoren, J. (eds.) 2016 Workshop on Automatic Machine Learning. JMLR Workshop and Conference Proceedings, vol. 64, pp. 31–40 (2016)

51. Kim, Y.B., Stratos, K., Sarikaya, R., Jeong, M.: New transfer learning techniques for disparate label sets. In: 53rd Annual Meeting of the Association for Computational Linguistics and 7th International Joint Conference on Natural Language Processing (ACL/IJCNLP), pp. 473–482. ACL, Stroudsburg (2015)

52. Kingma, D.P., Welling, M.: Auto-encoding variational Bayes. In: Proceedings of the International Conference on Learning Representations (ICLR). arXiv:1312.6114, Banff, AB, Canada (2014)

53. Kraska, T., Talwalkar, A., Duchi, J.C., Griffith, R., Franklin, M.J., Jordan, M.I.: MLbase: a distributed machine-learning system. In: 6th Biennial Conference on Innovative Data Systems Research (CIDR) (2013)

54. Kucherbaev, P., Daniel, F., Tranquillini, S., Marchese, M.: Crowdsourcing processes: a survey of approaches and opportunities. IEEE Internet Comput. **20**(2), 50–56 (2016)

55. Lampouras, G., Vlachos, A.: Imitation learning for language generation from unaligned data. In: 26th International Conference on Computational Linguistics (COLING), pp. 1101–1112. The COLING 2016 Organizing Committee, Osaka (2016)

56. Lang, T., Toussaint, M., Kersting, K.: Exploration in relational domains for model-based reinforcement learning. J. Mach. Learn. Res. **13**, 3725–3768 (2012)

57. Lecun, Y., Bengio, Y., Hinton, G.: Deep learning. Nature **521**(7553), 436–444 (2015)

58. Lewis, D.D., Gale, W.: A sequential algorithm for training text classifiers. In: Croft, B.W., van Rijsbergen, C.J. (eds.) SIGIR '94, pp. 3–12. Springer, London (1994). https://doi.org/10.1007/978-1-4471-2099-5_1

59. Lin, L.J.: Self-improving reactive agents based on reinforcement learning, planning and teaching. Mach. Learn. **8**(3), 293–321 (1992)

60. Lindauer, M.T., Hoos, H.H., Hutter, F., Schaub, T.: AutoFolio: an automatically configured algorithm selector. J. Artif. Intell. Res. **53**, 745–778 (2015)

61. Liu, Z., Heer, J.: The effects of interactive latency on exploratory visual analysis. IEEE Trans. Vis. Comput. Graph. **20**(12), 2122–2131 (2014)

62. Long, M., Cao, Y., Wang, J., Jordan, M.I.: Learning transferable features with deep adaptation networks. In: Bach, F., Blei, D. (eds.) 32nd International Conference on Machine Learning (ICML). JMLR: Workshop and Conference Proceedings, vol. 37, pp. 97–105 (2015)

63. Lu, J., Yang, J., Batra, D., Parikh, D.: Hierarchical question-image co-attention for visual question answering. In: Lee, D.D., Sugiyama, M., Luxburg, U.V., Guyon, I., Garnett, R. (eds.) Advances In Neural Information Processing Systems 29 (NIPS), pp. 289–297 (2016)

64. Mikolov, T., Sutskever, I., Chen, K., Corrado, G.S., Dean, J.: Distributed representations of words and phrases and their compositionality. In: Burges, C., Bottou, L., Welling, M., Ghahramani, Z., Weinberger, K. (eds.) Advances in Neural Information Processing Systems 26 (NIPS), pp. 3111–3119 (2013)
65. Mladenov, M., Kleinhans, L., Kersting, K.: Lifted inference for convex quadratic programs. In: 31st AAAI Conference on Artificial Intelligence (AAAI), pp. 2350–2356. AAAI Press, Palo Alto (2017)
66. Mnih, V., et al.: Human-level control through deep reinforcement learning. Nature 518(7540), 529–533 (2015)
67. Natarajan, S., Joshi, S., Tadepalli, P., Kersting, K., Shavlik, J.: Imitation learning in relational domains: a functional-gradient boosting approach. In: 22nd International Joint Conference on Artificial Intelligence (IJCAI), pp. 1414–1420. AAAI Press, Menlo Park (2011)
68. Ng, A.Y., Russell, S.J.: Algorithms for inverse reinforcement learning. In: Langley, P. (ed.) 17th International Conference on Machine Learning (ICML), pp. 663–670. Morgan Kaufmann, San Francisco (2000)
69. Odom, P., Natarajan, S.: Actively interacting with experts: a probabilistic logic approach. In: Frasconi, P., Landwehr, N., Manco, G., Vreeken, J. (eds.) ECML PKDD 2016. LNCS (LNAI), vol. 9852, pp. 527–542. Springer, Cham (2016). https://doi.org/10.1007/978-3-319-46227-1_33
70. Olsson, F.: A literature survey of active machine learning in the context of natural language processing. SICS Technical report T2009:06, Swedish Institute of Computer Science (2009)
71. Oquab, M., Bottou, L., Laptev, I., Sivic, J.: Learning and transferring mid-level image representations using convolutional neural networks. In: IEEE Conference on Computer Vision and Pattern Recognition (CVPR), pp. 1717–1724. IEEE, New York (2014)
72. Pan, S.J., Yang, Q.: A survey on transfer learning. IEEE Trans. Knowl. Data Eng. 22(10), 1345–1359 (2010)
73. Papandreou, G., Chen, L.C., Murphy, K., Yuille, A.L.: Weakly-and semi-supervised learning of a deep convolutional network for semantic image segmentation. In: International Conference on Computer Vision (ICCV), pp. 1742–1750. IEEE, New York (2015)
74. Parikh, D., Grauman, K.: Interactively building a discriminative vocabulary of nameable attributes. In: IEEE Conference on Computer Vision and Pattern Recognition (CVPR), pp. 1681–1688. IEEE, New York (2011)
75. Park, Y., Cafarella, M.J., Mozafari, B.: Neighbor-sensitive hashing. Proc. VLDB Endow. 9(3), 144–155 (2015)
76. Pennington, J., Socher, R., Manning, C.D.: GloVe: global vectors for word representation. In: Empirical Methods in Natural Language Processing (EMNLP), pp. 1532–1543. ACL, Stroudsburg (2014)
77. Piot, B., Geist, M., Pietquin, O.: Bridging the gap between imitation learning and inverse reinforcement learning. IEEE Trans. Neural Netw. 28(8), 1814–1826 (2016)
78. Porter, R., Theiler, J., Hush, D.: Interactive machine learning in data exploitation. Comput. Sci. Eng. 15(5), 12–20 (2013)
79. Radlinski, F., Joachims, T.: Query chains: learning to rank from implicit feedback. In: 11th ACM SIGKDD International Conference on Knowledge Discovery and Data Mining, pp. 239–248. ACM, New York (2005)
80. de Raedt, L., Bruynooghe, M.: Interactive concept-learning and constructive induction by analogy. Mach. Learn. 8(2), 107–150 (1992)

81. Ranganath, R., Tang, L., Charlin, L., Blei, D.M.: Deep exponential families. In: Lebanon, G., Vishwanathan, S. (eds.) 18th International Conference on Artificial Intelligence and Statistics (AISTATS). JMLR Workshop and Conference Proceedings, vol. 38, pp. 762–771 (2015)

82. Ratner, A., De Sa, C., Wu, S., Selsam, D., Re, C.: Data programming: creating large training sets, quickly. In: Lee, D.D., Sugiyama, M., Luxburg, U.V., Guyon, I., Garnett, R. (eds.) Advances in Neural Information Processing Systems 29 (NIPS), pp. 3567–3575 (2016)

83. Recht, B., Ré, C., Wright, S.J., Niu, F.: Hogwild: A lock-free approach to parallelizing stochastic gradient descent. In: Shawe-Taylor, J., Zemel, R.S., Bartlett, P.L., Pereira, F., Weinberger, K.Q. (eds.) Advances in Neural Information Processing Systems 24 (NIPS), pp. 693–701. Curran Associates, Red Hook (2011)

84. Rosenberg, C., Hebert, M., Schneiderman, H.: Semi-supervised self-training of object detection models. In: 7th IEEE Workshops on Application of Computer Vision (WACV), pp. 29–36. IEEE, New York (2005)

85. Rothe, S., Schütze, H.: Word embedding calculus in meaningful ultradense subspaces. In: 54th Annual Meeting of the Association for Computational Linguistics (ACL), pp. 512–517. ACL, Stroudsburg (2016)

86. Rother, C., Kolmogorov, V., Blake, A.: "GrabCut": interactive foreground extraction using iterated graph cuts. ACM Trans. Graph. 23(3), 309–314 (2004)

87. Rush, A.M., Chopra, S., Weston, J.: A neural attention model for abstractive sentence summarization. In: Conference on Empirical Methods in Natural Language Processing (EMNLP), pp. 379–389. ACL, Stroudsburg (2015)

88. Schaal, S.: Learning from demonstration. In: Jordan, M.I., Petsche, T. (eds.) Advances in Neural Information Processing Systems 9 (NIPS), pp. 1040–1046. MIT Press, Cambridge (1997)

89. Schein, A.I., Popescul, A., Ungar, L.H., Pennock, D.M.: Methods and metrics for cold-start recommendations. In: 25th Annual International ACM SIGIR Conference on Research and Development in Information Retrieval, pp. 253–260. ACM, New York (2002)

90. Schmidhuber, J.: Deep learning in neural networks: an overview. Neural Netw. 61, 85–117 (2015)

91. Settles, B.: Active Learning. Synthesis Lectures on Artificial Intelligence and Machine Learning. Morgan & Claypool, San Rafael (2012)

92. Seung, H.S., Opper, M., Sompolinsky, H.: Query by committee. In: 5th Annual ACM Workshop on Computational Learning Theory (COLT), pp. 287–294. ACM, New York (1992)

93. Shaikhina, T., Lowe, D., Daga, S., Briggs, D., Higgins, R., Khovanova, N.: Machine learning for predictive modelling based on small data in biomedical engineering. IFAC-PapersOnLine 48(20), 469–474 (2015)

94. Shivaswamy, P., Joachims, T.: Coactive learning. J. Artif. Intell. Res. 53, 1–40 (2015)

95. Simpson, E., Roberts, S.: Bayesian methods for intelligent task assignment in crowdsourcing systems. In: Guy, T., Kárný, M., Wolpert, D. (eds.) Decision Making: Uncertainty, Imperfection, Deliberation and Scalability. SCI, vol. 538, pp. 1–32. Springer, Cham (2015). https://doi.org/10.1007/978-3-319-15144-1_1

96. Simpson, E., Roberts, S., Psorakis, I., Smith, A.: Dynamic Bayesian combination of multiple imperfect classifiers. In: Guy, T., Karny, M., Wolpert, D. (eds.) Decision Making and Imperfection. SCI, vol. 474, pp. 1–35. Springer, Berlin, Heidelberg (2013). https://doi.org/10.1007/978-3-642-36406-8_1

97. Stecher, J., Janssen, F., Frederik, F.: Shorter rules are better, aren't they? In: Calders, T., Ceci, M., Malerba, D. (eds.) DS 2016. LNCS (LNAI), vol. 9956, pp. 279–294. Springer, Cham (2016). https://doi.org/10.1007/978-3-319-46307-0_18
98. Subramanian, K., Amor, H.B., Isbell, C.L., Thomaz, A.L. (eds.): Proceedings of the IJCAI 2016 Workshop on Interactive Machine Learning: Connecting Humans and Machines, New York (2016). https://sites.google.com/site/ijcai2016iml/
99. Sutton, R.S., Barto, A.G.: Reinforcement Learning: An Introduction. MIT Press, Cambridge (1998)
100. Tatbul, N.: Load shedding. In: Liu, L., Özsu, M.T. (eds.) Encyclopedia of Database Systems, pp. 1632–1636. Springer, New York (2009). https://doi.org/10.1007/978-1-4899-7993-3_211-2
101. Thornton, C., Hutter, F., Hoos, H.H., Leyton-Brown, K.: Auto-WEKA: combined selection and hyperparameter optimization of classification algorithms. In: 19th ACM SIGKDD International Conference on Knowledge Discovery and Data Mining, pp. 847–855. ACM, New York (2013)
102. Titov, I., Khoddam, E.: Unsupervised induction of semantic roles within a reconstruction-error minimization framework. In: Conference of the North American Chapter of the Association for Computational Linguistics: Human Language Technologies (NAACL-HLT), pp. 1–10. ACL, Stroudsburg (2015)
103. Tomanek, K., Olsson, F.: A web survey on the use of active learning to support annotation of text data. In: NAACL HLT 2009 Workshop on Active Learning for Natural Language Processing, pp. 45–48. ACL, Stroudsburg (2009)
104. Wang, S.I., Liang, P., Manning, C.D.: Learning language games through interaction. In: 54th Annual Meeting of the Association for Computational Linguistics (ACL), pp. 2368–2378. ACL, Stroudsburg (2016)
105. Welinder, P., Branson, S., Belongie, S., Perona, P.: The multidimensional wisdom of crowds. In: 23rd International Conference on Neural Information Processing Systems (NIPS), pp. 2424–2432. Curran Associates, Red Hook (2010)
106. Wilson, A.G., Kim, B., Herland, W. (eds.): Proceedings of the NIPS 2016 Workshop on Interpretable Machine Learning for Complex Systems, Barcelona, Spain (2016). https://sites.google.com/site/nips2016interpretml/
107. Yang, Z., Cohen, W., Salakhutdinov, R.: Revisiting semi-supervised learning with graph embeddings. In: Balcan, M.F., Weinberger, K.Q. (eds.) 33rd International Conference on Machine Learning (ICML). JMLR: Workshop and Conference Proceedings, vol. 48, pp. 40–48 (2016)
108. Yimam, S.M., Biemann, C., Eckart de Castilho, R., Gurevych, I.: Automatic annotation suggestions and custom annotation layers in WebAnno. In: 52nd Annual Meeting of the Association for Computational Linguistics (ACL): System Demonstrations, pp. 91–96. ACL, Stroudsburg (2014)
109. Zeiler, M.D., Fergus, R.: Visualizing and understanding convolutional networks. In: Fleet, D., Pajdla, T., Schiele, B., Tuytelaars, T. (eds.) ECCV 2014. LNCS, vol. 8689, pp. 818–833. Springer, Cham (2014). https://doi.org/10.1007/978-3-319-10590-1_53
110. Zhou, S., Chen, Q., Wang, X.: Active deep networks for semi-supervised sentiment classification. In: 23rd International Conference on Computational Linguistics (COLING), pp. 1515–1523. Tsinghua University Press, Beijing (2010)

Language Technology for Digital Linguistics: Turning the Linguistic Survey of India into a Rich Source of Linguistic Information

Lars Borin[1], Shafqat Mumtaz Virk[1(✉)], and Anju Saxena[2]

[1] Språkbanken, University of Gothenburg, Gothenburg, Sweden
`lars.borin@svenska.gu.se`, `virk.shafqat@gmail.com`
[2] Linguistics and Philology, Uppsala University, Uppsala, Sweden
`anju.saxena@lingfil.uu.se`

Abstract. We present our work aiming at turning the linguistic material available in Grierson's classical *Linguistic Survey of India* (LSI) from a printed discursive textual description into a formally structured digital language resource, a database suitable for a broad array of linguistic investigations of the languages of South Asia. While doing so, we develop state-of-the-art language technology for automatically extracting the relevant grammatical information from the text of the LSI, and interactive linguistic information visualization tools for better analysis and comparisons of languages based on their structural and functional features.

1 Introduction

South Asia (also "India[n subcontinent]")[1] with its rich and diverse linguistic tapestry of hundreds of languages, including many from four major language families (Indo-European, Dravidian, Austroasiatic and Tibeto-Burman), and a long history of intensive language contact, provides rich empirical data for studies of linguistic genealogy, linguistic typology, and language contact.

South Asia is often referred to as a *linguistic area*, a region where, due to close contact and widespread multilingualism, languages have influenced one another to the extent that both related and unrelated languages are more similar on many linguistic levels than we would expect. However, with some rare exceptions (e.g., [12]) most studies are largely impressionistic, drawing examples from a few languages [5]. To the best of our knowledge, there has not been any large-scale investigation – involving a substantial number of South Asian languages – of the claim that South Asia constitutes a linguistic area. Any such study will require access to linguistic information about a large number – hundreds – of languages in an easily comparable format. Traditionally such information is

---

[1] In linguistic works, South Asia is defined as the seven countries Pakistan, India, Nepal, Bhutan, Bangladesh, Sri Lanka, and the Maldives, plus some immediately adjacent areas (e.g., Tibet).

© Springer Nature Switzerland AG 2018
A. Gelbukh (Ed.): CICLing 2017, LNCS 10761, pp. 550–563, 2018.
https://doi.org/10.1007/978-3-319-77113-7_42

manually extracted from written descriptive grammars, i.e., linguistically trained individuals read grammatical descriptions and record the relevant information in a formally structured format.

In recent times there have been efforts to build large-scale typological databases of linguistic features intended to be useful for wider linguistic studies. Examples of such databases include the World Atlas of Language Structures (WALS),[2] the Atlas of Pidgin and Creole Language Structures (APiCS),[3] the South American Indigenous Language Structures,(SAILS)[4] and the Phonetics Information base and Lexicon (PHOIBLE).[5] Collecting this data is very labor-intensive and makes the information-gathering process slow and also inconsistent, in that many data points will be missing from the resulting databases.

In this paper we present some results of the methodology that we are developing for automatic extraction and visualization of linguistic information – using South Asian language data. This work forms a part of a larger linguistic project, aiming at investigating the hypothesis about South Asia as a linguistic area.

The information source for this project is Grierson's classical *Linguistic Survey of India* (LSI; [7]; see the next section). Because of the size of the LSI, we cannot extract useful information manually, so we have decided to investigate to which extent this could be automated using language technology. The language technology related aims of this project are: (1) to turn the linguistic material available in the LSI into a digital language resource, a database suitable for a broad array of linguistic investigations; (2) to develop state-of-the-art language technology for automatically extracting the relevant linguistic information from the text of the LSI – a novel application field, to our knowledge; (3) to develop interactive linguistic information visualization tools for better comparison and easier analysis of languages based on their functional and structural characteristics. The methodology and tools that are being developed in this project will be beneficial for others working in typological, genealogical, historical, and other related areas of linguistics. Notably, we expect its utility to extend beyond the LSI; hopefully the resulting information extraction method will be general enough to be applicable to the vast number of descriptive grammars of languages from all over the world written in English over the last century or so.

2   Background: *Linguistic Survey of India*

The LSI presents a comprehensive survey of the languages spoken in South Asia conducted in the late nineteenth and the early twentieth century by the British government. Under the supervision of George A. Grierson, the survey resulted into a detailed report comprising 19 books comprising around 9500 pages in total. The survey covered 723 linguistic varieties representing major language

---

[2] wals.info.

[3] apics.org.

[4] sails.clld.org.

[5] phoible.org.

families and some unclassified languages, of almost the whole of nineteenth-century British-controlled India (modern Pakistan, India, Bangladesh, and parts of Burma). For each major variety it provides (1) a grammatical sketch (including a description of the sound system); (2) a core word list; and (3) text specimens (including a glossed translation of the *Parable of the Prodigal Son*). The LSI grammar sketches provide basic linguistic information about the languages in a fairly standardized – although still discursive free-text – format. The focus is on the sound system and the morphology (nominal number and case inflection, verbal tense, aspect, and agreement inflection, etc.), but as we will see below in Sect. 4, there is also some syntactic information to be found in them. Crucially, the sketches provide information on some of the features that have been used in defining South Asia as a linguistic area, e.g. retroflexion, reduplication, compound verbs, word order, converbs/conjunctive participles, etc. This offers the possibility of a broad comparative study of South Asian languages.

3 Data Preparation

3.1 Preprocessing

As a first step, we are in the process of digitizing all LSI volumes dealing with the four main South Asian language families (16 out of the 19 books). This part of the work is almost completed. Since OCR software deals poorly with the complex typography and multitude of languages of the language examples and language specimens in the LSI, the digitization is accomplished by an initial scanning and OCR step, followed by a manual correction step, so-called double keying, done by a commercial provider. During the latter, we deliberately chose not to represent the many diacritic characters appearing in the text in their original shape, but rather replace them with unique character combinations easily entered using an ordinary QWERTY keyboard (e.g. X$V instead of X^v). However, we want these characters restored back to their original shapes in the text that we will be working with. Also, there was a lot of metadata present on each page, in the form of page headers and footers, that we wanted to separate from the language descriptions. So a natural first step was to do some cleaning and preprocessing. Using a set of regular expressions, and mostly relying on a search and replace strategy, both these tasks were completed. Though the process overall went smoothly, there are still some known issues, having to do with rendering superscript characters and characters with complex combinations of diacritics.

3.2 Text Processing and Annotation

The amount of text that has been digitized so far is well in excess of one million words. In order to be able to explore this amount of data – which is not feasible to do manually – we have strived to use existing language tools to the greatest extent possible, even if these tools were not designed explicitly for the kind of large-scale comparative linguistic investigations that we have in mind, but rather for more traditional corpus-linguistic studies.

The text data, i.e., grammar sketches excluding tabular data (e.g., inflection tables) and text specimens, have been imported and made searchable using Korp, a versatile open-source corpus infrastructure developed and maintained by our group [1].[6] Currently, the LSI "corpus" comprises about 1.3 MW, and contains data about around 550 linguistic varieties that we identified during the pre-processing step. The comparative dictionary and the tabular data from the grammar sketches still remain to be processed in a similar way.

Korp is a modular system with three main components: a (server-side) back-end, a (web-interface) front-end, and a configurable corpus import and export pipeline. The back-end offers a number of search functions and corpus statistics through a REST web service API. As the main corpus search engine, it uses Corpus Workbench [6]. The front-end provides options to search at simple, extended, and advanced levels in addition to providing a comparison facility between different search results, as well as various visualization options.

The corpus pipeline is a major component and can be used to import, annotate, and export the corpus to other formats. For annotations, it relies heavily on external annotation tools such as segmenters, POS taggers, and parsers. The vanilla distribution comes preconfigured with tools for Swedish. For our purposes, we have instead plugged in the English Stanford dependency parser [11] for lexical and syntactical annotations, but we are still relying on the default sentence and paragraph segmentation tools provided with the Korp distribution as we achieved reasonable performance also for English text. We have added the following word and text level annotations to the LSI data:

Word-level annotations: lemma, part of speech (POS), named-entity information, normalized word-form, dependency relation.

Text-level annotations: LSI volume/part number, language family, language name, ISO 639-3 language code, longitude, latitude, LSI classification, Ethnologue classification, Glottolog classification, page number, page source URL, paragraph and sentence level segmentation.

While most of the annotations are self-explanatory, there are a few which may need some explanation. The *normalized word form* is the form produced by removing the diacritics and replacing phonetic characters with their closest English-alphabet counterpart. The purpose is to make it easy to search the corpus, since the LSI consistently renders language names and glosses in a kind of phonetic transcription which will most likely be unfamiliar to many users, as well as hard to enter using a standard national keyboard. Thus, the normalization allows the user to search for, e.g., *Bihārī* using the search string "Bihari", or "bihari" (with case-sensitivity disabled).

The text-level annotations mostly represent the metadata which were collected from different sources[7] in addition to the LSI volumes themselves, and are maintained as part of the corpus. The page source URL, for example, is a

---

[6] https://spraakbanken.gu.se/eng/korp-info.

[7] For instance, location data come mainly from the Glottolog: http://glottolog.org.

link to the corresponding page image in University of Chicago's *Digital South Asia Library.*[8]

Figure 1 shows a screenshot of the Korp front-end displaying results of a simple corpus query in Korp's KWIC (Key Word In Context) view. The box to the right of the KWIC sentences shows annotations (*Word* and *Text* level attributes) and metadata for the selected word.

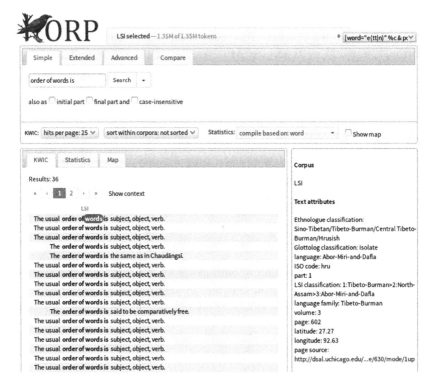

Fig. 1. Korp KWIC view resulting from searching the LSI for the string "order of words is"

In addition to these annotations, we have used the Stanford English Named Entity recognizer together with GeoNames[9] to extract all locations and their coordinates from within the description of a particular LSI language. The Korp front-end provides functionality for displaying locations on a map. All the proper names found in the query results are looked up in a database of locations, which also contains the coordinates. These locations then are displayed on a map by the front-end. See Fig. 2, where the map resulting from a text-level attribute search using the expression `language begins with` "konkani" is displayed. For

---

[8] http://dsal.uchicago.edu/books/lsi/ (Page images, no text search available.).

[9] http://www.geonames.org/.

identifying locations, the Korp front-end thus relies on POS tag information and a database of locations, a solution which is not perfect, but which gives reasonable accuracy for the purpose of providing a quick overview.

Locations: 44

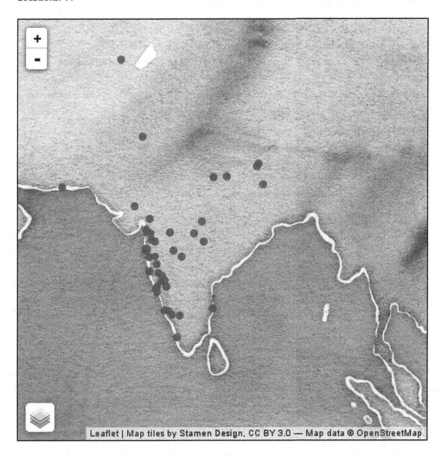

Fig. 2. Korp map view resulting from a text-level attribute search in the LSI for language begins with "konkani"

4   Grammatical Feature Extraction

After having cleaned the LSI data and stored it in a structured way, the next step is to extract information about particular grammatical features of LSI languages. The extracted grammatical feature values are to be used to investigate genetic relations and areal influences among the LSI languages during the later stages of the project. However, in the present paper our focus will be more on methodological development than on linguistic analysis of the results.

We have identified an initial list of about 100 features potentially retrievable from the LSI, that we judge to be most relevant for the purposes of the project. In this paper, however, the information extraction part is restricted to the following five features, for most of which tentative gold values could be manually extracted from the LSI comparative vocabulary, a much simpler task compared to retrieving them from the grammar sketches.

1. Apos: What is the order of adjective and noun in the NP?
2. NLPos: What is the order of numeral and noun in the NP?
3. NLBase: What numeral base the language has (decimal, vigesimal, or some other)?
4. Reflexive: Is reflexive indicated by a verbal affix, or by a pronoun?
5. Aagr: Does an adjectival modifier in the NP agree with its head noun?

4.1 Automatic Feature Extraction

The automatic feature extraction procedure consists of the following two steps:

(1) Using the standard search API of Korp, retrieve a set of potential sentences from the language descriptions by searching for a particular text string (representing a particular feature) and by limiting the search to 'within sentence'. The resulting sentences are further processed as described below to extract the feature values.

(2) Parse the sentences retrieved from step 1 using the Stanford dependency parser [10], and further process the resulting dependencies using a set of hand-built rules to extract the required information and to formulate the feature values.

Suppose that we are interested in extracting information about the order of adjective and noun in the Siyin[10] language. With step 4.1, we can extract all sentences containing the lemma "noun" or "adjective" from the grammar sketch of that particular language. One of the extracted sentences will be:

The adjectives follow the noun they qualify.

When we parse this sentence with the Stanford dependency parser, it will return the following dependencies:

```
det(adjectives-2, The-1)
nsubj(follow-3, adjectives-2)
root(ROOT-0, follow-3)
det(noun-5, the-4)
dobj(follow-3, noun-5)
nsubj(qualify-7, they-6)
acl:relcl(noun-5, qualify-7)
```

---

[10] A Tibeto-Burman language spoken in southern Tedim township, Chin State, Burma.

Algorithm 1. Extract Adjective Noun Order

```
 1: procedure EXTRACTADJECTIVENOUNORDER(parse)
 2:     for <every rel,arg1,arg2 in parse> do
 3:         AdjectiveSubjFollow ← False
 4:         NounObjFollow ← False
 5:         AdjectiveSubjPrecede ← False
 6:         NounObjPrecede ← False
 7:         if rel = nsubj then
 8:             if lemmatize(arg1) = follow ∧ lemmatize(arg2) = adjective then
 9:                 AdjectiveSubjFollow ← True
10:             else if lemmatize(arg1) = precede ∧ lemmatize(arg2) = adjective then
11:                 AdjectiveSubjPrecede ← True
12:             end if
13:         else if rel = dobj then
14:             if lemmatize(arg1) = follow ∧ lemmatize(arg2) = noun then
15:                 NounObjFollow ← True
16:             else if lemmatize(arg1) = precede ∧ lemmatize(arg2) = noun then
17:                 NounObjPrecede ← True
18:             end if
19:         end if
20:     end for
21:     if AdjectiveSubjFollow = True ∧ NounObjFollow = True then
22:         return NA
23:     else if AdjectiveSubPrecede = True ∧ NounObjPrecede = False then
24:         return AN
25:     end if
26: end procedure
```

These dependencies can be processed further with a set of rules to extract the required information. Algorithm 1 shows a simplified version of a procedure that can be used to extract noun-adjective order information from the dependencies given above.

The algorithm checks for a particular relation type (line 7 and 13) first, and then checks for the contents of arguments of the relation (lines 8–12 and 14–18) to adjust appropriate boolean parameters. For the above given dependencies, two boolean parameters $AdjectiveSubjFollow$ and $NounObjFollow$ will get the values $True$, and based on this the algorithm will formulate and return the value NA (the fact that adjectives follow nouns) for the adjective-noun-order feature (lines 21–25). Note that the algorithm shows how to extract noun-adjective order, and takes care of a couple of relation types and argument contents only. A full implementation will take care of other possible verbal predicates (e.g. *come*, *placed*, etc.), relation types, and argument contents including anaphora relations as described below.

The sentence containing the description of a particular feature may have anaphoric expressions referring to an antecedent or subsequent expression. For example, the sentence *They follow the nouns they qualify* may appear instead of *The adjectives follow the nouns they qualify* in the description, with the antecedent expression *adjectives* appearing somewhere else. To extract feature values from such sentences properly, such anaphoric relations need to be resolved. There exist many anaphora resolution systems [14,18], and a classical solution will involve using a state-of-the art system for this purpose. However, at the current stage of our experiments, we have chosen to employ a simple rule-based strategy to resolve such co-reference relations and extract feature values/description. The main idea is to investigate the context with a particular window size to resolve such co-references (if any). For example, if an argument in a relation is a pronoun (e.g. *they*, *it*, or *them*), we just investigate the arguments of one or more preceding or following sentences, and if those arguments contain the potentially linked entity (e.g. *nouns* or *adjectives*, etc.), we just assume that they are related to each other. This procedure can easily be incorporated in the above given algorithm with an extra if condition, but for simplicity reasons, we have excluded it. It is worth mentioning here that the rule-based anaphoric-resolution solution was chosen not only for its simplicity, but we also observed in experiments that in many cases this simple strategy was actually able to relate the arguments, whereas the Stanford anaphora resolution system [14] failed to do so.

Table 1 shows the number of languages for which we were able to extract the feature values using the strategy described above.

Table 1. Number of languages for which feature values were automatically extracted

Feature	Number of languages
Apos	66
NLPos	15
NLBase	31
Reflexive	39
Aagr	12

This simple dependency parsing and rule based approach allowed us to get off the ground quickly, but it has serious limitations. This strategy will very strictly match particular sentence structures and contents of arguments. This probably will not cover all possible ways the same information could have been encoded unless one designs rules rich enough to catch all possible verbal predicates and their argument strings – which does not seem realistically achievable. To address the limitations of this strategy, we plan to experiment with open information extraction based techniques, as well as with utilizing resources such as WordNet in order to handle alternative ways of expressing the same content.

4.2   Evaluation

Table 2 shows how accurately the proposed feature extraction methodology was able to extract different feature values. For each feature, the accuracy value was computed using the following simple formula:

$$Accuracy = \frac{N_{correct}}{N_{extracted}}$$

where $N_{correct}$ is the number of languages for which the feature value was correctly extracted, and $N_{extracted}$ is the total number of languages for which the feature value was extracted. To decide if an extracted value is correct or not, it was compared to the gold value which was retrieved manually by a human expert from the comparative vocabulary (most cases) or from the language descriptions.

Table 2. Evaluation results

Feature	Accuracy (%)
Apos	0.818
NLPos	1.0
NLBase	0.823
Reflexive	0.739
Aagr	0.857

At present, we do not have the information needed to compute the traditional evaluation metrics of precision, recall, and f-measure, since we do not know how many grammar sketches contain information about a particular feature. However, we plan to evaluate the proposed methodology on other available gold data sets (e.g. the SAILS[11] dataset) for which it will be possible to compute the above mentioned evaluation metrics.

5   Visualization of Extracted Grammatical Features for Linguistic Research

As mentioned earlier, the work presented here is part of a larger effort to design and deploy language-technology based e-science tools for research in large-scale comparative linguistics. There are indications that data visualization and visual analytics will play a crucial role in this connection (e.g., [2,3,9,15–17]). Consequently, one of the objectives of this project is to develop state-of-the-art tools for visualization of the extracted grammatical features in a way that makes it easy to observe the genetic relations between multiple languages and the areal influences of languages on each other. For this purpose, we have developed an

---

[11] sails.clld.org.

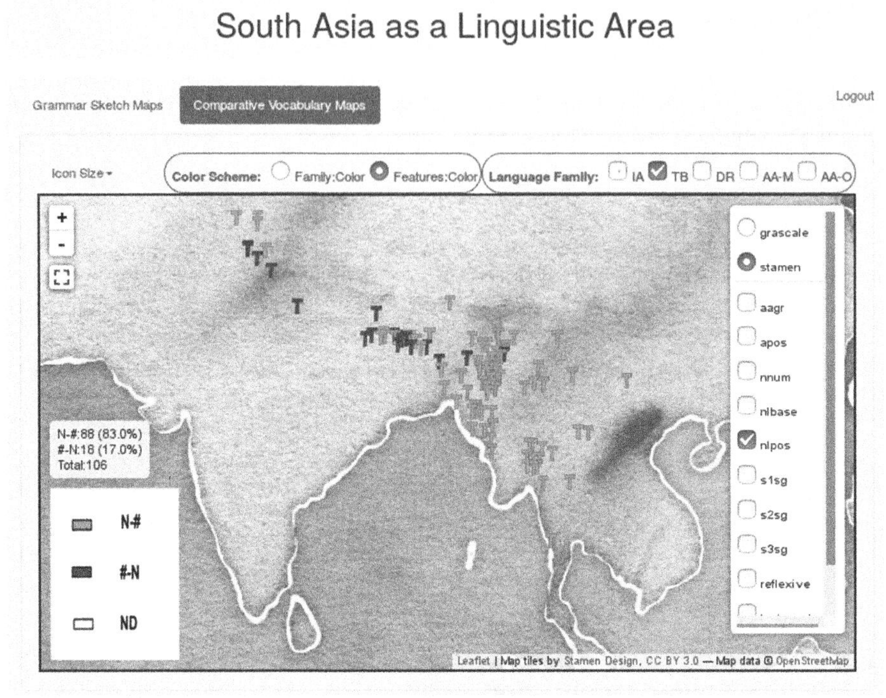

South Asia as a Linguistic Area

Fig. 3. Map showing numeral position w.r.t. the noun in Tibeto-Burman languages

interactive mapping solution where the users can choose to view values of particular feature(s) of the languages belonging to one or more families. Further, we provide switchable shape/color combinations for visualizing and differentiating family/feature characteristics. Figure 3 shows a snapshot visualizing the feature NLPos (i.e. the position of cardinal numerals w.r.t. the noun) for the languages belonging to the Tibeto-Burman family. As can be noticed, we have selected feature values to be encoded by the color, while the symbol 'T' shows that these languages belong to the Tibeto-Burman family. The user can select multiple families and multiple features at the same time by checking the appropriate check-boxes. From this map, it can be observed that the numerals mostly follow the noun (indicated by yellow/lighter color) in the languages spoken in the southeast, while it mostly precedes (indicated by purple/darker color) in the languages spoken in the nortwest. Such an interactive mapping facility provides a useful way to show the genetic relations and areal influences among languages spoken in different geographical areas and belonging to different language families.

In addition to this, we are aiming to convert the grammatical sketches into a typological database and store and visualize them appropriately. For this purpose, we have chosen cross-linguistic linked data (CLLD),[12] which is an open source python-based framework for storing and visualizing linguistic data from multiple languages. Previously, it has been used to store a number of well-known linguistic data sets including WALS [4], Glottolog [8], APiCS online [13], etc. It provides a number of useful mapping and navigation facilities in addition to storing the linguistic data using linked data principles. Figure 4 shows a snapshot of the LSI dataset as seen through the CLLD web interface.

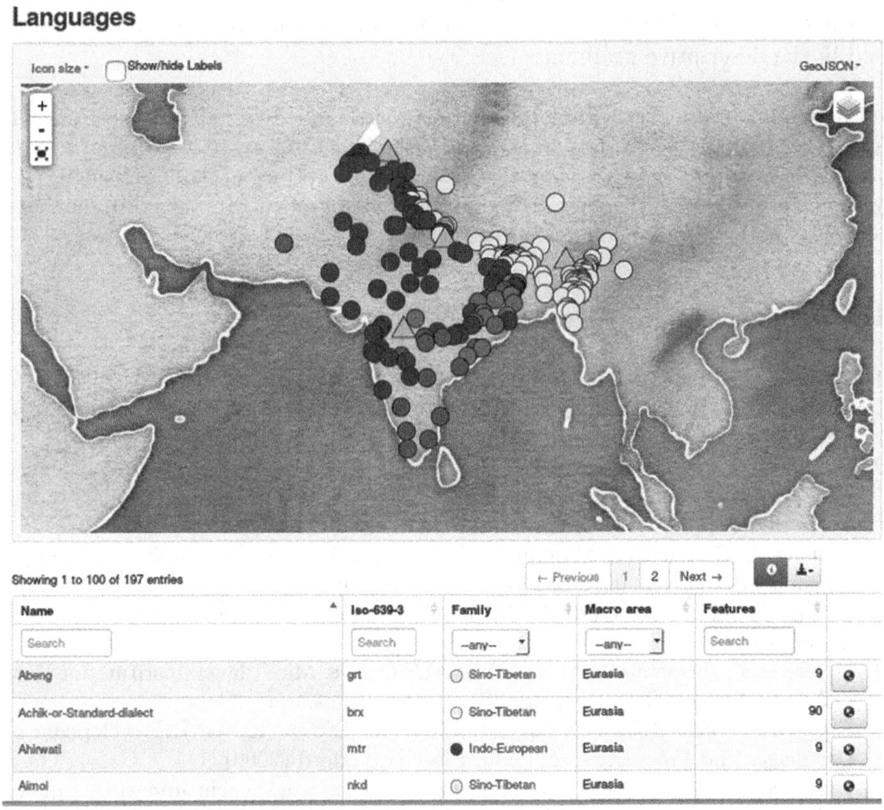

Fig. 4. The LSI dataset in CLLD

6   Conclusions and Future Work

Turning the LSI into a structured digital resource will provide a rich empirical foundation for large-scale comparative studies of genealogical, typological and areal linguistic relationships in South Asia. Today, the tendency is increasingly that such studies are not conducted manually, but need to draw on extensive

---

[12] http://clld.org/.

digitized language resources and state-of-the-art computational tools. This is the goal and motivations of our on-going work with the LSI. In addition to this, we aim to contribute to the methodological development of large-scale comparative linguistics drawing on digital language resources.

We are also planning to take into account the phonological and other related information present in the extensive tabular data found in the grammar sketches as well as the parallel annotated data present in the LSI text specimens. Further, we would like to experiment with open information extraction based techniques to improve the proposed feature extraction methodology, as well as including information from WordNet in order to cater for alternative ways of expression used in the descriptive grammars.

Acknowledgments. The work presented here was funded by the Swedish Research Council as part of the project *South Asia as a linguistic area? Exploring big-data methods in areal and genetic linguistics* (2015–2019, contract no. 421-2014-969), and by the University of Gothenburg as part of its funding of the Språkbanken language technology and digital humanities infrastructure.

References

1. Borin, L., Forsberg, M., Roxendal, J.: Korp – the corpus infrastructure of Språkbanken. In: Proceedings of LREC 2012, pp. 474–478. ELRA, Istanbul (2012). http://www.lrec-conf.org/proceedings/lrec2012/pdf/248_Paper.pdf
2. Broadwell, P.M., Tangherlini, T.R.: TrollFinder: geo-semantic exploration of a very large corpus of Danish folklore. In: The Third Workshop on Computational Models of Narrative, pp. 50–57. ELRA, Istanbul (2012)
3. Chuang, J., Ramage, D., Manning, C.D., Heer, J.: Interpretation and trust: designing model-driven visualizations for text analysis. In: ACM Human Factors in Computing Systems (CHI) (2012). http://vis.stanford.edu/papers/designing-model-driven-vis
4. Dryer, M.S., Haspelmath, M. (eds.): WALS Online. Max Planck Institute for Evolutionary Anthropology, Leipzig (2013). http://wals.info/
5. Ebert, K.: South Asia as a linguistic area. In: Brown, K. (ed.) Encyclopedia of Languages and Linguistics, 2nd edn. Elsevier, Oxford (2006)
6. Evert, S., Hardie, A.: Twenty-first century corpus workbench: updating a query architecture for the new millennium. In: Proceedings of the Corpus Linguistics 2011 Conference. University of Birmingham, Birmingham (2011)
7. Grierson, G.A.: A Linguistic Survey of India, vol. I-XI. Government of India, Central Publication Branch, Calcutta (1903–1927)
8. Hammarström, H., Forkel, R., Haspelmath, M., Bank, S.: Glottolog 2.7. Jena: Max Planck Institute for the Science of Human History (2016). http://glottolog.org
9. Havre, S., Hetzler, B., Nowell, L.: ThemeRiver: visualizing theme changes over time. IEEE Symposium on Information Visualization 2000. InfoVis 2000, pp. 115–123. IEEE, Salt Lake City (2000)
10. Klein, D., Manning, C.D.: Accurate unlexicalized parsing. In: Proceedings of ACL 2003, pp. 423–430. ACL, Sapporo (2003). http://dx.doi.org/10.3115/1075096.1075150

11. Manning, C.D., Surdeanu, M., Bauer, J., Finkel, J., Bethard, S.J., McClosky, D.: The Stanford CoreNLP natural language processing toolkit. In: ACL System Demonstrations, pp. 55–60. ACL, Portland (2014). http://www.aclweb.org/anthology/P/P14/P14-5010

12. Masica, C.P.: Defining a Linguistic Area: South Asia. Chicago University Press, Chicago (1976)

13. Michaelis, S.M., Maurer, P., Haspelmath, M., Huber, M. (eds.): APiCS Online. Max Planck Institute for Evolutionary Anthropology, Leipzig (2013). http://apics-online.info/

14. Recasens, M., Marneffe, M.C.D., Potts, C.: The life and death of discourse entities: identifying singleton mentions. In: Proceedings of NAACL-HLT 2013. ACL, Atlanta (2013)

15. Schilit, B.N., Kolak, O.: Exploring a digital library through key ideas. In: Proceedings of JCDL 2008, pp. 177–186. ACM, Pittsburgh (2008)

16. Smith, D.A.: Detecting and browsing events in unstructured text. In: SIGIR 2002. ACM, Tampere (2002)

17. Sun, G.D., Wu, Y.C., Liang, R.H., Liu, S.X.: A survey of visual analytics techniques and applications: state-of-the-art research and future challenges. J. Comput. Sci. Technol. 28(5), 852–867 (2013). http://dx.doi.org/10.1007/s11390-013-1383-8

18. Versley, Y., Moschitti, A., Poesio, M., Yang, X.: Coreference systems based on kernels methods. In: Proceedings of COLING 2008. ACL, Manchester (2008)

Classifying World Englishes from a Lexical Perspective: A Corpus-Based Approach

Frank Z. Xing[1($\boxtimes$)], Danyuan Ho[2], Diyana Hamzah[2], and Erik Cambria[1]

[1] School of Computer Science and Engineering, Nanyang Technological University, Singapore, Singapore
{zxing001,cambria}@ntu.edu.sg
[2] Temasek Laboratories, Nanyang Technological University, Singapore, Singapore
{dyho,diyana}@ntu.edu.sg

Abstract. The spread of English has led to the emergence of new English varieties worldwide. Existing quantitative approaches made use of several linguistic criteria, particularly morphological and syntactical features, in investigating variations across English varieties. Taking an alternative lexical perspective to the classification of World Englishes, this paper adopts a corpus-based approach in investigating the lexical frequency across 20 regional English varieties. Specifically, the lexical items in focus include culture-bound terms and words that have undergone semantic shift. The English varieties are categorized following a series of filtering rules and normalization techniques, and a hierarchical cluster of the varieties is subsequently formalized. Our findings generally corroborate with Kachru's *Three Circle of English* model, with subtle differences to the *Inner Circle-Outer Circle* groupings. The taxonomy of the English varieties additionally reveals geographical and cultural correlations.

Keywords: English · World Englishes · Lexicology · Word frequency
Classification · Corpus linguistics

1 Introduction

The colonial expansion of the British Empire in the 19th century and the rise of the United States as a major economic and military powerhouse in the 20th century have led to the spread of English across the globe. Today, it has become the leading world language, and is recognized as the official language in over 60 countries. English has gradually nativized into new localized varieties, and currently co-exists as an indigenized language alongside other local languages in several countries [2]. The implantation of English in novel linguistic settings resulted in the emergence of New Englishes, or non-native English varieties in former British colonies, which have distinctive linguistic forms and structures. These

© Springer Nature Switzerland AG 2018
A. Gelbukh (Ed.): CICLing 2017, LNCS 10761, pp. 564–575, 2018.
https://doi.org/10.1007/978-3-319-77113-7_43

new English varieties differ from standard English in several aspects including, inter alia, pronunciation, grammar, syntax and lexicon [7].

Comparisons between the lexicon of World Englishes have revealed that the lexis of new English varieties contain idiosyncratic words that are absent in historically input English (i.e., British English), including loanwords (from non-English languages), calques (or loan translations) and English-derived words that have undergone morphological derivation and semantic shift, among others [1,5]. However, such studies have been centered around qualitative approaches, which simply identified the general lexical similarities and/or differences between varieties. Meanwhile, those adopting quantitative approaches primarily focused on other linguistic aspects, in particular morphological and syntactical features [6,10,11,13,18]. This paper hence explores cross-varietal lexical variations using a corpus-based, quantitative approach, which can potentially provide new insights into the taxonomic relations between varieties.

The remainder of this paper is organized as follows: Sect. 2 discusses previous works investigating variations in World Englishes; the methodological design is outlined in Sect. 3; classification results are interpreted from multi-aspects in Sect. 4, followed by a hierarchical classification of the varieties and comparisons to findings of existing research; finally, Sect. 5 concludes with a brief summary of the findings.

2   Related Works

Scholars have proposed various theoretical models in the classification of World Englishes to explain the differences between varieties [9,16,21]. The *Three Circles of English* model [7], an influential account of the spread of English, groups English varieties into three concentric circles: *Inner Circle*, *Outer Circle* and *Expanding Circle*. The Inner Circle includes the traditional bases of English, like the UK, the US, Canada, Australia and New Zealand, where English is now the native language of the region's inhabitants. The Outer Circle came about as a result of the spread of English to Asia and Africa through extended colonization by Inner Circle powers, where English now serves as a chief institutional language and an important second language. The Outer Circle consists of former UK and US colonies, such as Singapore, Hong Kong and the Philippines. The Expanding Circle represents regions with no history of colonization by Inner Circle countries, like Japan and China, where English varieties are used in limited contexts, typically in business and scientific domains.

In the attempt of providing empirical support for the theoretical models, computer scientists employed quantitative approaches to discover the similarities and differences in linguistic patterns between the varieties. Nam et al. [18] investigated the use of syntactic alternations (more specifically, verb complication patterns of 'give') in the written texts of Indian English, Pakistani English and British English. Statistical analysis of the data involved the use of Akaike Information Criterion (AIC), which was employed to select model from a group of multi-nominal logistic regression. Hundt [6] looked into the use of progressive passive (which is formed by the combination of the verb 'be', the progressive auxiliary 'being' and a past participle) in Inner and Outer Circle English varieties. A frequency analysis was conducted on the data adapted from the International Corpus of Englishes (ICE). Biermeier [1] conducted a corpus-based study (from ICE) on word formation patterns across seven varieties of English.

To determine the trends in word coinage in the targeted varieties, the researcher analyzed the frequency data of lexical tokens through the lens of various word-formation categories such as compounding, conversion and affixation. Szmrecsanyi et al. [23] performed a large-scale investigation of 76 morphosyntactic variations (adapted from the morphosyntactic database of *Handbook of Varieties of English*) in 46 English varieties, which included not only native vernacular and traditional English dialects (L1 varieties), but also English as Second Language variants (L2 varieties) and English-based pidgins and creoles. A typological classification of these variants was performed using a series of quantitative analysis techniques (multidimensional scaling, clustering, and principle component analysis). These studies revealed common findings - i.e., the cross-varietal variations defy the neat *Inner circle-Outer circle* groupings, and the linguistic variations investigated can bring about subtle differences in the classification of English varieties.

Our paper looks into the variations exhibited by national varieties of English, which could consist of both standardized and localized forms used within the locality. An example is Singapore English, which can been described as a sociolect continuum consisting of Singapore Standard English and Singapore Colloquial English (or Singlish) on opposite ends of the spectrum [19]. This study approaches the categorization of English varieties from a lexical, rather than syntactic and morphosyntactic (as in previous works), perspective. Specifically, our approach investigated lexical frequency, or the frequency of word occurrences, in the different English varieties in order to establish their taxonomic relation via lexical distance and clustering.

3   Method

3.1   Dataset and Wordlist

The dataset used in this study is obtained from The Corpus of Global Web-Based English (GloWbE), which has a huge volume of available English texts. Firstly released in 2013 by Brigham Young University, GloWbE has now over 1.9 billion words from webpages and blogs from 20 countries, as well as data on word frequency. Our wordlist consists of 109,582 English words compiled and corrected in 1991 from lists obtained from the Interociter bulletin board maintained by SIL International[1]. Inflected forms are included in the wordlist, such as pluralized nouns and verbs in their past, present and progressive forms. In the analysis, we used the country codes provided in GloWbE (Table 1) to represent the varieties of English in their respective countries.

---

[1] http://www-01.sil.org/linguistics/wordlists/english/.

Table 1. Countries and Regions featured in GloWbe and their respective codes

Country	Code	Country	Code
Australia	AU	Malaysia	MY
Bangladesh	BD	New Zealand	NZ
Canada	CA	Nigeria	NG
Ghana	GH	Philippines	PH
Great Britain	GB	Pakistan	PK
Hong Kong	HK	Singapore	SG
India	IN	Sri Lanka	LK
Ireland	IE	South Africa	ZA
Jamaica	JM	Tanzania	TZ
Kenya	KE	United States	US

We have identified two types of lexical items that systematically exhibit variations in word frequency across English varieties, namely culture-bound terms and words that have undergone semantic shift.

Culture-Bound Terms. Culture-bound terms are words that are unique to the local culture. They include borrowings from indigenous languages that describe the local fauna and flora, and words for distinctive items and customs [21]. These words are likely to be used more frequently in the region of origin than other areas. For example, 'kangaroo', a hopping marsupial that is the cultural symbol of Australia, is likely to appear more frequently in Australian English than other English varieties. Similarly, 'ackee', a fruit eaten in Jamaica and West Africa, is less likely to appear in English varieties of other regions. However, the lexical frequency of certain words appears to be high in regions sharing cultural commonalities. For instance, given the sophisticated food culture of Hong Kong and Singapore and the fastidious attitude of their inhabitants towards food, 'restaurant' is used much more frequently in the English varieties spoken in these regions than others.

Semantic Shift. Words can undergo semantic shift, or a change in meaning, when used in new language settings. The divergence between the meanings of a particular word in different English varieties could be caused by the retention of the original meaning in one variety and its loss, and possibly the acquisition of a new meaning, in another [5]. For example, 'fall' is used to describe someone tumbling over in several English varieties, but can additionally refer to the autumn season in North American English. Similarly, in India, Malaysia and Singapore, 'auntie' can be used to refer to female elders, in addition its original meaning of the sister of one's parents (common in other English varieties). Semantic change is expected to affect the lexical frequency of words, since words

with broader meanings are likely to appear more frequently than those with
narrower meanings.

3.2   Dissimilarity

Measure of dissimilarity, or the overall difference in lexical frequencies, is a good
indicator of differences between two English varieties. To assess cross-varietal
lexical variations, we used the aforementioned types of lexical items (i.e., culture-
bound terms and words that have undergone semantic shift) as the main components of the variance of relative frequency.

In the analysis, we employed the variance S^2 to calculate the pairwise dissimilarity values. $F_{1j}, F_{2j}, \ldots, F_{kj}$ is used to denote word frequency in different
countries, where $k = 20$ is the total number of countries investigated and j is
the index of valid words. The variance can be represented as:

$$S^2(m, n) = \sum_{j=1} (\frac{F_{mj} - F_{nj}}{\bar{F}_j})^2$$

where $\bar{F}_j$ is the weighted average of F_{mj} and F_{nj} given by the chosen database.
We extracted the frequency distribution of the target words in our wordlist
from GloWbE in the format shown in Table 2. Each table contains the absolute
frequency, or the count of appearance of a particular word (FREQ), and its
relative frequency per million words (PerMil). In the case of the word 'red', for
instance, the values are $\bar{F} = 131.77, F_1 = 131.17, \cdots, F_{20} = 165.99$.

Table 2. Frequency distribution of the word 'red' in GloWbE

Region	All	US	CA	GB	$\cdots$	JM
FREQ	248332	50736	19776	58148	$\cdots$	6568
PerMil	131.77	131.17	146.74	150.01	$\cdots$	165.99

3.3   Filtering Rules and Penalty Items

Filtering of Hapax legomenona. *Hapax legomenon* refers to words that are
rarely used and have very low lexical frequency. Based on the lexical frequency
values provided in GloWbE, we observed that these words are mainly culture-
bound terms. The lexical frequency of such words are likely to be high in certain
varieties but similarly low in others. In addition, the semantics of these words
tend to be much less variant across varieties as compared to those of more
lexically frequent words. Due to their low lexical frequency and semantic uni-
formity, *hapax legomenon* is an unsuitable measure to investigate cross-varietal
variations, and are hence excluded from our analysis.

For this study, words that have an overall FREQ of less than 20 are considered as *hapax legomenon*, and are removed from the data. Examples include
'concessioners' (FREQ = 1), 'minitower' (FREQ = 1), 'squalled' (FREQ = 7)
and 'banjax' (FREQ = 9).

Stoplist. The stoplist is made up of words with very high average relative frequency. These words are terms that commonly appear across English varieties, and they show a remarkably low dissimilarity values when calculated using our algorithm in Sect. 3.2. We consider this dissimilarity to be random. Due to their high average relative frequency, these words demonstrated resistance to the frequency regulation process performed in the later stage, and are hence omitted from our analysis.

For this study, words having an average relative frequency value of over 200 were removed from the dataset prior to the analysis. This includes words like 'the', 'even' and 'men', which have average relative frequency values of 56371.98, 1193.94 and 329.11 respectively.

Frequency Regulation. After *hapax legomenona* and stoplist words were discarded, only words that exhibit moderate lexical frequency remained, of which those with a larger cross-varietal frequency variance are of a higher importance than those with a more evenly distributed frequency. This applies to words with relatively low, as well as high, absolute frequency. In addition, absolute frequency also has an effect on statistics. As absolute frequency decreases, sample size decreases and arbitrariness increases. The estimated frequency would have a wider confidence interval as a result. Therefore, we added a penalty item $-\beta|\bar{F}_j/\bar{F}_{All}|^2$ to the relative frequency of word j. This formula harmonizes two considerations: (1) This regulation is more relevant for words with high relative frequency $\bar{F}_j$; and (2) for words with lower relative frequency, a larger portion of low frequency variance words are trimmed. We thus use a parabola function to present the penalty item.

Parameter β is empirically set. From Fig. 1, we observe that β pulls the rank of word frequency (x-axis) towards negative. Using different β values, we found that frequency penalty item starts to dominate the rank measure from $\beta = 0.1$. In other words, the frequency distribution is overly distorted. As such, we decided on $\beta = 0.05$ and discarded all the words that fall into the negative side of x-axis plane.

Orthography. The spelling of words can differ across varieties. For instance, certain words are spelled differently in American English and British English (e.g., 'color', 'organize', 'theater', 'traveler' *versus* 'colour', 'organise', 'theatre', 'traveller'). In this study, English words that exhibit orthographic variants were normalized as a preprocessing step. Such a measure is necessary to prevent semantically similar words with different cross-varietal spellings from being processed as separate entries, which will in turn affect the lexical frequency values.

4   Results

After filtering, we sorted the wordlist by the regulated measure mentioned in Sect. 3. We then take the remaining words as lexical representatives of the different English varieties. In our two stage interpretation, we provide a descriptive

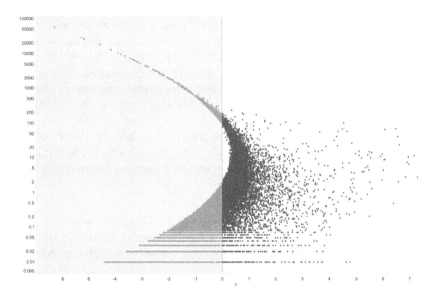

Fig. 1. The effect of adding penalty items to different words

analysis of the top 100 extracted words followed by a quantitative analysis of the entire wordlist in the latter part of this section.

4.1 Word Feature

Table 3 presents the top 100 most frequently recurring words in the dataset, which includes:

(i) Core English words (e.g., terminologies)
(ii) Culture-bound terms, consisting of (1) demonyms (e.g., 'nigerian', 'jamaican'), (2) geographic places (e.g., 'manila', 'nairobi'), and (3) loan-words that denote flora or fauna (e.g., 'carabao', a type of animal), distinctive items (e.g., 'pimento', a type of spice) and concepts (e.g., 'salaam', a greeting borrowed from Arabic)
(iii) Miscellaneous words (including grammatical loanwords)

4.2 Classification of Varieties

Hierarchical clustering was performed on the frequency data that was obtained from prior analysis. We assume that the frequency difference for the distribution of each word between two countries represents the overall lexical difference between the two varieties. For each country, we use a vector $(f_{m1}, f_{m2}, \ldots, f_{mj} \ldots)$ to denote its lexical distribution, where f_{mi} is the i-th word's relative frequency in the country m. We then calculate the "distance" $D = S^2(m, n)$, between every two countries.

Table 3. Top 100 words with the highest frequency

Core English words	gleaner	deejay	patois	safaris	maroons	chieftaincy
	breadfruits	wailer	wailers	crater	greasers	mattocks
	jamb	albinos	arsenic	artiste	tusker	mainlanders
	logwood	tings	breadfruit	deejays	acclimatization	
Demonyms	tanzanians	zanzibar	tanzanian	ghanaians	jamaicans	jamaican
	kenyans	malaysians	sinhalese	nigerians	filipinos	philippine
	malays	swahili	filipino	malaysian	nigerian	malay
	caribbean	philippines	afrikaans	singhalese	barbados	tagalog
	cantonese					
Geographic places	tamale	tanzania	jamaica	natron	ghana	volta
	colombo	penang	ceylon	nairobi	manila	kenya
	nigeria	malaysia	trinidad	pretoria	lahore	
Loanwords	salaam	reggae	jute	kopjes	obeah	wildebeest
	bazar	rastafarian	pimento	dhow	pap	ganja
	carabao	imam				
Miscellaneous words	es	whiles	orang	kl	arroyo	tun
	sutta	tented	atty	jackroll	sri	yaw
	kong	sundowns	bogle	dagoba	deles	moi
	delly	lambie	daystar			

The color map in Fig. 2 reveals two distinct groups of English varieties that corroborated with the classifications in Kachru's *Three Circle of English* model. The varieties denoted by the lighter color appears to represent those from Inner Circle countries (CA, GB, IE, AU and NZ), while the varieties denoted by the darker color group are those from Outer Circle countries. Interestingly, the calculated distance between IN (Indian English) and the varieties of the Inner Circle countries is smaller than between IN and varieties of the Outer Circle countries. Several reasons may explain the lack of a lexical distinction between Indian English and the Inner Circle varieties: British English is still held as the 'idealized linguistic norm' in India's education institutions [8], and English in India is largely confined to the educated, urban population [12].

The calculated distance between varieties of the lighter color group showed higher consistency than those of the darker color group. This indicates that the Inner Circle countries share more commonalities in lexical usage than the Outer Circle countries. The higher homogeneity of varieties of the Inner Circle group can be explained by the majority of the inhabitants of these countries being descendants of British emigrants and hence, the continuous transmission of English without major linguistic changes [17]. On the other hand, the lower homogeneity of varieties of the Outer Circle group may be indicative of the greater influence of indigenous languages and cultures in the development of

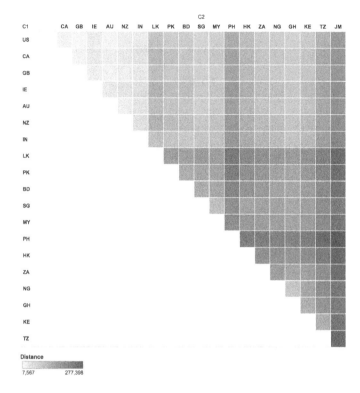

Fig. 2. Distance map showing the calculated distance between English varieties

these English. Additionally, disparate language contact settings might be present in the formation of these varieties, in which English interacted with distinct indigenous languages in various regions [5,17].

Following distance calculations, hierarchy clustering was conducted by continually merging two nearest country according to the distance map. Countries merged together are regarded as a distinct cluster, from which the distance to another country is the average distance from every member to it. This merging process continued until all the countries have been merged into their respective clusters, ultimately yielding a hierarchical tree of the 20 English varieties investigated, as shown in Fig. 3.

The visual representation of the hierarchy clustering in Fig. 3 shows how the individual varieties largely correlate with the geographical proximity and cultural affinity of the countries. Since English originated from the Great Britain, its variety (GB) is positioned to the left end, with which the lexical distance of the other varieties is compared. The varieties of New Zealand (NZ) and Australia (AU) are first merged with GB, since the inhabitants of these countries principally originated from the British Isles. The next to merge are CA and US, the varieties of Canada and the United States, which are also members of the Anglosphere. Canada English and American English (in this case, North American English)

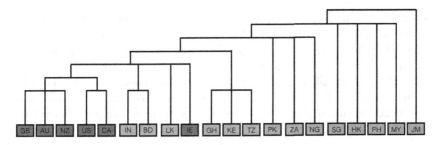

Fig. 3. Hierarchical classification of the 20 English varieties from a lexical perspective, where L1 and L2 varieties (based on Kachru's model) are labeled red and green accordingly (Color figure online)

are more *lexically* related to each other than to British English. English varieties of the Indian subcontinent, such varieties of India (IN), Bangladesh (BD) and Sri Lanka (LK), form another distinct cluster. These countries were once part of British India, except for Sri Lanka which was a separate Crown Colony geographically close to India. Another distinct cluster consists of varieties of Kenya (KE), Tanzania (TZ) and Ghana (GH), which are former British colonies in Africa.

The taxonomic relation between varieties found in this study is similar to the early Strevens' *Model of English in the World* [22], which indicates the genealogical relationships of the English varieties through a tree diagram overlaid on the world map. In the model, American English and British English formed the main branches, and the different English varieties, classified as leaf nodes, are grouped according to their geographical and historical affinities.

In our study, no strong lexical relationship is found among varieties of Singapore (SG), Malaysia (MY) and Hong Kong (HK), even though linguists attested typological similarities between Singapore English and Malaysian English [20], and between Singapore English and Hong Kong English [15]. Our findings suggested a tight cluster for Inner Circle varieties, but not for Outer Circle varieties.

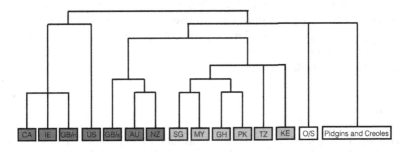

Fig. 4. Hierarchical agglomerative clustering result based on morphosyntactic variations, where GB/n and GB/s refer to North and South England respectively, and O/S refers to Orkney & Shetland; adapted from Szmrecsanyi et al. [23]

This is unlike the findings in [23], which indicated a distinct split between American varieties and British varieties. As shown in Fig. 4, Szmrecsanyi et al. found that British varieties are more closely related to Commonwealth Englishes than American varieties. Noting that morphosyntactic features are more resistant to language change, the differences in findings may be attributed to other predictors of variance in World Englishes besides geographical proximity [23]. Nonetheless, the two dendrograms generally reveal an alignment with Kachru's *Inner Circle-Outer Circle* distinction.

5   Conclusion

This study provides a novel account of the classification of World Englishes from a lexical perspective via hierarchical clustering of 20 English varieties based on data from a large-scale corpus. A descriptive analysis of the top 100 frequently recurring words revealed three distinct lexical categories. Culture-bound terms made up the bulk of the extracted terms, which included borrowing of words into the English variety to describe distinctive concepts and entities. The remaining words consists of English words describing certain terminologies or subject-matters, and miscellaneous words (including grammatical loanwords).

The classification of the English varieties revealed the presence of two distinct groups that corroborates with Kachru's *Inner Circle-Outer Circle* contrast. From hierarchical clustering, it is observed that English varieties share a relationship that largely correlates to their geographical and cultural proximities. The varieties in Africa form a distinctive cluster. Indian English, Sri Lankan English and Bangladeshi English form another cluster, and these varieties appear to be more lexically similar to that of the Inner Circle countries as compared to other Asian varieties like Hong Kong English and Singapore English. This classification is coherent with Strevens' *Model of English in the World*, which maps the relation of English varieties based on their geographical and historical affinities.

Our findings, however, did not show Singapore English, Malaysia English and Hong Kong English forming a distinctive cluster despite being typologically similar to each other. There could be other factors (for instance, language contact situation, depth of penetration of English in society and effects of language policies) contributing to the difference between these English varieties, which might be worth exploring in our future works. Additionally, using a more recent wordlist can potentially provide a more accurate classification of the English varieties as they are today.

References

1. Biermeier, T.: Word-formation in New Englishes. Properties and trends. In: World Englishes: Problems - Properties - Prospects: Selected Papers from the 13th IAWE Conference. John Benjamins, Amsterdam (2009)
2. Crystal, D.: English as a Global Language. Cambridge University Press, Cambridge (2012)

3. Cruse, A.: Lexical Semantics. Cambridge University Press, Cambridge (1986)
4. Davies, M., Fuchs, R.: Expanding horizons in the study of World Englishes with the 1.9 billion word Global Web-based English Corpus (GloWbE). Engl. World-Wide **36**(1), 1–28 (2015)
5. Görlach, M.: More Englishes: New Studies in Varieties of English 1988–1994. John Benjamins Publishing, Amsterdam (1995)
6. Hundt, M.: A case study on the progressive passive. In: World Englishes Problems, Properties and Prospects: Selected Papers from the 13th IAWE Conference. John Benjamins Publishing (2009)
7. Kachru, B.: The Other Tongue: English Across Cultures. University of Illinois Press, Urbana and Chicago (1992)
8. Kachru, B.: English in South Asia. In: Burchfield, R. (ed.) The Cambridge History of the English Language, vol. 5, pp. 497–553. Cambridge University Press, Cambridge (1994)
9. Kirkpatrick, A.: World Englishes Paperback with Audio CD: Implications for International Communication and English Language Teaching. Cambridge University Press, Cambridge (2007)
10. Kortmann, B., Schneider, E., Burridge, K., Mesthrie, R., Upton, C. (eds.): A Handbook of Varieties of English. 2 vols. Mouton de Gruyter, Berlin, New York (2004)
11. Kortmann, B., Szmrecsanyi, B.: Parameters of morphosyntactic variation in World Englishes: prospects and limitations of searching for universals. In: Linguistic Universals and Language Variation, vol. 1, pp. 264–290 (2011)
12. Krishnaswamy, N., Burde, A.S.: The Politics of Indians English. Linguistic Colonialism and the Expanding English Empire. Oxford University Press, Delhi (1998)
13. Labov, W.: Principles of Linguistic Change, vol. 1: Internal Factors. Basil Blackwell, Oxford (1994)
14. Leimgruber, J.R.E.: The trouble with World Englishes. Engl. Today **29**, 3–7 (2013)
15. Lim, L.: Not just an Outer Circle, Asian English. In: World Englishes – Problems, Properties and Prospects: Selected Papers from the 13th IAWE Conference, vol. 40. John Benjamins Publishing (2009)
16. McArthur, T.: The English Languages. Cambridge University Press, Cambridge (1998)
17. Mufwene, S.S.: The Ecology of Language Evolution. Cambridge University Press, Cambridge (2001)
18. Nam, C.F.H., Mukherjee, S., Schilk, M., Mukherjee, J.: Statistical analysis of varieties of English. J. R. Stat. Soc. Ser. A (Stat. Soc.) **176**, 777–793 (2013)
19. Platt, J.T.: Postcolonial English: the Singapore English speech continuum and its basilect 'Singlish' as a 'creoloid'. Anthropol. Linguist. **17**(7), 363–374 (1975)
20. Platt, J., Ho, M.L.: Singapore and Malaysia, vol. 4. John Benjamins Publishing, Amsterdam (1983)
21. Schneider, E.W.: Postcolonial English: Varieties Around the World. Cambridge University Press, Cambridge (2007)
22. Strevens, P.: Teaching English as an International Language: From Practice to Principle. Pergamon Press, Oxford (1980)
23. Szmrecsanyi, B., Kortmann, B.: The morphosyntax of varieties of English worldwide: a quantitative perspective. Ling. Int. Rev. General Linguist. **119**, 1643–1663 (2009)
24. Vicentini, A.: The economy principle in language, Notes and Observations from early modern English grammars. Mots Palabras Words **3**, 37–57 (2003)

Towards a Map of the Syntactic
Similarity of Languages

Alina Maria Ciobanu[1,2], Liviu P. Dinu[1,2(✉)], and Andrea Sgarro[2,3]

[1] Faculty of Mathematics and Computer Science, University of Bucharest,
Bucharest, Romania
`ldinu@fmi.unibuc.ro`
[2] Human Language Technologies Research Center, University of Bucharest,
Bucharest, Romania
[3] Department of Mathematics and Geosciences, University of Trieste, Trieste, Italy

Abstract. In this paper we propose a computational method for determining the syntactic similarity between languages. We investigate multiple approaches and metrics, showing that the results are consistent across methods. We report results on 16 languages belonging to various language families. The analysis that we conduct is adaptable to any languages, as far as resources are available.

1 Introduction

Language similarity and language classification are some of the most addressed research problems in historical and comparative linguistics, attracting not only academic scholars, but also the general public.

Traditionally, historical linguists have used the classical comparative method to (re)construct language phylogeny [17]. Another popular method used for identifying the language classification tree is Greenberg's method [18]. In spite of the fact that linguistic literature abounds in claims of classification of natural languages, [27] notes that the computational historical linguistics did not receive much attention until the beginning of the 1990s, and argues for the necessity of development of quantitative and computational methods in this field. In [31] the authors note a "tremendous increase in the use of computational techniques for estimating evolutionary histories of languages" in the last decade, and survey the attempts to use computational methods in historical linguistics. In spite of the diversity of the used datasets, the linguistic objects used by the large majority of methods employed in (computational) historical linguistics were the lexical items – "the most externally accessible elements" [25]. Instead, syntactic approaches were much less present in language classification.

However, the last decades have witnessed pioneering work on language phylogeny based on syntax. One of the most remarkable works in this direction is [25], which proposes a method for classifying the languages based on syntactic characters.[1]

---

[1] The authors also present a brief history of the syntactic approaches and acknowledge the work of [30], a pioneer in this field and a forerunner of their approach.

A. Gelbukh (Ed.): CICLing 2017, LNCS 10761, pp. 576–590, 2018.
https://doi.org/10.1007/978-3-319-77113-7_44

We want to provide a map of the syntactic similarity of the natural languages (i.e., to compute a quantitative degree of syntactic similarity between any two languages). We take word order into account, but we aim at going beyond the standard analysis of linguistic typology, which considers the normal order in which subject, object and verb occur in an "unemotional" statement, [3,8,10] and rather attempt to determine the overall quantitative degree of similarity between the sequences of POS tags between a given sentence in a source language L_1 and its translations in the target language L_2.

Contributions. We present a large-scale study on the syntax of 16 languages from 5 language families. To the best of our knowledge, this is one of the first endeavors of quantifying the syntactic similarities of the languages (that is, to provide a similarity score for a pair of languages).

Our Approach. We propose a computational method for determining the syntactic similarity between the natural languages. We proceed as follows: given a sentence-aligned multilingual parallel corpus, we measure the similarity between parallel sentences at a syntactic level: given a sentence S in a source language L_1 and its translation T in a target language L_2, we are interested in quantifying the similarity between POS(S) and POS(T), where by POS(S) we denote the string obtained from S by labeling each word with its corresponding POS tag. Further, for a finer-grained analysis of the syntactic similarity, and to take into account the syntactic structure of the sentences, we apply a similarity measure between the syntax trees of the sentences, labeled with POS tags.[2] Our results show that the metrics we proposed lead to consistent language classifications, a fact which vindicates the importance and soundness of a "multiple" approach as ours.

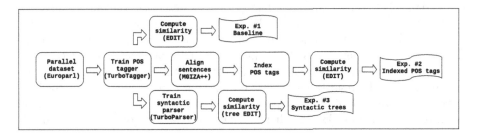

Fig. 1. Our methodology for computing the syntactic similarity.

---

[2] We use dependency parsing [21], so we rely on dependency trees to compute the syntactic similarity.

2   Methodology

Our goal is to asses the syntactic similarity between languages. To this end, we train a state-of-the-art POS tagger and dependency parser [26] for each language.[3] As training data we use a collection of treebanks with consistent dependency annotations [33] using the universal POS tagset of [35].

Having the parallel sequences of POS tags and the syntax trees, we investigate multiple methods of determining the syntactic similarity between the languages: a coarse-grained baseline, a finer-grained method that uses indexed POS tags (both methods quantifying word order distribution), and a method that leverages the syntactic structure of the sentences, using the syntax trees. To obtain an overall degree of similarity between two languages L_1 and L_2, we proceed as follows: for each sentence s_i from L_1 and its translation t_i in L_2, we compute the similarity score between s_i and t_i, and then we compute the average of these scores. The metrics that we propose are normalized in the $[0, 1]$ interval, where a similarity degree of 1 means that the input sequences are identical, while the lower the similarity, the more different the sequences are. The three methods are represented in Fig. 1.

2.1   Data

We run our experiments on *Europarl* [20], a multilingual parallel corpus extracted from the proceedings of the European Parliament, a very popular choice for experiments involving parallel texts [12,13].[4] We extract sentences in 16 languages belonging to various language families: Spanish (Es), French (Fr), Italian (It), Portuguese (Pt), Romanian (Ro) (the Romance family), Danish (Da), German (De), English (En), Dutch (Nl), Swedish (Sv) (the Germanic family), Bulgarian (Bg), Czech (Cs) (the Slavic family), Estonian (Et), Finnish (Fi), Hungarian (Hu) (the Finno-Ugric family), Greek (El) (the Hellenic family).

This dataset enables us to conduct a thorough analysis on the syntactic similarities between closer and more remotely related languages. Our dataset of parallel sentences contains 143,908 sentences for each language, but we trim the dataset to the maximum sentence length of 20. We only select sentences that we are able to identify in all 16 languages, because our experiments are designed for parallel corpora. While for very popular languages the number of sentences is much bigger than 143,908, for other languages the number of sentences is much

---

[3] Tagging and parsing accuracy for each language: Bg: 0.96,0.87; Cs: 0.98,0.82; Da: 0.95,0.82; De: 0.92,0.67; El: 0.96,0.82; En: 0.94,0.85; Es: 0.95,0.59; Et: 0.94,0.80; Fi: 0.93,0.75; Fr: 0.96,0.61; Hu: 0.92,0.79; It: 0.97,0.47; Nl: 0.89,0.77; Pt: 0.96,0.83; Ro: 0.95,0.82; Sv: 0.95,0.84.

[4] We believe that our investigation is not negatively influenced by the choice of corpus because we are consistent across all experiments in terms of text gender and we report results obtained solely by comparison between languages on the same dataset. In future work, we intend to apply the proposed methods on other datasets as well (for example, the EUR-Lex Corpus [2]).

lower, so we reduce the datasets for all languages to keep only parallel sentences in all 16 languages. After length pruning, we obtain about 62,000 sentences.[5]

Table 1. Example of raw and indexed POS tags for a German-French parallel sentence.

DE:	*Ich*	*bin*	*Ihnen*	*für*	*Ihre*	*konstruktiven*	*Beiträge*	*sehr*	*dankbar*
(1)	PRON	VERB	PRON	ADP	DET	ADJ	NOUN	ADP	ADJ
(2)	$PRON_0$	$VERB_1$	$PRON_2$	ADP_3	DET_4	ADJ_5	$NOUN_6$	ADP_7	ADJ_8
FR:	*Je*	*vous*	*suis*	*très*	*reconnaissante*	*pour*	*votre*	*apport*	*constructif*
(1)	PRON	PRON	VERB	ADV	ADJ	ADP	DET	NOUN	ADJ
(2)	$PRON_0$	$PRON_2$	$VERB_1$	ADV_7	ADJ_8	ADP_3	DET_4	$NOUN_6$	ADJ_5

3 Experiments

In this section we describe our approach for computing the syntactic similarity and report the similarity degrees obtained on the Europarl dataset.

3.1 Baseline (Exp. #1)

For each sentence from the dataset, we obtain the sequence of POS tags by applying a POS tagger [26]. Given the parallel sequences of POS tags, we employ the edit distance to determine degrees of syntactic similarity between the languages.[6] The edit distance [22] counts the minimum number of operations (insertion, deletion and substitution) required to transform one string into another. We use a normalized version of this metric, dividing the edit distance by the length of the longest string, and we convert it to similarity by subtracting it from 1. In Fig. 2 we report the results for this experiment, above the main diagonal of the matrix.

3.2 Indexed POS Tags (Exp. #2)

Although sequences of POS tags have been shown to display language-specific characteristics [28], we believe that such a similarity metric is not very reliable on its own, because we might obtain false positives when computing the edit distance on sequences of POS tags – that is, words that have the same POS tag and occupy the same position in the parallel sequences of POS tags, but are not actually the equivalent words.

---

[5] While the effect of translation cannot be denied, we rely on the fact that the interpreters/translators of Europarl are native speakers of the target language, which reduces the impact of the source language on translations significantly (as opposed to the translations performed by language learners, for example).

[6] We repeated Exp. #1 and #2 using the rank distance [7] instead of the edit distance, and there were no significant differences in the results.

As an example, consider the sentence from Table 1. Comparing the sequences of POS tags from (1) in DE and FR – that is, non-indexed POS tags – we observe that we have a match on the last position (ADJ). However, the actual words on the last position do not match (*dankbar* and *constructif*), so it is a *false positive* match. To overcome this drawback, we introduce an alignment step in our method: we align the parallel sentences using MGIZA++ [16], the multi-threaded version of the GIZA++ tool [34]. Thus, based on the alignment of the corresponding words, we are able to identify the matching tokens and to index the matching POS tags. Comparing the sequences of POS tags from (2) – that is, indexed POS tags – we observe that even though we have adjectives on the last position in both languages, we are able to tell they do not match based on the POS tags' indices. Having the POS tags indexed, we apply the edit distance again. For the sentence alignment phase, unaligned words received an index such that no word from the other sentence would match it (so that it would be penalized accordingly when computing the distance). We report the results in Fig. 2, below the main diagonal of the matrix.

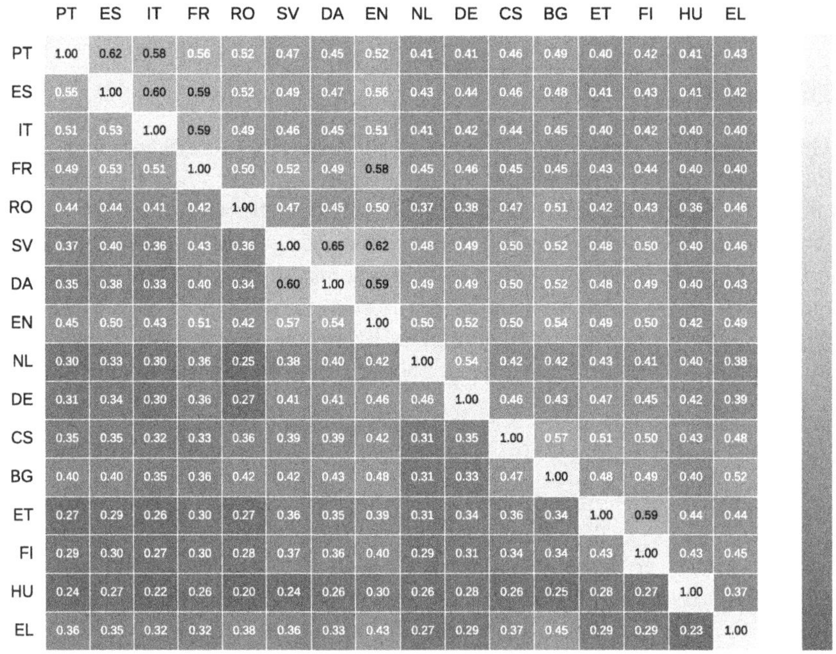

	PT	ES	IT	FR	RO	SV	DA	EN	NL	DE	CS	BG	ET	FI	HU	EL
PT	1.00	0.62	0.58	0.56	0.52	0.47	0.45	0.52	0.41	0.41	0.46	0.49	0.40	0.42	0.41	0.43
ES	0.55	1.00	0.60	0.59	0.52	0.49	0.47	0.56	0.43	0.44	0.46	0.48	0.41	0.43	0.41	0.42
IT	0.51	0.53	1.00	0.59	0.49	0.46	0.45	0.51	0.41	0.42	0.44	0.45	0.40	0.42	0.40	0.40
FR	0.49	0.53	0.51	1.00	0.50	0.52	0.49	0.58	0.45	0.46	0.45	0.45	0.43	0.44	0.40	0.40
RO	0.44	0.44	0.41	0.42	1.00	0.47	0.45	0.50	0.37	0.38	0.47	0.51	0.42	0.43	0.36	0.46
SV	0.37	0.40	0.36	0.43	0.36	1.00	0.65	0.62	0.48	0.49	0.50	0.52	0.48	0.50	0.40	0.46
DA	0.35	0.38	0.33	0.40	0.34	0.60	1.00	0.59	0.49	0.49	0.50	0.52	0.48	0.49	0.40	0.43
EN	0.45	0.50	0.43	0.51	0.42	0.57	0.54	1.00	0.50	0.52	0.50	0.54	0.49	0.50	0.42	0.49
NL	0.30	0.33	0.30	0.36	0.25	0.38	0.40	0.42	1.00	0.54	0.42	0.42	0.43	0.41	0.40	0.38
DE	0.31	0.34	0.30	0.36	0.27	0.41	0.41	0.46	0.46	1.00	0.46	0.43	0.47	0.45	0.42	0.39
CS	0.35	0.35	0.32	0.33	0.36	0.39	0.39	0.42	0.31	0.35	1.00	0.57	0.51	0.50	0.43	0.48
BG	0.40	0.40	0.35	0.36	0.42	0.42	0.43	0.48	0.31	0.33	0.47	1.00	0.48	0.49	0.40	0.52
ET	0.27	0.29	0.26	0.30	0.27	0.36	0.35	0.39	0.31	0.34	0.36	0.34	1.00	0.59	0.44	0.44
FI	0.29	0.30	0.27	0.30	0.28	0.37	0.36	0.40	0.29	0.31	0.34	0.34	0.43	1.00	0.43	0.45
HU	0.24	0.27	0.22	0.26	0.20	0.24	0.26	0.30	0.26	0.28	0.26	0.25	0.28	0.27	1.00	0.37
EL	0.36	0.35	0.32	0.32	0.38	0.36	0.33	0.43	0.27	0.29	0.37	0.45	0.29	0.29	0.23	1.00

(Legend: 0.90, 0.75, 0.60, 0.45, 0.30)

Fig. 2. Heatmap of syntactic similarity measured on sequences of POS tags (Exp. #1 – above the main diagonal) and on indexed sequences of POS tags (Exp. #2 – below the main diagonal).

3.3   Syntax Trees (Exp. #3)

The third metric that we apply on our parallel dataset is based on the dependency trees of the sentences [26]. We employ the tree edit distance [39] using the implementation of [1]. Similar to the version on strings, the tree edit distance represent the minimum number of operations (insertion, deletion or substitution) required to transform one tree into the other. The tree edit distance between trees T_1 and T_2 is normalized dividing the edit distance by $\max(N(T_1), N(T_2))$, where $N(T)$ is the number of vertices of tree T. We compute the tree edit distance between the dependency trees of the sentences, labeled with their POS tags. We do not use the labeled dependencies, but only the tree structure extracted from the dependency parsing (see an example in Fig. 3).

We chose dependency parsing because dependency is a 1:1 correspondence: every node in the syntactic tree corresponds to an element (a word) from the sentence. Thus, dependency structure serves our purpose better than constituency structure because having more nodes in the tree corresponding to a single word from the sentence would skew the values of the syntactic measures that we apply on parallel syntactic trees. We report the results in Fig. 4.

4   Discussion

Our initial goal was to provide a map of the syntactic similarity of the natural languages. To this end, we chose a multilingual parallel corpus in which a sentence, provided in a source language, is translated by professional translators into all the other target languages, and we attempted to determine a quantitative method to compare sequences of POS tags of the source sentence and of its translations in the target languages. The three experiments described in the previous section report the results of our experiments. Thus, we deem that, given a source language, the values reported on the line or column corresponding to it in the heatmaps represent the syntactic similarity with the other investigated languages.

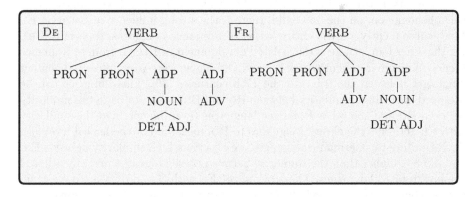

Fig. 3. Syntax trees with POS tag labels for sentences from Table 1.

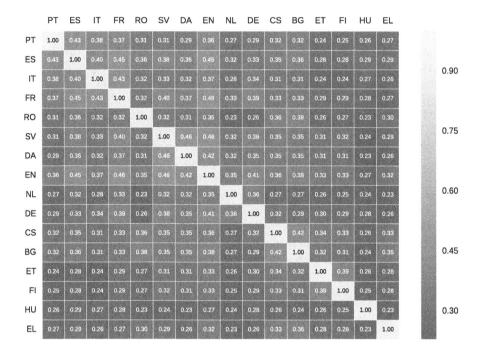

Fig. 4. Heatmap of syntactic similarity measured on syntax trees (Exp. #3).

4.1 Experimental Results and Interpretation

In Fig. 2 we report the results for Exp. #1 and #2. With a few exceptions, each language is most similar to the languages belonging to the same language family. For example, for Romance languages, in the first two experiments PT is closest to ES, followed by IT, FR, EN and RO, EN being the only one to break this rule. This observation stands for the other Romance languages as well. Thus, for any of FR, PT, IT, ES, the similarity between them and EN is higher than the similarity between them and RO. We identify two possible explanations for this phenomenon: on the one hand, there is the strong influence of Latin on EN (and, subsequently, on the entire kernel of Romance languages, especially FR). On the other hand, there is the isolated development of RO, far from te Romance kernel, in an environment where BG and Old Slavic were very influential contact languages between the 9th and the 17th centuries [36]. This influence can be observed in the high similarity between RO and BG, comparable to the similarity between RO and the other Romance languages (even higher than the similarity with FR and IT – the closest languages to RO, measured at the lexical level [6]).

Regarding the Germanic languages, we observe a high similarity between EN, DA and SV, higher than the similarity between these languages and any other of the investigated languages. The grouping of EN with SV and DA is, to a certain extent, consistent with previous results: [25] state that a similar grouping of En and Norwegian is in accordance with the historical influence of the Scandinavian languages on English, that is disclosed at the lexical level as well.

We remark that the similarity between Sv and Da is the highest of all the language pairs, followed by En-Sv and Pt-Es. In contrast, the lowest similarity is between Ro and Hu. In addition, Hu is the most distant language from all the others. Nl is most similar to De, followed by the other Germanic languages. Regarding En, we observe that, besides the high degrees of similarity with Sv and Da, the next languages, in terms of similarity, are Fr and Es (and, surprisingly, Bg). Among the Slavic languages, we note that Bg is most similar to Cs (and vice-versa in Exp. #1, while in Exp. #2 Cs is surpassed by En). Among the Finno-Ugric languages, we observe the highest degree of similarity between Fi and Et. As a general observation, the values reported for Exp. #1 are higher than those reported for Exp. #2.

For Exp. #3, we note that, for some of the languages, the dependency parsing accuracy was less than 75% (e.g., De, Es, Fr, It; see Sect. 2.1). We acknowledge the fact that the performance of the dependency parsing might influence the results.

We observe smaller values of similarity than in the previous experiments. Generally speaking, our previous observations still stand, with a few notable differences. For the Romance family, the languages from the Romance kernel (Fr, It, Es, Pt) are close to languages from the same kernel, but En plays a more important role. For example, for Fr, En is the closest language. Ro is drawn closer to the Slavic family, Cs and Bg being about as similar to Romanian as the other Romance languages. The cluster formed by Sv, Da and En has the same behavior as in the previous experiments. Thus, the highest degree of similarity belongs to the Sv-Da language pair, followed by Sv-En, En-Fr, Es-Fr, Pt-Es. Thus, we observe much higher similarities between En and the Romance languages. De is the closest language to En, and Nl is closest to De. Bg and Cs remain the closest languages, but their next similar language is Ro. Ro has a completely different rank of similarity than in Exp. #1 and #2, being closest to Bg, Cs, Es, En, followed by the other Romance languages.

We ran the experiments Exp. #1 – Exp. #3 again, but having the datasets split in multiple subsets: sentences with 1–5 words, 6–10 words, 11–15 words, 16–20 words. The results do not vary significantly. Removing outliers does not influence the reported values significantly as well. We compute the quartiles $Q1$, $Q2$ and $Q3$ with regard to the length of the sentences and we use the interquartile range $IQR = Q3 - Q1$ to find outliers. We consider outliers the observations that fall below $Q1 - 1.5(IQR)$ or above $Q3 + 1.5(IQR)$.

4.2 Language Clustering

The next step in our investigation was to analyze whether the degrees of syntactic similarity that we obtained can be used in language clustering. Grammar has been previously used in historical linguistics, as an indicator of the ancient relationship between the languages, but not so much directly, for phylogeny reconstruction [11]. A natural step was to apply clustering algorithms on the similarity matrices resulted from our three experiments. We used hierarchical clustering (the Ward method). The dendrograms obtained for Exp. #1 and #2 have an almost identical topology. Thus, due to space constraints, we only report one of them – for Exp. #2 – in Fig. 5a. We note that Romance languages are clustered together, as well as the Germanic (with two distinct clusters: DA-SV-EN and NL-DE), Slavic and Finno-Ugric languages. EL is linked to the Slavic family. Regarding the membership to language families, we obtain a fairly good classification of the languages. However, we note that the Romance families is exterior to the Germanic family, and the Finno-Ugric family is much closer to the Germanic family, having the same parent, which is generally not the case in linguistic phylogeny (Finno-Ugric languages are usually external).

Applying the same clustering algorithm on the results of Exp. #3 produces the dendrogram from Fig. 5b. We observe an essential difference from the previous dendrogram: the Romance languages are clustered together with the Germanic languages, having the same parent, as it is generally unanimously accepted. The only problem that arises is that Romanian does not belong to the Romance group, it migrates towards the Slavic group, together with BG and CS, which is also linked to EL. The grouping of RO with the Slavic languages (BG and CS) is not singular in this study. [12] obtained a similar grouping of the languages. A possible explanation could be the cultural evolution of Romanian,[7] factor with possible influence in its linguistic structure [10].

In Exp. #3, the Finno-Ugric languages are exterior to Romance and Germanic languages, having the same parent as the Slavic languages (and not being exterior to the entire group of Indo-European languages, as in the standard phylogenetic tree).

4.3   Comparison with Previous Work

The linguistic problem that we tackle, namely the syntactic language similarity, involves a certain "vagueness of reported values" (also noted by [12] in the problem of semantic language classification), as there isn't a gold standard that we can compare our results to. To overcome this drawback, we performed several experiments to compare our method with previous methods and results.

Comparison with Longobardi and Guardiano [25]. One of the very few studies that report quantitative measures to capture the syntactic similarity of languages is [25]. According to the authors, syntax has not been traditionally used as historical evidence for language relatedness mainly for two reasons: (1) the syntactic structures to be compared are difficult to identify; and (2) the

---

[7] Developing of Romanian far from the big Romance kernel made the contact of Romanian with Romance languages difficult until the 18th century. Instead, from the 9th to the 17th century there was a significant cultural influence of the South Slavic languages (especially Old Slavic), due in part to the exclusive use of Old Church Slavonic for religious purposes, which lead to giving South Slavic "the status of a cultural superstrate language" [36].

syntax is not as variable as the lexicon, and thus it might not provide enough variation to be able to draw conclusions. However, their results do not support this claim. The authors identified a series of syntactic parameters, and for each language they marked the presence or absence of each parameter with + or − (or 0, if the parameter cannot be accounted for in a language). Using a metric derived from the Hamming distance, the authors produced a distance matrix and a dendrogram. As the languages used in this study are not exactly the same as ours, we computed the correlation between the results reported on the subset of languages that are common in our studies. We obtained a high correlation (over 80%) for the Romance languages. In their experiments, much like in ours, Ro is, to some extent, exterior to the Romance kernel. In both experiments, Bg is close to Ro, but in our experiment Bg is also close to En, while in their experiment it is not so. Another similarity is that De and En are very close, as expected. Also, El is close to Bg in their experiments as well as in ours. As for the similarity with En, the previous remark still stands: our experiments draw El closer to En than the experiment of Longobardi and Guardiano.

Comparison with BLEU. So far, we have applied our syntactic metrics on different languages, to compute the similarity between them. In this section we describe a different use of our methods, as evaluation measures for the machine translation output. The values of syntactic similarity that we have obtained might be used as indicators of the quality of machine translation: a higher similarity between the reference sentence and the translated sentence indicates a better translation.

The most popular machine translation evaluation metrics rely on lexical feature to asses the performance of the machine translation systems. This approach on its own has some drawbacks, since there are other very important factors that contribute to a good translation, besides the lexicon. Syntax is one such factor. Initial attempts for including syntactic features in machine translation evaluation have been made by [9,23]. We further analyze and quantify the usefulness of the proposed syntactic metrics in the evaluation of the machine translation systems. To this end, we analyze the correlation of our syntactic metrics with BLEU, one of the most popular machine translation evaluation measures that relies on lexical features.

BLEU captures lexical similarity, while our methods capture syntactic similarity. Thus, we expect to not have a high correlation between them. On a random subset of 1,000 from Europarl, we took the English sentences and translated them with Bing[8] and Google Translate[9] in the other 14 languages. We then computed BLEU and our measures between the translated sentences and the reference translations (the sentences from Europarl, in all 14 languages except for English). We computed the Pearson correlation coefficient between the methods. For both translation systems, correlation is about 0.70 between our methods and BLEU.

---

[8] www.bing.com/translator.
[9] www.translate.google.com.

To have a more informative comparison, we then computed the BLEU score on the same subset of 1,000 sentences translated with Bing and Europarl, but using sequences of POS tags instead of words. Thus, we compared our method with a "syntactic version" of the BLEU measure, on both sequences of raw POS tags and indexed POS tags. In other words, we applied Exp. #1 and Exp. #2 using the BLEU metric instead of the edit distance. Computing the Pearson correlation between the two versions (BLEU vs edit), for each language, we notice that the results are highly correlated. We obtain an average correlation of 0.88 for Exp. #1 and 0.90 for Exp. #2 on the sentences translated with Google Translate, and 0.87 for Exp. #1 and 0.89 for Exp. #2 on the sentences translated with Bing.

Comparison with Human Judgement. We analyze the correlation of our syntactic metrics with human judgments for automatically translated sentences. To determine how well the syntactic metrics capture the syntactic differences between translated sentences and reference sentences, we evaluate the correlation between the syntactic metrics and human judgments. We conduct this experiment on the dataset of the 2012 Workshop on Statistical Machine Translation [4]. We used an experimental setup similar to [9]: we extracted the test English-German dataset and selected 7 out of the 15 systems that participated in the translation task: DFKI [37], JHU [15], KIT [32], UK [38] and the anonymized systems OnlineA, OnlineB, OnlineC. After computing the syntactic similarity between the automatically translated sentences (for each system) and the reference sentences, we evaluated our methods at system level.

To evaluate the syntactic metrics at system level, we computed Spearman's rank correlation coefficient between the human judgments and the average syntactic similarity (for the entire dataset), converting the syntactic similarities and the human judgments into rankings. The average results for the syntactic metrics and the values of Spearman's rank correlation coefficient are reported in Table 2. We obtained the highest correlation 0.75, which shows that the proposed syntactic metrics are able to capture the syntactic features of the translated sentences and to serve as evaluation measure for machine translation (in combination with metrics that capture other aspects of the translation quality).

5   Related Work

The syntactic similarity has been used mainly for the evaluation of the machine translation systems, to asses the similarity between the candidate and the reference translation. [5] proved the potential of syntax-based language models in statistical machine translation. They showed that the evaluation measures based on word n-grams (such as BLEU and NIST) have certain limitations in terms of assessing the fluency and the semantic accuracy of the candidate translation. The evaluation metrics that rely on lexical features cannot capture the structural similarity [9]. To overcome this drawback, the focus has shifted to structural (syntactic or semantic) machine translation evaluation metrics. [9,23]

have developed several syntactically-motivated evaluation metrics, based on the syntax trees (or subtrees) of the candidate and the reference translation.

An application of the cross-language syntactic similarity in historical linguistics has been emphasized by [25]: while most of the previous methods focused on lexical comparisons to identify relatedness indicators, the authors proposed a novel approach to study the genetic relationships between languages by focusing on syntactic structures. [24] further investigated the syntactic similarity of the languages by analogy with the patterns of genomic variation in Europe. [29] proposed an aggregate measure of syntactic distances using part-of-speech trigrams

Table 2. Average syntactic similarity for each machine translation system. The last column reports Spearman's correlation coefficient between syntactic similarity and human judgment, evaluated at system level.

Exp	DFKI	JHU	KIT	OnlineA	OnlineB	OnlineC	UK	ρ
#1	0.53	0.54	0.55	0.56	0.55	0.55	0.52	0.71
#2	0.36	0.38	0.40	0.42	0.40	0.40	0.36	0.75
#3	0.28	0.30	0.32	0.33	0.32	0.32	0.28	0.67

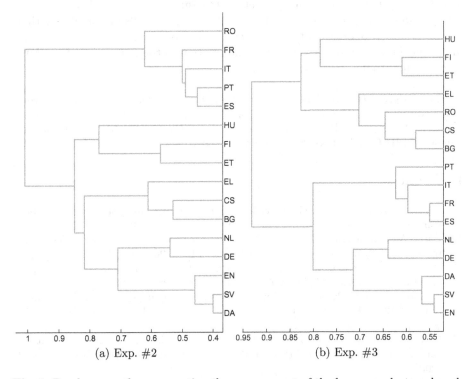

(a) Exp. #2   (b) Exp. #3

Fig. 5. Dendrograms for representing the arrangement of the language clusters, based on two of the three proposed syntactic metrics.

as markers of syntactic differences and measuring the overall distance between different corpora as the distance between frequency vectors of part-of-speech trigrams. However, they only apply the measure on a single language (English spoken by native vs. English spoken by Finnish immigrants), without providing a cross-language analysis. [28] studied mother tongue interference (the transfer of linguistic rules from L_1 to L_2) for non-native English speakers. They found that several syntactic features are informative in reconstructing phylogenies and in native language identification. [19] recently proposed a socio-linguistics study on the syntactic variation with respect to demographic variables, showing that several age and gender-specific variations hold across languages. [14] conducted a thorough analysis on the word order freedom in dependency corpora, showing the reliability of the conditional entropy measures. They reported results of 34 languages independently, without performing a cross-language analysis.

6   Conclusions

In this paper we proposed a methodology for computing the syntactic similarity of natural languages. We used Europarl, a large dataset of parallel sentences, from which we extracted sentences in 16 languages from various language families. Results on the underlying geometric structure are comparable and significantly stable across the various methods with different levels of granularity that we employed. We showed that the syntactic similarity correlates well with human judgment, promising to provide useful information in the evaluation of machine translation output.

Acknowledgments. We thank the anonymous reviewers for their helpful and constructive comments. This work was supported by a grant of the Romanian National Authority for Scientific Research and Innovation, CNCS/CCCDI UEFISCDI, project number PN-III-P2-2.1-53BG/2016, within PNCDI III.

References

1. Augsten, N., Böhlen, M., Gamper, J.: Approximate matching of hierarchical data using pq-grams. In: Proceedings of VLDB 2005, pp. 301–312 (2005)
2. Baisa, V., Michelfeit, J., Medved, M., Jakubícek, M.: European union language resources in sketch engine. In: Proceedings of LREC 2016, pp. 2799–2803 (2016)
3. Bortolussi, L., Sgarro, A., Dinu, L.P.: Measures of fuzzy disarray in linguistic typology. In: Proceedings of IPMU 2008, pp. 167–172 (2008)
4. Callison-Burch, C., Koehn, P., Monz, C., Post, M., Soricut, R., Specia, L.: Findings of the 2012 workshop on statistical machine translation. In: Proceedings of WMT 2012, pp. 10–51 (2012)
5. Charniak, E., Knight, K., Yamada, K.: Syntax-based language models for statistical machine translation. In: Proceedings of the 9th Machine Translation Summit (2003)
6. Ciobanu, A.M., Dinu, L.P.: An etymological approach to cross-language orthographic similarity. Application on Romanian. In: Proceedings of EMNLP 2014, pp. 1047–1058 (2014)

7. Dinu, A., Dinu, L.P.: On the syllabic similarities of romance languages. In: Gelbukh, A. (ed.) CICLing 2005. LNCS, vol. 3406, pp. 785–788. Springer, Heidelberg (2005). https://doi.org/10.1007/978-3-540-30586-6_88

8. Dryer, M.S.: 81 order of subject, object, and verb. In: The World Atlas of Language Structures, pp. 330–333 (2005)

9. Duma, M., Vertan, C., Menzel, W.: A new syntactic metric for evaluation of machine translation. In: Proceedings of the ACL Student Research Workshop, pp. 130–135 (2013)

10. Dunn, M., Greenhill, S., Levinson, S., Gray, R.: Evolved structure of language shows lineage-specific trends in word-order universals. Nature **473**(7345), 79–82 (2011)

11. Dunn, M., Terrill, A., Reesink, G., Foley, R., Levinson, S.: Structural phylogenetics and the reconstruction of ancient language history. Science **309**(5743), 2072–2075 (2005)

12. Eger, S., Hoenen, A., Mehler, A.: Language classification from bilingual word embedding graphs. In: Proceedings of COLING 2016, Technical Papers, pp. 3507–3518 (2016)

13. Eger, S., Schenk, N., Mehler, A.: Towards semantic language classification: inducing and clustering semantic association networks from Europarl. In: Proceedings of *SEM 2015, pp. 127–136 (2015)

14. Futrell, R., Mahowald, K., Gibson, E.: Quantifying word order freedom in dependency corpora. In: Proceedings of Depling 2015, pp. 91–100 (2015)

15. Ganitkevitch, J., Cao, Y., Weese, J., Post, M., Callison-Burch, C.: Joshua 4.0: packing, PRO, and paraphrases. In: Proceedings of WMT 2012, pp. 283–291 (2012)

16. Gao, Q., Vogel, S.: Parallel implementations of word alignment tool. In: Proceedings of SETQA-NLP 2008, pp. 49–57 (2008)

17. Gray, R., Atkinson, Q.: Language tree divergences support the Anatolian theory of Indo-European origin. Nature **426**, 435–439 (2003)

18. Greenberg, J.H.: Language in the Americas. Stanford University Press, Stanford (1987)

19. Johannsen, A., Hovy, D., Søgaard, A.: Cross-lingual syntactic variation over age and gender. In: Proceedings of CoNLL 2015, pp. 103–112 (2015)

20. Koehn, P.: Europarl: a parallel corpus for statistical machine translation. In: Proceedings of the 10th Machine Translation Summit, pp. 79–86 (2005)

21. Kübler, S., McDonald, R., Nivre, J.: Dependency parsing. Synth. Lect. Hum. Lang. Technol. **1**(1), 1–127 (2009)

22. Levenshtein, V.I.: Binary codes capable of correcting deletions, insertions, and reversals. Sov. Phys. Dokl. **10**, 707–710 (1965)

23. Liu, D., Gildea, D.: Syntactic features for evaluation of machine translation. In: Proceedings of the ACL Workshop on Intrinsic and Extrinsic Evaluation Measures for Machine Translation and/or Summarization, pp. 25–32 (2005)

24. Longobardi, G., et al.: Across language families: genome diversity mirrors linguistic variation within Europe. Am. J. Phys. Anthropol. **157**(4), 630–640 (2015)

25. Longobardi, G., Guardiano, C.: Evidence for syntax as a signal of historical relatedness. Lingua **119**(11), 1679–1706 (2009)

26. Martins, A.F.T., Almeida, M.B., Smith, N.A.: Turning on the turbo: fast third-order non-projective turbo parsers. In: Proceedings of ACL 2013, Short Papers, vol. 2, pp. 617–622 (2013)

27. McMahon, A., McMahon, R.: Finding families: quantitative methods in language classification. Trans. Philol. Soc. **101**(1), 7–55 (2003)

28. Nagata, R., Whittaker, E.: Reconstructing an Indo-European family tree from non-native English texts. In: Proceedings of ACL 2013, Long Papers, vol. 1, pp. 1137–1147 (2013)
29. Nerbonne, J., Wiersma, W.: A measure of aggregate syntactic distance. In: Proceedings of the Workshop on Linguistic Distances, pp. 82–90 (2006)
30. Nichols, J.: Linguistic Diversity in Space and Time. University of Chicago Press, Chicago (1992)
31. Nichols, J., Warnow, T.: Tutorial on computational linguistic phylogeny. Lang. Linguist. Compass **2**(5), 760–820 (2008)
32. Niehues, J., Zhang, Y., Mediani, M., Herrmann, T., Cho, E., Waibel, A.: The Karlsruhe Institute of Technology translation systems for the WMT 2012. In: Proceedings of WMT 2012, pp. 349–355 (2012)
33. Nivre, J.: Towards a universal grammar for natural language processing. In: Gelbukh, A. (ed.) CICLing 2015. LNCS, vol. 9041, pp. 3–16. Springer, Cham (2015). https://doi.org/10.1007/978-3-319-18111-0_1
34. Och, F.J., Ney, H.: A systematic comparison of various statistical alignment models. Comput. Linguist. **29**(1), 19–51 (2003)
35. Petrov, S., Das, D., McDonald, R.T.: A universal part-of-speech tagset. In: Proceedings of LREC 2012, pp. 2089–2096 (2012)
36. Schulte, K.: Loanwords in Romanian. In: Loanwords in the World's Languages: A Comparative Handbook, pp. 230–259 (2009)
37. Vilar, D.: DFKI's SMT system for WMT 2012. In: Proceedings of WMT 2012, pp. 382–387 (2012)
38. Zeman, D.: Data issues of the multilingual translation matrix. In: Proceedings of WMT 2012, pp. 395–400 (2012)
39. Zhang, K., Shasha, D.: Simple fast algorithms for the editing distance between trees and related problems. SIAM J. Comput. **18**(6), 1245–1262 (1989)

Romanian Word Production: An Orthographic Approach Based on Sequence Labeling

Liviu P. Dinu[1,2(✉)] and Alina Maria Ciobanu[1,2]

[1] Faculty of Mathematics and Computer Science, University of Bucharest,
Bucharest, Romania
`liviu.p.dinu@gmail.com, ldinu@fmi.unibuc.ro`
[2] Human Language Technologies Research Center, University of Bucharest,
Bucharest, Romania

Abstract. Languages borrow words from one another for various reasons. How the borrowing process takes place, how new words enter a recipient language are key questions of historical linguistics. In this paper, we propose a multilingual method for word form production based on the orthography of the words. For borrowed words, we investigate the derivation from a donor language into a recipient language. We also address the problem of genetic cognates derivation. We experiment with Romanian as a recipient language and we investigate borrowings from multiple donor languages. The advantages of the proposed method are that it does not use any external knowledge, except for the training word pairs, and it does not require the phonetic transcriptions of the input words.

1 Introduction

Natural languages are living eco-systems, they are constantly in contact and, by consequence, they borrow from one another. The process in which words enter one language from another is called *linguistic borrowing*. A borrowed word, also called *loanword*, is defined as a "lexical item (a word) which has been 'borrowed' from another language, a word which originally was not part of the vocabulary of the recipient language but was adopted from some other language and made part of the borrowing language's vocabulary" [6].

The unprecedented contact between languages in today's context of high mobility and the explosion of communication tools conduct to an inherent enrichment of languages by borrowings.[1] *Why* and *how* the borrowing process takes place are fundamental questions which, by their nature, invite to experimental perspective [7]. To answer the first question, [6] notes that 'Languages borrow words from other languages primarily because of need and prestige". Further,

---

[1] A dictionary of recent word which entered Romanian [12] counts 4,853 new words entered after 1965, most of them entering the language after 1990.

ⓒ Springer Nature Switzerland AG 2018
A. Gelbukh (Ed.): CICLing 2017, LNCS 10761, pp. 591–603, 2018.
https://doi.org/10.1007/978-3-319-77113-7_45

the author states that the result of the borrowing process depends on numerous factors, such as the length and intensity of the contact and the extent to which the populations in question are bilingual. Hence, the outcome of the contact between two populations is a challenging and interesting research problem.

1.1 Related Work

Historical linguistics addresses two main questions [28]: (i) how are languages related? and (ii) how do languages change across space and time? Traditionally, both problems were investigated with comparative linguistics instruments [6] and required a manual process. Modern approaches impose the use and development of quantitative and computational methods in this field [2,18,25], or even cross-disciplinary methods (such as those borrowed from biology). In this paper we focus on the second question. More specifically, we investigate how words enter a recipient language from a donor language. Most of the previous approaches to word form production relied on phonetic transcriptions [13,17,19]. Nowadays, computers are able to learn these changes from pairs of known related words. [3] proposed such a method for cognate production. Aligning the related words to extract orthographic changes from one language to another has proven very effective, when applied to both the orthographic [15] and the phonetic form of the words [21]. Another contribution belongs to [26], who introduced an algorithm for cognate production based on edit distance alignment and the identification of orthographic cues when words enter a new language. Other probabilistic approaches to word form production belong to [4,5,16].

1.2 Problem Formulation/Our Approach

We propose a method for word form production based on the orthography of the words, building on the idea that orthographic changes represent sound correspondences to a fairly large extent [11]. We account for the type of relationship between the words, making a clear distinction between borrowed words (also known as loanwords: words that entered a recipient language from a donor language) and genetic cognates (words in different languages having the same etymology and a common ancestor). Our goal is to perform word form production without using any external resources (e.g., a lexicon or a dataset in the recipient language). We use sequence labeling, an approach that has been proven useful in generating transliterations [1,14].

For borrowed words, we investigate the derivation of a word from a donor language into a recipient language. Given the form of a word u in a donor language L_1, we intend to develop a methods that predicts the form v of the word u in a recipient language L_2, in the hypotheses that the word v will be derived in L_2 from the word u (through a borrowing process).

For cognate production, we begin with cognate pairs in two related language L_1 and L_2, and investigate the derivation of a word from L_2 starting with words from L_1 and viceversa. We also conduct a comparison between recipient languages: given a donor language whose words were borrowed in multiple recipient languages, we compare the performance of the system for each recipient language. We also include here the information given by the common ancestor.

We conduct our experiments on Romanian as recipient language, which borrows words from 20 donor languages. In other words, we want to see how well our approach models the form of foreign words which have been borrowed by Romanian, and which donor language was modeled better by our approach. Why Romanian? Romanian is a Romance language, belonging to the Italic branch of the Indo-European language family. It is surrounded by speakers of non-Romance languages, namely Hungarian and several Slavic languages (Ukrainian, Bulgarian, Serbian), and throughout history, Romania has also had Poland, Czechoslovakia, USSR, Ottoman Empire and the Austro-Hungarian Empire as neighbours. Thus, its relationship with the big Romance kernel was difficult. While the contact with the Romance languages was difficult, the contact with various other languages was much easier. For example, from the 9th to the 17th century there was a significant cultural influence of the South Slavic languages (especially Old Slavic), due in part to the exclusive use of Old Church Slavonic for religious purposes, which lead to giving South Slavic "the status of a cultural superstrate language" [29]. Due to the expansion of the Ottoman Empire in the Balkans, Romanian was in contact with Turkish via multiple channels: politics, administration, commerce, military. Starting with the 19th century, Romania's orientation towards Western European culture and lifestyle led to an increased contact between Romanian and predominantly written French, Italian and German, which allowed a large numbers of loanwords to enter the language [29].

Thus, Romanian is very interesting to study not only in terms of its similarities with other natural languages [9], but also regarding the borrowings, and especially the form in which borrowings from various languages entered Romanian. Our study aims to investigate, among others, what words have a better chance of being correctly predicted: those that entered the language via the phylogenetic branch, or those for which the contact was due to cultural, geographic or political circumstances?

2   Word Form Production: Methodology

From the alignment of the related words in the training set we learn orthographic cues and patterns for the changes in spelling, and we attempt to infer the form, in the recipient language, of the words related to the input words in the test set.

We use a word form production method that is based on sequence labeling (assigning a sequence of labels to a sequence of tokens). This approach has been proven useful for cognate production [8] and for generating transliterations [1,14]. From the alignment of the related words in the training set we learn orthographic cues and patterns for the changes in spelling. For the words in the test set, using the trained system, we infer the form of their related words in the recipient language. In this section we describe the proposed approach.

2.1 Orthographic Alignment

To align pairs of words we employ the Needleman-Wunsch global alignment algorithm [27], with words as input sequences and a very simple substitution matrix, which gives equal scores to all substitutions, disregarding diacritics (e.g., we ensure that *e* and *è* are matched). For example, for the Romanian word *traducător* (meaning *translator*) borrowed from the Italian word *traduttore*, the alignment is as follows:

```
t r a d u t - t o r e
t r a d u c ă t o r -
```

2.2 Sequence Labeling

In our case, the words in the donor language are the sequences, and the characters are the tokens. Our purpose is to obtain, for each input word, a sequence of characters that compose its related word. To this end, we use first- and second-order conditional random fields (CRFs) [22].

For each character in the donor word (after the alignment), the label is the character which occurs on the same position in the recipient word. In the case of insertions, because there is no input character in the donor language to which we could associate the inserted character as label, we add it to the previous label. We account for affixes separately: for each input word we add two more characters B and E, marking the beginning and the end of the word. The characters that are inserted in the recipient word at the beginning or at the end of the word are associated to these special characters. In order to reduce the number of labels, for input tokens that are identical to their labels we replace the label with *. Thus, for the previous example, the labels are as follows:

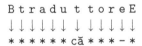

3   Experimental Setup

In this section we describe the experimental setup used to assess the performance and to analyze the proposed method for word form production.

3.1 Datasets

We use a dataset of word-etymons and cognate pairs [9], from which we extract Romanian words having etymons in 20 languages.[2] The dataset was built from

---

[2] Romanian borrowed words from over 40 languages. In our experiments, we use the top 20 languages in terms of number of borrowed words, so that we have enough training data. See Table 1 for the complete list of languages.

an aggregation of machine-readable dictionaries[3] that contain information about the origin of the words. The information about the etymology of the words was extracted from those definitions. The dataset of cognates was built similarly, using resources with etymological information for the other languages as well. For a proper comparison, we use datasets of equal length (800 word pairs) for all languages.

For each borrowing, the donor word (the etymon, or the ancestor) is provided, together with the donor language. The dataset is structured as a list of word pairs having the form: $w_1(L_1) \rightarrow w_2(L_2)$, where word w_2 entered L_2 from the L_1 word w_1. Example: *victoria (Latin)* $\rightarrow$ *victorie (Romanian)*.

3.2 Task Setup

We split each of the datasets in three subsets for training, development and testing with a 3:1:1 ratio. We use the CRF implementation provided by the Mallet toolkit for machine learning [24]. As features we use n-grams of characters from the input word around the current token, in a window of size w, where $n \in \{1, ..., w\}$. For parameter tuning, we perform a grid search for the number of iterations in $\{1, 5, 10, 25, 50, 100\}$ and for the size of the window w in $\{1, 2, 3\}$.

We use a "majority class" type of baseline that does not take context into account, as described in [8].

3.3 Evaluation Measures

Following previous work in this area [3,5], we use the evaluation measures listed below to assess the performance of our method:

Coverage. The coverage (also known as *top n accuracy*) is a relaxed metric which computes the percentage of input words for which the n-best output list contains the correct cognate pair (the gold standard). We use n $\in \{1, 5, 10\}$ for a better understanding of our system's performance. The practical importance of analyzing the top n results is that we offer a filter to narrow down the possible forms of the output words to a low-dimensional list, that linguists can analyze, aiming to identify the correct form of the word. Note that the coverage for n = 1 is the well-known measure accuracy.

Average Edit Distance. To assess how close the production is to the correct form of the words, we report the edit distance between the produced words and the gold standard. The edit distance [23] counts the minimum number of operations (insertion, deletion and substitution) required to transform one string into another.

4 Experiments

In this section we describe the experiments we conducted on word form production, having Romanian as a recipient language. We report the results of our

---

[3] https://dexonline.ro.

method and we analyze and compare the performance of the system in different scenarios.

4.1   Experiment #1: Borrowings

In this experiment we aim to produce the orthographic form of borrowings in Romanian. We use *etymon-word* pairs: given a word in a donor language L1, we aim to produce the form in which this word entered a recipient language L2. We use Romanian as the recipient language and experiment with a wide range of donor languages (20), covering all the European language families.

Table 1. Exp. #1.1: Word form production for borrowings, using lemmas as input. The **Language** column indicates the donor language. The recipient language is, in all cases, Romanian. We report the average edit distance between the produced form and the correct form of the borrowing (**edit**) un-normalized (and between parentheses the normalized version) and the coverage (**cov** for n $\in$ {1, 5, 10}) for the baseline and for the method presented in this paper.

Language	Baseline				This work			
	EDIT	COV_1	COV_5	COV_{10}	EDIT	COV_1	COV_5	COV_{10}
English	2.04 (0.23)	0.02	0.16	0.25	1.33 (0.15)	0.36	0.56	0.61
French	2.16 (0.24)	0.06	0.25	0.35	1.42 (0.15)	0.32	0.63	0.70
Italian	2.60 (0.32)	0.00	0.17	0.23	1.62 (0.23)	0.35	0.47	0.53
Latin	2.75 (0.34)	0.00	0.08	0.17	1.76 (0.22)	0.28	0.48	0.55
Neo-Greek	2.39 (0.29)	0.08	0.17	0.25	1.82 (0.24)	0.25	0.53	0.58
Old Slavic	2.34 (0.33)	0.08	0.18	0.23	1.84 (0.27)	0.17	0.39	0.47
German	2.36 (0.32)	0.07	0.23	0.26	2.00 (0.29)	0.26	0.41	0.45
Turkish	1.88 (0.27)	0.11	0.17	0.21	2.01 (0.29)	0.23	0.37	0.41
Bulgarian	2.33 (0.34)	0.06	0.20	0.21	2.22 (0.33)	0.15	0.23	0.28
Ruthenian	2.33 (0.35)	0.09	0.19	0.25	2.31 (0.35)	0.11	0.18	0.21
Russian	2.24 (0.33)	0.09	0.19	0.23	2.33 (0.33)	0.13	0.20	0.25
Albanian	2.60 (0.42)	0.06	0.11	0.12	2.35 (0.38)	0.08	0.20	0.25
Serbian	2.43 (0.37)	0.01	0.19	0.21	2.38 (0.36)	0.11	0.23	0.27
Polish	2.49 (0.38)	0.04	0.12	0.15	2.43 (0.36)	0.08	0.13	0.19
Portuguese	2.95 (0.52)	0.00	0.03	0.08	2.50 (0.43)	0.07	0.30	0.33
Slavic	2.88 (0.42)	0.05	0.11	0.17	2.66 (0.41)	0.12	0.27	0.31
Provencal	3.01 (0.49)	0.01	0.04	0.07	2.70 (0.44)	0.05	0.17	0.21
Hungarian	2.80 (0.43)	0.05	0.16	0.21	2.73 (0.42)	0.05	0.19	0.21
Spanish	3.22 (0.53)	0.02	0.06	0.11	3.06 (0.50)	0.05	0.12	0.15
Greek	4.36 (0.49)	0.01	0.08	0.08	4.28 (0.48)	0.05	0.15	0.15

Experiment #1.1: Lemmas as Input. First, we experiment using lemmas (dictionary word forms) as input. The results for this experiment are listed in Table 1. We observe that the best results are obtained from French and English donor words. The top 8 ranked languages are those with which Romanian had the most intense cultural collaboration, either more recently (English, for example) or in the past (in the period of the "languages" of Romanian, when the Italian and French influence was remarkable) or by continuous contact (with Turkish). The word production performance is lower even for related langauges (as Portuguese and Spanish); these languages are more remote from Romania, from a geographical point of view, and this might have made the contact between languages more difficult. What is more, for Spanish, for example, there has never been a significant Spanish cultural influence on Romanian.

The method we propose outperforms the baseline significantly for all languages except for Turkish, as measured by the edit distance. Our assumption was that context is very relevant in word form production and our hypothesis is confirmed by the difference in performance between our method and the baseline (since the baseline does not take context into account).

We repeat **Exp. #1** with a modified version of the dataset, in which we discard diacritics, in order to see if (and how) diacritics influence the learning process. The results are slightly improved when diacritics are not taken into account. The coverage values (n = 10) for both versions of the dataset, for the CRF system, are represented in Fig. 1.

Table 2. Exp. #1.2: Word form production for borrowings, using stems as input. The **Language** column indicates the donor language. The recipient language is, in all cases, Romanian. For our method, we mark with * the results for which the difference to **Exp. #1.1** is statistically significant (pairwise T-test, $p < 0.05$).

Language	Baseline				This work			
	EDIT	COV_1	COV_5	COV_{10}	EDIT	COV_1	COV_5	COV_{10}
French	1.65 (0.20)	0.27	0.44	0.49	1.43 (0.17)	0.41	0.56	0.61
English	1.94 (0.22)	0.18	0.34	0.39	1.48 (0.17)	0.38	0.51	0.56
Portuguese	2.13 (0.45)	0.06	0.20	0.23	1.67 (0.36)*	0.18	0.33	0.38
Italian	1.83 (0.27)	0.25	0.32	0.33	1.70 (0.26)	0.31	0.45	0.45
German	1.99 (0.29)	0.17	0.31	0.33	1.75 (0.27)*	0.28	0.49	0.51
Russian	2.18 (0.33)	0.12	0.20	0.29	1.97 (0.32)*	0.20	0.29	0.33
Turkish	1.96 (0.31)	0.13	0.23	0.25	2.12 (0.33)	0.20	0.30	0.33
Spanish	2.53 (0.46)	0.13	0.21	0.23	2.26 (0.44)*	0.15	0.23	0.26
Hungarian	2.48 (0.43)	0.08	0.16	0.21	2.38 (0.42)*	0.12	0.20	0.24

Experiment #1.2: Stems as Input. In this experiment we use stems instead of lemmas. Both lemmatization and stemming reduce the form of inflected or derived words to a common base word form, but stemming does this in a more drastic manner. That is, it reduces the words to much shorter forms than lemmatization. Furthermore, stemming might also remove prefixes from the words, which lemmatization does not. We believe that this might make a difference. We are interested to see if training and testing the system on stems, instead of lemmas, leads to better results. We use the Snowball Stemmer[4], which provides stemmers for 9 of our 20 donor languages. The results for this experiment are reported in Table 2. We observe that stemming does not improve the performance of the system. This shows that Romanian is a complex language, and foreign influences, in the case of new words entering the languages, occur in the root of the words as well. Thus, the root is not necessarily easier to produce than the entire word, including affixes.

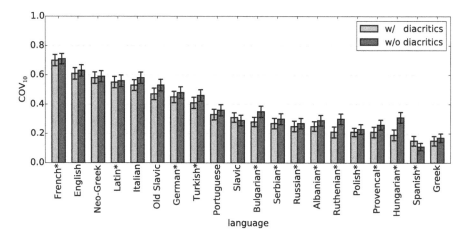

Fig. 1. cov_{10} for **Exp. #1.1**, with and without diacritics. The error bars represent standard deviation, estimated by 10,000 iterations of bootstrap resampling with replacement [20]. We mark with * the languages for which the difference between the two versions of the dataset is statistically significant (pairwise T-test, $p < 0.05$).

4.2   Experiment #2: Cognates

Further, we address cognate production. This task is very similar to word form production for borrowings (**Exp. #1**). The only difference is that instead of using a dataset of *etymon-word* pairs, we use a dataset of *cognate* pairs. We analyze *genetic cognates*, i.e. pairs of words with a common ancestor. In the recent years, there has been a significant interest in identifying cognates using computational methods [10], but very few studies addressed the automatic production of the cognate pairs. Our purpose is to determine whether the system behaves differently, in terms of performance, for cognates and for borrowings. To this end, we conduct the following two experiments:

---

[4] http://snowball.tartarus.org.

Table 3. Exp. #2.1: Word form production for cognates. The **Language** column indicates the donor language. The recipient language is, in all cases, Romanian. For this experiment, we also perform the experiment in the reverse direction (from recipient to donor). The arrows ($\rightarrow$, $\leftarrow$) indicate the direction of the production process. For our method, we mark with * the results for which the differences to **Exp. #1.1** are statistically significant (Mann-Whitney U-test, $p < 0.05$).

Language		Baseline				This work			
		EDIT	COV_1	COV_5	COV_{10}	edit	COV_1	COV_5	COV_{10}
Spanish	$\rightarrow$	1.71 (0.19)	0.10	0.34	0.41	0.91 (0.11)*	0.49	0.75	0.80
	$\leftarrow$	1.60 (0.18)	0.12	0.19	0.22	0.94 (0.12)*	0.48	0.66	0.70
Italian	$\rightarrow$	2.05 (0.23)	0.06	0.22	0.29	1.24 (0.15)	0.36	0.61	0.70
	$\leftarrow$	1.85 (0.21)	0.08	0.19	0.22	1.25 (0.14)*	0.43	0.56	0.64
Turkish	$\rightarrow$	1.30 (0.17)	0.27	0.32	0.35	1.28 (0.17)*	0.33	0.56	0.60
	$\leftarrow$	1.33 (0.18)	0.29	0.45	0.50	1.29 (0.17)*	0.41	0.61	0.65
Portuguese	$\rightarrow$	1.54 (0.18)	0.17	0.42	0.50	1.29 (0.16)*	0.38	0.55	0.61
	$\leftarrow$	1.77 (0.21)	0.17	0.27	0.31	1.36 (0.16)*	0.32	0.51	0.55
English	$\rightarrow$	2.01 (0.27)	0.01	0.14	0.19	1.30 (0.18)	0.35	0.62	0.70
	$\leftarrow$	2.08 (0.27)	0.03	0.13	0.18	1.47 (0.20)*	0.38	0.50	0.56

Table 4. Exp. #2.2: Word form production for parallel lists of cognates. The **Language** column indicates here the recipient language. The donor language is Latin in (4a) and French in (4b).

(a) Cognates with Latin ancestors

Language	Baseline				This work			
	EDIT	COV_1	COV_5	COV_{10}	EDIT	COV_1	COV_5	COV_{10}
Italian	1.14 (0.14)	0.44	0.52	0.54	0.98 (0.12)	0.48	0.63	0.69
Romanian	2.37 (0.31)	0.03	0.22	0.40	1.06 (0.16)	0.47	0.61	0.66
Spanish	1.67 (0.21)	0.16	0.39	0.44	1.16 (0.15)	0.45	0.60	0.63
Portuguese	2.93 (0.33)	0.07	0.16	0.17	2.62 (0.30)	0.20	0.28	0.33

(b) Cognates with French ancestors

Language	Baseline				This work			
	EDIT	COV_1	COV_5	COV_{10}	EDIT	COV_1	COV_5	COV_{10}
Romanian	1.90 (0.21)	0.14	0.26	0.28	0.86 (0.12)	0.46	0.73	0.77
Turkish	2.15 (0.27)	0.15	0.23	0.29	1.56 (0.19)	0.38	0.53	0.56

Experiment #2.1: Cognate Production. In the first experiment we use cognates between Romanian and 5 languages: Spanish, Italian, Turkish, Portuguese and English. We are interested in investigating if, for a given pair of languages, having one word from a cognate pair, we can automatically determine the orthographic form of its cognate. The results for this experiment are reported in Table 3. For all five languages, the system performs, in both directions (i.e., from L_1 to L_2 and from L_2 to L_1) better than for deriving modern word forms from their ancestors (**Exp. # 1**). In Table 5 we report several examples of our system on cognate production.

Experiment 2.2: Cognate Derivation. Further, we take into account the common ancestor of the cognate pairs and investigate in which language the production is better. To this end, we extract two datasets: one dataset of Latin words that entered Romanian, Spanish, Portuguese and Italian (Table 4a), and another dataset of French words that entered Romanian and Turkish (Table 4b). We observe that for the first dataset (Latin ancestors) the performance is best for Italian, followed by Romanian and Spanish.

We perform the one-way ANOVA F-test ($p < 0.05$) with the null hypothesis $\mathcal{H}_0$: $\text{EDIT}_{Ro} = \text{EDIT}_{Es} = \text{EDIT}_{Pt} = \text{EDIT}_{It}$, where EDIT_L is the average edit distance between the produced and the correct word form, on the test set, for language L. Since the p-value is less than 0.05, we reject the null hypothesis. Further pairwise T-tests show that difference is statistically significant ($p < 0.05$) for Italian-Portuguese, Portuguese-Romanian and Portuguese-Spanish. For the second dataset (French ancestors), the system was able to learn orthographic patterns much better for Romanian than for Turkish. A pairwise T-test shows that the difference between the two languages is statistically significant ($p < 0.05$).

Table 5. Examples of cognate production output. **Language** column indicates the donor language. The recipient language is, in all cases, Romanian. The arrows ($\rightarrow$, $\leftarrow$) indicate the direction of the production process. In the **Output** column, we emphasize the true cognate (in bold).

Language	Cognate pair		Output (5-best list)
Italian	*millenario - milenar*	$\rightarrow$	***milenar**, milenarium, millenar, millenarium, milenariu*
		$\leftarrow$	*milenario, milenare, milenarro, **millenario**, milanario*
Spanish	*petrificado - petrificat*	$\rightarrow$	***petrificat**, petrificatum, petrificatus, petrificart, petrificant*
		$\leftarrow$	***petrificado**, petrificados, petrificacio, petrificacin, petrificada*
Portuguese	*hipnose - hipnoză*	$\rightarrow$	***hipnoză**, hipnosiune, ipnoz, ipnosiune, hipnos*
		$\leftarrow$	*hipnoză, hipnos, **hipnose**, hipnoser, hipnos*
Turkish	*otokrasi - autocrație*	$\rightarrow$	*autocrasie, **autocrație**, otocrasie, autocracie, otocrație*
		$\leftarrow$	*otokrasyon, **otokrasi**, otokrasiyon, otokrațyon, otokrasyalamak*

5   Conclusions

In this paper, we proposed an automatic method for word form production, based on the orthography of the words. We applied our method using Romanian as the recipient language and we experimented with multiple donor languages. Our results are encouraging. They show that even though the rules for adapting borrowed words to the recipient language are sometimes not uniform, there are certain patterns and regularities that allow the production of n-best lists of output words, to be further analyzed by linguists. We conclude that languages are grouped, in the ranking, rather by their cultural influence on Romanian, than by the language families. We emphasize the difference in behavior between learning and producing borrowings, given their etymons (ancestors), and learning and producing genetic cognates. The direction of the cognate production does not seem to influence the results. Even when the output sequence doesn't match the true cognate, it might be a valid word in the recipient language. Sometimes, the produced sequences represent older forms of the words used today or, for nouns, the feminine form of the word. We observe that learning patterns from cognates leads to much better results than learning patterns from borrowings. We plan to study further means of improving our method with the purpose of obtaining high quality word form production without external knowledge.

Acknowledgments. We thank the anonymous reviewers for their helpful and constructive comments. This work was supported by a grant of the Romanian National Authority for Scientific Research and Innovation, CNCS/CCCDI UEFISCDI, project number PN-III-P2-2.1-53BG/2016, within PNCDI III.

References

1. Ammar, W., Dyer, C., Smith, N.A.: Transliteration by sequence labeling with lattice encodings and reranking. In: Proceedings of the 4th Named Entity Workshop, pp. 66–70 (2012)
2. Atkinson, Q.D.: The descent of words. Proc. Natl. Acad. Sci. **110**(11), 4159–4160 (2013)
3. Beinborn, L., Zesch, T., Gurevych, I.: Cognate production using character-based machine translation. In: Proceedings of IJCNLP 2013, pp. 883–891 (2013)
4. Bouchard-Côté, A., Griffiths, T.L., Klein, D.: Improved reconstruction of protolanguage word forms. In: Proceedings of NAACL 2009, pp. 65–73 (2009)
5. Bouchard-Côté, A., Liang, P., Griffiths, T., Klein, D.: A probabilistic approach to diachronic phonology. In: Proceedings of EMNLP-CoNLL 2007, pp. 887–896 (2007)
6. Campbell, L.: Historical Linguistics. An Introduction. MIT Press, Cambridge (1998)
7. Chitoran, I.: The nature of historical change. In: Cohn, A.C., Fougeron, C., Huffman, M.K. (eds.) The Oxford Handbook of Laboratory Phonology. Oxford University Press, Oxford (2011)
8. Ciobanu, A.M.: Sequence labeling for cognate production. Procedia Comput. Sci. **96**, 1391–1399 (2016). Knowledge-Based and Intelligent Information and Engineering Systems: Proceedings of KES 2016

9. Ciobanu, A.M., Dinu, L.P.: An etymological approach to cross-language orthographic similarity. Application on Romanian. In: Proceedings of EMNLP 2014, pp. 1047–1058 (2014)

10. Ciobanu, A.M., Dinu, L.P.: Automatic detection of cognates using orthographic alignment. In: Proceedings of ACL 2014, Volume 2: Short Papers, pp. 99–105 (2014)

11. Delmestri, A., Cristianini, N.: String similarity measures and PAM-like matrices for cognate identification. Buchar. Work. Pap. Linguist. **12**(2), 71–82 (2010)

12. Dimitrescu, F.: Dictionar de Cuvinte Recente. Ed. Logos (1997)

13. Eastlack, C.L.: Iberochange: a program to simulate systematic sound change in Ibero-romance. Comput. Humanit. **11**, 81–88 (1977)

14. Ganesh, S., Harsha, S., Pingali, P., Verma, V.: Statistical transliteration for cross language information retrieval using HMM alignment model and CRF. In: Proceedings of the 2nd Workshop on Cross Lingual Information Access (2008)

15. Gomes, L., Pereira Lopes, J.G.: Measuring spelling similarity for cognate identification. In: Antunes, L., Pinto, H.S. (eds.) EPIA 2011. LNCS (LNAI), vol. 7026, pp. 624–633. Springer, Heidelberg (2011). https://doi.org/10.1007/978-3-642-24769-9_45

16. Hall, D., Klein, D.: Finding cognate groups using phylogenies. In: Proceedings of the 48th Annual Meeting of the Association for Computational Linguistics, pp. 1030–1039 (2010)

17. Hartman, S.L.: A universal alphabet for experiments in comparative phonology. Comput. Humanit. **15**, 75–82 (1981)

18. Heggarty, P.: Beyond lexicostatistics: how to get more out of "Word List" comparisons. In: Quantitative Approaches to Linguistic Diversity: Commemorating the Centenary of the Birth of Morris Swadesh, pp. 113–137. Benjamins (2012)

19. Hewson, J.: Comparative reconstruction on the computer. In: Proceedings of ICHL 1974, pp. 191–197 (1974)

20. Koehn, P.: Statistical significance tests for machine translation evaluation. In: Proceedings of EMNLP 2004, pp. 388–395 (2004)

21. Kondrak, G.: A new algorithm for the alignment of phonetic sequences. In: Proceedings of NAACL 2000, pp. 288–295 (2000)

22. Lafferty, J.D., McCallum, A., Pereira, F.C.N.: Conditional random fields: probabilistic models for segmenting and labeling sequence data. In: Proceedings of ICML 2001, pp. 282–289 (2001)

23. Levenshtein, V.I.: Binary codes capable of correcting deletions, insertions, and reversals. Sov. Phys. Dokl. **10**, 707–710 (1965)

24. McCallum, A.K.: MALLET: A Machine Learning for Language Toolkit (2002). http://mallet.cs.umass.edu

25. McMahon, A., Heggarty, P., McMahon, R., Slaska, N.: Swadesh sublists and the benefits of borrowing: an Andean case study. Trans. Philol. Soc. **103**(2), 147–170 (2005)

26. Mulloni, A.: Automatic prediction of cognate orthography using support vector machines. In: Proceedings of the ACL Student Research Workshop, pp. 25–30 (2007)

27. Needleman, S.B., Wunsch, C.D.: A general method applicable to the search for similarities in the amino acid sequence of two proteins. J. Mol. Biol. **48**(3), 443–453 (1970)

28. Rama, T., Borin, L.: Comparative evaluation of string similarity measures for automatic language classification. In: Mikros, G.K., Macutek, J. (eds.) Sequences in Language and Text. De Gruyter Mouton, Berlin (2014)
29. Schulte, K.: Loanwords in Romanian. In: Haspelmath, M. (ed.) Loanwords in the World's Languages: A Comparative Handbook, pp. 230–259. De Gruyter Mouton, Berlin (2009)

Author Index

Printed in the United States
By Bookmasters